SCHAUM'S
outlines

Signals and Systems

SCHAUM'S.
outlines

Signals and Systems

———————————————————————— *Fourth Edition*

Hwei P. Hsu, PhD

Schaum's Outline Series

Mc
Graw
Hill
Education

New York   Chicago   San Francisco   Athens   London   Madrid
Mexico City   Milan   New Delhi   Singapore   Sydney   Toronto

HWEI P. HSU received his BS from National Taiwan University and his MS and PhD from Case Institute of Technology. He has published several books, including *Schaum's Outline of Analog and Digital Communications* and *Schaum's Outline of Probability, Random Variables, and Random Processes.*

2 3 4 5 6 7 8 9   LOV   24 23 22

ISBN   978-1-260-45424-6
MHID   1-260-45424-X

e-ISBN   978-1-260-45425-3
e-MHID   1-260-45425-8

McGraw-Hill Education books are available at special quantity discounts to use as premiums and sales promotions or for use in corporate training programs. To contact a representative, please visit the Contact Us pages at www.mhprofessional.com.

Preface to The Second Edition

The purpose of this book, like its previous edition, is to provide the concepts and theory of signals and systems needed in almost all electrical engineering fields and in many other engineering and science disciplines as well.

In the previous edition the book focused strictly on deterministic signals and systems. This new edition expands the contents of the first edition by adding two chapters dealing with random signals and the response of linear systems to random inputs. The background material on probability needed for these two chapters is included in Appendix B.

I wish to express my appreciation to Ms. Kimberly Eaton and Mr. Charles Wall of the McGraw-Hill Schaum Series for inviting me to revise the book.

HWEI P. HSU
Shannondell at Valley Forge, Audubon, Pennsylvania

Preface to The First Edition

The concepts and theory of signals and systems are needed in almost all electrical engineering fields and in many other engineering and scientific disciplines as well. They form the foundation for further studies in areas such as communication, signal processing, and control systems.

This book is intended to be used as a supplement to all textbooks on signals and systems or for self-study. It may also be used as a textbook in its own right. Each topic is introduced in a chapter with numerous solved problems. The solved problems constitute an integral part of the text.

Chapter 1 introduces the mathematical description and representation of both continuous-time and discrete-time signals and systems. Chapter 2 develops the fundamental input-output relationship for linear time-invariant (LTI) systems and explains the unit impulse response of the system and convolution operation. Chapters 3 and 4 explore the transform techniques for the analysis of LTI systems. The Laplace transform and its application to continuous-time LTI systems are considered in Chapter 3. Chapter 4 deals with the z-transform and its application to discrete-time LTI systems. The Fourier analysis of signals and systems is treated in Chapters 5 and 6. Chapter 5 considers the Fourier analysis of continuous-time signals and systems, while Chapter 6 deals with discrete-time signals and systems. The final chapter, Chapter 7, presents the state space or state variable concept and analysis for both discrete-time and continuous-time systems. In addition, background material on matrix analysis needed for Chapter 7 is included in Appendix A.

I am grateful to Professor Gordon Silverman of Manhattan College for his assistance, comments, and careful review of the manuscript. I also wish to thank the staff of the McGraw-Hill Schaum Series, especially John Aliano for his helpful comments and suggestions and Maureen Walker for her great care in preparing this book. Last, I am indebted to my wife, Daisy, whose understanding and constant support were necessary factors in the completion of this work.

HWEI P. HSU
Montville, New Jersey

To the Student

To understand the material in this text, the reader is assumed to have a basic knowledge of calculus, along with some knowledge of differential equations and the first circuit course in electrical engineering.

This text covers both continuous-time and discrete-time signals and systems. If the course you are taking covers only continuous-time signals and systems, you may study parts of Chapters 1 and 2 covering the continuous-time case, Chapters 3 and 5, and the second part of Chapter 7. If the course you are taking covers only discrete-time signals and systems, you may study parts of Chapters 1 and 2 covering the discrete-time case, Chapters 4 and 6, and the first part of Chapter 7.

To really master a subject, a continuous interplay between skills and knowledge must take place. By studying and reviewing many solved problems and seeing how each problem is approached and how it is solved, you can learn the skills of solving problems easily and increase your store of necessary knowledge. Then, to test and reinforce your learned skills, it is imperative that you work out the supplementary problems (hints and answers are provided). I would like to emphasize that there is no short cut to learning except by "doing."

Contents

* The laptop icon next to an exercise indicates that the exercise is also available as a video with step-by-step instructions. These videos are available on the Schaums.com website by following the instructions on the inside front cover.

CHAPTER 1

Signals and Systems

1.1 Introduction

The concept and theory of signals and systems are needed in almost all electrical engineering fields and in many other engineering and scientific disciplines as well. In this chapter we introduce the mathematical description and representation of signals and systems and their classifications. We also define several important basic signals essential to our studies.

1.2 Signals and Classification of Signals

A *signal* is a function representing a physical quantity or variable, and typically it contains information about the behavior or nature of the phenomenon. For instance, in an *RC* circuit the signal may represent the voltage across the capacitor or the current flowing in the resistor. Mathematically, a signal is represented as a function of an independent variable t. Usually t represents time. Thus, a signal is denoted by $x(t)$.

A. Continuous-Time and Discrete-Time Signals:

A signal $x(t)$ is a *continuous-time* signal if t is a continuous variable. If t is a discrete variable—that is, $x(t)$ is defined at discrete times—then $x(t)$ is a *discrete-time* signal. Since a discrete-time signal is defined at discrete times, a discrete-time signal is often identified as a *sequence* of numbers, denoted by $\{x_n\}$ or $x[n]$, where n = integer. Illustrations of a continuous-time signal $x(t)$ and of a discrete-time signal $x[n]$ are shown in Fig. 1-1.

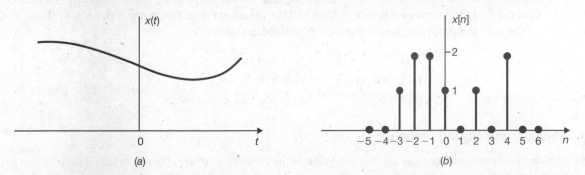

Fig. 1-1 Graphical representation of (a) continuous-time and (b) discrete-time signals.

A discrete-time signal $x[n]$ may represent a phenomenon for which the independent variable is inherently discrete. For instance, the daily closing stock market average is by its nature a signal that evolves at discrete points in time (that is, at the close of each day). On the other hand a discrete-time signal $x[n]$ may be obtained by *sampling* a continuous-time signal $x(t)$ such as

$$x(t_0), x(t_1), \ldots, x(t_n), \ldots$$

or in a shorter form as

$$x[0], x[1], \ldots, x[n], \ldots$$

or

$$x_0, x_1, \ldots, x_n, \ldots$$

where we understand that

$$x_n = x[n] = x(t_n)$$

and x_n's are called *samples* and the time interval between them is called the *sampling interval*. When the sampling intervals are equal (uniform sampling), then

$$x_n = x[n] = x(nT_s)$$

where the constant T_s is the sampling interval.

A discrete-time signal $x[n]$ can be defined in two ways:

1. We can specify a rule for calculating the nth value of the sequence. For example,

$$x[n] = x_n = \begin{cases} \left(\dfrac{1}{2}\right)^n & n \geq 0 \\ 0 & n < 0 \end{cases}$$

or

$$\{x_n\} = \left\{ 1, \frac{1}{2}, \frac{1}{4}, \ldots, \left(\frac{1}{2}\right)^n, \ldots \right\}$$

2. We can also explicitly list the values of the sequence. For example, the sequence shown in Fig. 1-1(b) can be written as

$$\{x_n\} = \{\ldots, 0, 0, 1, 2, 2, 1, 0, 1, 0, 2, 0, 0, \ldots\}$$
$$\uparrow$$

or

$$\{x_n\} = \{1, 2, 2, 1, 0, 1, 0, 2\}$$
$$\uparrow$$

We use the arrow to denote the $n = 0$ term. We shall use the convention that if no arrow is indicated, then the first term corresponds to $n = 0$ and all the values of the sequence are zero for $n < 0$.

The sum and product of two sequences are defined as follows:

$$\{c_n\} = \{a_n\} + \{b_n\} \rightarrow c_n = a_n + b_n$$
$$\{c_n\} = \{a_n\}\{b_n\} \quad \rightarrow c_n = a_n b_n$$
$$\{c_n\} = \alpha\{a_n\} \quad\quad \rightarrow c_n = \alpha a_n \quad\quad \alpha = \text{constant}$$

B. Analog and Digital Signals:

If a continuous-time signal $x(t)$ can take on any value in the continuous interval (a, b), where a may be $-\infty$ and b may be $+\infty$, then the continuous-time signal $x(t)$ is called an *analog* signal. If a discrete-time signal $x[n]$ can take on only a finite number of distinct values, then we call this signal a *digital* signal.

C. Real and Complex Signals:

A signal $x(t)$ is a *real* signal if its value is a real number, and a signal $x(t)$ is a *complex* signal if its value is a complex number. A general complex signal $x(t)$ is a function of the form

$$x(t) = x_1(t) + jx_2(t) \tag{1.1}$$

where $x_1(t)$ and $x_2(t)$ are real signals and $j = \sqrt{-1}$.

Note that in Eq. (1.1) t represents either a continuous or a discrete variable.

D. Deterministic and Random Signals:

Deterministic signals are those signals whose values are completely specified for any given time. Thus, a deterministic signal can be modeled by a known function of time t. *Random* signals are those signals that take random values at any given time and must be characterized statistically. Random signals will be discussed in Chaps. 8 and 9.

E. Even and Odd Signals:

A signal $x(t)$ or $x[n]$ is referred to as an *even* signal if

$$x(-t) = x(t)$$
$$x[-n] = x[n] \tag{1.2}$$

A signal $x(t)$ or $x[n]$ is referred to as an *odd* signal if

$$x(-t) = -x(t)$$
$$x[-n] = -x[n] \tag{1.3}$$

Examples of even and odd signals are shown in Fig. 1-2.

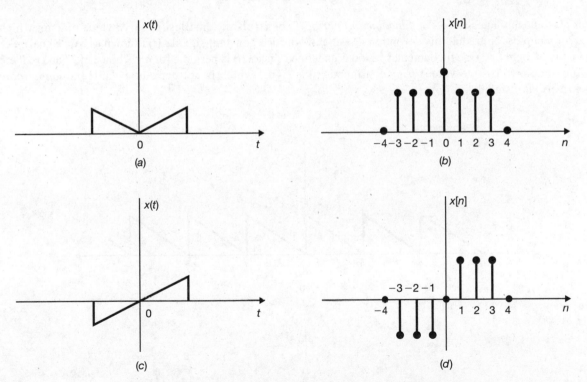

Fig. 1-2 Examples of even signals (*a* and *b*) and odd signals (*c* and *d*).

Any signal $x(t)$ or $x[n]$ can be expressed as a sum of two signals, one of which is even and one of which is odd. That is,

$$x(t) = x_e(t) + x_o(t)$$
$$x[n] = x_e[n] + x_o[n] \tag{1.4}$$

Where

$$x_e(t) = \frac{1}{2}\{x(t) + x(-t)\} \qquad \text{even part of } x(t)$$

$$x_e[n] = \frac{1}{2}\{x[n] + x[-n]\} \qquad \text{even part of } x[n]$$

$$(1.5)$$

$$x_o(t) = \frac{1}{2}\{x(t) - x(-t)\} \qquad \text{odd part of } x(t)$$

$$x_o[n] = \frac{1}{2}\{x[n] - x[-n]\} \qquad \text{odd part of } x[n]$$

$$(1.6)$$

Note that the product of two even signals or of two odd signals is an even signal and that the product of an even signal and an odd signal is an odd signal (Prob. 1.7).

F. Periodic and Nonperiodic Signals:

A continuous-time signal $x(t)$ is said to be *periodic with period T* if there is a positive nonzero value of T for which

$$x(t + T) = x(t) \qquad \text{all } t \tag{1.7}$$

An example of such a signal is given in Fig. 1-3(*a*). From Eq. (1.7) or Fig. 1-3(*a*) it follows that

$$x(t + mT) = x(t) \tag{1.8}$$

for all t and any integer m. The *fundamental period* T_0 of $x(t)$ is the smallest positive value of T for which Eq. (1.7) holds. Note that this definition does not work for a constant signal $x(t)$ (known as a dc signal). For a constant signal $x(t)$ the fundamental period is undefined since $x(t)$ is periodic for *any* choice of T (and so there is no smallest positive value). Any continuous-time signal which is not periodic is called a *nonperiodic* (or *aperiodic*) signal.

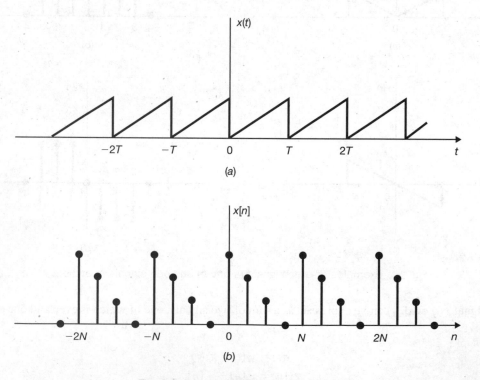

Fig. 1-3 Examples of periodic signals.

Periodic discrete-time signals are defined analogously. A sequence (discrete-time signal) $x[n]$ is *periodic with period N* if there is a positive integer N for which

$$x[n + N] = x[n] \qquad \text{all } n \tag{1.9}$$

An example of such a sequence is given in Fig. 1-3(*b*). From Eq. (1.9) and Fig. 1-3(*b*) it follows that

$$x[n + mN] = x[n] \tag{1.10}$$

for all n and any integer m. The fundamental period N_0 of $x[n]$ is the smallest positive integer N for which Eq. (1.9) holds. Any sequence which is not periodic is called a *nonperiodic* (or *aperiodic*) sequence.

Note that a sequence obtained by uniform sampling of a periodic continuous-time signal may not be periodic (Probs. 1.12 and 1.13). Note also that the sum of two continuous-time periodic signals may not be periodic but that the sum of two periodic sequences is always periodic (Probs. 1.14 and 1.15).

G. Energy and Power Signals:

Consider $v(t)$ to be the voltage across a resistor R producing a current $i(t)$. The instantaneous power $p(t)$ per ohm is defined as

$$p(t) = \frac{v(t)i(t)}{R} = i^2(t) \tag{1.11}$$

Total energy E and average power P on a per-ohm basis are

$$E = \int_{-\infty}^{\infty} i^2(t)\, dt \quad \text{joules} \tag{1.12}$$

$$P = \lim_{T \to \infty} \frac{1}{T} \int_{-T/2}^{T/2} i^2(t)\, dt \quad \text{watts} \tag{1.13}$$

For an arbitrary continuous-time signal $x(t)$, the *normalized energy content E* of $x(t)$ is defined as

$$E = \int_{-\infty}^{\infty} |x(t)|^2\, dt \tag{1.14}$$

The *normalized average power P* of $x(t)$ is defined as

$$P = \lim_{T \to \infty} \frac{1}{T} \int_{-T/2}^{T/2} |x(t)|^2\, dt \tag{1.15}$$

Similarly, for a discrete-time signal $x[n]$, the normalized energy content E of $x[n]$ is defined as

$$E = \sum_{n=-\infty}^{\infty} |x[n]|^2 \tag{1.16}$$

The normalized average power P of $x[n]$ is defined as

$$P = \lim_{N \to \infty} \frac{1}{2N + 1} \sum_{n=-N}^{N} |x[n]|^2 \tag{1.17}$$

Based on definitions (1.14) to (1.17), the following classes of signals are defined:

1. $x(t)$ (or $x[n]$) is said to be an *energy* signal (or sequence) if and only if $0 < E < \infty$, and so $P = 0$.
2. $x(t)$ (or $x[n]$) is said to be a *power* signal (or sequence) if and only if $0 < P < \infty$, thus implying that $E = \infty$.
3. Signals that satisfy neither property are referred to as neither energy signals nor power signals.

Note that a periodic signal is a power signal if its energy content per period is finite, and then the average power of this signal need only be calculated over a period (Prob. 1.18).

1.3   Basic Continuous-Time Signals

A.   The Unit Step Function:

The *unit step* function $u(t)$, also known as the *Heaviside unit* function, is defined as

$$u(t) = \begin{cases} 1 & t > 0 \\ 0 & t < 0 \end{cases} \tag{1.18}$$

which is shown in Fig. 1-4 (*a*). Note that it is discontinuous at $t = 0$ and that the value at $t = 0$ is undefined. Similarly, the shifted unit step function $u(t - t_0)$ is defined as

$$u(t - t_0) = \begin{cases} 1 & t > t_0 \\ 0 & t < t_0 \end{cases} \tag{1.19}$$

which is shown in Fig. 1-4 (*b*)

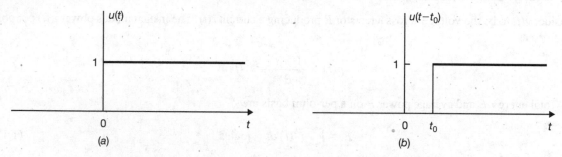

Fig. 1-4 (*a*) Unit step function; (*b*) shifted unit step function.

B.   The Unit Impulse Function:

The *unit impulse* function $\delta(t)$, also known as the *Dirac delta* function, plays a central role in system analysis. Traditionally, $\delta(t)$ is often defined as the limit of a suitably chosen conventional function having unity area over an infinitesimal time interval as shown in Fig. 1-5 and possesses the following properties:

$$\delta(t) = \begin{cases} 0 & t \neq 0 \\ \infty & t = 0 \end{cases}$$

$$\int_{-\varepsilon}^{\varepsilon} \delta(t)\, dt = 1$$

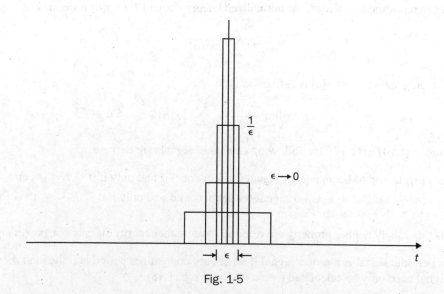

Fig. 1-5

But an ordinary function which is everywhere 0 except at a single point must have the integral 0 (in the Riemann integral sense). Thus, $\delta(t)$ cannot be an ordinary function and mathematically it is defined by

$$\int_{-\infty}^{\infty} \phi(t)\delta(t)\,dt = \phi(0) \tag{1.20}$$

where $\phi(t)$ is any regular function continuous at $t = 0$.

An alternative definition of $\delta(t)$ is given by

$$\int_{a}^{b} \phi(t)\delta(t)\,dt = \begin{cases} \phi(0) & a < 0 < b \\ 0 & a < b < 0 \quad \text{or} \quad 0 < a < b \\ \text{undefined} & a = 0 \quad \text{or} \quad b = 0 \end{cases} \tag{1.21}$$

Note that Eq. (1.20) or (1.21) is a symbolic expression and should not be considered an ordinary Riemann integral. In this sense, $\delta(t)$ is often called a *generalized function* and $\phi(t)$ is known as a *testing function*. A different class of testing functions will define a different generalized function (Prob. 1.24). Similarly, the delayed delta function $\delta(t - t_0)$ is defined by

$$\int_{-\infty}^{\infty} \phi(t)\delta(t - t_0)\,dt = \phi(t_0) \tag{1.22}$$

where $\phi(t)$ is any regular function continuous at $t = t_0$. For convenience, $\delta(t)$ and $\delta(t - t_0)$ are depicted graphically as shown in Fig. 1-6.

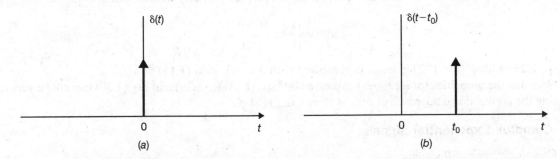

Fig. 1-6 (a) Unit impulse function; (b) shifted unit impulse function.

Some additional properties of $\delta(t)$ are

$$\delta(at) = \frac{1}{|a|}\delta(t) \tag{1.23}$$

$$\delta(-t) = \delta(t) \tag{1.24}$$

$$x(t)\delta(t) = x(0)\delta(t) \tag{1.25}$$

if $x(t)$ is continuous at $t = 0$.

$$x(t)\delta(t - t_0) = x(t_0)\delta(t - t_0) \tag{1.26}$$

if $x(t)$ is continuous at $t = t_0$.

Using Eqs. (1.22) and (1.24), any continuous-time signal $x(t)$ can be expressed as

$$x(t) = \int_{-\infty}^{\infty} x(\tau)\delta(t - \tau)\,d\tau \tag{1.27}$$

Generalized Derivatives:

If $g(t)$ is a generalized function, its nth generalized derivative $g^{(n)}(t) = d^n g(t)/dt^n$ is defined by the following relation:

$$\int_{-\infty}^{\infty} \phi(t) g^{(n)}(t)\, dt = (-1)^n \int_{-\infty}^{\infty} \phi^{(n)}(t) g(t)\, dt \tag{1.28}$$

where $\phi(t)$ is a testing function which can be differentiated an arbitrary number of times and vanishes outside some fixed interval and $\phi^{(n)}(t)$ is the nth derivative of $\phi(t)$. Thus, by Eqs. (1.28) and (1.20) the derivative of $\delta(t)$ can be defined as

$$\int_{-\infty}^{\infty} \phi(t) \delta'(t)\, dt = -\phi'(0) \tag{1.29}$$

where $\phi(t)$ is a testing function which is continuous at $t = 0$ and vanishes outside some fixed interval and $\phi'(0) = d\phi(t)/dt\big|_{t=0}$. Using Eq. (1.28), the derivative of $u(t)$ can be shown to be $\delta(t)$ (Prob. 1.28); that is,

$$\delta(t) = u'(t) = \frac{du(t)}{dt} \tag{1.30}$$

Then the unit step function $u(t)$ can be expressed as

$$u(t) = \int_{-\infty}^{t} \delta(\tau)\, d\tau \tag{1.31}$$

Note that the unit step function $u(t)$ is discontinuous at $t = 0$; therefore, the derivative of $u(t)$ as shown in Eq. (1.30) is not the derivative of a function in the ordinary sense and should be considered a generalized derivative in the sense of a generalized function. From Eq. (1.31) we see that $u(t)$ is undefined at $t = 0$ and

$$u(t) = \begin{cases} 1 & t > 0 \\ 0 & t < 0 \end{cases}$$

by Eq. (1.21) with $\phi(t) = 1$. This result is consistent with the definition (1.18) of $u(t)$.

Note that the properties (or identities) expressed by Eqs. (1.23) to (1.26) and Eq. (1.30) can not be verified by using the conventional approach of $\delta(t)$ as shown in Fig. 1-5.

C. Complex Exponential Signals:

The *Complex exponential* signal

$$x(t) = e^{j\omega_0 t} \tag{1.32}$$

is an important example of a complex signal. Using Euler's formula, this signal can be defined as

$$x(t) = e^{j\omega_0 t} = \cos \omega_0 t + j \sin \omega_0 t \tag{1.33}$$

Thus, $x(t)$ is a complex signal whose real part is $\cos \omega_0 t$ and imaginary part is $\sin \omega_0 t$. An important property of the complex exponential signal $x(t)$ in Eq. (1.32) is that it is periodic. The fundamental period T_0 of $x(t)$ is given by (Prob. 1.9)

$$T_0 = \frac{2\pi}{\omega_0} \tag{1.34}$$

Note that $x(t)$ is periodic for any value of ω_0.

General Complex Exponential Signals:

Let $s = \sigma + j\omega$ be a complex number. We define $x(t)$ as

$$x(t) = e^{st} = e^{(\sigma + j\omega)t} = e^{\sigma t}(\cos \omega t + j \sin \omega t) \tag{1.35}$$

Then signal $x(t)$ in Eq. (1.35) is known as a *general complex exponential* signal whose real part $e^{\sigma t} \cos \omega t$ and imaginary part $e^{\sigma t} \sin \omega t$ are exponentially increasing ($\sigma > 0$) or decreasing ($\sigma < 0$) sinusoidal signals (Fig. 1-7).

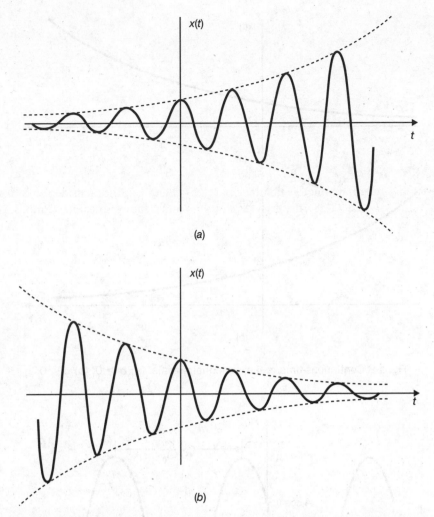

Fig. 1-7 (*a*) Exponentially increasing sinusoidal signal; (*b*) exponentially decreasing sinusoidal signal.

Real Exponential Signals:

Note that if $s = \sigma$ (a real number), then Eq. (1.35) reduces to a *real exponential* signal

$$x(t) = e^{\sigma t} \tag{1.36}$$

As illustrated in Fig. 1-8, if $\sigma > 0$, then $x(t)$ is a growing exponential; and if $\sigma < 0$, then $x(t)$ is a decaying exponential.

D. Sinusoidal Signals:

A continuous-time *sinusoidal* signal can be expressed as

$$x(t) = A \cos(\omega_0 t + \theta) \tag{1.37}$$

where A is the *amplitude* (real), ω_0 is the *radian frequency* in radians per second, and θ is the *phase angle* in radians. The sinusoidal signal $x(t)$ is shown in Fig. 1-9, and it is periodic with fundamental period

$$T_0 = \frac{2\pi}{\omega_0} \tag{1.38}$$

The reciprocal of the fundamental period T_0 is called the *fundamental frequency f_0*:

$$f_0 = \frac{1}{T_0} \quad \text{hertz (Hz)} \tag{1.39}$$

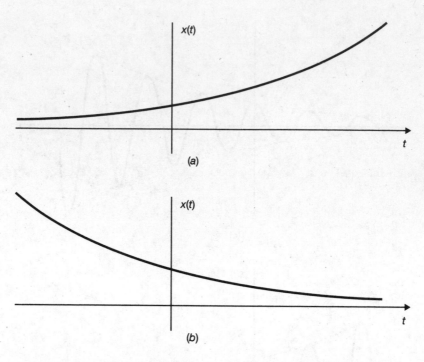

Fig. 1-8 Continuous-time real exponential signals. (a) $\sigma > 0$; (b) $\sigma < 0$.

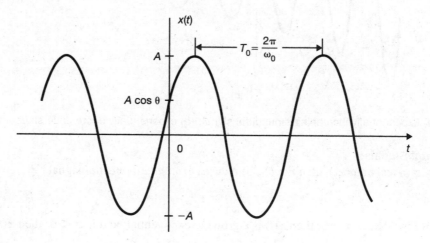

Fig. 1-9 Continuous-time sinusoidal signal.

From Eqs. (1.38) and (1.39) we have

$$\omega_0 = 2\pi f_0 \qquad\qquad (1.40)$$

which is called the *fundamental angular frequency*. Using Euler's formula, the sinusoidal signal in Eq. (1.37) can be expressed as

$$A\cos(\omega_0 t + \theta) = A\operatorname{Re}\{e^{j(\omega_0 t + \theta)}\} \qquad\qquad (1.41)$$

where "Re" denotes "real part of." We also use the notation "Im" to denote "imaginary part of." Then

$$A\operatorname{Im}\{e^{j(\omega_0 t + \theta)}\} = A\sin(\omega_0 t + \theta) \qquad\qquad (1.42)$$

1.4   Basic Discrete-Time Signals

A.   The Unit Step Sequence:

The *unit step* sequence $u[n]$ is defined as

$$u[n] = \begin{cases} 1 & n \geq 0 \\ 0 & n < 0 \end{cases} \tag{1.43}$$

which is shown in Fig. 1-10 (a). Note that the value of $u[n]$ at $n = 0$ is defined [unlike the continuous-time step function $u(t)$ at $t = 0$] and equals unity. Similarly, the shifted unit step sequence $u[n - k]$ is defined as

$$u[n - k] = \begin{cases} 1 & n \geq k \\ 0 & n < k \end{cases} \tag{1.44}$$

which is shown in Fig. 1-10 (b).

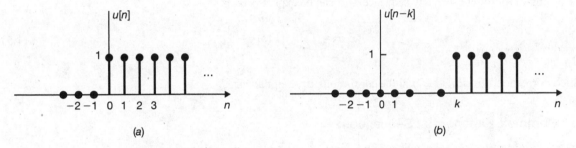

(a)   (b)

Fig. 1-10 (a) Unit step sequence; (b) shifted unit step sequence.

B.   The Unit Impulse Sequence:

The *unit impulse* (or *unit sample*) sequence $\delta[n]$ is defined as

$$\delta[n] = \begin{cases} 1 & n = 0 \\ 0 & n \neq 0 \end{cases} \tag{1.45}$$

which is shown in Fig. 1-11(a). Similarly, the shifted unit impulse (or sample) sequence $\delta[n - k]$ is defined as

$$\delta[n - k] = \begin{cases} 1 & n = k \\ 0 & n \neq k \end{cases} \tag{1.46}$$

which is shown in Fig. 1-11(b).

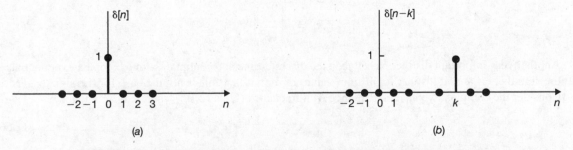

(a)   (b)

Fig. 1-11 (a) Unit impulse (sample) sequence; (b) shifted unit impulse sequence.

Unlike the continuous-time unit impulse function $\delta(t)$, $\delta[n]$ is defined without mathematical complication or difficulty. From definitions (1.45) and (1.46) it is readily seen that

$$x[n]\delta[n] = x[0]\delta[n] \tag{1.47}$$

$$x[n]\delta[n - k] = x[k]\delta[n - k] \tag{1.48}$$

which are the discrete-time counterparts of Eqs. (1.25) and (1.26), respectively. From definitions (1.43) to (1.46), $\delta[n]$ and $u[n]$ are related by

$$\delta[n] = u[n] - u[n - 1] \tag{1.49}$$

$$u[n] = \sum_{k=-\infty}^{n} \delta[k] = \sum_{k=0}^{\infty} \delta[n - k] \tag{1.50}$$

which are the discrete-time counterparts of Eqs. (1.30) and (1.31), respectively.

Using definition (1.46), any sequence $x[n]$ can be expressed as

$$x[n] = \sum_{k=-\infty}^{\infty} x[k]\delta[n - k] \tag{1.51}$$

which corresponds to Eq. (1.27) in the continuous-time signal case.

C. Complex Exponential Sequences:

The *complex exponential* sequence is of the form

$$x[n] = e^{j\Omega_0 n} \tag{1.52}$$

Again, using Euler's formula, $x[n]$ can be expressed as

$$x[n] = e^{j\Omega_0 n} = \cos \Omega_0 n + j \sin \Omega_0 n \tag{1.53}$$

Thus, $x[n]$ is a complex sequence whose real part is $\cos \Omega_0 n$ and imaginary part is $\sin \Omega_0 n$.

Periodicity of $e^{j\Omega_0 n}$:

In order for $e^{j\Omega_0 n}$ to be periodic with period N (> 0), Ω_0 must satisfy the following condition (Prob. 1.11):

$$\frac{\Omega_0}{2\pi} = \frac{m}{N} \qquad m = \text{positive integer} \tag{1.54}$$

Thus, the sequence $e^{j\Omega_0 n}$ is not periodic for any value of Ω_0. It is periodic only if $\Omega_0/2\pi$ is a rational number. Note that this property is quite different from the property that the continuous-time signal $e^{j\omega_0 t}$ is periodic for any value of ω_0. Thus, if Ω_0 satisfies the periodicity condition in Eq. (1.54), $\Omega_0 \neq 0$, and N and m have no factors in common, then the fundamental period of the sequence $x[n]$ in Eq. (1.52) is N_0 given by

$$N_0 = m\left(\frac{2\pi}{\Omega_0}\right) \tag{1.55}$$

Another very important distinction between the discrete-time and continuous-time complex exponentials is that the signals $e^{j\omega_0 t}$ are all distinct for distinct values of ω_0 but that this is not the case for the signals $e^{j\Omega_0 n}$.

Consider the complex exponential sequence with frequency $(\Omega_0 + 2\pi k)$, where k is an integer:

$$e^{j(\Omega_0 + 2\pi k)n} = e^{j\Omega_0 n}e^{j2\pi kn} = e^{j\Omega_0 n} \tag{1.56}$$

since $e^{j2\pi kn} = 1$. From Eq. (1.56) we see that the complex exponential sequence at frequency Ω_0 is the same as that at frequencies $(\Omega_0 \pm 2\pi)$, $(\Omega_0 \pm 4\pi)$, and so on. Therefore, in dealing with discrete-time exponentials, we need only consider an interval of length 2π in which to choose Ω_0. Usually, we will use the interval $0 \leq \Omega_0 < 2\pi$ or the interval $-\pi \leq \Omega_0 < \pi$.

General Complex Exponential Sequences:

The most general complex exponential sequence is often defined as

$$x[n] = C\alpha^n \tag{1.57}$$

where C and α are, in general, complex numbers. Note that Eq. (1.52) is the special case of Eq. (1.57) with $C = 1$ and $\alpha = e^{j\Omega_0}$.

Real Exponential Sequences:

If C and α in Eq. (1.57) are both real, then $x[n]$ is a real exponential sequence. Four distinct cases can be identified: $\alpha > 1$, $0 < \alpha < 1$, $-1 < \alpha < 0$, and $\alpha < -1$. These four real exponential sequences are shown in Fig. 1-12. Note that if $\alpha = 1$, $x[n]$ is a constant sequence, whereas if $\alpha = -1$, $x[n]$ alternates in value between $+C$ and $-C$.

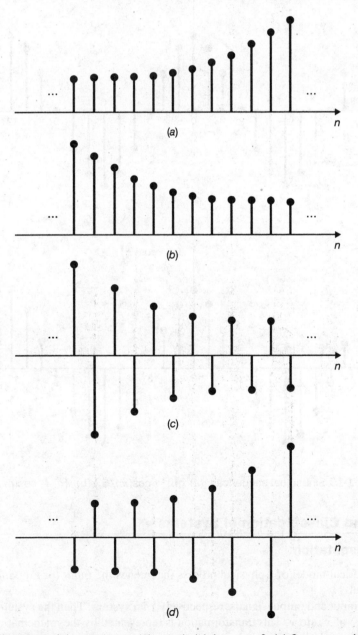

Fig. 1-12 Real exponential sequences. (a) $\alpha > 1$; (b) $1 > \alpha > 0$; (c) $0 > \alpha > -1$; (d) $\alpha < -1$.

D. Sinusoidal Sequences:

A *sinusoidal* sequence can be expressed as

$$x[n] = A \cos(\Omega_0 n + \theta) \tag{1.58}$$

If n is dimensionless, then both Ω_0 and θ have units of radians. Two examples of sinusoidal sequences are shown in Fig. 1-13. As before, the sinusoidal sequence in Eq. (1.58) can be expressed as

$$A \cos(\Omega_0 n + \theta) = A \operatorname{Re}\{e^{j(\Omega_0 n + \theta)}\} \tag{1.59}$$

As we observed in the case of the complex exponential sequence in Eq. (1.52), the same observations [Eqs. (1.54) and (1.56)] also hold for sinusoidal sequences. For instance, the sequence in Fig. 1-13(*a*) is periodic with fundamental period 12, but the sequence in Fig. 1-13(*b*) is not periodic.

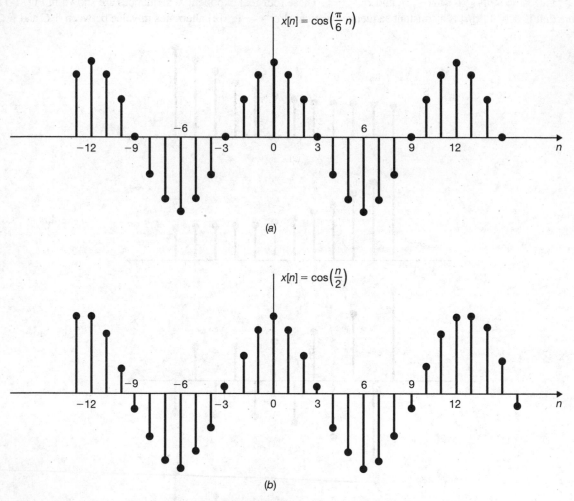

Fig. 1-13 Sinusoidal sequences. (*a*) $x[n] = \cos(\pi n/6)$; (*b*) $x[n] = \cos(n/2)$.

1.5 Systems and Classification of Systems

A. System Representation:

A *system* is a mathematical model of a physical process that relates the *input* (or *excitation*) signal to the *output* (or *response*) signal.

Let x and y be the input and output signals, respectively, of a system. Then the system is viewed as a *transformation* (or *mapping*) of x into y. This transformation is represented by the mathematical notation

$$y = \mathbf{T}x \tag{1.60}$$

where **T** is the *operator* representing some well-defined rule by which x is transformed into y. Relationship (1.60) is depicted as shown in Fig. 1-14(a). Multiple input and/or output signals are possible, as shown in Fig. 1-14(b). We will restrict our attention for the most part in this text to the single-input, single-output case.

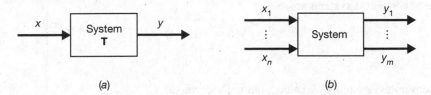

(a)　　　　　　　　　　　　　　　　　　(b)

Fig. 1-14　System with single or multiple input and output signals.

B.　Deterministic and Stochastic Systems:

If the input and output signals x and y are deterministic signals, then the system is called a deterministic system. If the input and output signals x and y are random signals, then the system is called a *stochastic* system.

C.　Continuous-Time and Discrete-Time Systems:

If the input and output signals x and y are continuous-time signals, then the system is called a *continuous-time system* [Fig. 1-15(a)]. If the input and output signals are discrete-time signals or sequences, then the system is called a *discrete-time system* [Fig. 1-15(b)].

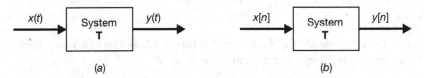

(a)　　　　　　　　　　　　　　　　　　(b)

Fig. 1-15　(a) Continuous-time system; (b) discrete-time system.

Note that in a continuous-time system the input $x(t)$ and output $y(t)$ are often expressed by a differential equation (see Prob. 1.32) and in a discrete-time system the input $x[n]$ and output $y[n]$ are often expressed by a difference equation (see Prob. 1.37).

D.　Systems with Memory and without Memory

A system is said to be *memoryless* if the output at any time depends on only the input at that same time. Otherwise, the system is said to have *memory*. An example of a memoryless system is a resistor R with the input $x(t)$ taken as the current and the voltage taken as the output $y(t)$. The input-output relationship (Ohm's law) of a resistor is

$$y(t) = Rx(t) \tag{1.61}$$

An example of a system with memory is a capacitor C with the current as the input $x(t)$ and the voltage as the output $y(t)$; then

$$y(t) = \frac{1}{C}\int_{-\infty}^{t} x(\tau)\, d\tau \tag{1.62}$$

A second example of a system with memory is a discrete-time system whose input and output sequences are related by

$$y[n] = \sum_{k=-\infty}^{n} x[k] \tag{1.63}$$

E.   Causal and Noncausal Systems:

A system is called *causal* if its output at the present time depends on only the present and/or past values of the input. Thus, in a causal system, it is not possible to obtain an output before an input is applied to the system. A system is called *noncausal* (or *anticipative*) if its output at the present time depends on future values of the input. Example of noncausal systems are

$$y(t) = x(t + 1) \tag{1.64}$$

$$y[n] = x[-n] \tag{1.65}$$

Note that all memoryless systems are causal, but not vice versa.

F.   Linear Systems and Nonlinear Systems:

If the operator $\mathbf{T}$ in Eq. (1.60) satisfies the following two conditions, then $\mathbf{T}$ is called a *linear operator* and the system represented by a linear operator $\mathbf{T}$ is called a *linear system*:

1.   Additivity:
Given that $\mathbf{T}x_1 = y_1$ and $\mathbf{T}x_2 = y_2$, then

$$\mathbf{T}\{x_1 + x_2\} = y_1 + y_2 \tag{1.66}$$

for any signals x_1 and x_2.

2.   Homogeneity (or *Scaling*):

$$\mathbf{T}\{\alpha x\} = \alpha y \tag{1.67}$$

for any signals x and any scalar α.

Any system that does not satisfy Eq. (1.66) and/or Eq. (1.67) is classified as a *nonlinear* system. Eqs. (1.66) and (1.67) can be combined into a single condition as

$$\mathbf{T}\{\alpha_1 x_1 + \alpha_2 x_2\} = \alpha_1 y_1 + \alpha_2 y_2 \tag{1.68}$$

where α_1 and α_2 are arbitrary scalars. Eq. (1.68) is known as the *superposition property*. Examples of linear systems are the resistor [Eq. (1.61)] and the capacitor [Eq. (1.62)]. Examples of nonlinear systems are

$$y = x^2 \tag{1.69}$$

$$y = \cos x \tag{1.70}$$

Note that a consequence of the homogeneity (or scaling) property [Eq. (1.67)] of linear systems is that *a zero input yields a zero output*. This follows readily by setting $\alpha = 0$ in Eq. (1.67). This is another important property of linear systems.

G.   Time-Invariant and Time-Varying Systems:

A system is called *time-invariant* if a time shift (delay or advance) in the input signal causes the same time shift in the output signal. Thus, for a continuous-time system, the system is time-invariant if

$$\mathbf{T}\{x(t - \tau)\} = y(t - \tau) \tag{1.71}$$

for any real value of τ. For a discrete-time system, the system is time-invariant (or *shift-invariant*) if

$$\mathbf{T}\{x[n - k]\} = y[n - k] \tag{1.72}$$

for any integer k. A system which does not satisfy Eq. (1.71) (continuous-time system) or Eq. (1.72) (discrete-time system) is called a *time-varying* system. To check a system for time-invariance, we can compare the shifted output with the output produced by the shifted input (Probs. 1.33 to 1.39).

H. Linear Time-Invariant Systems:

If the system is linear and also time-invariant, then it is called a *linear time-invariant* (LTI) system.

I. Stable Systems:

A system is *bounded-input/bounded-output* (BIBO) *stable* if for any bounded input x defined by

$$|x| \leq k_1 \tag{1.73}$$

the corresponding output y is also bounded defined by

$$|y| \leq k_2 \tag{1.74}$$

where k_1 and k_2 are finite real constants. An *unstable* system is one in which not all bounded inputs lead to bounded output. For example, consider the system where output $y[n]$ is given by $y[n] = (n + 1)u[n]$, and input $x[n] = u[n]$ is the unit step sequence. In this case the input $u[n] = 1$, but the output $y[n]$ increases without bound as n increases.

J. Feedback Systems:

A special class of systems of great importance consists of systems having *feedback*. In a *feedback system*, the output signal is fed back and added to the input to the system as shown in Fig. 1-16.

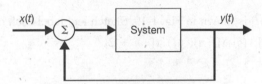

Fig. 1-16 Feedback system.

SOLVED PROBLEMS

Signals and Classification of Signals

1.1. A continuous-time signal $x(t)$ is shown in Fig. 1-17. Sketch and label each of the following signals.

(*a*) $x(t - 2)$; (*b*) $x(2t)$; (*c*) $x(t/2)$; (*d*) $x(-t)$

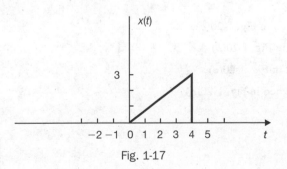

Fig. 1-17

(*a*) $x(t - 2)$ is sketched in Fig. 1-18(*a*).

(*b*) $x(2t)$ is sketched in Fig. 1-18(*b*).

(*c*) $x(t/2)$ is sketched in Fig. 1-18(*c*).

(*d*) $x(-t)$ is sketched in Fig. 1-18(*d*).

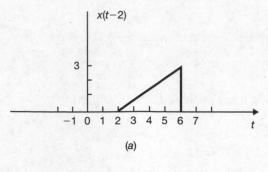

(a)

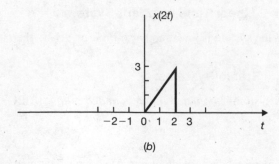

(b)

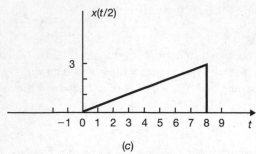

(c)

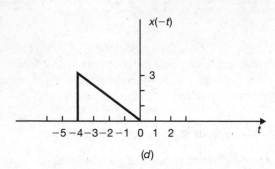

(d)

Fig. 1-18

1.2. A discrete-time signal $x[n]$ is shown in Fig. 1-19. Sketch and label each of the following signals.

(a) $x[n - 2]$; (b) $x[2n]$; (c) $x[-n]$; (d) $x[-n + 2]$

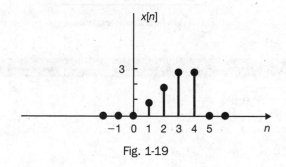

Fig. 1-19

(a) $x[n - 2]$ is sketched in Fig. 1-20(a).

(b) $x[2n]$ is sketched in Fig. 1-20(b).

(c) $x[-n]$ is sketched in Fig. 1-20(c).

(d) $x[-n + 2]$ is sketched in Fig. 1-20(d).

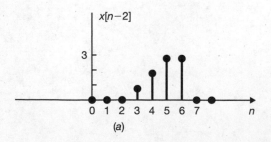

(a)

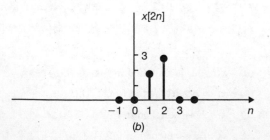

(b)

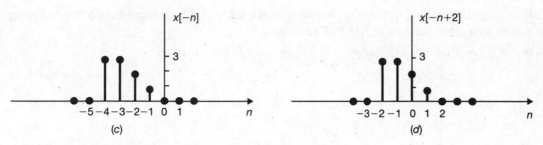

Fig. 1-20

1.3. Given the continuous-time signal specified by

$$x(t) = \begin{cases} 1 - |t| & -1 \le t \le 1 \\ 0 & \text{otherwise} \end{cases}$$

determine the resultant discrete-time sequence obtained by uniform sampling of $x(t)$ with a sampling interval of (a) 0.25 s, (b) 0.5 s, and (c) 1.0 s.

It is easier to take the graphical approach for this problem. The signal $x(t)$ is plotted in Fig. 1-21(a). Figs. 1-21(b) to (d) give plots of the resultant sampled sequences obtained for the three specified sampling intervals.

(a) $T_s = 0.25$ s. From Fig. 1-21(b) we obtain

$$x[n] = \{\dots, 0, 0.25, 0.5, 0.75, 1, 0.75, 0.5, 0.25, 0, \dots\}$$
$$\uparrow$$

(b) $T_s = 0.5$ s. From Fig. 1-21(c) we obtain

$$x[n] = \{\dots, 0, 0.5, 1, 0.5, 0, \dots\}$$
$$\uparrow$$

(c) $T_s = 1$ s. From Fig. 1-21(d) we obtain

$$x[n] = \{\dots, 0, 1, 0, \dots\} = \delta[n]$$
$$\uparrow$$

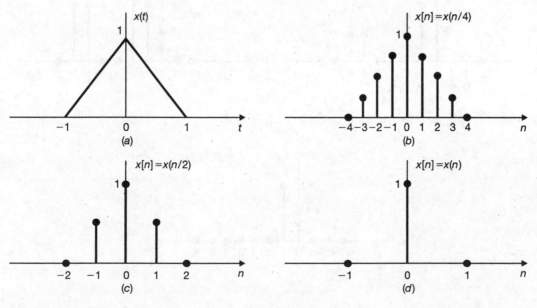

Fig. 1-21

1.4. Using the discrete-time signals $x_1[n]$ and $x_2[n]$ shown in Fig. 1-22, represent each of the following signals by a graph and by a sequence of numbers.

(*a*) $y_1[n] = x_1[n] + x_2[n]$; (*b*) $y_2[n] = 2x_1[n]$; (*c*) $y_3[n] = x_1[n]x_2[n]$

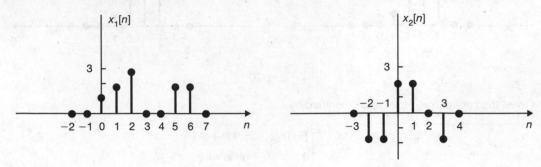

Fig. 1-22

(*a*) $y_1[n]$ is sketched in Fig. 1-23(*a*). From Fig. 1-23(*a*) we obtain

$$y_1[n] = \{\ldots, 0, -2, -2, 3, 4, 3, -2, 0, 2, 2, 0, \ldots\}$$
$$\uparrow$$

(*b*) $y_2[n]$ is sketched in Fig. 1-23(*b*). From Fig. 1-23(*b*) we obtain

$$y_2[n] = \{\ldots, 0, 2, 4, 6, 0, 0, 4, 4, 0, \ldots\}$$
$$\uparrow$$

(*c*) $y_3[n]$ is sketched in Fig. 1-23(*c*). From Fig. 1-23(*c*) we obtain

$$y_3[n] = \{\ldots, 0, 2, 4, 0, \ldots\}$$
$$\uparrow$$

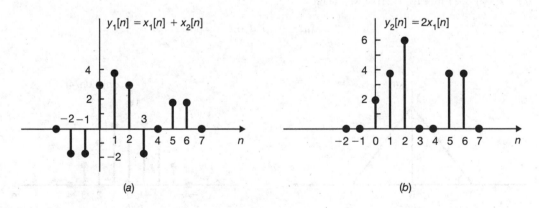

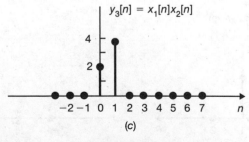

Fig. 1-23

1.5. Sketch and label the even and odd components of the signals shown in Fig. 1-24.

Using Eqs. (1.5) and (1.6), the even and odd components of the signals shown in Fig. 1-24 are sketched in Fig. 1-25.

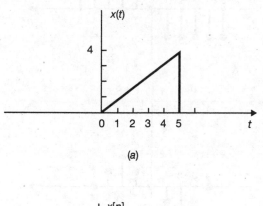

(a)

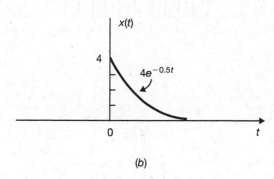

(b)

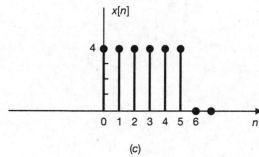

(c)

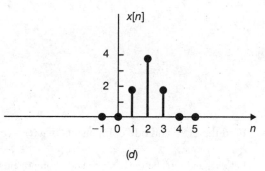

(d)

Fig. 1-24

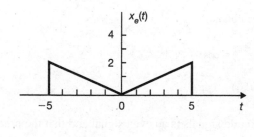

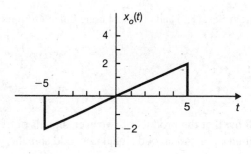

(a)

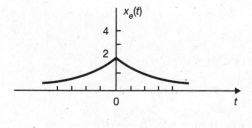

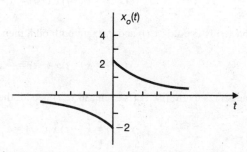

(b)

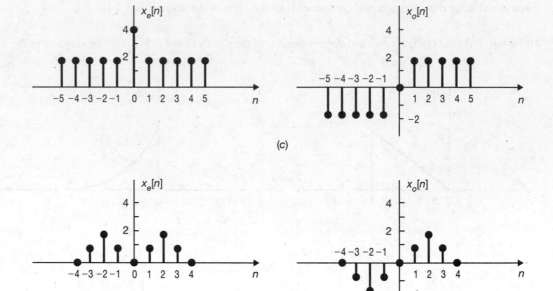

Fig. 1-25

1.6. Find the even and odd components of $x(t) = e^{jt}$.

Let $x_e(t)$ and $x_o(t)$ be the even and odd components of e^{jt}, respectively.

$$e^{jt} = x_e(t) + x_o(t)$$

From Eqs. (1.5) and (1.6) and using Euler's formula, we obtain

$$x_e(t) = \frac{1}{2}(e^{jt} + e^{-jt}) = \cos t$$

$$x_o(t) = \frac{1}{2}(e^{jt} - e^{-jt}) = j\sin t$$

1.7. Show that the product of two even signals or of two odd signals is an even signal and that the product of an even and an odd signal is an odd signal.

Let $x(t) = x_1(t)x_2(t)$. If $x_1(t)$ and $x_2(t)$ are both even, then

$$x(-t) = x_1(-t)x_2(-t) = x_1(t)x_2(t) = x(t)$$

and $x(t)$ is even. If $x_1(t)$ and $x_2(t)$ are both odd, then

$$x(-t) = x_1(-t)x_2(-t) = -x_1(t)\,[-x_2(t)] = x_1(t)x_2(t) = x(t)$$

and $x(t)$ is even. If $x_1(t)$ is even and $x_2(t)$ is odd, then

$$x(-t) = x_1(-t)\,x_2(-t) = x_1(t)\,[-x_2(t)] = -x_1(t)x_2(t) = -x(t)$$

and $x(t)$ is odd. Note that in the above proof, variable t represents either a continuous or a discrete variable.

1.8. Show that

    (*a*)   If $x(t)$ and $x[n]$ are even, then

$$\int_{-a}^{a} x(t)\, dt = 2\int_{0}^{a} x(t)\, dt \tag{1.75a}$$

$$\sum_{n=-k}^{k} x[n] = x[0] + 2\sum_{n=1}^{k} x[n] \tag{1.75b}$$

    (*b*)   If $x(t)$ and $x[n]$ are odd, then

$$x(0) = 0 \qquad \text{and} \qquad x[0] = 0 \tag{1.76}$$

$$\int_{-a}^{a} x(t)\, dt = 0 \qquad \text{and} \qquad \sum_{n=-k}^{k} x[n] = 0 \tag{1.77}$$

    (*a*)   We can write

$$\int_{-a}^{a} x(t)\, dt = \int_{-a}^{0} x(t)\, dt + \int_{0}^{a} x(t)\, dt$$

Letting $t = -\lambda$ in the first integral on the right-hand side, we get

$$\int_{-a}^{0} x(t)\, dt = \int_{a}^{0} x(-\lambda)(-d\lambda) = \int_{0}^{a} x(-\lambda)\, d\lambda$$

Since $x(t)$ is even, that is, $x(-\lambda) = x(\lambda)$, we have

$$\int_{0}^{a} x(-\lambda)\, d\lambda = \int_{0}^{a} x(\lambda)\, d\lambda = \int_{0}^{a} x(t)\, dt$$

Hence,

$$\int_{-a}^{a} x(t)\, dt = \int_{0}^{a} x(t)\, dt + \int_{0}^{a} x(t)\, dt = 2\int_{0}^{a} x(t)\, dt$$

Similarly,

$$\sum_{n=-k}^{k} x[n] = \sum_{n=-k}^{-1} x[n] + x[0] + \sum_{n=1}^{k} x[n]$$

Letting $n = -m$ in the first term on the right-hand side, we get

$$\sum_{n=-k}^{-1} x[n] = \sum_{m=1}^{k} x[-m]$$

Since $x[n]$ is even, that is, $x[-m] = x[m]$, we have

$$\sum_{m=1}^{k} x[-m] = \sum_{m=1}^{k} x[m] = \sum_{n=1}^{k} x[n]$$

Hence,

$$\sum_{n=-k}^{k} x[n] = \sum_{n=1}^{k} x[n] + x[0] + \sum_{n=1}^{k} x[n] = x[0] + 2\sum_{n=1}^{k} x[n]$$

    (*b*)   Since $x(t)$ and $x[n]$ are odd, that is, $x(-t) = -x(t)$ and $x[-n] = -x[n]$, we have

$$x(-0) = -x(0) \qquad \text{and} \qquad x[-0] = -x[0]$$

Hence,

$$x(-0) = x(0) = -x(0) \Rightarrow x(0) = 0$$

$$x[-0] = x[0] = -x[0] \Rightarrow x[0] = 0$$

Similarly,

$$\int_{-a}^{a} x(t)\, dt = \int_{-a}^{0} x(t)\, dt + \int_{0}^{a} x(t)\, dt = \int_{0}^{a} x(-\lambda)\, d\lambda + \int_{0}^{a} x(t)\, dt$$

$$= -\int_{0}^{a} x(\lambda)\, d\lambda + \int_{0}^{a} x(t)\, dt = -\int_{0}^{a} x(t)\, dt + \int_{0}^{a} x(t)\, dt = 0$$

and

$$\sum_{n=-k}^{k} x[n] = \sum_{n=-k}^{-1} x[n] + x[0] + \sum_{n=1}^{k} x[n] = \sum_{m=1}^{k} x[-m] + x[0] + \sum_{n=1}^{k} x[n]$$

$$= -\sum_{m=1}^{k} x[m] + x[0] + \sum_{n=1}^{k} x[n] = -\sum_{n=1}^{k} x[n] + x[0] + \sum_{n=1}^{k} x[n]$$

$$= x[0] = 0$$

in view of Eq. (1.76).

1.9. Show that the complex exponential signal

$$x(t) = e^{j\omega_0 t}$$

is periodic and that its fundamental period is $2\pi/\omega_0$.

By Eq. (1.7), $x(t)$ will be periodic if

$$e^{j\omega_0(t + T)} = e^{j\omega_0 t}$$

Since

$$e^{j\omega_0(t + T)} = e^{j\omega_0 t} e^{j\omega_0 T}$$

we must have

$$e^{j\omega_0 T} = 1 \tag{1.78}$$

If $\omega_0 = 0$, then $x(t) = 1$, which is periodic for any value of T. If $\omega_0 \neq 0$, Eq. (1.78) holds if

$$\omega_0 T = m2\pi \quad \text{or} \quad T = m\frac{2\pi}{\omega_0} \quad m = \text{positive integer}$$

Thus, the fundamental period T_0, the smallest positive T, of $x(t)$ is given by $2\pi/\omega_0$.

1.10. Show that the sinusoidal signal

$$x(t) = \cos(\omega_0 t + \theta)$$

is periodic and that its fundamental period is $2\pi/\omega_0$.

The sinusoidal signal $x(t)$ will be periodic if

$$\cos[\omega_0(t + T) + \theta] = \cos(\omega_0 t + \theta)$$

We note that

$$\cos[\omega_0(t+T)+\theta] = \cos[\omega_0 t + \theta + \omega_0 T] = \cos(\omega_0 t + \theta)$$

if

$$\omega_0 T = m2\pi \qquad \text{or} \qquad T = m\frac{2\pi}{\omega_0} \qquad m = \text{positive integer}$$

Thus, the fundamental period T_0 of $x(t)$ is given by $2\pi/\omega_0$.

1.11. Show that the complex exponential sequence

$$x[n] = e^{j\Omega_0 n}$$

is periodic only if $\Omega_0/2\pi$ is a rational number.

By Eq. (1.9), $x[n]$ will be periodic if

$$e^{j\Omega_0(n+N)} = e^{j\Omega_0 n}e^{j\Omega_0 N} = e^{j\Omega_0 n}$$

or

$$e^{j\Omega_0 N} = 1 \tag{1.79}$$

Equation (1.79) holds only if

$$\Omega_0 N = m2\pi \qquad m = \text{positive integer}$$

or

$$\frac{\Omega_0}{2\pi} = \frac{m}{N} = \text{rational numbers} \tag{1.80}$$

Thus, $x[n]$ is periodic only if $\Omega_0/2\pi$ is a rational number.

1.12. Let $x(t)$ be the complex exponential signal

$$x(t) = e^{j\omega_0 t}$$

with radian frequency ω_0 and fundamental period $T_0 = 2\pi/\omega_0$. Consider the discrete-time sequence $x[n]$ obtained by uniform sampling of $x(t)$ with sampling interval T_s. That is,

$$x[n] = x(nT_s) = e^{j\omega_0 nT_s}$$

Find the condition on the value of T_s so that $x[n]$ is periodic.

If $x[n]$ is periodic with fundamental period N_0, then

$$e^{j\omega_0(n+N_0)T_s} = e^{j\omega_0 nT_s}e^{j\omega_0 N_0 T_s} = e^{j\omega_0 nT_s}$$

Thus, we must have

$$e^{j\omega_0 N_0 T_s} = 1 \Rightarrow \omega_0 N_0 T_s = \frac{2\pi}{T_0}N_0 T_s = m2\pi \qquad m = \text{positive integer}$$

or

$$\frac{T_s}{T_0} = \frac{m}{N_0} = \text{rational number} \tag{1.81}$$

Thus, $x[n]$ is periodic if the ratio T_s/T_0 of the sampling interval and the fundamental period of $x(t)$ is a rational number.

Note that the above condition is also true for sinusoidal signals $x(t) = \cos(\omega_0 t + \theta)$.

1.13. Consider the sinusoidal signal

$$x(t) = \cos 15t$$

(a) Find the value of sampling interval T_s such that $x[n] = x(nT_s)$ is a periodic sequence.

(b) Find the fundamental period of $x[n] = x(nT_s)$ if $T_s = 0.1\pi$ seconds.

(a) The fundamental period of $x(t)$ is $T_0 = 2\pi/\omega_0 = 2\pi/15$. By Eq. (1.81), $x[n] = x(nT_s)$ is periodic if

$$\frac{T_s}{T_0} = \frac{T_s}{2\pi/15} = \frac{m}{N_0} \tag{1.82}$$

where m and N_0 are positive integers. Thus, the required value of T_s is given by

$$T_s = \frac{m}{N_0} T_0 = \frac{m}{N_0} \frac{2\pi}{15} \tag{1.83}$$

(b) Substituting $T_s = 0.1\pi = \pi/10$ in Eq. (1.82), we have

$$\frac{T_s}{T_0} = \frac{\pi/10}{2\pi/15} = \frac{15}{20} = \frac{3}{4}$$

Thus, $x[n] = x(nT_s)$ is periodic. By Eq. (1.82)

$$N_0 = m \frac{T_0}{T_s} = m \frac{4}{3}$$

The smallest positive integer N_0 is obtained with $m = 3$. Thus, the fundamental period of $x[n] = x(0.1\pi n)$ is $N_0 = 4$.

1.14. Let $x_1(t)$ and $x_2(t)$ be periodic signals with fundamental periods T_1 and T_2, respectively. Under what conditions is the sum $x(t) = x_1(t) + x_2(t)$ periodic, and what is the fundamental period of $x(t)$ if it is periodic?

Since $x_1(t)$ and $x_2(t)$ are periodic with fundamental periods T_1 and T_2, respectively, we have

$$x_1(t) = x_1(t + T_1) = x_1(t + mT_1) \qquad m = \text{positive integer}$$
$$x_2(t) = x_2(t + T_2) = x_2(t + kT_2) \qquad k = \text{positive integer}$$

Thus,

$$x(t) = x_1(t + mT_1) + x_2(t + kT_2)$$

In order for $x(t)$ to be periodic with period T, one needs

$$x(t + T) = x_1(t + T) + x_2(t + T) = x_1(t + mT_1) + x_2(t + kT_2)$$

Thus, we must have

$$mT_1 = kT_2 = T \tag{1.84}$$

or

$$\frac{T_1}{T_2} = \frac{k}{m} = \text{rational number} \tag{1.85}$$

In other words, the sum of two periodic signals is periodic only if the ratio of their respective periods can be expressed as a rational number. Then the fundamental period is the least common multiple of T_1 and T_2, and it is given by Eq. (1.84) if the integers m and k are relative prime. If the ratio T_1/T_2 is an irrational number, then the signals $x_1(t)$ and $x_2(t)$ do not have a common period and $x(t)$ cannot be periodic.

1.15. Let $x_1[n]$ and $x_2[n]$ be periodic sequences with fundamental periods N_1 and N_2, respectively. Under what conditions is the sum $x[n] = x_1[n] + x_2[n]$ periodic, and what is the fundamental period of $x[n]$ if it is periodic?

Since $x_1[n]$ and $x_2[n]$ are periodic with fundamental periods N_1 and N_2, respectively, we have

$$x_1[n] = x_1[n + N_1] = x_1[n + mN_1] \qquad m = \text{positive integer}$$
$$x_2[n] = x_2[n + N_2] = x_2[n + kN_2] \qquad k = \text{positive integer}$$

Thus,

$$x[n] = x_1[n + mN_1] + x_2[n + kN_2]$$

In order for $x[n]$ to be periodic with period N, one needs

$$x[n + N] = x_1[n + N] + x_2[n + N] = x_1[n + mN_1] + x_2[n + kN_2]$$

Thus, we must have

$$mN_1 = kN_2 = N \qquad\qquad\qquad (1.86)$$

Since we can always find integers m and k to satisfy Eq. (1.86), it follows that the sum of two periodic sequences is also periodic and its fundamental period is the least common multiple of N_1 and N_2.

1.16. Determine whether or not each of the following signals is periodic. If a signal is periodic, determine its fundamental period.

(a) $x(t) = \cos\left(t + \dfrac{\pi}{4}\right)$ (b) $x(t) = \sin\dfrac{2\pi}{3}t$

(c) $x(t) = \cos\dfrac{\pi}{3}t + \sin\dfrac{\pi}{4}t$ (d) $x(t) = \cos t + \sin\sqrt{2}t$

(e) $x(t) = \sin^2 t$ (f) $x(t) = e^{j[(\pi/2)t - 1]}$

(g) $x[n] = e^{j(\pi/4)n}$ (h) $x[n] = \cos\dfrac{1}{4}n$

(i) $x[n] = \cos\dfrac{\pi}{3}n + \sin\dfrac{\pi}{4}n$ (j) $x[n] = \cos^2\dfrac{\pi}{8}n$

(a) $x(t) = \cos\left(t + \dfrac{\pi}{4}\right) = \cos\left(\omega_0 t + \dfrac{\pi}{4}\right) \to \omega_0 = 1$

$x(t)$ is periodic with fundamental period $T_0 = 2\pi/\omega_0 = 2\pi$.

(b) $x(t) = \sin\dfrac{2\pi}{3}t \to \omega_0 = \dfrac{2\pi}{3}$

$x(t)$ is periodic with fundamental period $T_0 = 2\pi/\omega_0 = 3$.

(c) $x(t) = \cos\dfrac{\pi}{3}t + \sin\dfrac{\pi}{4}t = x_1(t) + x_2(t)$

where $x_1(t) = \cos(\pi/3)t = \cos\omega_1 t$ is periodic with $T_1 = 2\pi/\omega_1 = 6$ and $x_2(t) = \sin(\pi/4)t = \sin\omega_2 t$ is periodic with $T_2 = 2\pi/\omega_2 = 8$. Since $T_1/T_2 = \frac{6}{8} = \frac{3}{4}$ is a rational number, $x(t)$ is periodic with fundamental period $T_0 = 4T_1 = 3T_2 = 24$.

(d) $x(t) = \cos t + \sin \sqrt{2} t = x_1(t) + x_2(t)$

where $x_1(t) = \cos t = \cos \omega_1 t$ is periodic with $T_1 = 2\pi/\omega_1 = 2\pi$ and $x_2(t) = \sin \sqrt{2} t = \sin \omega_2 t$ is periodic with $T_2 = 2\pi/\omega_2 = \sqrt{2} \pi$. Since $T_1/T_2 = \sqrt{2}$ is an irrational number, $x(t)$ is nonperiodic.

(e) Using the trigonometric identity $\sin^2 \theta = \frac{1}{2}(1 - \cos 2\theta)$, we can write

$$x(t) = \sin^2 t = \frac{1}{2} - \frac{1}{2}\cos 2t = x_1(t) + x_2(t)$$

where $x_1(t) = \frac{1}{2}$ is a dc signal with an arbitrary period and $x_2(t) = -\frac{1}{2}\cos 2t = -\frac{1}{2}\cos \omega_2 t$ is periodic with $T_2 = 2\pi/\omega_2 = \pi$. Thus, $x(t)$ is periodic with fundamental period $T_0 = \pi$.

(f) $x(t) = e^{j[(\pi/2)t - 1]} = e^{-j}e^{j(\pi/2)t} = e^{-j}e^{j\omega_0 t} \to \omega_0 = \frac{\pi}{2}$

$x(t)$ is periodic with fundamental period $T_0 = 2\pi/\omega_0 = 4$.

(g) $x[n] = e^{j(\pi/4)n} = e^{j\Omega_0 n} \to \Omega_0 = \frac{\pi}{4}$

Since $\Omega_0/2\pi = \frac{1}{8}$ is a rational number, $x[n]$ is periodic, and by Eq. (1.55) the fundamental period is $N_0 = 8$.

(h) $x[n] = \cos \frac{1}{4}n = \cos \Omega_0 n \to \Omega_0 = \frac{1}{4}$

Since $\Omega_0/2\pi = 1/8\pi$ is not a rational number, $x[n]$ is nonperiodic.

(i) $x[n] = \cos \frac{\pi}{3}n + \sin \frac{\pi}{4}n = x_1[n] + x_2[n]$

where

$$x_1[n] = \cos\frac{\pi}{3}n = \cos \Omega_1 n \to \Omega_1 = \frac{\pi}{3}$$

$$x_2[n] = \sin\frac{\pi}{4}n = \cos \Omega_2 n \to \Omega_2 = \frac{\pi}{4}$$

Since $\Omega_1/2\pi = \frac{1}{6}$ (= rational number), $x_1[n]$ is periodic with fundamental period $N_1 = 6$, and since $\Omega_2/2\pi = \frac{1}{8}$ (= rational number), $x_2[n]$ is periodic with fundamental period $N_2 = 8$. Thus, from the result of Prob. 1.15, $x[n]$ is periodic and its fundamental period is given by the least common multiple of 6 and 8, that is, $N_0 = 24$.

(j) Using the trigonometric identity $\cos^2 \theta = \frac{1}{2}(1 + \cos 2\theta)$, we can write

$$x[n] = \cos^2 \frac{\pi}{8}n = \frac{1}{2} + \frac{1}{2}\cos \frac{\pi}{4}n = x_1[n] + x_2[n]$$

where $x_1[n] = \frac{1}{2} = \frac{1}{2}(1)^n$ is periodic with fundamental period $N_1 = 1$ and $x_2[n] = \frac{1}{2}\cos(\pi/4)n = \frac{1}{2}\cos \Omega_2 n \to \Omega_2 = \pi/4$. Since $\Omega_2/2\pi = \frac{1}{8}$ (= rational number), $x_2[n]$ is periodic with fundamental period $N_2 = 8$. Thus, $x[n]$ is periodic with fundamental period $N_0 = 8$ (the least common multiple of N_1 and N_2).

1.17. Show that if $x(t + T) = x(t)$, then

$$\int_\alpha^\beta x(t)\, dt = \int_{\alpha+T}^{\beta+T} x(t)\, dt \tag{1.87}$$

$$\int_0^T x(t)\, dt = \int_a^{a+T} x(t)\, dt \tag{1.88}$$

for any real α, β, and a.

If $x(t + T) = x(t)$, then letting $t = \tau - T$, we have

$$x(\tau - T + T) = x(\tau) = x(\tau - T)$$

and

$$\int_\alpha^\beta x(t)\, dt = \int_{\alpha+T}^{\beta+T} x(\tau - T)\, d\tau = \int_{\alpha+T}^{\beta+T} x(\tau)\, d\tau = \int_{\alpha+T}^{\beta+T} x(t)\, dt$$

Next, the right-hand side of Eq. (1.88) can be written as

$$\int_a^{a+T} x(t)\,dt = \int_a^0 x(t)\,dt + \int_0^{a+T} x(t)\,dt$$

By Eq. (1.87) we have

$$\int_a^0 x(t)\,dt = \int_{a+T}^T x(t)\,dt$$

Thus,

$$\int_a^{a+T} x(t)\,dt = \int_{a+T}^T x(t)\,dt + \int_0^{a+T} x(t)\,dt$$

$$= \int_0^{a+T} x(t)\,dt + \int_{a+T}^T x(t)\,dt = \int_0^T x(t)\,dt$$

1.18. Show that if $x(t)$ is periodic with fundamental period T_0, then the normalized average power P of $x(t)$ defined by Eq. (1.15) is the same as the average power of $x(t)$ over any interval of length T_0, that is,

$$P = \frac{1}{T_0} \int_0^{T_0} |x(t)|^2\,dt \tag{1.89}$$

By Eq. (1.15)

$$P = \lim_{T \to \infty} \frac{1}{T} \int_{-T/2}^{T/2} |x(t)|^2\,dt$$

Allowing the limit to be taken in a manner such that T is an integral multiple of the fundamental period, $T = kT_0$, the total normalized energy content of $x(t)$ over an interval of length T is k times the normalized energy content over one period. Then

$$P = \lim_{k \to \infty} \left[\frac{1}{kT_0} k \int_0^{T_0} |x(t)|^2\,dt \right] = \frac{1}{T_0} \int_0^{T_0} |x(t)|^2\,dt$$

1.19. The following equalities are used on many occasions in this text. Prove their validity.

(a) $\displaystyle\sum_{n=0}^{N-1} \alpha^n = \begin{cases} \dfrac{1-\alpha^N}{1-\alpha} & \alpha \neq 1 \\ N & \alpha = 1 \end{cases}$ $\qquad\qquad$ (1.90)

(b) $\displaystyle\sum_{n=0}^{\infty} \alpha^n = \frac{1}{1-\alpha}$ $\qquad |\alpha| < 1$ $\qquad\qquad$ (1.91)

(c) $\displaystyle\sum_{n=k}^{\infty} \alpha^n = \frac{\alpha^k}{1-\alpha}$ $\qquad |\alpha| < 1$ $\qquad\qquad$ (1.92)

(d) $\displaystyle\sum_{n=0}^{\infty} n\alpha^n = \frac{\alpha}{(1-\alpha)^2}$ $\qquad |\alpha| < 1$ $\qquad\qquad$ (1.93)

(a) Let

$$S = \sum_{n=0}^{N-1} \alpha^n = 1 + \alpha + \alpha^2 + \cdots + \alpha^{N-1} \tag{1.94}$$

Then

$$\alpha S = \alpha \sum_{n=0}^{N-1} \alpha^n = \alpha + \alpha^2 + \alpha^3 + \cdots + \alpha^N \tag{1.95}$$

Subtracting Eq. (1.95) from Eq. (1.94), we obtain

$$(1 - \alpha) S = 1 - \alpha^N$$

Hence if $\alpha \neq 1$, we have

$$S = \sum_{n=0}^{N-1} \alpha^n = \frac{1 - \alpha^N}{1 - \alpha} \tag{1.96}$$

If $\alpha = 1$, then by Eq. (1.94)

$$\sum_{n=0}^{N-1} \alpha^n = 1 + 1 + 1 + \cdots + 1 = N$$

(b) For $|\alpha| < 1$, $\lim_{N \to \infty} \alpha^N = 0$. Then by Eq. (1.96) we obtain

$$\sum_{n=0}^{\infty} \alpha^n = \lim_{N \to \infty} \sum_{n=0}^{N-1} \alpha^n = \lim_{N \to \infty} \frac{1 - \alpha^N}{1 - \alpha} = \frac{1}{1 - \alpha}$$

(c) Using Eq. (1.91), we obtain

$$\sum_{n=k}^{\infty} \alpha^n = \alpha^k + \alpha^{k+1} + \alpha^{k+2} + \cdots$$

$$= \alpha^k(1 + \alpha + \alpha^2 + \cdots) = \alpha^k \sum_{n=0}^{\infty} \alpha^n = \frac{\alpha^k}{1 - \alpha}$$

(d) Taking the derivative of both sides of Eq. (1.91) with respect to α, we have

$$\frac{d}{d\alpha}\left(\sum_{n=0}^{\infty} \alpha^n\right) = \frac{d}{d\alpha}\left(\frac{1}{1 - \alpha}\right) = \frac{1}{(1 - \alpha)^2}$$

and

$$\frac{d}{d\alpha}\left(\sum_{n=0}^{\infty} \alpha^n\right) = \sum_{n=0}^{\infty} \frac{d}{d\alpha} \alpha^n = \sum_{n=0}^{\infty} n\alpha^{n-1} = \frac{1}{\alpha} \sum_{n=0}^{\infty} n\alpha^n$$

Hence,

$$\frac{1}{\alpha} \sum_{n=0}^{\infty} n\alpha^n = \frac{1}{(1 - \alpha)^2} \quad \text{or} \quad \sum_{n=0}^{\infty} n\alpha^n = \frac{\alpha}{(1 - \alpha)^2}$$

1.20. Determine whether the following signals are energy signals, power signals, or neither.

(a) $x(t) = e^{-at}u(t), \quad a > 0$ $\qquad$ (b) $x(t) = A\cos(\omega_0 t + \theta)$

(c) $x(t) = tu(t)$ $\qquad$ (d) $x[n] = (-0.5)^n u[n]$

(e) $x[n] = u[n]$ $\qquad$ (f) $x[n] = 2e^{j3n}$

(a) $E = \int_{-\infty}^{\infty} |x(t)|^2 \, dt = \int_0^{\infty} e^{-2at} \, dt = \frac{1}{2a} < \infty$

Thus, $x(t)$ is an energy signal.

(b) The sinusoidal signal $x(t)$ is periodic with $T_0 = 2\pi/\omega_0$. Then by the result from Prob. 1.18, the average power of $x(t)$ is

$$P = \frac{1}{T_0} \int_0^{T_0} [x(t)]^2 \, dt = \frac{\omega_0}{2\pi} \int_0^{2\pi/\omega_0} A^2 \cos^2(\omega_0 t + \theta) \, dt$$

$$= \frac{A^2 \omega_0}{2\pi} \int_0^{2\pi/\omega_0} \frac{1}{2}[1 + \cos(2\omega_0 t + 2\theta)] \, dt = \frac{A^2}{2} < \infty$$

Thus, $x(t)$ is a power signal. Note that periodic signals are, in general, power signals.

(c) $E = \lim\limits_{T \to \infty} \int_{-T/2}^{T/2} |x(t)|^2 \, dt = \lim\limits_{T \to \infty} \int_0^{T/2} t^2 \, dt = \lim\limits_{T \to \infty} \dfrac{(T/2)^3}{3} = \infty$

$P = \lim\limits_{T \to \infty} \dfrac{1}{T} \int_{-T/2}^{T/2} |x(t)|^2 \, dt = \lim\limits_{T \to \infty} \dfrac{1}{T} \int_0^{T/2} t^2 \, dt = \lim\limits_{T \to \infty} \dfrac{1}{T} \dfrac{(T/2)^3}{3} = \infty$

Thus, $x(t)$ is neither an energy signal nor a power signal.

(d)   By definition (1.16) and using Eq. (1.91), we obtain

$$E = \sum_{n=-\infty}^{\infty} |x[n]|^2 = \sum_{n=0}^{\infty} 0.25^n = \dfrac{1}{1 - 0.25} = \dfrac{4}{3} < \infty$$

Thus, $x[n]$ is an energy signal.

(e)   By definition (1.17)

$$P = \lim_{N \to \infty} \dfrac{1}{2N+1} \sum_{n=-N}^{N} |x[n]|^2$$

$$= \lim_{N \to \infty} \dfrac{1}{2N+1} \sum_{n=0}^{N} 1^2 = \lim_{N \to \infty} \dfrac{1}{2N+1} (N+1) = \dfrac{1}{2} < \infty$$

Thus, $x[n]$ is a power signal.

(f)   Since $|x[n]| = |2e^{j3n}| = 2|e^{j3n}| = 2$,

$$P = \lim_{N \to \infty} \dfrac{1}{2N+1} \sum_{n=-N}^{N} |x[n]|^2 = \lim_{N \to \infty} \dfrac{1}{2N+1} \sum_{n=-N}^{N} 2^2$$

$$= \lim_{N \to \infty} \dfrac{1}{2N+1} 4(2N+1) = 4 < \infty$$

Thus, $x[n]$ is a power signal.

Basic Signals

1.21. Show that

$$u(-t) = \begin{cases} 0 & t > 0 \\ 1 & t < 0 \end{cases} \tag{1.97}$$

Let $\tau = -t$. Then by definition (1.18)

$$u(-t) = u(\tau) = \begin{cases} 1 & \tau > 0 \\ 0 & \tau < 0 \end{cases}$$

Since $\tau > 0$ and $\tau < 0$ imply, respectively, that $t < 0$ and $t > 0$, we obtain

$$u(-t) = \begin{cases} 0 & t > 0 \\ 1 & t < 0 \end{cases}$$

which is shown in Fig. 1-26.

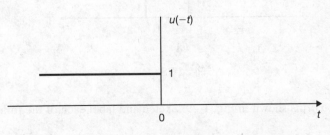

Fig. 1-26

1.22. A continuous-time signal $x(t)$ is shown in Fig. 1-27. Sketch and label each of the following signals.

    (*a*) $x(t)u(1-t)$;   (*b*) $x(t)[u(t)-u(t-1)]$;   (*c*) $x(t)\delta(t-\frac{3}{2})$

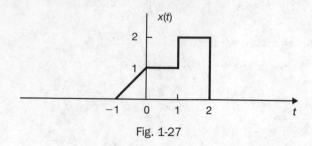

Fig. 1-27

    (*a*)   By definition (1.19)

$$u(1-t)=\begin{cases}1 & t<1\\0 & t>1\end{cases}$$

and $x(t)u(1-t)$ is sketched in Fig. 1-28(*a*).

    (*b*)   By definitions (1.18) and (1.19)

$$u(t)-u(t-1)=\begin{cases}1 & 0<t\le1\\0 & \text{otherwise}\end{cases}$$

and $x(t)[u(t)-u(t-1)]$ is sketched in Fig. 1-28(*b*).

    (*c*)   By Eq. (1.26)

$$x(t)\delta\left(t-\frac{3}{2}\right)=x\left(\frac{3}{2}\right)\delta\left(t-\frac{3}{2}\right)=2\delta\left(t-\frac{3}{2}\right)$$

which is sketched in Fig. 1-28(*c*).

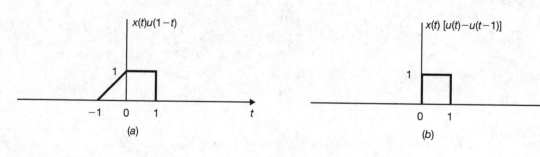

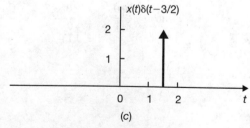

Fig. 1-28

1.23. A discrete-time signal $x[n]$ is shown in Fig. 1-29. Sketch and label each of the following signals.

    (*a*) $x[n]u[1-n]$;   (*b*) $x[n]\{u[n+2]-u[n]\}$;   (*c*) $x[n]\delta[n-1]$

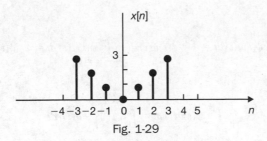

Fig. 1-29

(a) By definition (1.44)

$$u[1-n] = \begin{cases} 1 & n \le 1 \\ 0 & n > 1 \end{cases}$$

and $x[n]u[1-n]$ is sketched in Fig. 1-30(a).

(b) By definitions (1.43) and (1.44)

$$u[n+2] - u[n] = \begin{cases} 1 & -2 \le n < 0 \\ 0 & \text{otherwise} \end{cases}$$

and $x[n]\{u[n+2] - u[n]\}$ is sketched in Fig. 1-30(b).

(c) By definition (1.48)

$$x[n]\delta[n-1] = x[1]\delta[n-1] = \delta[n-1] = \begin{cases} 1 & n=1 \\ 0 & n \ne 1 \end{cases}$$

which is sketched in Fig. 1-30(c).

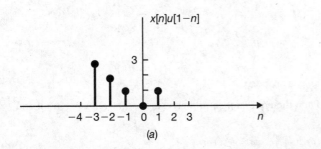

(a)

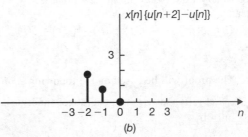

(b)

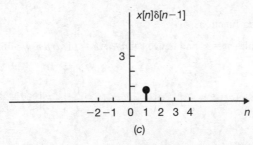

(c)

Fig. 1-30

1.24. The unit step function $u(t)$ can be defined as a generalized function by the following relation:

$$\int_{-\infty}^{\infty} \phi(t)u(t)\, dt = \int_{0}^{\infty} \phi(t)\, dt \qquad (1.98)$$

where $\phi(t)$ is a testing function which is integrable over $0 < t < \infty$. Using this definition, show that

$$u(t) = \begin{cases} 1 & t > 0 \\ 0 & t < 0 \end{cases}$$

Rewriting Eq. (1.98) as

$$\int_{-\infty}^{\infty} \phi(t)u(t)\,dt = \int_{-\infty}^{0} \phi(t)u(t)\,dt + \int_{0}^{\infty} \phi(t)u(t)\,dt = \int_{0}^{\infty} \phi(t)\,dt$$

we obtain

$$\int_{-\infty}^{0} \phi(t)u(t)\,dt = \int_{0}^{\infty} \phi(t)[1-u(t)]\,dt$$

This can be true only if

$$\int_{-\infty}^{0} \phi(t)u(t)\,dt = 0 \quad \text{and} \quad \int_{0}^{\infty} \phi(t)[1-u(t)]\,dt = 0$$

These conditions imply that

$$\phi(t)u(t) = 0, t < 0 \quad \text{and} \quad \phi(t)[1-u(t)] = 0, t > 0$$

Since $\phi(t)$ is arbitrary, we have

$$u(t) = 0, t < 0 \quad \text{and} \quad 1-u(t) = 0, t > 0$$

that is,

$$u(t) = \begin{cases} 1 & t > 0 \\ 0 & t < 0 \end{cases}$$

1.25. Verify Eqs. (1.23) and (1.24); that is,

(a) $\delta(at) = \dfrac{1}{|a|}\delta(t)$; (b) $\delta(-t) = \delta(t)$

The proof will be based on the following *equivalence* property:

Let $g_1(t)$ and $g_2(t)$ be generalized functions. Then the equivalence property states that $g_1(t) = g_2(t)$ if and only if

$$\int_{-\infty}^{\infty} \phi(t)\,g_1(t)\,dt = \int_{-\infty}^{\infty} \phi(t)\,g_2(t)\,dt \tag{1.99}$$

for all suitably defined testing functions $\phi(t)$.

(a) With a change of variable, $at = \tau$, and hence $t = \tau/a$, $dt = (1/a)\,d\tau$, we obtain the following equations:
  If $a > 0$,

$$\int_{-\infty}^{\infty} \phi(t)\delta(at)\,dt = \frac{1}{a}\int_{-\infty}^{\infty} \phi\!\left(\frac{\tau}{a}\right)\delta(\tau)\,d\tau = \frac{1}{a}\phi\!\left(\frac{\tau}{a}\right)\bigg|_{\tau=0} = \frac{1}{|a|}\phi(0)$$

  If $a < 0$,

$$\int_{-\infty}^{\infty} \phi(t)\delta(at)\,dt = \frac{1}{a}\int_{\infty}^{-\infty} \phi\!\left(\frac{\tau}{a}\right)\delta(\tau)\,d\tau = -\frac{1}{a}\int_{-\infty}^{\infty} \phi\!\left(\frac{\tau}{a}\right)\delta(\tau)\,d\tau$$

$$= -\frac{1}{a}\phi\!\left(\frac{\tau}{a}\right)\bigg|_{\tau=0} = \frac{1}{|a|}\phi(0)$$

  Thus, for any a

$$\int_{-\infty}^{\infty} \phi(t)\delta(at)\,dt = \frac{1}{|a|}\phi(0)$$

Now, using Eq. (1.20) for $\phi(0)$, we obtain

$$\int_{-\infty}^{\infty} \phi(t)\delta(at)\,dt = \frac{1}{|a|}\phi(0) = \frac{1}{|a|}\int_{-\infty}^{\infty} \phi(t)\delta(t)\,dt$$

$$= \int_{-\infty}^{\infty} \phi(t)\frac{1}{|a|}\delta(t)\,dt$$

for any $\phi(t)$. Then, by the equivalence property (1.99), we obtain

$$\delta(at) = \frac{1}{|a|}\delta(t)$$

(b) Setting $a = -1$ in the above equation, we obtain

$$\delta(-t) = \frac{1}{|-1|}\delta(t) = \delta(t)$$

which shows that $\delta(t)$ is an even function.

1.26. (a) Verify Eq. (1.26):

$$x(t)\delta(t - t_0) = x(t_0)\delta(t - t_0)$$

if $x(t)$ is continuous at $t = t_0$.

(b) Verify Eq. (1.25):

$$x(t)\delta(t) = x(0)\delta(t)$$

if $x(t)$ is continuous at $t = 0$.

(a) If $x(t)$ is continuous at $t = t_0$, then by definition (1.22) we have

$$\int_{-\infty}^{\infty} \phi(t)[x(t)\delta(t - t_0)]\,dt = \int_{-\infty}^{\infty} [\phi(t)x(t)]\delta(t - t_0)\,dt = \phi(t_0)x(t_0)$$

$$= x(t_0)\int_{-\infty}^{\infty} \phi(t)\delta(t - t_0)\,dt$$

$$= \int_{-\infty}^{\infty} \phi(t)[x(t_0)\delta(t - t_0)]\,dt$$

for all $\phi(t)$ which are continuous at $t = t_0$. Hence, by the equivalence property (1.99) we conclude that

$$x(t)\delta(t - t_0) = x(t_0)\delta(t - t_0)$$

(b) Setting $t_0 = 0$ in the above expression, we obtain

$$x(t)\delta(t) = x(0)\delta(t)$$

1.27. Show that

(a) $t\delta(t) = 0$

(b) $\sin t\delta(t) = 0$

(c) $\cos t\delta(t - \pi) = -\delta(t - \pi)$

Using Eqs. (1.25) and (1.26), we obtain

(a) $t\delta(t) = (0)\delta(t) = 0$

(b) $\sin t\delta(t) = (\sin 0)\delta(t) = (0)\delta(t) = 0$

(c) $\cos t\delta(t - \pi) = (\cos \pi)\,\delta(t - \pi) = (-1)\delta(t - \pi) = -\delta(t - \pi)$

1.28. Verify Eq. (1.30):

$$\delta(t) = u'(t) = \frac{du(t)}{dt}$$

From Eq. (1.28) we have

$$\int_{-\infty}^{\infty} \phi(t) u'(t)\, dt = -\int_{-\infty}^{\infty} \phi'(t) u(t)\, dt \qquad\qquad (1.100)$$

where $\phi(t)$ is a testing function which is continuous at $t = 0$ and vanishes outside some fixed interval. Thus, $\phi'(t)$ exists and is integrable over $0 < t < \infty$ and $\phi(\infty) = 0$. Then using Eq. (1.98) or definition (1.18), we have

$$\int_{-\infty}^{\infty} \phi(t) u'(t)\, dt = -\int_{0}^{\infty} \phi'(t)\, dt = -\phi(t)\Big|_{0}^{\infty} = -[\phi(\infty) - \phi(0)]$$

$$= \phi(0) = \int_{-\infty}^{\infty} \phi(t)\delta(t)\, dt$$

Since $\phi(t)$ is arbitrary and by equivalence property (1.99), we conclude that

$$\delta(t) = u'(t) = \frac{du(t)}{dt}$$

1.29. Show that the following properties hold for the derivative of $\delta(t)$:

(a) $\displaystyle\int_{-\infty}^{\infty} \phi(t)\delta'(t)\, dt = -\phi'(0)$ where $\phi'(0) = \dfrac{d\phi(t)}{dt}\bigg|_{t=0}$ $\qquad\qquad (1.101)$

(b) $t\delta'(t) = -\delta(t)$ $\qquad\qquad\qquad\qquad\qquad\qquad\qquad\qquad\qquad\quad (1.102)$

(a)   Using Eqs. (1.28) and (1.20), we have

$$\int_{-\infty}^{\infty} \phi(t)\delta'(t)\, dt = -\int_{-\infty}^{\infty} \phi'(t)\delta(t)\, dt = -\phi'(0)$$

(b)   Using Eqs. (1.101) and (1.20), we have

$$\int_{-\infty}^{\infty} \phi(t)\,[t\delta'(t)]\, dt = \int_{-\infty}^{\infty} [t\phi(t)]\delta'(t)\, dt = -\frac{d}{dt}\,[t\phi(t)]\Big|_{t=0}$$

$$= -[\phi(t) + t\phi'(t)]\big|_{t=0} = -\phi(0)$$

$$= -\int_{-\infty}^{\infty} \phi(t)\delta(t)\, dt = \int_{-\infty}^{\infty} \phi(t)[-\delta(t)]\, dt$$

Thus, by the equivalence property (1.99) we conclude that

$$t\delta'(t) = -\delta(t)$$

1.30. Evaluate the following integrals:

(a) $\displaystyle\int_{-1}^{1} (3t^2 + 1)\delta(t)\, dt$

(b) $\displaystyle\int_{1}^{2} (3t^2 + 1)\delta(t)\, dt$

(c) $\displaystyle\int_{-\infty}^{\infty} (t^2 + \cos \pi t)\delta(t - 1)\, dt$

(d) $\displaystyle\int_{-\infty}^{\infty} e^{-t}\delta(2t - 2)\, dt$

(e) $\displaystyle\int_{-\infty}^{\infty} e^{-t}\delta'(t)\, dt$

(a) By Eq. (1.21), with $a = -1$ and $b = 1$, we have

$$\int_{-1}^{1}(3t^2 + 1)\delta(t)\,dt = (3t^2 + 1)\Big|_{t=0} = 1$$

(b) By Eq. (1.21), with $a = 1$ and $b = 2$, we have

$$\int_{1}^{2}(3t^2 + 1)\delta(t)\,dt = 0$$

(c) By Eq. (1.22)

$$\int_{-\infty}^{\infty}(t^2 + \cos \pi t)\delta(t-1)\,dt = (t^2 + \cos \pi t)\Big|_{t=1}$$
$$= 1 + \cos \pi = 1 - 1 = 0$$

(d) Using Eqs. (1.22) and (1.23), we have

$$\int_{-\infty}^{\infty}e^{-t}\delta(2t - 2)\,dt = \int_{-\infty}^{\infty}e^{-t}\delta[2(t-1)]\,dt$$
$$= \int_{-\infty}^{\infty}e^{-t}\frac{1}{|2|}\delta(t-1)\,dt = \frac{1}{2}e^{-t}\Big|_{t=1} = \frac{1}{2e}$$

(e) By Eq. (1.29)

$$\int_{-\infty}^{\infty}e^{-t}\delta'(t)\,dt = -\frac{d}{dt}(e^{-t})\Big|_{t=0} = e^{-t}\Big|_{t=0} = 1$$

1.31. Find and sketch the first derivatives of the following signals:

(a) $x(t) = u(t) - u(t - a)$, $a > 0$

(b) $x(t) = t[u(t) - u(t - a)]$, $a > 0$

(c) $x(t) = \operatorname{sgn} t = \begin{cases} 1 & t > 0 \\ -1 & t < 0 \end{cases}$

(a) Using Eq. (1.30), we have

$$u'(t) = \delta(t) \quad \text{and} \quad u'(t - a) = \delta(t - a)$$

Then

$$x'(t) = u'(t) - u'(t - a) = \delta(t) - \delta(t - a)$$

Signals $x(t)$ and $x'(t)$ are sketched in Fig. 1-31(a).

(b) Using the rule for differentiation of the product of two functions and the result from part (a), we have

$$x'(t) = [u(t) - u(t - a)] + t[\delta(t) - \delta(t - a)]$$

But by Eqs. (1.25) and (1.26)

$$t\delta(t) = (0)\delta(t) = 0 \quad \text{and} \quad t\delta(t - a) = a\delta(t - a)$$

Thus,

$$x'(t) = u(t) - u(t - a) - a\delta(t - a)$$

Signals $x(t)$ and $x'(t)$ are sketched in Fig. 1-31(b).

(c) $x(t) = \operatorname{sgn} t$ can be rewritten as

$$x(t) = \operatorname{sgn} t = u(t) - u(-t)$$

Then using Eq. (1.30), we obtain

$$x'(t) = u'(t) - u'(-t) = \delta(t) - [-\delta(t)] = 2\delta(t)$$

Signals $x(t)$ and $x'(t)$ are sketched in Fig. 1-31(c).

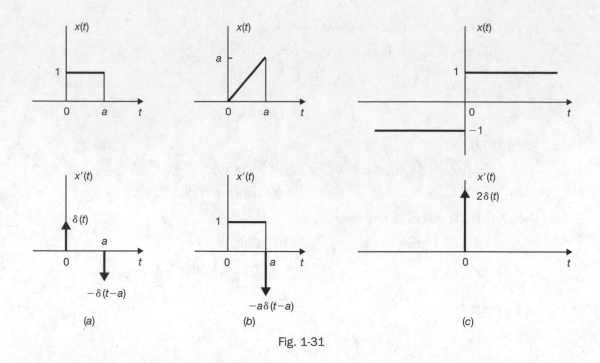

Fig. 1-31

Systems and Classification of Systems

1.32. Consider the *RC* circuit shown in Fig. 1-32. Find the relationship between the input $x(t)$ and the output $y(t)$

    (*a*)   If $x(t) = v_s(t)$ and $y(t) = v_c(t)$.

    (*b*)   If $x(t) = v_s(t)$ and $y(t) = i(t)$.

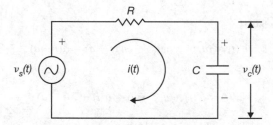

Fig. 1-32 *RC* circuit.

    (*a*)   Applying Kirchhoff's voltage law to the *RC* circuit in Fig. 1-32, we obtain

$$v_s(t) = Ri(t) + v_c(t) \tag{1.103}$$

The current $i(t)$ and voltage $v_c(t)$ are related by

$$i(t) = C\frac{dv_c(t)}{dt} \tag{1.104}$$

Letting $v_s(t) = x(t)$ and $v_c(t) = y(t)$ and substituting Eq. (1.04) into Eq. (1.103), we obtain

$$RC\frac{dy(t)}{dt} + y(t) = x(t)$$

or

$$\frac{dy(t)}{dt} + \frac{1}{RC}y(t) = \frac{1}{RC}x(t) \tag{1.105}$$

Thus, the input-output relationship of the *RC* circuit is described by a first-order linear differential equation with constant coefficients.

(*b*)   Integrating Eq. (1.104), we have

$$v_c(t) = \frac{1}{C}\int_{-\infty}^{t} i(\tau)\,d\tau \tag{1.106}$$

Substituting Eq. (1.106) into Eq. (1.103) and letting $v_s(t) = x(t)$ and $i(t) = y(t)$, we obtain

$$Ry(t) + \frac{1}{C}\int_{-\infty}^{t} y(\tau)\,d\tau = x(t)$$

or

$$y(t) + \frac{1}{RC}\int_{-\infty}^{t} y(\tau)\,d\tau = \frac{1}{R}x(t)$$

Differentiating both sides of the above equation with respect to *t*, we obtain

$$\frac{dy(t)}{dt} + \frac{1}{RC}y(t) = \frac{1}{R}\frac{dx(t)}{dt} \tag{1.107}$$

Thus, the input-output relationship is described by another first-order linear differential equation with constant coefficients.

1.33. Consider the capacitor shown in Fig. 1-33. Let input $x(t) = i(t)$ and output $y(t) = v_c(t)$.

(*a*)   Find the input-output relationship.

(*b*)   Determine whether the system is (*i*) memoryless, (*ii*) causal, (*iii*) linear, (*iv*) time-invariant, or (*v*) stable.

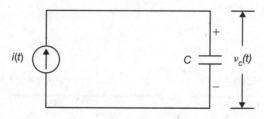

Fig. 1-33

(*a*)   Assume the capacitance *C* is constant. The output voltage $y(t)$ across the capacitor and the input current $x(t)$ are related by [Eq. (1.106)]

$$y(t) = \mathbf{T}\{x(t)\} = \frac{1}{C}\int_{-\infty}^{t} x(\tau)\,d\tau \tag{1.108}$$

(*b*)   (*i*)   From Eq. (1.108) it is seen that the output $y(t)$ depends on the past and the present values of the input. Thus, the system is not memoryless.

(*ii*)   Since the output $y(t)$ does not depend on the future values of the input, the system is causal.

(*iii*)   Let $x(t) = \alpha_1 x_1(t) + \alpha_2 x_2(t)$. Then

$$\begin{aligned} y(t) = \mathbf{T}\{x(t)\} &= \frac{1}{C}\int_{-\infty}^{t} [\alpha_1 x_1(\tau) + \alpha_2 x_2(\tau)]\,d\tau \\ &= \alpha_1\left[\frac{1}{C}\int_{-\infty}^{t} x_1(\tau)\,d\tau\right] + \alpha_2\left[\frac{1}{C}\int_{-\infty}^{t} x_2(\tau)\,d\tau\right] \\ &= \alpha_1 y_1(t) + \alpha_2 y_2(t) \end{aligned}$$

Thus, the superposition property (1.68) is satisfied and the system is linear.

(iv)   Let $y_1(t)$ be the output produced by the shifted input current $x_1(t) = x(t - t_0)$.

Then

$$y_1(t) = \mathbf{T}\{x(t - t_0)\} = \frac{1}{C}\int_{-\infty}^{t} x(\tau - t_0)\,d\tau$$

$$= \frac{1}{C}\int_{-\infty}^{t-t_0} x(\lambda)\,d\lambda = y(t - t_0)$$

Hence, the system is time-invariant.

(v)   Let $x(t) = k_1 u(t)$, with $k_1 \neq 0$. Then

$$y(t) = \frac{1}{C}\int_{-\infty}^{t} k_1 u(\tau)\,d\tau = \frac{k_1}{C}\int_0^t d\tau = \frac{k_1}{C} t u(t) = \frac{k_1}{C} r(t) \tag{1.109}$$

where $r(t) = tu(t)$ is known as the *unit ramp* function (Fig. 1-34). Since $y(t)$ grows linearly in time without bound, the system is not BIBO stable.

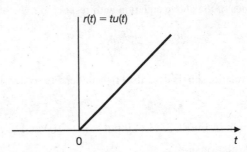

Fig. 1-34 Unit ramp function.

1.34. Consider the system shown in Fig. 1-35. Determine whether it is *(a)* memoryless, *(b)* causal, *(c)* linear, *(d)* time-invariant, or *(e)* stable.

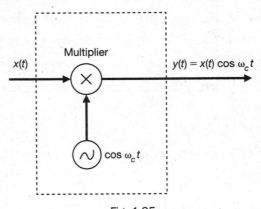

Fig. 1-35

(a)   From Fig. 1-35 we have

$$y(t) = \mathbf{T}\{x(t)\} = x(t)\cos \omega_c t$$

Since the value of the output $y(t)$ depends on only the present values of the input $x(t)$, the system is memoryless.

(b)   Since the output $y(t)$ does not depend on the future values of the input $x(t)$, the system is causal.

(c)   Let $x(t) = \alpha_1 x(t) + \alpha_2 x(t)$. Then

$$y(t) = \mathbf{T}\{x(t)\} = [\alpha_1 x_1(t) + \alpha_2 x_2(t)]\cos \omega_c t$$

$$= \alpha_1 x_1(t)\cos \omega_c t + \alpha_2 x_2(t)\cos \omega_c t$$

$$= \alpha_1 y_1(t) + \alpha_2 y_2(t)$$

Thus, the superposition property (1.68) is satisfied and the system is linear.

(*d*) Let $y_1(t)$ be the output produced by the shifted input $x_1(t) = x(t - t_0)$. Then

$$y_1(t) = \mathbf{T}\{x(t - t_0)\} = x(t - t_0) \cos \omega_c t$$

But

$$y(t - t_0) = x(t - t_0) \cos \omega_c(t - t_0) \neq y_1(t)$$

Hence, the system is not time-invariant.

(*e*) Since $|\cos \omega_c t| \le 1$, we have

$$|y(t)| = |x(t) \cos \omega_c t| \le |x(t)|$$

Thus, if the input $x(t)$ is bounded, then the output $y(t)$ is also bounded and the system is BIBO stable.

1.35. A system has the input-output relation given by

$$y = \mathbf{T}\{x\} = x^2 \tag{1.110}$$

Show that this system is nonlinear.

$$\mathbf{T}\{x_1 + x_2\} = (x_1 + x_2)^2 = x_1^2 + x_2^2 + 2x_1 x_2$$
$$\neq \mathbf{T}\{x_1\} + \mathbf{T}\{x_2\} = x_1^2 + x_2^2$$

Thus, the system is nonlinear.

1.36. The discrete-time system shown in Fig. 1-36 is known as the *unit delay* element. Determine whether the system is (*a*) memoryless, (*b*) causal, (*c*) linear, (*d*) time-invariant, or (*e*) stable.

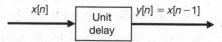

Fig. 1-36 Unit delay element

(*a*) The system input-output relation is given by

$$y[n] = \mathbf{T}\{x[n]\} = x[n - 1] \tag{1.111}$$

Since the output value at n depends on the input values at $n - 1$, the system is not memoryless.

(*b*) Since the output does not depend on the future input values, the system is causal.

(*c*) Let $x[n] = \alpha_1 x_1[n] + \alpha_2 x_2[n]$. Then

$$y[n] = \mathbf{T}\{\alpha_1 x_1[n] + \alpha_2 x_2[n]\} = \alpha_1 x_1[n - 1] + \alpha_2 x_2[n - 1]$$
$$= \alpha_1 y_1[n] + \alpha_2 y_2[n]$$

Thus, the superposition property (1.68) is satisfied and the system is linear.

(*d*) Let $y_1[n]$ be the response to $x_1[n] = x[n - n_0]$. Then

$$y_1[n] = \mathbf{T}\{x_1[n]\} = x_1[n - 1] = x[n - 1 - n_0]$$

and

$$y[n - n_0] = x[n - n_0 - 1] = x[n - 1 - n_0] = y_1[n]$$

Hence, the system is time-invariant.

(*e*) Since

$$|y[n]| = |x[n - 1]| \le k \qquad \text{if } |x[n]| \le k \text{ for all } n$$

the system is BIBO stable.

1.37. Find the input-output relation of the feedback system shown in Fig. 1-37.

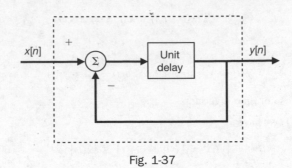

Fig. 1-37

From Fig. 1-37 the input to the unit delay element is $x[n] - y[n]$. Thus, the output $y[n]$ of the unit delay element is [Eq. (1.111)]

$$y[n] = x[n-1] - y[n-1]$$

Rearranging, we obtain

$$y[n] + y[n-1] = x[n-1] \tag{1.112}$$

Thus, the input-output relation of the system is described by a first-order difference equation with constant coefficients.

1.38. A system has the input-output relation given by

$$y[n] = \mathbf{T}\{x[n]\} = nx[n] \tag{1.113}$$

Determine whether the system is (*a*) memoryless, (*b*) causal, (*c*) linear, (*d*) time-invariant, or (*e*) stable.

(*a*)   Since the output value at n depends on only the input value at n, the system is memoryless.

(*b*)   Since the output does not depend on the future input values, the system is causal.

(*c*)   Let $x[n] = \alpha_1 x_1[n] + \alpha_2 x_2[n]$. Then

$$y[n] = \mathbf{T}\{x[n]\} = n\{\alpha_1 x_1[n] + \alpha_2 x_2[n]\}$$
$$= \alpha_1 n x_1[n] + \alpha_2 n x_2[n] = \alpha_1 y_1[n] + \alpha_2 y_2[n]$$

Thus, the superposition property (1.68) is satisfied and the system is linear.

(*d*)   Let $y_1[n]$ be the response to $x_1[n] = x[n - n_0]$. Then

$$y_1[n] = \mathbf{T}\{x[n - n_0]\} = nx[n - n_0]$$

But $$y[n - n_0] = (n - n_0) x[n - n_0] \neq y_1[n]$$

Hence, the system is not time-invariant.

(*e*)   Let $x[n] = u[n]$. Then $y[n] = nu[n]$. Thus, the bounded unit step sequence produces an output sequence that grows without bound (Fig. 1-38) and the system is not BIBO stable.

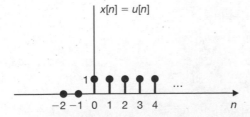
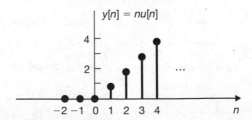

Fig. 1-38

1.39. A system has the input-output relation given by

$$y[n] = \mathbf{T}\{x[n]\} = x[k_0 n] \tag{1.114}$$

where k_0 is a positive integer. Is the system time-invariant?

Let $y_1[n]$ be the response to $x_1[n] = x[n - n_0]$. Then

$$y_1[n] = \mathbf{T}\{x_1[n]\} = x_1[k_0 n] = x[k_0 n - n_0]$$

But
$$y[n - n_0] = x[k_0(n - n_0)] \neq y_1[n]$$

Hence, the system is not time-invariant unless $k_0 = 1$. Note that the system described by Eq. (1.114) is called a *compressor*. It creates the output sequence by selecting every k_0th sample of the input sequence. Thus, it is obvious that this system is time-varying.

1.40. Consider the system whose input-output relation is given by the linear equation

$$y = ax + b \tag{1.115}$$

where x and y are the input and output of the system, respectively, and a and b are constants. Is this system linear?

If $b \neq 0$, then the system is not linear because $x = 0$ implies $y = b \neq 0$. If $b = 0$, then the system is linear.

1.41. The system represented by $\mathbf{T}$ in Fig. 1-39 is known to be time-invariant. When the inputs to the system are $x_1[n]$, $x_2[n]$, and $x_3[n]$, the outputs of the system are $y_1[n]$, $y_2[n]$, and $y_3[n]$ as shown. Determine whether the system is linear.

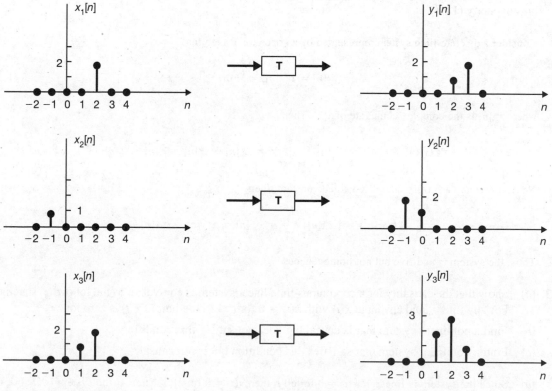

Fig. 1-39

From Fig. 1-39 it is seen that

$$x_3[n] = x_1[n] + x_2[n - 2]$$

Thus, if **T** is linear, then

$$\mathbf{T}\{x_3[n]\} = \mathbf{T}\{x_1[n]\} + \mathbf{T}\{x_2[n - 2]\} = y_1[n] + y_2[n - 2]$$

which is shown in Fig. 1-40. From Figs. 1-39 and 1-40 we see that

$$y_3[n] \neq y_1[n] + y_2[n - 2]$$

Hence, the system is not linear.

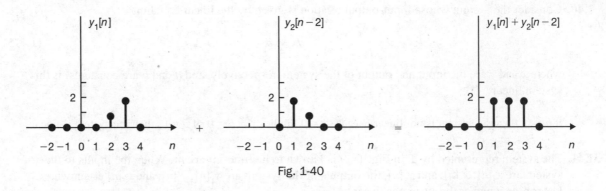

Fig. 1-40

1.42. Give an example of a system that satisfies the condition of additivity (1.66) but not the condition of homogeneity (1.67).

Consider a discrete-time system represented by an operator **T** such that

$$y[n] = \mathbf{T}\{x[n]\} = x^*[n] \tag{1.116}$$

where $x^*[n]$ is the complex conjugate of $x[n]$. Then

$$\mathbf{T}\{x_1[n] + x_2[n]\} = \{x_1[n] + x_2[n]\}^* = x_1^*[n] + x_2^*[n] = y_1[n] + y_2[n]$$

Next, if α is any arbitrary complex-valued constant, then

$$\mathbf{T}\{\alpha x[n]\} = \{\alpha x[n]\}^* = \alpha^* x^*[n] = \alpha^* y[n] \neq \alpha y[n]$$

Thus, the system is additive but not homogeneous.

1.43. (a) Show that the causality for a continuous-time linear system is equivalent to the following statement: For any time t_0 and any input $x(t)$ with $x(t) = 0$ for $t \leq t_0$, the output $y(t)$ is zero for $t \leq t_0$.

(b) Find a nonlinear system that is causal but does not satisfy this condition.

(c) Find a nonlinear system that satisfies this condition but is not causal.

(a) Since the system is linear, if $x(t) = 0$ for all t, then $y(t) = 0$ for all t. Thus, if the system is causal, then $x(t) = 0$ for $t \leq t_0$ implies that $y(t) = 0$ for $t \leq t_0$. This is the necessary condition. That this condition is also sufficient is shown as follows: let $x_1(t)$ and $x_2(t)$ be two inputs of the system and let $y_1(t)$ and $y_2(t)$ be the corresponding outputs. If $x_1(t) = x_2(t)$ for $t \le t_0$, or $x(t) = x_1(t) - x_2(t) = 0$ for $t \le t_0$, then $y_1(t) = y_2(t)$ for $t \le t_0$, or $y(t) = y_1(t) - y_2(t) = 0$ for $t \le t_0$.

(b) Consider the system with the input-output relation

$$y(t) = x(t) + 1$$

This system is nonlinear (Prob. 1.40) and causal since the value of $y(t)$ depends on only the present value of $x(t)$. But with $x(t) = 0$ for $t \le t_0$, $y(t) = 1$ for $t \le t_0$.

(c) Consider the system with the input-output relation

$$y(t) = x(t)x(t + 1)$$

It is obvious that this system is nonlinear (see Prob. 1.35) and noncausal since the value of $y(t)$ at time t depends on the value of $x(t + 1)$ of the input at time $t + 1$. Yet $x(t) = 0$ for $t \le t_0$ implies that $y(t) = 0$ for $t \le t_0$.

1.44. Let **T** represent a continuous-time LTI system. Then show that

$$\mathbf{T}\{e^{st}\} = \lambda e^{st} \tag{1.117}$$

where s is a complex variable and λ is a complex constant.

Let $y(t)$ be the output of the system with input $x(t) = e^{st}$. Then

$$\mathbf{T}\{e^{st}\} = y(t)$$

Since the system is time-invariant, we have

$$\mathbf{T}\{e^{s(t + t_0)}\} = y(t + t_0)$$

for arbitrary real t_0. Since the system is linear, we have

$$\mathbf{T}\{e^{s(t + t_0)}\} = \mathbf{T}\{e^{st} e^{st_0}\} = e^{st_0}\mathbf{T}\{e^{st}\} = e^{st_0}y(t)$$

Hence,

$$y(t + t_0) = e^{st_0}y(t)$$

Setting $t = 0$, we obtain

$$y(t_0) = y(0)e^{st_0} \tag{1.118}$$

Since t_0 is arbitrary, by changing t_0 to t, we can rewrite Eq. (1.118) as

$$y(t) = y(0) e^{st} = \lambda e^{st}$$

or

$$\mathbf{T}\{e^{st}\} = \lambda e^{st}$$

where $\lambda = y(0)$.

1.45. Let **T** represent a discrete-time LTI system. Then show that

$$\mathbf{T}\{z^n\} = \lambda z^n \tag{1.119}$$

where z is a complex variable and λ is a complex constant.

Let $y[n]$ be the output of the system with input $x[n] = z^n$. Then

$$\mathbf{T}\{z^n\} = y[n]$$

Since the system is time-invariant, we have

$$\mathbf{T}\{z^{n+n_0}\} = y[n + n_0]$$

for arbitrary integer n_0. Since the system is linear, we have

$$\mathbf{T}\{z^{n+n_0}\} = \mathbf{T}\{z^n z^{n_0}\} = z^{n_0}\mathbf{T}\{z^n\} = z^{n_0}y[n]$$

Hence, $$y[n + n_0] = z^{n_0}y[n]$$

Setting $n = 0$, we obtain

$$y[n_0] = y[0]z^{n_0} \tag{1.120}$$

Since n_0 is arbitrary, by changing n_0 to n, we can rewrite Eq. (1.120) as

$$y[n] = y[0]z^n = \lambda z^n$$

or $$\mathbf{T}\{z^n\} = \lambda z^n$$

where $\lambda = y[0]$.

In mathematical language, a function $x(\cdot)$ satisfying the equation

$$\mathbf{T}\{x(\cdot)\} = \lambda x(\cdot) \tag{1.121}$$

is called an *eigenfunction* (or *characteristic function*) of the operator $\mathbf{T}$, and the constant λ is called the *eigenvalue* (or *characteristic value*) corresponding to the eigenfunction $x(\cdot)$. Thus, Eqs. (1.117) and (1.119) indicate that the complex exponential functions are eigenfunctions of any LTI system.

SUPPLEMENTARY PROBLEMS

1.46. Express the signals shown in Fig. 1-41 in terms of unit step functions.

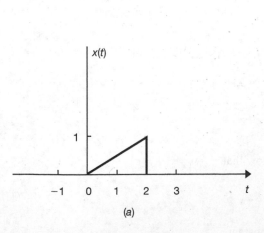
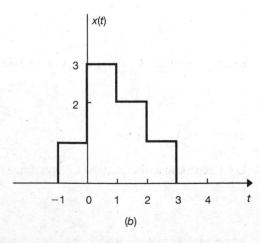

Fig. 1-41

1.47. Express the sequences shown in Fig. 1-42 in terms of unit step sequences.

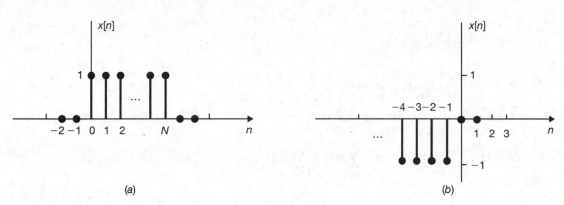

(a)

(b)

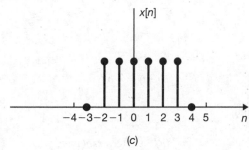

(c)

Fig. 1-42

1.48. Determine the even and odd components of the following signals:

(a) $x(t) = u(t)$

(b) $x(t) = \sin\left(\omega_0 t + \dfrac{\pi}{4}\right)$

(c) $x[n] = e^{j(\Omega_0 n + \pi/2)}$

(d) $x[n] = \delta[n]$

1.49. Let $x(t)$ be an arbitrary signal with even and odd parts denoted by $x_e(t)$ and $x_o(t)$, respectively. Show that

$$\int_{-\infty}^{\infty} x^2(t)\, dt = \int_{-\infty}^{\infty} x_e^2(t)\, dt + \int_{-\infty}^{\infty} x_o^2(t)\, dt$$

1.50. Let $x[n]$ be an arbitrary sequence with even and odd parts denoted by $x_e[n]$ and $x_o[n]$, respectively. Show that

$$\sum_{n=-\infty}^{\infty} x^2[n] = \sum_{n=-\infty}^{\infty} x_e^2[n] + \sum_{n=-\infty}^{\infty} x_o^2[n]$$

1.51. Determine whether or not each of the following signals is periodic. If a signal is periodic, determine its fundamental period.

(a) $x(t) = \cos\left(2t + \dfrac{\pi}{4}\right)$

(b) $x(t) = \cos^2 t$

(c) $x(t) = (\cos 2\pi t)u(t)$

(d) $x(t) = e^{j\pi t}$

(e) $x[n] = e^{j[(n/4)-\pi]}$

(f) $x[n] = \cos\left(\dfrac{\pi n^2}{8}\right)$

(g) $x[n] = \cos\left(\dfrac{n}{2}\right)\cos\left(\dfrac{\pi n}{4}\right)$

(h) $x[n] = \cos\left(\dfrac{\pi n}{4}\right) + \sin\left(\dfrac{\pi n}{8}\right) - 2\cos\left(\dfrac{\pi n}{2}\right)$

1.52. Show that if $x[n]$ is periodic with period N, then

(a) $\displaystyle\sum_{k=n_0}^{n} x[k] = \sum_{k=n_0+N}^{n+N} x[k];$ (b) $\displaystyle\sum_{k=0}^{N} x[k] = \sum_{k=n_0}^{n_0+N} x[k]$

1.53. (a) What is $\delta(2t)$?

(b) What is $\delta[2n]$?

1.54. Show that

$$\delta'(-t) = -\delta'(t)$$

1.55. Evaluate the following integrals:

(a) $\displaystyle\int_{-\infty}^{t} (\cos\tau)u(\tau)\,d\tau$ (b) $\displaystyle\int_{-\infty}^{t} (\cos\tau)\delta(\tau)\,d\tau$

(c) $\displaystyle\int_{-\infty}^{\infty} (\cos t)u(t-1)\delta(t)\,dt$ (d) $\displaystyle\int_{0}^{2\pi} t\sin\dfrac{t}{2}\delta(\pi-t)\,dt$

1.56. Consider a continuous-time system with the input-output relation

$$y(t) = \mathbf{T}\{x(t)\} = \frac{1}{T}\int_{t-T/2}^{t+T/2} x(\tau)\,d\tau$$

Determine whether this system is (a) linear, (b) time-invariant, (c) causal.

1.57. Consider a continuous-time system with the input-output relation

$$y(t) = \mathbf{T}\{x(t)\} = \sum_{k=-\infty}^{\infty} x(t)\delta(t - kT_s)$$

Determine whether this system is (a) linear, (b) time-invariant.

1.58. Consider a discrete-time system with the input-output relation

$$y[n] = \mathbf{T}\{x[n]\} = x^2[n]$$

Determine whether this system is (a) linear, (b) time-invariant.

1.59. Give an example of a system that satisfies the condition of homogeneity (1.67) but not the condition of additivity (1.66).

1.60 Give an example of a linear time-varying system such that with a periodic input the corresponding output is not periodic.

1.61. A system is called *invertible* if we can determine its input signal x uniquely by observing its output signal y. This is illustrated in Fig. 1-43. Determine if each of the following systems is invertible. If the system is invertible, give the inverse system.

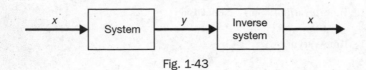

Fig. 1-43

(a) $y(t) = 2x(t)$

(b) $y(t) = x^2(t)$

(c) $y(t) = \int_{-\infty}^{t} x(\tau)\, d\tau$

(d) $y[n] = \sum_{k=-\infty}^{n} x[k]$

(e) $y[n] = nx[n]$

ANSWERS TO SUPPLEMENTARY PROBLEMS

1.46. (a) $x(t) = \frac{t}{2}[u(t) - u(t-2)]$

(b) $x(t) = u(t+1) + 2u(t) - u(t-1) - u(t-2) - u(t-3)$

1.47. (a) $x[n] = u[n] - u[n-(N+1)]$

(b) $x[n] = -u[-n-1]$

(c) $x[n] = u[n+2] - u[n-4]$

1.48. (a) $x_e(t) = \frac{1}{2},\, x_o(t) = \frac{1}{2}\operatorname{sgn} t$

(b) $x_e(t) = \frac{1}{\sqrt{2}}\cos \omega_0 t,\, x_o(t) = \frac{1}{\sqrt{2}}\sin \omega_0 t$

(c) $x_e[n] = j\cos \Omega_0 n,\, x_o[n] = -\sin \Omega_0 n$

(d) $x_e[n] = \delta[n],\, x_o[n] = 0$

1.49. *Hint:*   Use the results from Prob. 1.7 and Eq. (1.77).

1.50. *Hint:*   Use the results from Prob. 1.7 and Eq. (1.77).

1.51. (a) Periodic, period = π (b) Periodic, period = π

(c) Nonperiodic   (d) Periodic, period = 2

(e) Nonperiodic   (f) Periodic, period = 8

(g) Nonperiodic   (h) Periodic, period = 16

1.52. *Hint:*   See Prob. 1.17.

1.53. (a) $\delta(2t) = \frac{1}{2}\delta(t)$

(b) $\delta[2n] = \delta[n]$

1.54. *Hint:* Use Eqs. (1.101) and (1.99).

1.55. (a) $\sin t$

(b) 1 for $t > 0$ and 0 for $t < 0$; not defined for $t = 0$

(c) 0

(d) π

1.56. (*a*) Linear; (*b*) Time-invariant; (*c*) Noncausal

1.57. (*a*) Linear; (*b*) Time-varying

1.58. (*a*) Nonlinear; (*b*) Time-invariant

1.59. Consider the system described by

$$y(t) = \mathbf{T}\{x(t)\} = \left[\int_a^b [x(\tau)]^2 \, d\tau \right]^{1/2}$$

1.60 $y[n] = \mathbf{T}\{x[n]\} = nx[n]$

1.61. (*a*) Invertible; $x(t) = \dfrac{1}{2} y(t)$

 (*b*) Not invertible

 (*c*) Invertible; $x(t) = \dfrac{dy(t)}{dt}$

 (*d*) Invertible; $x[n] = y[n] - y[n-1]$

 (*e*) Not invertible

CHAPTER 2

Linear Time-Invariant Systems

2.1 Introduction

Two most important attributes of systems are linearity and time-invariance. In this chapter we develop the fundamental input-output relationship for systems having these attributes. It will be shown that the input-output relationship for LTI systems is described in terms of a convolution operation. The importance of the convolution operation in LTI systems stems from the fact that knowledge of the response of an LTI system to the unit impulse input allows us to find its output to any input signals. Specifying the input-output relationships for LTI systems by differential and difference equations will also be discussed.

2.2 Response of a Continuous-Time LTI System and the Convolution Integral

A. Impulse Response:

The *impulse response* $h(t)$ of a continuous-time LTI system (represented by $\mathbf{T}$) is defined to be the response of the system when the input is $\delta(t)$, that is,

$$h(t) = \mathbf{T}\{\delta(t)\} \qquad (2.1)$$

B. Response to an Arbitrary Input:

From Eq. (1.27) the input $x(t)$ can be expressed as

$$x(t) = \int_{-\infty}^{\infty} x(\tau)\delta(t - \tau)\, d\tau \qquad (2.2)$$

Since the system is linear, the response $y(t)$ of the system to an arbitrary input $x(t)$ can be expressed as

$$y(t) = \mathbf{T}\{x(t)\} = \mathbf{T}\left\{\int_{-\infty}^{\infty} x(\tau)\delta(t - \tau)\, d\tau\right\}$$
$$= \int_{-\infty}^{\infty} x(\tau)\mathbf{T}\{\delta(t - \tau)\}\, d\tau \qquad (2.3)$$

Since the system is time-invariant, we have

$$h(t - \tau) = \mathbf{T}\{\delta(t - \tau)\} \qquad (2.4)$$

Substituting Eq. (2.4) into Eq. (2.3), we obtain

$$y(t) = \int_{-\infty}^{\infty} x(\tau)h(t - \tau)\, d\tau \qquad (2.5)$$

Equation (2.5) indicates that a continuous-time LTI system is completely characterized by its impulse response $h(t)$.

C. Convolution Integral:

Equation (2.5) defines the *convolution* of two continuous-time signals $x(t)$ and $h(t)$ denoted by

$$y(t) = x(t) * h(t) = \int_{-\infty}^{\infty} x(\tau) h(t - \tau) \, d\tau \qquad (2.6)$$

Equation (2.6) is commonly called the *convolution integral*. Thus, we have the fundamental result that *the output of any continuous-time LTI system is the convolution of the input x(t) with the impulse response h(t) of the system*. Fig. 2-1 illustrates the definition of the impulse response $h(t)$ and the relationship of Eq. (2.6).

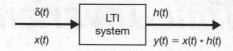

Fig. 2-1 Continuous-time LTI system.

D. Properties of the Convolution Integral:

The convolution integral has the following properties.

1. Commutative:

$$x(t) * h(t) = h(t) * x(t) \qquad (2.7)$$

2. Associative:

$$\{x(t) * h_1(t)\} * h_2(t) = x(t) * \{h_1(t) * h_2(t)\} \qquad (2.8)$$

3. Distributive:

$$x(t) * \{h_1(t)\} + h_2(t) = x(t) * h_1(t) + x(t) * h_2(t) \qquad (2.9)$$

E. Convolution Integral Operation:

Applying the commutative property (2.7) of convolution to Eq. (2.6), we obtain

$$y(t) = h(t) * x(t) = \int_{-\infty}^{\infty} h(\tau) x(t - \tau) \, d\tau \qquad (2.10)$$

which may at times be easier to evaluate than Eq. (2.6). From Eq. (2.6) we observe that the convolution integral operation involves the following four steps:

1. The impulse response $h(\tau)$ is time-reversed (that is, reflected about the origin) to obtain $h(-\tau)$ and then shifted by t to form $h(t - \tau) = h[-(\tau - t)]$, which is a function of τ with parameter t.
2. The signal $x(\tau)$ and $h(t - \tau)$ are multiplied together for all values of τ with t fixed at some value.
3. The product $x(\tau)h(t - \tau)$ is integrated over all τ to produce a single output value $y(t)$.
4. Steps 1 to 3 are repeated as t varies over $-\infty$ to ∞ to produce the entire output $y(t)$.

Examples of the above convolution integral operation are given in Probs. 2.4 to 2.6.

F. Step Response:

The *step response* $s(t)$ of a continuous-time LTI system (represented by $\mathbf{T}$) is defined to be the response of the system when the input is $u(t)$; that is,

$$s(t) = \mathbf{T}\{u(t)\} \qquad (2.11)$$

In many applications, the step response $s(t)$ is also a useful characterization of the system. The step response $s(t)$ can be easily determined by Eq. (2.10); that is,

$$s(t) = h(t) * u(t) = \int_{-\infty}^{\infty} h(\tau) u(t-\tau) \, d\tau = \int_{-\infty}^{t} h(\tau) \, d\tau \qquad (2.12)$$

Thus, the step response $s(t)$ can be obtained by integrating the impulse response $h(t)$. Differentiating Eq. (2.12) with respect to t, we get

$$h(t) = s'(t) = \frac{ds(t)}{dt} \qquad (2.13)$$

Thus, the impulse response $h(t)$ can be determined by differentiating the step response $s(t)$.

2.3 Properties of Continuous-Time LTI Systems

A. Systems with or without Memory:

Since the output $y(t)$ of a memoryless system depends on only the present input $x(t)$, then, if the system is also linear and time-invariant, this relationship can only be of the form

$$y(t) = Kx(t) \qquad (2.14)$$

where K is a (gain) constant. Thus, the corresponding impulse response $h(t)$ is simply

$$h(t) = K\delta(t) \qquad (2.15)$$

Therefore, if $h(t_0) \neq 0$ for $t_0 \neq 0$, then continuous-time LTI system has memory.

B. Causality:

As discussed in Sec. 1.5D, a causal system does not respond to an input event until that event actually occurs. Therefore, for a causal continuous-time LTI system, we have

$$h(t) = 0 \qquad t < 0 \qquad (2.16)$$

Applying the causality condition (2.16) to Eq. (2.10), the output of a causal continuous-time LTI system is expressed as

$$y(t) = \int_{0}^{\infty} h(\tau) x(t-\tau) \, d\tau \qquad (2.17)$$

Alternatively, applying the causality condition (2.16) to Eq. (2.6), we have

$$y(t) = \int_{-\infty}^{t} x(\tau) h(t-\tau) \, d\tau \qquad (2.18)$$

Equation (2.18) shows that the only values of the input $x(t)$ used to evaluate the output $y(t)$ are those for $\tau \leq t$. Based on the causality condition (2.16), any signal $x(t)$ is called *causal* if

$$x(t) = 0 \qquad t < 0 \qquad (2.19a)$$

and is called *anticausal* if

$$x(t) = 0 \qquad t > 0 \qquad (2.19b)$$

Then, from Eqs. (2.17), (2.18), and (2.19a), when the input $x(t)$ is causal, the output $y(t)$ of a causal continuous-time LTI system is given by

$$y(t) = \int_{0}^{t} h(\tau) x(t-\tau) \, d\tau = \int_{0}^{t} x(\tau) h(t-\tau) \, d\tau \qquad (2.20)$$

C. Stability:

The BIBO (bounded-input/bounded-output) stability of an LTI system (Sec. 1.5H) is readily ascertained from its impulse response. It can be shown (Prob. 2.13) that a continuous-time LTI system is BIBO stable if its impulse response is absolutely integrable; that is,

$$\int_{-\infty}^{\infty} |h(\tau)| \, d\tau < \infty \tag{2.21}$$

2.4 Eigenfunctions of Continuous-Time LTI Systems

In Chap. 1 (Prob. 1.44) we saw that the eigenfunctions of continuous-time LTI systems represented by $\mathbf{T}$ are the complex exponentials e^{st}, with s a complex variable. That is,

$$\mathbf{T}\{e^{st}\} = \lambda e^{st} \tag{2.22}$$

where λ is the eigenvalue of $\mathbf{T}$ associated with e^{st}. Setting $x(t) = e^{st}$ in Eq. (2.10), we have

$$y(t) = \mathbf{T}\{e^{st}\} = \int_{-\infty}^{\infty} h(\tau) \, e^{s(t-\tau)} \, d\tau = \left[\int_{-\infty}^{\infty} h(\tau) \, e^{-s\tau} \, d\tau\right] e^{st}$$

$$= H(s) \, e^{st} = \lambda e^{st} \tag{2.23}$$

where

$$\lambda = H(s) = \int_{-\infty}^{\infty} h(\tau) \, e^{-s\tau} \, d\tau \tag{2.24}$$

Thus, the eigenvalue of a continuous-time LTI system associated with the eigenfunction e^{st} is given by $H(s)$, which is a complex constant whose value is determined by the value of s via Eq. (2.24). Note from Eq. (2.23) that $y(0) = H(s)$ (see Prob. 1.44).

The above results underlie the definitions of the Laplace transform and Fourier transform, which will be discussed in Chaps. 3 and 5.

2.5 Systems Described by Differential Equations

A. Linear Constant-Coefficient Differential Equations:

A general Nth-order linear constant-coefficient differential equation is given by

$$\sum_{k=0}^{N} a_k \frac{d^k y(t)}{dt^k} = \sum_{k=0}^{M} b_k \frac{d^k x(t)}{dt^k} \tag{2.25}$$

where coefficients a_k and b_k are real constants. The order N refers to the highest derivative of $y(t)$ in Eq. (2.25). Such differential equations play a central role in describing the input-output relationships of a wide variety of electrical, mechanical, chemical, and biological systems. For instance, in the RC circuit considered in Prob. 1.32, the input $x(t) = v_s(t)$ and the output $y(t) = v_c(t)$ are related by a first-order constant-coefficient differential equation [Eq. (1.105)]

$$\frac{dy(t)}{dt} + \frac{1}{RC} y(t) = \frac{1}{RC} x(t)$$

The general solution of Eq. (2.25) for a particular input $x(t)$ is given by

$$y(t) = y_p(t) + y_h(t) \tag{2.26}$$

where $y_p(t)$ is a *particular solution* satisfying Eq. (2.25) and $y_h(t)$ is a *homogeneous solution* (or *complementary solution*) satisfying the homogeneous differential equation

$$\sum_{k=0}^{N} a_k \frac{d^k y_h(t)}{dt^k} = 0 \tag{2.27}$$

The exact form of $y_h(t)$ is determined by N auxiliary conditions. Note that Eq. (2.25) does not completely specify the output $y(t)$ in terms of the input $x(t)$ unless auxiliary conditions are specified. In general, a set of auxiliary conditions are the values of

$$y(t), \frac{dy(t)}{dt}, \dots, \frac{d^{N-1}y(t)}{dt^{N-1}}$$

at some point in time.

B. Linearity:

The system specified by Eq. (2.25) will be linear only if all of the auxiliary conditions are zero (see Prob. 2.21). If the auxiliary conditions are not zero, then the response $y(t)$ of a system can be expressed as

$$y(t) = y_{zi}(t) + y_{zs}(t) \tag{2.28}$$

where $y_{zi}(t)$, called the *zero-input response*, is the response to auxiliary conditions, and $y_{zs}(t)$, called the *zero-state response,* is the response of a linear system with zero auxiliary conditions. This is illustrated in Fig. 2-2.

Note that $y_{zi}(t) \neq y_h(t)$ and $y_{zs}(t) \neq y_p(t)$ and that in general $y_{zi}(t)$ contains $y_h(t)$ and $y_{zs}(t)$ contains both $y_h(t)$ and $y_p(t)$ (see Prob. 2.20).

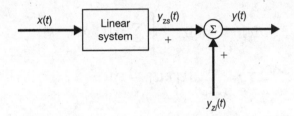

Fig. 2-2 Zero-state and zero-input responses.

C. Causality:

In order for the linear system described by Eq. (2.25) to be causal we must assume the condition of *initial rest* (or an *initially relaxed condition*). That is, if $x(t) = 0$ for $t \leq t_0$, then assume $y(t) = 0$ for $t \leq t_0$ (see Prob. 1.43). Thus, the response for $t > t_0$ can be calculated from Eq. (2.25) with the initial conditions

$$y(t_0) = \frac{dy(t_0)}{dt} = \dots = \frac{d^{N-1}y(t_0)}{dt^{N-1}} = 0$$

where

$$\frac{d^k y(t_0)}{dt^k} = \frac{d^k y(t)}{dt^k}\bigg|_{t=t_0}.$$

Clearly, at initial rest $y_{zi}(t) = 0$.

D. Time-Invariance:

For a linear causal system, initial rest also implies time-invariance (Prob. 2.22).

E. Impulse Response:

The impulse response $h(t)$ of the continuous-time LTI system described by Eq. (2.25) satisfies the differential equation

$$\sum_{k=0}^{N} a_k \frac{d^k h(t)}{dt^k} = \sum_{k=0}^{M} b_k \frac{d^k \delta(t)}{dt^k} \tag{2.29}$$

with the initial rest condition. Examples of finding impulse responses are given in Probs. 2.23 to 2.25. In later chapters, we will find the impulse response by using transform techniques.

2.6 Response of a Discrete-Time LTI System and Convolution Sum

A. Impulse Response:

The *impulse response* (or *unit sample response*) $h[n]$ of a discrete-time LTI system (represented by **T**) is defined to be the response of the system when the input is $\delta[n]$; that is,

$$h[n] = \mathbf{T}\{\delta[n]\} \tag{2.30}$$

B. Response to an Arbitrary Input:

From Eq. (1.51) the input $x[n]$ can be expressed as

$$x[n] = \sum_{k=-\infty}^{\infty} x[k]\,\delta[n-k] \tag{2.31}$$

Since the system is linear, the response $y[n]$ of the system to an arbitrary input $x[n]$ can be expressed as

$$y[n] = \mathbf{T}\{x[n]\} = \mathbf{T}\left\{\sum_{k=-\infty}^{\infty} x[k]\,\delta[n-k]\right\}$$
$$= \sum_{k=-\infty}^{\infty} x[k]\,\mathbf{T}\{\delta[n-k]\} \tag{2.32}$$

Since the system is time-invariant, we have

$$h[n-k] = \mathbf{T}\{\delta[n-k]\} \tag{2.33}$$

Substituting Eq. (2.33) into Eq. (2.32), we obtain

$$y[n] = \sum_{k=-\infty}^{\infty} x[k]h[n-k] \tag{2.34}$$

Equation (2.34) indicates that a discrete-time LTI system is completely characterized by its impulse response $h[n]$.

C. Convolution Sum:

Equation (2.34) defines the *convolution* of two sequences $x[n]$ and $h[n]$ denoted by

$$y[n] = x[n] * h[n] = \sum_{k=-\infty}^{\infty} x[k]h[n-k] \tag{2.35}$$

Equation (2.35) is commonly called the *convolution sum*. Thus, again, we have the fundamental result that *the output of any discrete-time LTI system is the convolution of the input $x[n]$ with the impulse response $h[n]$ of the system*.

Fig. 2-3 illustrates the definition of the impulse response $h[n]$ and the relationship of Eq. (2.35).

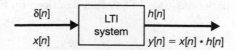

Fig. 2-3 Discrete-time LTI system.

D.   Properties of the Convolution Sum:

The following properties of the convolution sum are analogous to the convolution integral properties shown in Sec. 2.3.

1.   Commutative:

$$x[n] * h[n] = h[n] * x[n] \tag{2.36}$$

2.   Associative:

$$\{x[n] * h_1[n]\} * h_2[n] = x[n] * \{h_1[n] * h_2[n]\} \tag{2.37}$$

3.   Distributive:

$$x[n] * \{h_1[n]\} + h_2[n]\} = x[n] * h_1[n] + x[n] * h_2[n] \tag{2.38}$$

E.   Convolution Sum Operation:

Again, applying the commutative property (2.36) of the convolution sum to Eq. (2.35), we obtain

$$y[n] = h[n] * x[n] = \sum_{k=-\infty}^{\infty} h[k]x[n-k] \tag{2.39}$$

which may at times be easier to evaluate than Eq. (2.35). Similar to the continuous-time case, the convolution sum [Eq. (2.35)] operation involves the following four steps:

1.   The impulse response $h[k]$ is time-reversed (that is, reflected about the origin) to obtain $h[-k]$ and then shifted by n to form $h[n-k] = h[-(k-n)]$, which is a function of k with parameter n.
2.   Two sequences $x[k]$ and $h[n-k]$ are multiplied together for all values of k with n fixed at some value.
3.   The product $x[k]h[n-k]$ is summed over all k to produce a single output sample $y[n]$.
4.   Steps 1 to 3 are repeated as n varies over $-\infty$ to ∞ to produce the entire output $y[n]$.

Examples of the above convolution sum operation are given in Probs. 2.28 and 2.30.

F.   Step Response:

The *step response* $s[n]$ of a discrete-time LTI system with the impulse response $h[n]$ is readily obtained from Eq. (2.39) as

$$s[n] = h[n] * u[n] = \sum_{k=-\infty}^{\infty} h[k]u[n-k] = \sum_{k=-\infty}^{n} h[k] \tag{2.40}$$

From Eq. (2.40) we have

$$h[n] = s[n] - s[n-1] \tag{2.41}$$

Equations (2.40) and (2.41) are the discrete-time counterparts of Eqs. (2.12) and (2.13), respectively.

2.7   Properties of Discrete-Time LTI Systems

A.   Systems with or without Memory:

Since the output $y[n]$ of a memoryless system depends on only the present input $x[n]$, then, if the system is also linear and time-invariant, this relationship can only be of the form

$$y[n] = Kx[n] \tag{2.42}$$

where K is a (gain) constant. Thus, the corresponding impulse response is simply

$$h[n] = K\delta[n] \tag{2.43}$$

Therefore, if $h[n_0] \neq 0$ for $n_0 \neq 0$, the discrete-time LTI system has memory.

B. Causality:

Similar to the continuous-time case, the causality condition for a discrete-time LTI system is

$$h[n] = 0 \qquad n < 0 \tag{2.44}$$

Applying the causality condition (2.44) to Eq. (2.39), the output of a causal discrete-time LTI system is expressed as

$$y[n] = \sum_{k=0}^{\infty} h[k]x[n-k] \tag{2.45}$$

Alternatively, applying the causality condition (2.44) to Eq. (2.35), we have

$$y[n] = \sum_{k=-\infty}^{n} x[k]h[n-k] \tag{2.46}$$

Equation (2.46) shows that the only values of the input $x[n]$ used to evaluate the output $y[n]$ are those for $k \leq n$. As in the continuous-time case, we say that any sequence $x[n]$ is called *causal* if

$$x[n] = 0 \qquad n < 0 \tag{2.47a}$$

and is called *anticausal* if

$$x[n] = 0 \qquad n \geq 0 \tag{2.47b}$$

Then, when the input $x[n]$ is causal, the output $y[n]$ of a causal discrete-time LTI system is given by

$$y[n] = \sum_{k=0}^{n} h[k]x[n-k] = \sum_{k=0}^{n} x[k]h[n-k] \tag{2.48}$$

C. Stability:

It can be shown (Prob. 2.37) that a discrete-time LTI system is BIBO stable if its impulse response is absolutely summable; that is,

$$\sum_{k=-\infty}^{\infty} |h[k]| < \infty \tag{2.49}$$

2.8   Eigenfunctions of Discrete-Time LTI Systems

In Chap. 1 (Prob. 1.45) we saw that the eigenfunctions of discrete-time LTI systems represented by **T** are the complex exponentials z^n, with z a complex variable. That is,

$$\mathbf{T}\{z^n\} = \lambda z^n \tag{2.50}$$

where λ is the eigenvalue of **T** associated with z^n. Setting $x[n] = z^n$ in Eq. (2.39), we have

$$y[n] = \mathbf{T}\{z^n\} = \sum_{k=-\infty}^{\infty} h[k]\, z^{n-k} = \left[\sum_{k=-\infty}^{\infty} h[k]\, z^{-k} \right] z^n$$

$$= H(z)\, z^n = \lambda z^n \tag{2.51}$$

where

$$\lambda = H(z) = \sum_{k=-\infty}^{\infty} h[k]\, z^{-k} \tag{2.52}$$

Thus, the eigenvalue of a discrete-time LTI system associated with the eigenfunction z^n is given by $H(z)$, which is a complex constant whose value is determined by the value of z via Eq. (2.52). Note from Eq. (2.51) that $y[0] = H(z)$ (see Prob. 1.45).

The above results underlie the definitions of the z-transform and discrete-time Fourier transform, which will be discussed in Chaps. 4 and 6.

2.9 Systems Described by Difference Equations

The role of differential equations in describing continuous-time systems is played by *difference equations* for discrete-time systems.

A. Linear Constant-Coefficient Difference Equations:

The discrete-time counterpart of the general differential equation (2.25) is the Nth-order linear constant-coefficient difference equation given by

$$\sum_{k=0}^{N} a_k y[n-k] = \sum_{k=0}^{M} b_k x[n-k] \tag{2.53}$$

where coefficients a_k and b_k are real constants. The order N refers to the largest delay of $y[n]$ in Eq. (2.53). An example of the class of linear constant-coefficient difference equations is given in Chap. 1 (Prob. 1.37). Analogous to the continuous-time case, the solution of Eq. (2.53) and all properties of systems, such as linearity, causality, and time-invariance, can be developed following an approach that directly parallels the discussion for differential equations. Again we emphasize that the system described by Eq. (2.53) will be causal and LTI if the system is initially at rest.

B. Recursive Formulation:

An alternate and simpler approach is available for the solution of Eq. (2.53). Rearranging Eq. (2.53) in the form

$$y[n] = \frac{1}{a_0} \left\{ \sum_{k=0}^{M} b_k x[n-k] - \sum_{k=1}^{N} a_k y[n-k] \right\} \tag{2.54}$$

we obtain a formula to compute the output at time n in terms of the present input and the previous values of the input and output. From Eq. (2.54) we see that the need for auxiliary conditions is obvious and that to calculate $y[n]$ starting at $n = n_0$, we must be given the values of $y[n_0 - 1], y[n_0 - 2], \ldots, y[n_0 - N]$ as well as the input $x[n]$ for $n \geq n_0 - M$. The general form of Eq. (2.54) is called a *recursive equation*, since it specifies a recursive procedure for determining the output in terms of the input and previous outputs. In the special case when $N = 0$, from Eq. (2.53) we have

$$y[n] = \frac{1}{a_0} \left\{ \sum_{k=0}^{M} b_k x[n-k] \right\} \tag{2.55}$$

which is a *nonrecursive equation*, since previous output values are not required to compute the present output. Thus, in this case, auxiliary conditions are not needed to determine $y[n]$.

C. Impulse Response:

Unlike the continuous-time case, the impulse response $h[n]$ of a discrete-time LTI system described by Eq. (2.53) or, equivalently, by Eq. (2.54) can be determined easily as

$$h[n] = \frac{1}{a_0}\left\{\sum_{k=0}^{M} b_k \,\delta[n-k] - \sum_{k=1}^{N} a_k \,h[n-k]\right\} \tag{2.56}$$

For the system described by Eq. (2.55) the impulse response $h[n]$ is given by

$$h[n] = \frac{1}{a_0}\sum_{k=0}^{M} b_k \,\delta[n-k] = \begin{cases} b_n/a_0 & 0 \le n \le M \\ 0 & \text{otherwise} \end{cases} \tag{2.57}$$

Note that the impulse response for this system has finite terms; that is, it is nonzero for only a finite time duration. Because of this property, the system specified by Eq. (2.55) is known as a *finite impulse response* (FIR) system. On the other hand, a system whose impulse response is nonzero for an infinite time duration is said to be an *infinite impulse response* (IIR) system. Examples of finding impulse responses are given in Probs. 2.44 and 2.45. In Chap. 4, we will find the impulse response by using transform techniques.

SOLVED PROBLEMS

Responses of a Continuous-Time LTI System and Convolution

2.1. Verify Eqs. (2.7) and (2.8); that is,

(a) $x(t) * h(t) = h(t) * x(t)$

(b) $\{x(t) * h_1(t)\} * h_2(t) = x(t) * \{h_1(t) * h_2(t)\}$

(a) By definition (2.6)

$$x(t) * h(t) = \int_{-\infty}^{\infty} x(\tau)h(t-\tau)\,d\tau$$

By changing the variable $t - \tau = \lambda$, we have

$$x(t) * h(t) = \int_{-\infty}^{\infty} x(t-\lambda)h(\lambda)\,d\lambda = \int_{-\infty}^{\infty} h(\lambda)x(t-\lambda)\,d\lambda = h(t) * x(t)$$

(b) Let $x(t) * h_1(t) = f_1(t)$ and $h_1(t) * h_2(t) = f_2(t)$. Then

$$f_1(t) = \int_{-\infty}^{\infty} x(\tau)h_1(t-\tau)\,d\tau$$

and

$$\{x(t) * h_1(t)\} * h_2(t) = f_1(t) * h_2(t) = \int_{-\infty}^{\infty} f_1(\sigma)h_2(t-\sigma)\,d\sigma$$

$$= \int_{-\infty}^{\infty} \left[\int_{-\infty}^{\infty} x(\tau)h_1(\sigma-\tau)\,d\tau\right]h_2(t-\sigma)\,d\sigma$$

Substituting $\lambda = \sigma - \tau$ and interchanging the order of integration, we have

$$\{x(t) * \mathrm{h}_1(t)\} * h_2(t) = \int_{-\infty}^{\infty} x(\tau)\left[\int_{-\infty}^{\infty} h_1(\lambda)h_2(t-\tau-\lambda)\,d\lambda\right]d\tau$$

Now, since

$$f_2(t) = \int_{-\infty}^{\infty} h_1(\lambda)h_2(t-\lambda)\,d\lambda$$

we have

$$f_2(t-\tau) = \int_{-\infty}^{\infty} h_1(\lambda)h_2(t-\tau-\lambda)\,d\lambda$$

Thus,
$$\{x(t)*h_1(t)\}*h_2(t) = \int_{-\infty}^{\infty} x(\tau)f_2(t-\tau)\,d\tau$$
$$= x(t)*f_2(t) = x(t)*\{h_1(t)*h_2(t)\}$$

2.2. Show that

(*a*) $x(t)*\delta(t) = x(t)$ (2.58)

(*b*) $x(t)*\delta(t-t_0) = x(t-t_0)$ (2.59)

(*c*) $x(t)*u(t) = \int_{-\infty}^{t} x(\tau)\,d\tau$ (2.60)

(*d*) $x(t)*u(t-t_0) = \int_{-\infty}^{t-t_0} x(\tau)\,d\tau$ (2.61)

(*a*) By definition (2.6) and Eq. (1.22) we have

$$x(t)*\delta(t) = \int_{-\infty}^{\infty} x(\tau)\,\delta(t-\tau)\,d\tau = x(\tau)\big|_{\tau=t} = x(t)$$

(*b*) By Eqs. (2.7) and (1.22) we have

$$x(t)*\delta(t-t_0) = \delta(t-t_0)*x(t) = \int_{-\infty}^{\infty} \delta(\tau-t_0)x(t-\tau)\,d\tau$$
$$= x(t-\tau)\big|_{\tau=t_0} = x(t-t_0)$$

(*c*) By Eqs. (2.6) and (1.19) we have

$$x(t)*u(t) = \int_{-\infty}^{\infty} x(\tau)u(t-\tau)\,d\tau = \int_{-\infty}^{t} x(\tau)\,d\tau$$

since $u(t-\tau) = \begin{cases} 1 & \tau < t \\ 0 & \tau > t \end{cases}$

(*d*) In a similar manner, we have

$$x(t)*u(t-t_0) = \int_{-\infty}^{\infty} x(\tau)u(t-\tau-t_0)\,d\tau = \int_{-\infty}^{t-t_0} x(\tau)\,d\tau$$

since $u(t-\tau-t_0) = \begin{cases} 1 & \tau < t-t_0 \\ 0 & \tau > t-t_0 \end{cases}$.

2.3. Let $y(t) = x(t)*h(t)$. Then show that

$$x(t-t_1)*h(t-t_2) = y(t-t_1-t_2) \tag{2.62}$$

By Eq. (2.6) we have

$$y(t) = x(t)*h(t) = \int_{-\infty}^{\infty} x(\tau)h(t-\tau)\,d\tau \tag{2.63a}$$

and
$$x(t-t_1)*h(t-t_2) = \int_{-\infty}^{\infty} x(\tau-t_1)h(t-\tau-t_2)\,d\tau \tag{2.63b}$$

Let $\tau - t_1 = \lambda$. Then $\tau = \lambda + t_1$ and Eq. (2.63b) becomes

$$x(t-t_1)*h(t-t_2) = \int_{-\infty}^{\infty} x(\lambda)h(t-t_1-t_2-\lambda)\,d\lambda \tag{2.63c}$$

62 CHAPTER 2 *Linear Time-Invariant Systems*

Comparing Eqs. (2.63a) and (2.63c), we see that replacing t in Eq. (2.63a) by $t - t_1 - t_2$, we obtain Eq. (2.63c). Thus, we conclude that

$$x(t - t_1) * h(t - t_2) = y(t - t_1 - t_2)$$

2.4. The input $x(t)$ and the impulse response $h(t)$ of a continuous time LTI system are given by

$$x(t) = u(t) \qquad h(t) = e^{-\alpha t} u(t), \alpha > 0$$

(a) Compute the output $y(t)$ by Eq. (2.6).

(b) Compute the output $y(t)$ by Eq. (2.10).

(a) By Eq. (2.6)

$$y(t) = x(t) * h(t) = \int_{-\infty}^{\infty} x(\tau)h(t - \tau)\, d\tau$$

Functions $x(\tau)$ and $h(t - \tau)$ are shown in Fig. 2-4(a) for $t < 0$ and $t > 0$. From Fig. 2-4(a) we see that for $t < 0$, $x(\tau)$ and $h(t - \tau)$ do not overlap, while for $t > 0$, they overlap from $\tau = 0$ to $\tau = t$. Hence, for $t < 0$, $y(t) = 0$. For $t > 0$, we have

$$y(t) = \int_0^t e^{-\alpha(t-\tau)}\, d\tau = e^{-\alpha t} \int_0^t e^{\alpha \tau}\, d\tau$$

$$= e^{-\alpha t} \frac{1}{\alpha} (e^{\alpha t} - 1) = \frac{1}{\alpha}(1 - e^{-\alpha t})$$

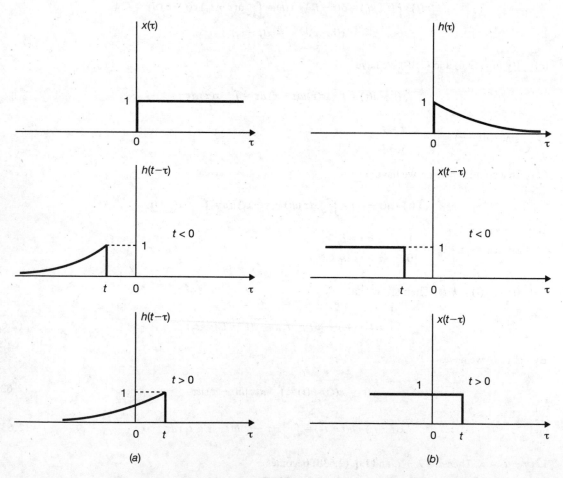

(a)   (b)

Fig. 2-4

Thus, we can write the output $y(t)$ as

$$y(t) = \frac{1}{\alpha}(1 - e^{-\alpha t})u(t) \tag{2.64}$$

(b) By Eq. (2.10)

$$y(t) = h(t) * x(t) = \int_{-\infty}^{\infty} h(\tau)x(t-\tau)\,d\tau$$

Functions $h(\tau)$ and $x(t-\tau)$ are shown in Fig. 2-4(b) for $t < 0$ and $t > 0$. Again from Fig. 2-4(b) we see that for $t < 0$, $h(\tau)$ and $x(t-\tau)$ do not overlap, while for $t > 0$, they overlap from $\tau = 0$ to $\tau = t$. Hence, for $t < 0$, $y(t) = 0$. For $t > 0$, we have

$$y(t) = \int_0^t e^{-\alpha \tau}d\tau = \frac{1}{\alpha}(1 - e^{-\alpha t})$$

Thus, we can write the output $y(t)$ as

$$y(t) = \frac{1}{\alpha}(1 - e^{-\alpha t})u(t) \tag{2.65}$$

which is the same as Eq. (2.64).

2.5. Compute the output $y(t)$ for a continuous-time LTI system whose impulse response $h(t)$ and the input $x(t)$ are given by

$$h(t) = e^{-\alpha t}u(t) \qquad x(t) = e^{\alpha t}u(-t) \quad \alpha > 0$$

By Eq. (2.6)

$$y(t) = x(t) * h(t) = \int_{-\infty}^{\infty} x(\tau)h(t-\tau)\,d\tau$$

Functions $x(\tau)$ and $h(t-\tau)$ are shown in Fig. 2-5 (a) for $t < 0$ and $t > 0$. From Fig. 2-5 (a) we see that for $t < 0$, $x(\tau)$ and $h(t-\tau)$ overlap from $\tau = -\infty$ to $\tau = t$, while for $t > 0$, they overlap from $\tau = -\infty$ to $\tau = 0$. Hence, for $t < 0$, we have

$$y(t) = \int_{-\infty}^{t} e^{\alpha \tau}\, e^{-\alpha(t-\tau)}d\tau = e^{-\alpha t}\int_{-\infty}^{t} e^{2\alpha \tau}\,d\tau = \frac{1}{2\alpha}e^{\alpha t} \tag{2.66a}$$

For $t > 0$, we have

$$y(t) = \int_{-\infty}^{0} e^{\alpha \tau}\, e^{-\alpha(t-\tau)}d\tau = e^{-\alpha t}\int_{-\infty}^{0} e^{2\alpha \tau}\,d\tau = \frac{1}{2\alpha}e^{-\alpha t} \tag{2.66b}$$

Combining Eqs. (2.66a) and (2.66b), we can write $y(t)$ as

$$y(t) = \frac{1}{2\alpha}e^{-\alpha |t|} \qquad \alpha > 0 \tag{2.67}$$

which is shown in Fig. 2-5(b).

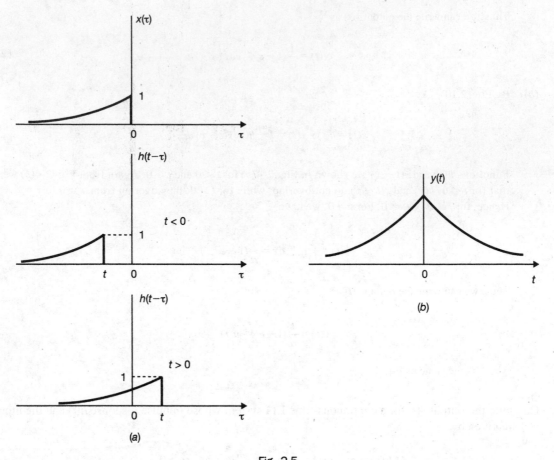

Fig. 2-5

2.6. Evaluate $y(t) = x(t) * h(t)$, where $x(t)$ and $h(t)$ are shown in Fig. 2-6, (a) by an analytical technique, and (b) by a graphical method.

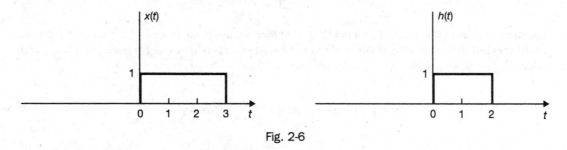

Fig. 2-6

(a) We first express $x(t)$ and $h(t)$ in functional form:

$$x(t) = u(t) - u(t - 3) \qquad h(t) = u(t) - u(t - 2)$$

Then, by Eq. (2.6) we have

$$y(t) = x(t) * h(t) = \int_{-\infty}^{\infty} x(\tau) h(t - \tau)\, d\tau$$

$$= \int_{-\infty}^{\infty} [u(\tau) - u(\tau - 3)][u(t - \tau) - u(t - \tau - 2)]\, d\tau$$

$$= \int_{-\infty}^{\infty} u(\tau) u(t - \tau)\, d\tau - \int_{-\infty}^{\infty} u(\tau) u(t - 2 - \tau)\, d\tau$$

$$\quad - \int_{-\infty}^{\infty} u(\tau - 3) u(t - \tau)\, d\tau + \int_{-\infty}^{\infty} u(\tau - 3) u(t - 2 - \tau)\, d\tau$$

Since

$$u(\tau)u(t-\tau) = \begin{cases} 1 & 0 < \tau < t, t > 0 \\ 0 & \text{otherwise} \end{cases}$$

$$u(\tau)u(t-2-\tau) = \begin{cases} 1 & 0 < \tau < t-2, t > 2 \\ 0 & \text{otherwise} \end{cases}$$

$$u(\tau-3)u(t-\tau) = \begin{cases} 1 & 3 < \tau < t, t > 3 \\ 0 & \text{otherwise} \end{cases}$$

$$u(\tau-3)u(t-2-\tau) = \begin{cases} 1 & 3 < \tau < t-2, t > 5 \\ 0 & \text{otherwise} \end{cases}$$

we can express $y(t)$ as

$$y(t) = \left(\int_0^t d\tau\right)u(t) - \left(\int_0^{t-2} d\tau\right)u(t-2)$$

$$- \left(\int_3^t d\tau\right)u(t-3) + \left(\int_3^{t-2} d\tau\right)u(t-5)$$

$$= tu(t) - (t-2)u(t-2) - (t-3)u(t-3) + (t-5)u(t-5)$$

which is plotted in Fig. 2-7.

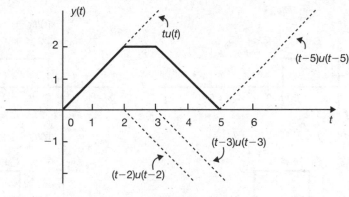

Fig. 2-7

(b)   Functions $h(\tau), x(\tau)$ and $h(t-\tau), x(\tau)h(t-\tau)$ for different values of t are sketched in Fig. 2-8. From Fig. 2-8 we see that $x(\tau)$ and $h(t-\tau)$ do not overlap for $t < 0$ and $t > 5$, and hence, $y(t) = 0$ for $t < 0$ and $t > 5$. For the other intervals, $x(\tau)$ and $h(t-\tau)$ overlap. Thus, computing the area under the rectangular pulses for these intervals, we obtain

$$y(t) = \begin{cases} 0 & t < 0 \\ t & 0 < t \le 2 \\ 2 & 2 < t \le 3 \\ 5-t & 3 < t \le 5 \\ 0 & 5 < t \end{cases}$$

which is plotted in Fig. 2-9.

2.7.   Let $h(t)$ be the triangular pulse shown in Fig. 2-10(a) and let $x(t)$ be the unit impulse train [Fig. 2-10(b)] expressed as

$$x(t) = \delta_T(t) = \sum_{n=-\infty}^{\infty} \delta(t - nT) \tag{2.68}$$

Determine and sketch $y(t) = h(t) * x(t)$ for the following values of T: (a) $T = 3$, (b) $T = 2$, (c) $T = 1.5$.

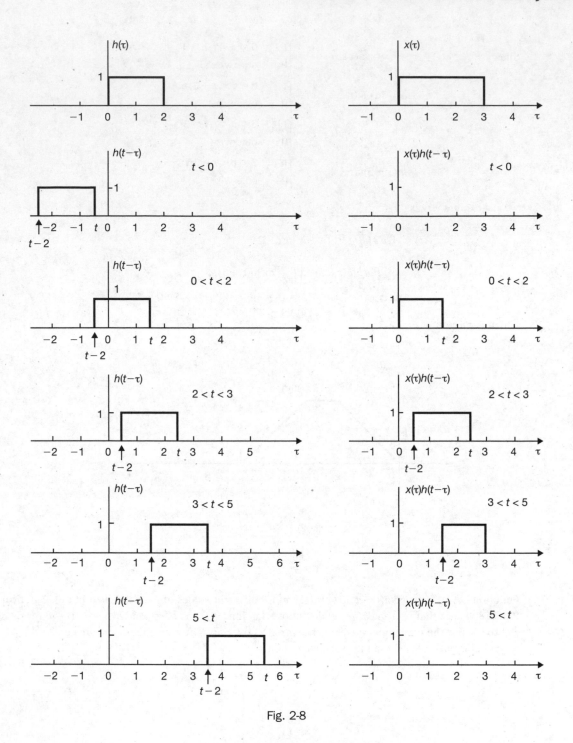

Fig. 2-8

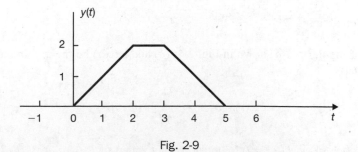

Fig. 2-9

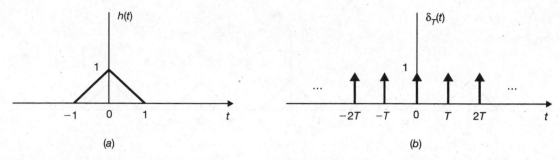

Fig. 2-10

Using Eqs. (2.59) and (2.9), we obtain

$$y(t) = h(t) * \delta_T(t) = h(t) * \left[\sum_{n=-\infty}^{\infty} \delta(t - nT) \right]$$

$$= \sum_{n=-\infty}^{\infty} h(t) * \delta(t - nT) = \sum_{n=-\infty}^{\infty} h(t - nT) \tag{2.69}$$

(a) For $T = 3$, Eq. (2.69) becomes

$$y(t) = \sum_{n=-\infty}^{\infty} h(t - 3n)$$

which is sketched in Fig. 2-11(a).

(b) For $T = 2$, Eq. (2.69) becomes

$$y(t) = \sum_{n=-\infty}^{\infty} h(t - 2n)$$

which is sketched in Fig. 2-11(b).

(c) For $T = 1.5$, Eq. (2.69) becomes

$$y(t) = \sum_{n=-\infty}^{\infty} h(t - 1.5n)$$

which is sketched in Fig. 2-11(c). Note that when $T < 2$, the triangular pulses are no longer separated and they overlap.

2.8. If $x_1(t)$ and $x_2(t)$ are both periodic signals with a common period T_0, the convolution of $x_1(t)$ and $x_2(t)$ does not converge. In this case, we define the *periodic convolution* of $x_1(t)$ and $x_2(t)$ as

$$f(t) = x_1(t) \otimes x_2(t) = \int_0^{T_0} x_1(\tau) x_2(t - \tau) \, d\tau \tag{2.70}$$

(a) Show that $f(t)$ is periodic with period T_0.

(b) Show that

$$f(t) = \int_a^{a + T_0} x_1(\tau) x_2(t - \tau) \, d\tau \tag{2.71}$$

for any a.

(c) Compute and sketch the periodic convolution of the square-wave signal $x(t)$ shown in Fig. 2-12 with itself.

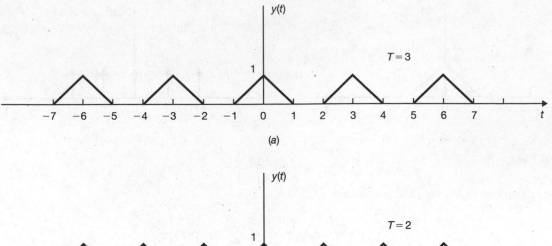

(a)

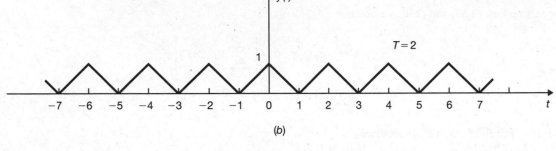

(b)

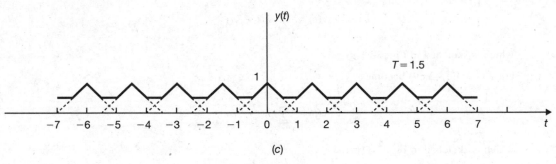

(c)

Fig. 2-11

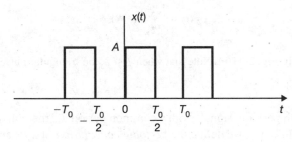

Fig. 2-12

(a)   Since $x_2(t)$ is periodic with period T_0, we have

$$x_2(t + T_0 - \tau) = x_2(t - \tau)$$

Then from Eq. (2.70) we have

$$f(t + T_0) = \int_0^{T_0} x_1(\tau)x_2(t + T_0 - \tau)\,d\tau$$

$$= \int_0^{T_0} x_1(\tau)x_2(t - \tau)\,d\tau = f(t)$$

Thus, $f(t)$ is periodic with period T_0.

(b) Since both $x_1(\tau)$ and $x_2(\tau)$ are periodic with the same period T_0, $x_1(\tau)x_2(t-\tau)$ is also periodic with period T_0. Then using property (1.88) (Prob. 1.17), we obtain

$$f(t) = \int_0^{T_0} x_1(\tau)x_2(t-\tau)\,d\tau = \int_a^{a+T_0} x_1(\tau)x_2(t-\tau)\,d\tau$$

for an arbitrary a.

(c) We evaluate the periodic convolution graphically. Signals $x(\tau)$, $x(t-\tau)$, and $x(\tau)x(t-\tau)$ are sketched in Fig. 2-13(a), from which we obtain

$$f(t) = \begin{cases} A^2 t & 0 < t \le T_0/2 \\ \\ -A^2(t - T_0) & T_0/2 < t \le T_0 \end{cases} \quad \text{and} \quad f(t + T_0) = f(t)$$

which is plotted in Fig. 2-13(b).

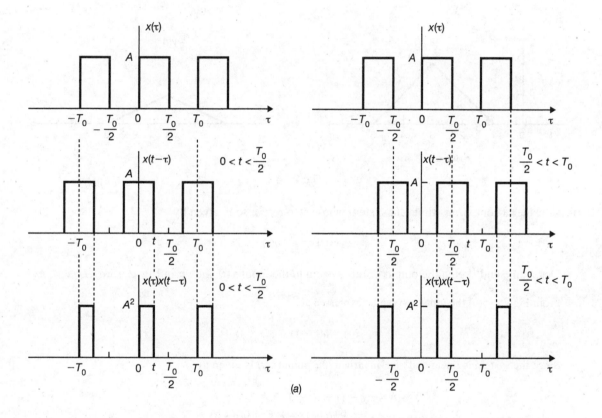

(a)

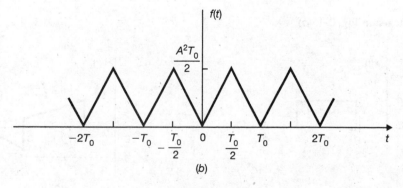

(b)

Fig. 2-13

Properties of Continuous-Time LTI Systems

2.9. The signals in Figs. 2-14(a) and (b) are the input $x(t)$ and the output $y(t)$, respectively, of a certain continuous-time LTI system. Sketch the output to the following inputs: (a) $x(t - 2)$; (b) $\frac{1}{2}x(t)$.

(a)   Since the system is time-invariant, the output will be $y(t - 2)$, which is sketched in Fig. 2-14(c).

(b)   Since the system is linear, the output will be $\frac{1}{2}y(t)$, which is sketched in Fig. 2-14(d).

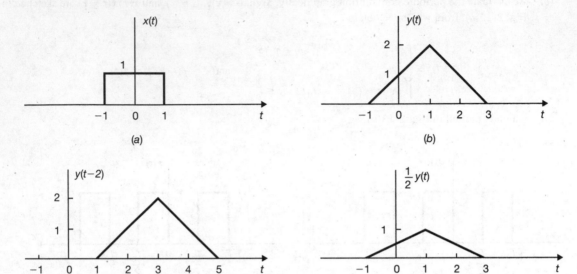

Fig. 2-14

2.10. Consider a continuous-time LTI system whose step response is given by

$$s(t) = e^{-t}u(t)$$

Determine and sketch the output of this system to the input $x(t)$ shown in Fig. 2-15(a).

From Fig. 2-15(a) the input $x(t)$ can be expressed as

$$x(t) = u(t - 1) - u(t - 3)$$

Since the system is linear and time-invariant, the output $y(t)$ is given by

$$y(t) = s(t - 1) - s(t - 3)$$
$$= e^{-(t-1)}u(t - 1) - e^{-(t-3)}u(t - 3)$$

which is sketched in Fig. 2-15(b).

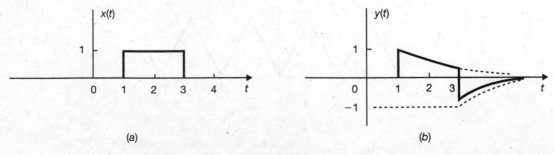

Fig. 2-15

2.11. Consider a continuous-time LTI system described by (see Prob. 1.56)

$$y(t) = \mathbf{T}\{x(t)\} = \frac{1}{T} \int_{t-T/2}^{t+T/2} x(\tau) \, d\tau \tag{2.72}$$

(a) Find and sketch the impulse response $h(t)$ of the system.

(b) Is this system causal?

(a) Equation (2.72) can be rewritten as

$$y(t) = \frac{1}{T} \int_{-\infty}^{t+T/2} x(\tau) \, d\tau - \frac{1}{T} \int_{-\infty}^{t-T/2} x(\tau) \, d\tau \tag{2.73}$$

Using Eqs. (2.61) and (2.9), Eq. (2.73) can be expressed as

$$y(t) = \frac{1}{T} x(t) * u\left(t + \frac{T}{2}\right) - \frac{1}{T} x(t) * u\left(t - \frac{T}{2}\right)$$

$$= x(t) * \frac{1}{T} \left[u\left(t + \frac{T}{2}\right) - u\left(t - \frac{T}{2}\right) \right] = x(t) * h(t) \tag{2.74}$$

Thus, we obtain

$$h(t) = \frac{1}{T} \left[u\left(t + \frac{T}{2}\right) - u\left(t - \frac{T}{2}\right) \right] = \begin{cases} 1/T & -T/2 < t \le T/2 \\ 0 & \text{otherwise} \end{cases} \tag{2.75}$$

which is sketched in Fig. 2-16.

(b) From Fig. 2-16 or Eq. (2.75) we see that $h(t) \ne 0$ for $t < 0$. Hence, the system is not causal.

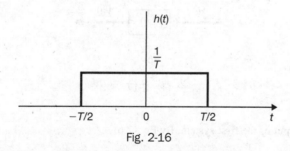

Fig. 2-16

2.12. Let $y(t)$ be the output of a continuous-time LTI system with input $x(t)$. Find the output of the system if the input is $x'(t)$, where $x'(t)$ is the first derivative of $x(t)$.

From Eq. (2.10)

$$y(t) = h(t) * x(t) = \int_{-\infty}^{\infty} h(\tau) x(t - \tau) \, d\tau$$

Differentiating both sides of the above convolution integral with respect to t, we obtain

$$y'(t) = \frac{d}{dt} \left[\int_{-\infty}^{\infty} h(\tau) x(t - \tau) \, d\tau \right] = \int_{-\infty}^{\infty} \frac{d}{dt} [h(\tau) x(t - \tau) \, d\tau]$$

$$= \int_{-\infty}^{\infty} h(\tau) x'(t - \tau) \, d\tau = h(t) * x'(t) \tag{2.76}$$

which indicates that $y'(t)$ is the output of the system when the input is $x'(t)$.

2.13. Verify the BIBO stability condition [Eq. (2.21)] for continuous-time LTI systems.

Assume that the input $x(t)$ of a continuous-time LTI system is bounded, that is,

$$|x(t)| \le k_1 \qquad \text{all } t \tag{2.77}$$

Then, using Eq. (2.10), we have

$$|y(t)| = \left| \int_{-\infty}^{\infty} h(\tau)x(t-\tau)\,d\tau \right| \le \int_{-\infty}^{\infty} |h(\tau)x(t-\tau)|\,d\tau$$

$$= \int_{-\infty}^{\infty} |h(\tau)||x(t-\tau)|\,d\tau \le k_1 \int_{-\infty}^{\infty} |h(\tau)|\,d\tau$$

since $|x(t-\tau)| \le k_1$ from Eq. (2.77). Therefore, if the impulse response is absolutely integrable, that is,

$$\int_{-\infty}^{\infty} |h(\tau)|\,d\tau = K < \infty$$

then $|y(t)| \le k_1 K = k_2$ and the system is BIBO stable.

2.14. The system shown in Fig. 2-17(a) is formed by connecting two systems *in cascade*. The impulse responses of the systems are given by $h_1(t)$ and $h_2(t)$, respectively, and

$$h_1(t) = e^{-2t}u(t) \qquad h_2(t) = 2e^{-t}u(t)$$

(a) Find the impulse response $h(t)$ of the overall system shown in Fig. 2-17(b).

(b) Determine if the overall system is BIBO stable.

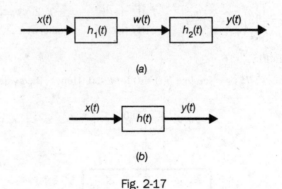

(a)

(b)

Fig. 2-17

(a) Let $w(t)$ be the output of the first system. By Eq. (2.6)

$$w(t) = x(t) * h_1(t) \tag{2.78}$$

Then we have

$$y(t) = w(t) * h_2(t) = [x(t) * h_1(t)] * h_2(t) \tag{2.79}$$

But by the associativity property of convolution (2.8), Eq. (2.79) can be rewritten as

$$y(t) = x(t) * [h_1(t) * h_2(t)] = x(t) * h(t) \tag{2.80}$$

Therefore, the impulse response of the overall system is given by

$$h(t) = h_1(t) * h_2(t) \tag{2.81}$$

Thus, with the given $h_1(t)$ and $h_2(t)$, we have

$$h(t) = \int_{-\infty}^{\infty} h_1(\tau)h_2(t-\tau)\,d\tau = \int_{-\infty}^{\infty} e^{-2\tau}u(\tau)\,2e^{-(t-\tau)}u(t-\tau)\,d\tau$$

$$= 2e^{-t}\int_{-\infty}^{\infty} e^{-\tau}u(\tau)u(t-\tau)\,d\tau = 2e^{-t}\left[\int_0^t e^{-\tau}\,d\tau\right]u(t)$$

$$= 2(e^{-t} - e^{-2t})u(t)$$

(b) Using the above $h(t)$, we have

$$\int_{-\infty}^{\infty} |h(\tau)| \, d\tau = 2\int_0^\infty (e^{-\tau} - e^{-2\tau}) \, d\tau = 2\left[\int_0^\infty e^{-\tau} \, d\tau - \int_0^\infty e^{-2\tau} \, d\tau\right]$$

$$= 2\left(1 - \frac{1}{2}\right) = 1 < \infty$$

Thus, the system is BIBO stable.

Eigenfunctions of Continuous-Time LTI Systems

2.15. Consider a continuous-time LTI system with the input-output relation given by

$$y(t) = \int_{-\infty}^{t} e^{-(t-\tau)} x(\tau) \, d\tau \tag{2.82}$$

(a) Find the impulse response $h(t)$ of this system.

(b) Show that the complex exponential function e^{st} is an eigenfunction of the system.

(c) Find the eigenvalue of the system corresponding to e^{st} by using the impulse response $h(t)$ obtained in part (a).

(a) From Eq. (2.82), definition (2.1), and Eq. (1.21) we get

$$h(t) = \int_{-\infty}^{t} e^{-(t-\tau)} \delta(\tau) \, d\tau = e^{-(t-\tau)}\Big|_{\tau=0} = e^{-t} \qquad t > 0$$

Thus, $\qquad\qquad\qquad\qquad h(t) = e^{-t}u(t) \tag{2.83}$

(b) Let $x(t) = e^{st}$. Then

$$y(t) = \int_{-\infty}^{t} e^{-(t-\tau)} e^{s\tau} \, d\tau = e^{-t}\int_{-\infty}^{t} e^{(s+1)\tau} \, d\tau$$

$$= \frac{1}{s+1}e^{st} = \lambda e^{st} \qquad \text{if Re } s > -1 \tag{2.84}$$

Thus, by definition (2.22) e^{st} is the eigenfunction of the system and the associated eigenvalue is

$$\lambda = \frac{1}{s+1} \tag{2.85}$$

(c) Using Eqs. (2.24) and (2.83), the eigenvalue associated with e^{st} is given by

$$\lambda = H(s) = \int_{-\infty}^{\infty} h(\tau)e^{-s\tau} \, d\tau = \int_{-\infty}^{\infty} e^{-\tau}u(\tau)e^{-s\tau} \, d\tau$$

$$= \int_0^\infty e^{-(s+1)\tau} \, d\tau = \frac{1}{s+1} \qquad \text{if Re } s > -1$$

which is the same as Eq. (2.85).

2.16. Consider the continuous-time LTI system described by

$$y(t) = \frac{1}{T}\int_{t-T/2}^{t+T/2} x(\tau) \, d\tau \tag{2.86}$$

(a) Find the eigenvalue of the system corresponding to the eigenfunction e^{st}.

(b) Repeat part (a) by using the impulse function $h(t)$ of the system.

(a) Substituting $x(\tau) = e^{s\tau}$ in Eq. (2.86), we obtain

$$y(t) = \frac{1}{T} \int_{t-T/2}^{t+T/2} e^{s\tau} \, d\tau$$

$$= \frac{1}{sT} (e^{sT/2} - e^{-sT/2}) e^{st} = \lambda e^{st}$$

Thus, the eigenvalue of the system corresponding to e^{st} is

$$\lambda = \frac{1}{sT}(e^{sT/2} - e^{-sT/2}) \tag{2.87}$$

(b) From Eq. (2.75) in Prob. 2.11 we have

$$h(t) = \frac{1}{T}\left[u\left(t + \frac{T}{2}\right) - u\left(t - \frac{T}{2}\right)\right] = \begin{cases} 1/T & -T/2 < t \le T/2 \\ 0 & \text{otherwise} \end{cases}$$

Using Eq. (2.24), the eigenvalue $H(s)$ corresponding to e^{st} is given by

$$H(s) = \int_{-\infty}^{\infty} h(\tau)\, e^{-s\tau}\, d\tau = \frac{1}{T}\int_{-T/2}^{T/2} e^{-s\tau}\, d\tau = \frac{1}{sT}(e^{sT/2} - e^{-sT/2})$$

which is the same as Eq. (2.87).

2.17. Consider a stable continuous-time LTI system with impulse response $h(t)$ that is real and even. Show that $\cos \omega t$ and $\sin \omega t$ are eigenfunctions of this system with the same real eigenvalue.

By setting $s = j\omega$ in Eqs. (2.23) and (2.24), we see that $e^{j\omega t}$ is an eigenfunction of a continuous-time LTI system and the corresponding eigenvalue is

$$\lambda = H(j\omega) = \int_{-\infty}^{\infty} h(\tau)\, e^{-j\omega\tau}\, d\tau \tag{2.88}$$

Since the system is stable, that is,

$$\int_{-\infty}^{\infty} |h(\tau)|\, d\tau < \infty$$

then

$$\int_{-\infty}^{\infty} \left| h(\tau)\, e^{-j\omega\tau} \right| d\tau = \int_{-\infty}^{\infty} |h(\tau)| \left| e^{-j\omega\tau} \right| d\tau = \int_{-\infty}^{\infty} |h(\tau)|\, d\tau < \infty$$

since $\left| e^{-j\omega\tau} \right| = 1$. Thus, $H(j\omega)$ converges for any ω. Using Euler's formula, we have

$$H(j\omega) = \int_{-\infty}^{\infty} h(\tau)\, e^{-j\omega\tau}\, d\tau = \int_{-\infty}^{\infty} h(\tau)(\cos \omega\tau - j \sin \omega\tau)\, d\tau$$

$$= \int_{-\infty}^{\infty} h(\tau) \cos \omega\tau \, d\tau - j\int_{-\infty}^{\infty} h(\tau) \sin \omega\tau \, d\tau \tag{2.89}$$

Since $\cos \omega\tau$ is an even function of τ and $\sin \omega\tau$ is an odd function of τ, and if $h(t)$ is real and even, then $h(\tau)$ $\cos \omega\tau$ is even and $h(\tau) \sin \omega\tau$ is odd. Then by Eqs. (1.75a) and (1.77), Eq. (2.89) becomes

$$H(j\omega) = 2\int_0^{\infty} h(\tau)\cos \omega\tau \, d\tau \tag{2.90}$$

Since $\cos \omega\tau$ is an even function of ω, changing ω to $-\omega$ in Eq. (2.90) and changing j to $-j$ in Eq. (2.89), we have

$$H(-j\omega) = H(j\omega)^* = 2\int_0^{\infty} h(\tau)\cos(-\omega\tau)\, d\tau$$

$$= 2\int_0^{\infty} h(\tau)\cos \omega\tau \, d\tau = H(j\omega) \tag{2.91}$$

Thus, we see that the eigenvalue $H(j\omega)$ corresponding to the eigenfunction $e^{j\omega t}$ is real. Let the system be represented by **T**. Then by Eqs. (2.23), (2.24), and (2.91) we have

$$\mathbf{T}\{e^{j\omega t}\} = H(j\omega)\, e^{j\omega t} \tag{2.92a}$$

$$\mathbf{T}\{e^{-j\omega t}\} = H(-j\omega)\, e^{-j\omega t} = H(j\omega)\, e^{-j\omega t} \tag{2.92b}$$

Now, since **T** is linear, we get

$$\mathbf{T}\{\cos \omega t\} = \mathbf{T}\left\{\frac{1}{2}(e^{j\omega t} + e^{-j\omega t})\right\} = \frac{1}{2}\mathbf{T}\{e^{j\omega t}\} + \frac{1}{2}\mathbf{T}\{e^{-j\omega t}\}$$

$$= H(j\omega)\left\{\frac{1}{2}(e^{j\omega t} + e^{-j\omega t})\right\} = H(j\omega)\cos \omega t \tag{2.93a}$$

and

$$\mathbf{T}\{\sin \omega t\} = \mathbf{T}\left\{\frac{1}{2j}(e^{j\omega t} - e^{-j\omega t})\right\} = \frac{1}{2j}\mathbf{T}\{e^{j\omega t}\} - \frac{1}{2j}\mathbf{T}\{e^{-j\omega t}\}$$

$$= H(j\omega)\left\{\frac{1}{2j}(e^{j\omega t} - e^{-j\omega t})\right\} = H(j\omega)\sin \omega t \tag{2.93b}$$

Thus, from Eqs. (2.93a) and (2.93b) we see that $\cos \omega t$ and $\sin \omega t$ are the eigenfunctions of the system with the same real eigenvalue $H(j\omega)$ given by Eq. (2.88) or (2.90).

Systems Described by Differential Equations

2.18. The continuous-time system shown in Fig. 2-18 consists of one integrator and one scalar multiplier. Write a differential equation that relates the output $y(t)$ and the input $x(t)$.

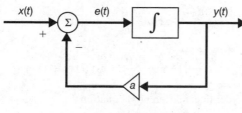

Fig. 2-18

Let the input of the integrator shown in Fig. 2-18 be denoted by $e(t)$. Then the input-output relation of the integrator is given by

$$y(t) = \int_{-\infty}^{t} e(\tau)\, d\tau \tag{2.94}$$

Differentiating both sides of Eq. (2.94) with respect to t, we obtain

$$\frac{dy(t)}{dt} = e(t) \tag{2.95}$$

Next, from Fig. 2-18 the input $e(t)$ to the integrator is given by

$$e(t) = x(t) - ay(t) \tag{2.96}$$

Substituting Eq. (2.96) into Eq. (2.95), we get

$$\frac{dy(t)}{dt} = x(t) - ay(t)$$

or

$$\frac{dy(t)}{dt} + ay(t) = x(t) \tag{2.97}$$

which is the required first-order linear differential equation.

2.19. The continuous-time system shown in Fig. 2-19 consists of two integrators and two scalar multipliers. Write a differential equation that relates the output $y(t)$ and the input $x(t)$.

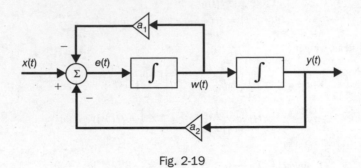

Fig. 2-19

Let $e(t)$ and $w(t)$ be the input and the output of the first integrator in Fig. 2-19, respectively. Using Eq. (2.95), the input to the first integrator is given by

$$e(t) = \frac{dw(t)}{dt} = -a_1 w(t) - a_2 y(t) + x(t) \qquad (2.98)$$

Since $w(t)$ is the input to the second integrator in Fig. 2-19, we have

$$w(t) = \frac{dy(t)}{dt} \qquad (2.99)$$

Substituting Eq. (2.99) into Eq. (2.98), we get

$$\frac{d^2 y(t)}{dt^2} = -a_1 \frac{dy(t)}{dt} - a_2 y(t) + x(t)$$

or

$$\frac{d^2 y(t)}{dt^2} + a_1 \frac{dy(t)}{dt} + a_2 y(t) = x(t) \qquad (2.100)$$

which is the required second-order linear differential equation.

Note that, in general, the order of a continuous-time LTI system consisting of the interconnection of integrators and scalar multipliers is equal to the number of integrators in the system.

2.20. Consider a continuous-time system whose input $x(t)$ and output $y(t)$ are related by

$$\frac{dy(t)}{dt} + ay(t) = x(t) \qquad (2.101)$$

where a is a constant.

(*a*)   Find $y(t)$ with the auxiliary condition $y(0) = y_0$ and

$$x(t) = Ke^{-bt} u(t) \qquad (2.102)$$

(*b*)   Express $y(t)$ in terms of the zero-input and zero-state responses.

(*a*)   Let

$$y(t) = y_p(t) + y_h(t)$$

where $y_p(t)$ is the particular solution satisfying Eq. (2.101) and $y_h(t)$ is the homogeneous solution which satisfies

$$\frac{dy_h(t)}{dt} + ay_h(t) = 0 \qquad (2.103)$$

Assume that

$$y_p(t) = Ae^{-bt} \qquad t > 0 \tag{2.104}$$

Substituting Eq. (2.104) into Eq. (2.101), we obtain

$$-bAe^{-bt} + aAe^{-bt} = Ke^{-bt}$$

from which we obtain $A = K/(a - b)$, and

$$y_p(t) = \frac{K}{a-b}e^{-bt} \qquad t > 0 \tag{2.105}$$

To obtain $y_h(t)$, we assume

$$y_h(t) = Be^{st}$$

Substituting this into Eq. (2.103) gives

$$sBe^{st} + aBe^{st} = (s+a)Be^{st} = 0$$

from which we have $s = -a$ and

$$y_h(t) = Be^{-at}$$

Combining $y_p(t)$ and $y_h(t)$, we get

$$y(t) = Be^{-at} + \frac{K}{a-b}e^{-bt} \qquad t > 0 \tag{2.106}$$

From Eq. (2.106) and the auxiliary condition $y(0) = y_0$, we obtain

$$B = y_0 - \frac{K}{a-b}$$

Thus, Eq. (2.106) becomes

$$y(t) = \left(y_0 - \frac{K}{a-b}\right)e^{-at} + \frac{K}{a-b}e^{-bt} \qquad t > 0 \tag{2.107}$$

For $t < 0$, we have $x(t) = 0$, and Eq. (2.101) becomes Eq. (2.103). Hence,

$$y(t) = Be^{-at} \qquad t < 0$$

From the auxiliary condition $y(0) = y_0$ we obtain

$$y(t) = y_0 e^{-at} \qquad t < 0 \tag{2.108}$$

(b) Combining Eqs. (2.107) and (2.108), $y(t)$ can be expressed in terms of $y_{zi}(t)$ (zero-input response) and $y_{zs}(t)$ (zero-state response) as

$$y(t) = y_0 e^{-at} + \frac{K}{a-b}(e^{-bt} - e^{-at})u(t)$$

$$= y_{zi}(t) + y_{zs}(t) \tag{2.109}$$

where

$$y_{zi}(t) = y_0 e^{-at} \tag{2.110a}$$

$$y_{zs}(t) = \frac{K}{a-b}(e^{-bt} - e^{-at})u(t) \tag{2.110b}$$

2.21. Consider the system in Prob. 2.20.

  (*a*)   Show that the system is not linear if $y(0) = y_0 \neq 0$.

  (*b*)   Show that the system is linear if $y(0) = 0$.

  (*a*)   Recall that a linear system has the property that zero input produces zero output (Sec. 1.5E). However, if we let $K = 0$ in Eq. (2.102), we have $x(t) = 0$, but from Eq. (2.109) we see that

$$y(t) = y_0 e^{-at} \neq 0 \qquad y_0 \neq 0$$

Thus, this system is nonlinear if $y(0) = y_0 \neq 0$.

  (*b*)   If $y(0) = 0$, the system is linear. This is shown as follows. Let $x_1(t)$ and $x_2(t)$ be two input signals, and let $y_1(t)$ and $y_2(t)$ be the corresponding outputs. That is,

$$\frac{dy_1(t)}{dt} + ay_1(t) = x_1(t) \tag{2.111}$$

$$\frac{dy_2(t)}{dt} + ay_2(t) = x_2(t) \tag{2.112}$$

with the auxiliary conditions

$$y_1(0) = y_2(0) = 0 \tag{2.113}$$

Consider

$$x(t) = \alpha_1 x_1(t) + \alpha_2 x_2(t)$$

where α_1 and α_2 are any complex numbers. Multiplying Eq. (2.111) by α_1 and Eq. (2.112) by α_2 and adding, we see that

$$y(t) = \alpha_1 y_1(t) + \alpha_2 y_2(t)$$

satisfies the differential equation

$$\frac{dy(t)}{dt} + ay(t) = x(t)$$

and also, from Eq. (2.113),

$$y(0) = \alpha_1 y_1(0) + \alpha_2 y_2(0) = 0$$

Therefore, $y(t)$ is the output corresponding to $x(t)$, and thus the system is linear.

2.22. Consider the system in Prob. 2.20. Show that the initial rest condition $y(0) = 0$ also implies that the system is time-invariant.

Let $y_1(t)$ be the response to an input $x_1(t)$ and

$$x_1(t) = 0 \qquad t \leq 0 \tag{2.114}$$

Then

$$\frac{dy_1(t)}{dt} + ay_1(t) = x_1(t) \tag{2.115}$$

and

$$y_1(0) = 0 \tag{2.116}$$

Now, let $y_2(t)$ be the response to the shifted input $x_2(t) = x_1(t - \tau)$. From Eq. (2.114) we have

$$x_2(t) = 0 \qquad t \leq \tau \tag{2.117}$$

Then $y_2(t)$ must satisfy

$$\frac{dy_2(t)}{dt} + ay_2(t) = x_2(t) \tag{2.118}$$

and
$$y_2(\tau) = 0 \tag{2.119}$$

Now, from Eq. (2.115) we have

$$\frac{dy_1(t-\tau)}{dt} + ay_1(t-\tau) = x_1(t-\tau) = x_2(t)$$

If we let $y_2(t) = y_1(t-\tau)$, then by Eq. (2.116) we have

$$y_2(\tau) = y_1(\tau-\tau) = y_1(0) = 0$$

Thus, Eqs. (2.118) and (2.119) are satisfied and we conclude that the system is time-invariant.

2.23 Consider the system in Prob. 2.20. Find the impulse response $h(t)$ of the system.

The impulse response $h(t)$ should satisfy the differential equation

$$\frac{dh(t)}{dt} + ah(t) = \delta(t) \tag{2.120}$$

The homogeneous solution $h_h(t)$ to Eq. (2.120) satisfies

$$\frac{dh_h(t)}{dt} + ah_h(t) = 0 \tag{2.121}$$

To obtain $h_h(t)$, we assume

$$h_h(t) = ce^{st}$$

Substituting this into Eq. (2.121) gives

$$sce^{st} + ace^{st} = (s+a)ce^{st} = 0$$

from which we have $s = -a$ and

$$h_h(t) = ce^{-at}u(t) \tag{2.122}$$

We predict that the particular solution $h_p(t)$ is zero since $h_p(t)$ cannot contain $\delta(t)$. Otherwise, $h(t)$ would have a derivative of $\delta(t)$ that is not part of the right-hand side of Eq. (2.120). Thus,

$$h(t) = ce^{-at}u(t) \tag{2.123}$$

To find the constant c, substituting Eq. (2.123) into Eq. (2.120), we obtain

$$\frac{d}{dt}[ce^{-at}u(t)] + ace^{-at}u(t) = \delta(t)$$

or
$$-ace^{-at}u(t) + ce^{-at}\frac{du(t)}{dt} + ace^{-at}u(t) = \delta(t)$$

Using Eqs. (1.25) and (1.30), the above equation becomes

$$ce^{-at}\frac{du(t)}{dt} = ce^{-at}\delta(t) = c\delta(t) = \delta(t)$$

so that $c = 1$. Thus, the impulse response is given by

$$h(t) = e^{-at}u(t) \tag{2.124}$$

2.24 Consider the system in Prob. 2.20 with $y(0) = 0$.

(a) Find the step response $s(t)$ of the system without using the impulse response $h(t)$.

(b) Find the step response $s(t)$ with the impulse response $h(t)$ obtained in Prob. 2.23.

(c) Find the impulse response $h(t)$ from $s(t)$.

(a) In Prob. 2.20

$$x(t) = Ke^{-bt}u(t)$$

Setting $K = 1$, $b = 0$, we obtain $x(t) = u(t)$ and then $y(t) = s(t)$. Thus, setting $K = 1$, $b = 0$, and $y(0) = y_0 = 0$ in Eq. (2.109), we obtain the step response

$$s(t) = \frac{1}{a}(1 - e^{-at})u(t) \tag{2.125}$$

(b) Using Eqs. (2.12) and (2.124) in Prob. 2.23, the step response $s(t)$ is given by

$$s(t) = \int_{-\infty}^{t} h(\tau)d\tau = \int_{-\infty}^{t} e^{-a\tau}u(\tau)d\tau$$

$$= \left[\int_{-\infty}^{t} e^{-a\tau}d\tau\right]u(t) = \frac{1}{a}(1 - e^{-at})u(t)$$

which is the same as Eq. (2.125).

(c) Using Eqs. (2.13) and (2.125), the impulse response $h(t)$ is given by

$$h(t) = s'(t) = \frac{d}{dt}\left[\frac{1}{a}(1 - e^{-at})u(t)\right]$$

$$= e^{-at}u(t) + \frac{1}{a}(1 - e^{-at})u'(t)$$

Using Eqs. (1.25) and (1.30), we have

$$\frac{1}{a}(1 - e^{-at})u'(t) = \frac{1}{a}(1 - e^{-at})\delta(t) = \frac{1}{a}(1-1)\delta(t) = 0$$

Thus, $$h(t) = e^{-at}u(t)$$

which is the same as Eq. (1.124).

2.25. Consider the system described by

$$y'(t) + 2y(t) = x(t) + x'(t) \tag{2.126}$$

Find the impulse response $h(t)$ of the system.

The impulse response $h(t)$ should satisfy the differential equation

$$h'(t) + 2h(t) = \delta(t) + \delta'(t) \tag{2.127}$$

The homogeneous solution $h_h(t)$ to Eq. (2.127) is [see Prob. 2.23 and Eq. (2.122)]

$$h_h(t) = c_1 e^{-2t}u(t)$$

Assuming the particular solution $h_p(t)$ of the form

$$h_p(t) = c_2\delta(t)$$

the general solution is

$$h(t) = c_1 e^{-2t} u(t) + c_2 \delta(t) \qquad (2.128)$$

The delta function $\delta(t)$ must be present so that $h'(t)$ contributes $\delta'(t)$ to the left-hand side of Eq. (1.127). Substituting Eq. (2.128) into Eq. (2.127), we obtain

$$-2c_1 e^{-2t} u(t) + c_1 e^{-2t} u'(t) + c_2 \delta'(t) + 2c_1 e^{-2t} u(t) + 2c_2 \delta(t)$$
$$= \delta(t) + \delta'(t)$$

Again, using Eqs. (1.25) and (1.30), we have

$$(c_1 + 2c_2) \delta(t) + c_2 \delta'(t) = \delta(t) + \delta'(t)$$

Equating coefficients of $\delta(t)$ and $\delta'(t)$, we obtain

$$c_1 + 2c_2 = 1 \qquad c_2 = 1$$

from which we have $c_1 = -1$ and $c_2 = 1$. Substituting these values in Eq. (2.128), we obtain

$$h(t) = -e^{-2t} u(t) + \delta(t) \qquad (2.129)$$

Responses of a Discrete-Time LTI System and Convolution

2.26 Verify Eqs. (2.36) and (2.37); that is,

(*a*) $x[n] * h[n] = h[n] * x[n]$

(*b*) $\{x[n] * h_1[n]\} * h_2[n] = x[n] * \{h_1[n] * h_2[n]\}$

(*a*) By definition (2.35)

$$x[n] * h[n] = \sum_{k=-\infty}^{\infty} x[k]h[n-k]$$

By changing the variable $n - k = m$, we have

$$x[n] * h[n] = \sum_{m=-\infty}^{\infty} x[n-m]h[m] = \sum_{m=-\infty}^{\infty} h[m]x[n-m] = h[n] * x[n]$$

(*b*) Let $x[n] * h_1[n] = f_1[n]$ and $h_1[n] * h_2[n] = f_2[n]$. Then

$$f_1[n] = \sum_{k=-\infty}^{\infty} x[k]h_1[n-k]$$

and

$$\{x[n] * h_1[n]\} * h_2[n] = f_1[n] * h_2[n] = \sum_{m=-\infty}^{\infty} f_1[m]h_2[n-m]$$

$$= \sum_{m=-\infty}^{\infty} \left[\sum_{k=-\infty}^{\infty} x[k]h_1[m-k] \right] h_2[n-m]$$

Substituting $r = m - k$ and interchanging the order of summation, we have

$$\{x[n] * h_1[n]\} * h_2[n] = \sum_{k=-\infty}^{\infty} x[k] \left(\sum_{r=-\infty}^{\infty} h_1[r]h_2[n-k-r] \right)$$

Now, since

$$f_2[n] = \sum_{r=-\infty}^{\infty} h_1[r] h_2[n-r]$$

we have

$$f_2[n-k] = \sum_{r=-\infty}^{\infty} h_1[r] h_2[n-k-r]$$

Thus,
$$\{x[n] * h_1[n]\} * h_2[n] = \sum_{k=-\infty}^{\infty} x[k] f_2[n-k]$$

$$= x[n] * f_2[n] = x[n] * \{h_1[n] * h_2[n]\}$$

2.27. Show that

(a) $x[n] * \delta[n] = x[n]$ (2.130)

(b) $x[n] * \delta[n - n_0] = x[n - n_0]$ (2.131)

(c) $x[n] * u[n] = \sum_{k=-\infty}^{\infty} x[k]$ (2.132)

(d) $x[n] * u[n - n_0] = \sum_{k=-\infty}^{n-n_0} x[k]$ (2.133)

(a) By Eq. (2.35) and property (1.46) of $\delta[n-k]$ we have

$$x[n] * \delta[n] = \sum_{k=-\infty}^{\infty} x[k] \delta[n-k] = x[n]$$

(b) Similarly, we have

$$x[n] * \delta[n - n_0] = \sum_{k=-\infty}^{\infty} x[k] \delta[n-k-n_0] = x[n-n_0]$$

(c) By Eq. (2.35) and definition (1.44) of $u[n-k]$ we have

$$x[n] * u[n] = \sum_{k=-\infty}^{\infty} x[k] u[n-k] = \sum_{k=-\infty}^{n} x[k]$$

(d) In a similar manner, we have

$$x[n] * u[n - n_0] = \sum_{k=-\infty}^{\infty} x[k] u[n-k-n_0] = \sum_{k=-\infty}^{n-n_0} x[k]$$

2.28 The input $x[n]$ and the impulse response $h[n]$ of a discrete-time LTI system are given by

$$x[n] = u[n] \qquad h[n] = \alpha^n u[n] \qquad 0 < \alpha < 1$$

(a) Compute the output $y[n]$ by Eq. (2.35).

(b) Compute the output $y[n]$ by Eq. (2.39).

(a) By Eq. (2.35) we have

$$y[n] = x[n] * h[n] = \sum_{k=-\infty}^{\infty} x[k] h[n-k]$$

Sequences $x[k]$ and $h[n-k]$ are shown in Fig. 2-20(a) for $n<0$ and $n>0$. From Fig. 2-20(a) we see that for $n<0$, $x[k]$ and $h[n-k]$ do not overlap, while for $n \geq 0$, they overlap from $k=0$ to $k=n$. Hence, for $n<0$, $y[n]=0$. For $n \geq 0$, we have

$$y[n] = \sum_{k=0}^{n} \alpha^{n-k}$$

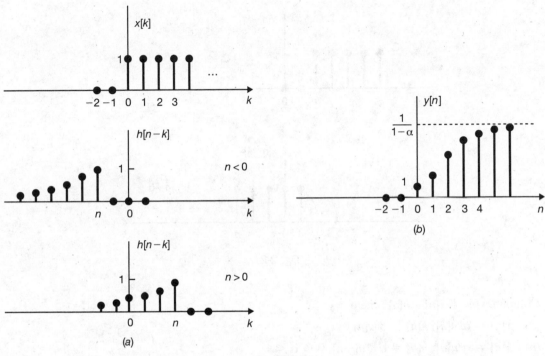

Fig. 2-20

Changing the variable of summation k to $m=n-k$ and using Eq. (1.90), we have

$$y[n] = \sum_{m=n}^{0} \alpha^m = \sum_{m=0}^{n} \alpha^m = \frac{1-\alpha^{n+1}}{1-\alpha}$$

Thus, we can write the output $y[n]$ as

$$y[n] = \left(\frac{1-\alpha^{n+1}}{1-\alpha} \right) u[n] \tag{2.134}$$

which is sketched in Fig. 2-20(b).

(b)   By Eq. (2.39)

$$y[n] = h[n] * x[n] = \sum_{k=-\infty}^{\infty} h[k]x[n-k]$$

Sequences $h[k]$ and $x[n-k]$ are shown in Fig. 2-21 for $n<0$ and $n>0$. Again from Fig. 2-21 we see that for $n<0$, $h[k]$ and $x[n-k]$ do not overlap, while for $n \geq 0$, they overlap from $k=0$ to $k=n$. Hence, for $n<0$, $y[n]=0$. For $n \geq 0$, we have

$$y[n] = \sum_{k=0}^{n} \alpha^k = \frac{1-\alpha^{n+1}}{1-\alpha}$$

Thus, we obtain the same result as shown in Eq. (2.134).

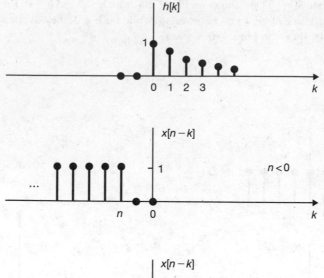

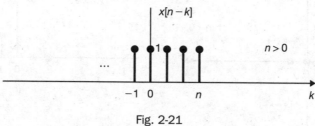

Fig. 2-21

2.29 Compute $y[n] = x[n] * h[n]$, where

    (a) $x[n] = \alpha^n u[n]$, $h[n] = \beta^n u[n]$

    (b) $x[n] = \alpha^n u[n]$, $h[n] = \alpha^{-n} u[-n]$, $0 < \alpha < 1$

    (a)   From Eq. (2.35) we have

$$y[n] = \sum_{k=-\infty}^{\infty} x[k]h[n-k] = \sum_{k=-\infty}^{\infty} \alpha^k u[k] \beta^{n-k} u[n-k]$$

$$= \sum_{k=-\infty}^{\infty} \alpha^k \beta^{n-k} u[k] u[n-k]$$

since

$$u[k]u[n-k] = \begin{cases} 1 & 0 \le k \le n \\ 0 & \text{otherwise} \end{cases}$$

we have

$$y[n] = \sum_{k=0}^{n} \alpha^k \beta^{n-k} = \beta^n \sum_{k=0}^{n} \left(\frac{\alpha}{\beta} \right)^k \qquad n \ge 0$$

Using Eq. (1.90), we obtain

$$y[n] = \begin{cases} \beta^n \dfrac{1 - (\alpha/\beta)^{n+1}}{1 - (\alpha/\beta)} u[n] & \alpha \ne \beta \\ \beta^n (n+1) u[n] & \alpha = \beta \end{cases} \qquad\qquad (2.135a)$$

or

$$y[n] = \begin{cases} \dfrac{1}{\beta - \alpha} (\beta^{n+1} - \alpha^{n+1}) u[n] & \alpha \ne \beta \\ \beta^n (n+1) u[n] & \alpha = \beta \end{cases} \qquad\qquad (2.135b)$$

(b)

$$y[n] = \sum_{k=-\infty}^{\infty} x[k]h[n-k] = \sum_{k=-\infty}^{\infty} \alpha^k u[k] \alpha^{-(n-k)} u[-(n-k)]$$

$$= \sum_{k=-\infty}^{\infty} \alpha^{-n} \alpha^{2k} u[k] u[k-n]$$

For $n \leq 0$, we have

$$u[k]u[k-n] = \begin{cases} 1 & 0 \leq k \\ 0 & \text{otherwise} \end{cases}$$

Thus, using Eq. (1.91), we have

$$y[n] = \alpha^{-n} \sum_{k=0}^{\infty} \alpha^{2k} = \alpha^{-n} \sum_{k=0}^{\infty} (\alpha^2)^k = \frac{\alpha^{-n}}{1-\alpha^2} \qquad n \leq 0 \qquad (2.136a)$$

For $n \geq 0$, we have

$$u[k]u[k-n] = \begin{cases} 1 & n \leq k \\ 0 & \text{otherwise} \end{cases}$$

Thus, using Eq. (1.92), we have

$$y[n] = \alpha^{-n} \sum_{k=n}^{\infty} (\alpha^2)^k = \alpha^{-n} \frac{\alpha^{2n}}{1-\alpha^2} = \frac{\alpha^n}{1-\alpha^2} \qquad n \geq 0 \qquad (2.136b)$$

Combining Eqs. (2.136a) and (2.136b), we obtain

$$y[n] = \frac{\alpha^{|n|}}{1-\alpha^2} \qquad \text{all } n \qquad (2.137)$$

which is sketched in Fig. 2-22.

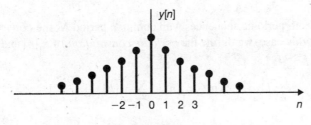

Fig. 2-22

2.30. Evaluate $y[n] = x[n] * h[n]$, where $x[n]$ and $h[n]$ are shown in Fig. 2-23, (*a*) by an analytical technique, and (*b*) by a graphical method.

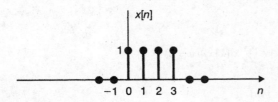
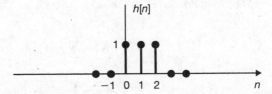

Fig. 2-23

(*a*)   Note that $x[n]$ and $h[n]$ can be expressed as

$$x[n] = \delta[n] + \delta[n-1] + \delta[n-2] + \delta[n-3]$$
$$h[n] = \delta[n] + \delta[n-1] + \delta[n-2]$$

Now, using Eqs. (2.38), (2.130), and (2.131), we have

$$x[n] * h[n] = x[n] * \{\delta[n] + \delta[n-1] + \delta[n-2]\}$$
$$= x[n] * \delta[n] + x[n] * \delta[n-1] + x[n] * \delta[n-2]$$
$$= x[n] + x[n-1] + x[n-2]$$

Thus,

$$y[n] = \delta[n] + \delta[n-1] + \delta[n-2] + \delta[n-3]$$
$$+ \delta[n-1] + \delta[n-2] + \delta[n-3] + \delta[n-4]$$
$$+ \delta[n-2] + \delta[n-3] + \delta[n-4] + \delta[n-5]$$

or

$$y[n] = \delta[n] + 2\delta[n-1] + 3\delta[n-2] + 3\delta[n-3] + 2\delta[n-4] + \delta[n-5]$$

or

$$y[n] = \{1, 2, 3, 3, 2, 1\}$$

(*b*)   Sequences $h[k]$, $x[k]$ and $h[n-k]$, $x[k]h[n-k]$ for different values of n are sketched in Fig. 2-24. From Fig. 2-24 we see that $x[k]$ and $h[n-k]$ do not overlap for $n < 0$ and $n > 5$, and hence, $y[n] = 0$ for $n < 0$ and $n > 5$. For $0 \leq n \leq 5$, $x[k]$ and $h[n-k]$ overlap. Thus, summing $x[k]h[n-k]$ for $0 \leq n \leq 5$, we obtain

$$y[0] = 1 \qquad y[1] = 2 \qquad y[2] = 3 \qquad y[3] = 3 \qquad y[4] = 2 \qquad y[5] = 1$$

or

$$y[n] = \{1, 2, 3, 3, 2, 1\}$$

which is plotted in Fig. 2-25.

2.31. If $x_1[n]$ and $x_2[n]$ are both periodic sequences with common period N, the convolution of $x_1[n]$ and $x_2[n]$ does not converge. In this case, we define the *periodic convolution* of $x_1[n]$ and $x_2[n]$ as

$$f[n] = x_1[n] \otimes x_2[n] = \sum_{k=0}^{N-1} x_1[k]x_2[n-k] \tag{2.138}$$

Show that $f[n]$ is periodic with period N.

Since $x_2[n]$ is periodic with period N, we have

$$x_2[(n-k) + N] = x_2[n-k]$$

Then from Eq. (2.138) we have

$$f[n+N] = \sum_{k=0}^{N-1} x_1[k]x_2[n+N-k] = \sum_{k=0}^{N-1} x_1[k]x_2[(n-k)+N]$$
$$= \sum_{k=0}^{N-1} x_1[k]x_2[(n-k)] = f[n]$$

Thus, $f[n]$ is periodic with period N.

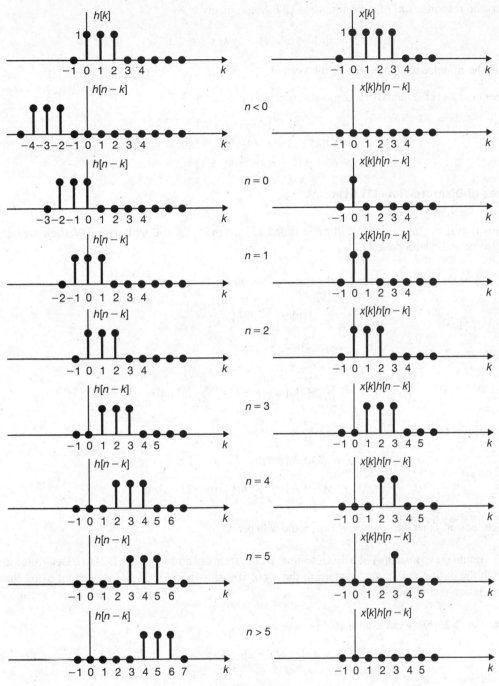

Fig. 2-24

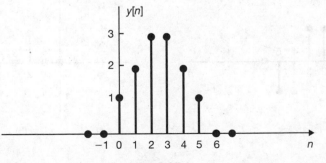

Fig. 2-25

2.32. The step response $s[n]$ of a discrete-time LTI system is given by

$$s[n] = \alpha^n u[n] \qquad 0 < \alpha < 1$$

Find the impulse response $h[n]$ of the system.

From Eq. (2.41) the impulse response $h[n]$ is given by

$$
\begin{aligned}
h[n] = s[n] - s[n-1] &= \alpha^n u[n] - \alpha^{n-1} u[n-1] \\
&= \{\delta[n] + \alpha^n u[n-1]\} - \alpha^{n-1} u[n-1] \\
&= \delta[n] - (1-\alpha)\alpha^{n-1} u[n-1]
\end{aligned}
$$

Properties of Discrete-Time LTI Systems

2.33. Show that if the input $x[n]$ to a discrete-time LTI system is periodic with period N, then the output $y[n]$ is also periodic with period N.

Let $h[n]$ be the impulse response of the system. Then by Eq. (2.39) we have

$$y[n] = \sum_{k=-\infty}^{\infty} h[k]x[n-k]$$

Let $n = m + N$. Then

$$y[m+N] = \sum_{k=-\infty}^{\infty} h[k]x[m+N-k] = \sum_{k=-\infty}^{\infty} h[k]x[(m-k)+N]$$

Since $x[n]$ is periodic with period N, we have

$$x[(m-k)+N] = x[m-k]$$

Thus,
$$y[m+N] = \sum_{k=-\infty}^{\infty} h[k]x[m-k] = y[m]$$

which indicates that the output $y[n]$ is periodic with period N.

2.34. The impulse response $h[n]$ of a discrete-time LTI system is shown in Fig. 2-26(*a*). Determine and sketch the output $y[n]$ of this system to the input $x[n]$ shown in Fig. 2-26(*b*) without using the convolution technique.

From Fig. 2-26(*b*) we can express $x[n]$ as

$$x[n] = \delta[n-2] - \delta[n-4]$$

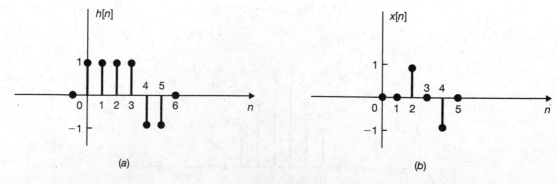

(a)   (b)

Fig. 2-26

Since the system is linear and time-invariant and by the definition of the impulse response, we see that the output $y[n]$ is given by

$$y[n] = h[n-2] - h[n-4]$$

which is sketched in Fig. 2-27.

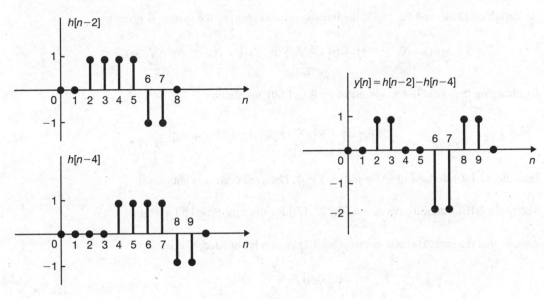

Fig. 2-27

2.35. A discrete-time system is causal if for every choice of n_0 the value of the output sequence $y[n]$ at $n = n_0$ depends on only the values of the input sequence $x[n]$ for $n \le n_0$ (see Sec. 1.5D). From this definition derive the causality condition (2.44) for a discrete-time LTI system; that is,

$$h[n] = 0 \qquad n < 0$$

From Eq. (2.39) we have

$$y[n] = \sum_{k=-\infty}^{\infty} h[k]x[n-k]$$

$$= \sum_{k=-\infty}^{-1} h[k]x[n-k] + \sum_{k=0}^{\infty} h[k]x[n-k] \qquad (2.139)$$

Note that the first summation represents a weighted sum of future values of $x[n]$. Thus, if the system is causal, then

$$\sum_{k=-\infty}^{-1} h[k]x[n-k] = 0$$

This can be true only if

$$h[n] = 0 \qquad n < 0$$

Now if $h[n] = 0$ for $n < 0$, then Eq. (2.139) becomes

$$y[n] = \sum_{k=0}^{\infty} h[k]x[n-k]$$

which indicates that the value of the output $y[n]$ depends on only the past and the present input values.

2.36. Consider a discrete-time LTI system whose input $x[n]$ and output $y[n]$ are related by

$$y[n] = \sum_{k=-\infty}^{n} 2^{k-n} x[k+1]$$

Is the system causal?

By definition (2.30) and Eq. (1.48) the impulse response $h[n]$ of the system is given by

$$h[n] = \sum_{k=-\infty}^{n} 2^{k-n} \delta[k+1] = \sum_{k=-\infty}^{n} 2^{-(n+1)} \delta[k+1] = 2^{-(n+1)} \sum_{k=-\infty}^{n} \delta[k+1]$$

By changing the variable $k + 1 = m$ and by Eq. (1.50), we obtain

$$h[n] = 2^{-(n+1)} \sum_{m=-\infty}^{n+1} \delta[m] = 2^{-(n+1)} u[n+1] \qquad (2.140)$$

From Eq. (2.140) we have $h[-1] = u[0] = 1 \neq 0$. Thus, the system is not causal.

2.37. Verify the BIBO stability condition [Eq. (2.49)] for discrete-time LTI systems.

Assume that the input $x[n]$ of a discrete-time LTI system is bounded, that is,

$$|x[n]| \leq k_1 \qquad \text{all } n \qquad (2.141)$$

Then, using Eq. (2.35), we have

$$|y[n]| = \left| \sum_{k=-\infty}^{\infty} h[k] x[n-k] \right| \leq \sum_{k=-\infty}^{\infty} |h[k]||x[n-k]| \leq k_1 \sum_{k=-\infty}^{\infty} |h[k]|$$

Since $|x[n-k]| \leq k_1$ from Eq. (2.141). Therefore, if the impulse response is absolutely summable, that is,

$$\sum_{k=-\infty}^{\infty} |h[k]| = K < \infty$$

we have

$$|y[n]| \leq k_1 K = k_2 < \infty$$

and the system is BIBO stable.

2.38. Consider a discrete-time LTI system with impulse response $h[n]$ given by

$$h[n] = \alpha^n u[n]$$

    (a)   Is this system causal?

    (b)   Is this system BIBO stable?

    (a)   Since $h[n] = 0$ for $n < 0$, the system is causal.

    (b)   Using Eq. (1.91) (Prob. 1.19), we have

$$\sum_{k=-\infty}^{\infty} |h[k]| = \sum_{k=-\infty}^{\infty} |\alpha^k u[n]| = \sum_{k=0}^{\infty} |\alpha|^k = \frac{1}{1-|\alpha|} \qquad |\alpha| < 1$$

Therefore, the system is BIBO stable if $|\alpha| < 1$ and unstable if $|\alpha| \geq 1$.

Systems Described by Difference Equations

2.39. The discrete-time system shown in Fig. 2-28 consists of one unit delay element and one scalar multiplier. Write a difference equation that relates the output $y[n]$ and the input $x[n]$.

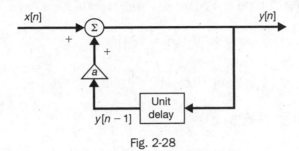

Fig. 2-28

In Fig. 2-28 the output of the unit delay element is $y[n - 1]$. Thus, from Fig. 2-28 we see that

$$y[n] = ay[n - 1] + x[n] \qquad (2.142)$$

or
$$y[n] - ay[n - 1] = x[n] \qquad (2.143)$$

which is the required first-order linear difference equation.

2.40. The discrete-time system shown in Fig. 2-29 consists of two unit delay elements and two scalar multipliers. Write a difference equation that relates the output $y[n]$ and the input $x[n]$.

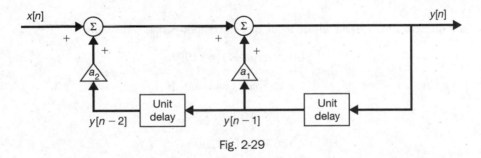

Fig. 2-29

In Fig. 2-29 the output of the first (from the right) unit delay element is $y[n - 1]$ and the output of the second (from the right) unit delay element is $y[n - 2]$. Thus, from Fig. 2-29 we see that

$$y[n] = a_1 y[n - 1] + a_2 y[n - 2] + x[n] \qquad (2.144)$$

or
$$y[n] - a_1 y[n - 1] - a_2 y[n - 2] = x[n] \qquad (2.145)$$

which is the required second-order linear difference equation.

Note that, in general, the order of a discrete-time LTI system consisting of the interconnection of unit delay elements and scalar multipliers is equal to the number of unit delay elements in the system.

2.41. Consider the discrete-time system in Fig. 2-30. Write a difference equation that relates the output $y[n]$ and the input $x[n]$.

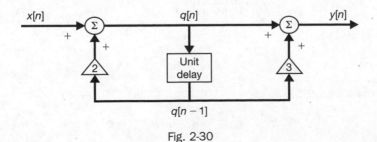

Fig. 2-30

Let the input to the unit delay element be $q[n]$. Then from Fig. 2-30 we see that

$$q[n] = 2q[n-1] + x[n] \tag{2.146a}$$

$$y[n] = q[n] + 3q[n-1] \tag{2.146b}$$

Solving Eqs. (2.146a) and (2.146b) for $q[n]$ and $q[n-1]$ in terms of $x[n]$ and $y[n]$, we obtain

$$q[n] = \frac{2}{5}y[n] + \frac{3}{5}x[n] \tag{2.147a}$$

$$q[n-1] = \frac{1}{5}y[n] - \frac{1}{5}x[n] \tag{2.147b}$$

Changing n to $(n-1)$ in Eq. (2.147a), we have

$$q[n-1] = \frac{2}{5}y[n-1] + \frac{3}{5}x[n-1] \tag{2.147c}$$

Thus, equating Eq. (2.147b) and Eq. (2.147c), we have

$$\frac{1}{5}y[n] - \frac{1}{5}x[n] = \frac{2}{5}y[n-1] + \frac{3}{5}x[n-1]$$

Multiplying both sides of the above equation by 5 and rearranging terms, we obtain

$$y[n] - 2y[n-1] = x[n] + 3x[n-1] \tag{2.148}$$

which is the required difference equation.

2.42. Consider a discrete-time system whose input $x[n]$ and output $y[n]$ are related by

$$y[n] - ay[n-1] = x[n] \tag{2.149}$$

where a is a constant. Find $y[n]$ with the auxiliary condition $y[-1] = y_{-1}$ and

$$x[n] = Kb^n u[n] \tag{2.150}$$

Let $y[n] = y_p[n] + y_h[n]$

where $y_p[n]$ is the particular solution satisfying Eq. (2.149) and $y_h[n]$ is the homogeneous solution which satisfies

$$y[n] - ay[n-1] = 0 \tag{2.151}$$

Assume that

$$y_p[n] = Ab^n \qquad n \geq 0 \tag{2.152}$$

Substituting Eq. (2.152) into Eq. (2.149), we obtain

$$Ab^n - aAb^{n-1} = Kb^n$$

from which we obtain $A = Kb/(b-a)$, and

$$y_p[n] = \frac{K}{b-a}b^{n+1} \qquad n \geq 0 \tag{2.153}$$

To obtain $y_h[n]$, we assume

$$y_h[n] = Bz^n$$

Substituting this into Eq. (2.151) gives

$$Bz^n - aBz^{n-1} = (z-a)Bz^{n-1} = 0$$

from which we have $z = a$ and

$$y_h[n] = Ba^n \tag{2.154}$$

Combining $y_p[n]$ and $y_h[n]$, we get

$$y[n] = Ba^n + \frac{K}{b-a}b^{n+1} \qquad n \geq 0 \tag{2.155}$$

In order to determine B in Eq. (2.155) we need the value of $y[0]$. Setting $n = 0$ in Eqs. (2.149) and (2.150), we have

$$y[0] - ay[-1] = y[0] - ay_{-1} = x[0] = K$$

or

$$y[0] = K + ay_{-1} \tag{2.156}$$

Setting $n = 0$ in Eq. (2.155), we obtain

$$y[0] = B + K\frac{b}{b-a} \tag{2.157}$$

Therefore, equating Eqs. (2.156) and (2.157), we have

$$K + ay_{-1} = B + K\frac{b}{b-a}$$

from which we obtain

$$B = ay_{-1} - K\frac{a}{b-a}$$

Hence, Eq. (2.155) becomes

$$y[n] = y_{-1}a^{n+1} + K\frac{b^{n+1} - a^{n+1}}{b-a} \qquad n \geq 0 \tag{2.158}$$

For $n < 0$, we have $x[n] = 0$, and Eq. (2.149) becomes Eq. (2.151). Hence,

$$y[n] = Ba^n \tag{2.159}$$

From the auxiliary condition $y[-1] = y_{-1}$, we have

$$y[-1] = y_{-1} = Ba^{-1}$$

from which we obtain $B = y_{-1}a$. Thus,

$$y[n] = y_{-1}a^{n+1} \qquad n < 0 \tag{2.160}$$

Combining Eqs. (2.158) and (2.160), $y[n]$ can be expressed as

$$y[n] = y_{-1}a^{n+1} + K\frac{b^{n+1} - a^{n+1}}{b-a}u[n] \tag{2.161}$$

Note that as in the continuous-time case (Probs. 2.21 and 2.22), the system described by Eq. (2.149) is not linear if $y[-1] \neq 0$. The system is causal and time-invariant if it is initially at rest; that is, $y[-1] = 0$. Note also that Eq. (2.149) can be solved recursively (see Prob. 2.43).

2.43. Consider the discrete-time system in Prob. 2.42. Find the output $y[n]$ when $x[n] = K\delta[n]$ and $y[-1] = y_{-1} = \alpha$.

We can solve Eq. (2.149) for successive values of $y[n]$ for $n \geq 0$ as follows: rearrange Eq. (2.149) as

$$y[n] = ay[n-1] + x[n] \tag{2.162}$$

Then

$$y[0] = ay[-1] + x[0] = a\alpha + K$$
$$y[1] = ay[0] + x[1] = a(a\alpha + K)$$
$$y[2] = ay[1] + x[2] = a^2(a\alpha + K)$$
$$\vdots$$
$$y[n] = ay[n-1] + x[n] = a^n(a\alpha + K) = a^{n+1}\alpha + a^n K \tag{2.163}$$

Similarly, we can also determine $y[n]$ for $n < 0$ by rearranging Eq. (2.149) as

$$y[n-1] = \frac{1}{a}\{y[n] - x[n]\} \tag{2.164}$$

Then

$$y[-1] = \alpha$$
$$y[-2] = \frac{1}{a}\{y[-1] - x[-1]\} = \frac{1}{a}\alpha = a^{-1}\alpha$$
$$y[-3] = \frac{1}{a}\{y[-2] - x[-2]\} = a^{-2}\alpha$$
$$\vdots$$
$$y[-n] = \frac{1}{a}\{y[-n+1] - x[-n+1]\} = a^{-n+1}\alpha \tag{2.165}$$

Combining Eqs. (2.163) and (2.165), we obtain

$$y[n] = a^{n+1}\alpha + Ka^n u[n] \tag{2.166}$$

2.44. Consider the discrete-time system in Prob. 2.43 for an initially at rest condition.

(a) Find in impulse response $h[n]$ of the system.
(b) Find the step response $s[n]$ of the system.
(c) Find the impulse response $h[n]$ from the result of part (b).

(a) Setting $K = 1$ and $y[-1] = \alpha = 0$ in Eq. (2.166), we obtain

$$h[n] = a^n u[n] \tag{2.167}$$

(b) Setting $K = 1$, $b = 1$, and $y[-1] = y_{-1} = 0$ in Eq. (2.161), we obtain

$$s[n] = \left(\frac{1 - a^{n+1}}{1 - a}\right)u[n] \tag{2.168}$$

(c) From Eqs. (2.41) and (2.168) the impulse response $h[n]$ is given by

$$h[n] = s[n] - s[n-1] = \left(\frac{1 - a^{n+1}}{1 - a}\right)u[n] - \left(\frac{1 - a^n}{1 - a}\right)u[n-1]$$

When $n = 0$,

$$h[0] = \left(\frac{1 - a}{1 - a}\right)u[0] = 1$$

When $n \geq 1$,

$$h[n] = \frac{1}{1 - a}[1 - a^{n+1} - (1 - a^n)] = \frac{a^n(1 - a)}{1 - a} = a^n$$

Thus,

$$h[n] = a^n u[n]$$

which is the same as Eq. (2.167).

2.45. Find the impulse response $h[n]$ for each of the causal LTI discrete-time systems satisfying the following difference equations and indicate whether each system is a FIR or an IIR system.

    (a) $y[n] = x[n] - 2x[n-2] + x[n-3]$

    (b) $y[n] + 2y[n-1] = x[n] + x[n-1]$

    (c) $y[n] - \frac{1}{2}y[n-2] = 2x[n] - x[n-2]$

    (a) By definition (2.56)

$$h[n] = \delta[n] - 2\delta[n-2] + \delta[n-3]$$

    or

$$h[n] = \{1, 0, -2, 1\}$$

Since $h[n]$ has only four terms, the system is a FIR system.

    (b) $h[n] = -2h[n-1] + \delta[n] + \delta[n-1]$

Since the system is causal, $h[-1] = 0$. Then

$$h[0] = -2h[-1] + \delta[0] + \delta[-1] = \delta[0] = 1$$
$$h[1] = -2h[0] + \delta[1] + \delta[0] = -2 + 1 = -1$$
$$h[2] = -2h[1] + \delta[2] + \delta[1] = -2(-1) = 2$$
$$h[3] = -2h[2] + \delta[3] + \delta[2] = -2(2) = -2^2$$
$$\vdots$$
$$h[n] = -2h[n-1] + \delta[n] + \delta[n-1] = (-1)^n 2^{n-1}$$

Hence,
$$h[n] = \delta[n] + (-1)^n 2^{n-1} u[n-1]$$

Since $h[n]$ has infinite terms, the system is an IIR system.

    (c) $h[n] = \frac{1}{2}h[n-2] + 2\delta[n] - \delta[n-2]$

Since the system is causal, $h[-2] = h[-1] = 0$. Then

$$h[0] = \frac{1}{2}h[-2] + 2\delta[0] - \delta[-2] = 2\delta[0] = 2$$

$$h[1] = \frac{1}{2}h[-1] + 2\delta[1] - \delta[-1] = 0$$

$$h[2] = \frac{1}{2}h[0] + 2\delta[2] - \delta[0] = \frac{1}{2}(2) = -1 = 0$$

$$h[3] = \frac{1}{2}h[1] + 2\delta[3] - \delta[1] = 0$$
$$\vdots$$

Hence,
$$h[n] = 2\delta[n]$$

Since $h[n]$ has only one term, the system is a FIR system.

SUPPLEMENTARY PROBLEMS

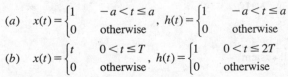

2.46. Compute the convolution $y(t) = x(t) * h(t)$ of the following pair of signals:

    (a) $x(t) = \begin{cases} 1 & -a < t \le a \\ 0 & \text{otherwise} \end{cases}$, $h(t) = \begin{cases} 1 & -a < t \le a \\ 0 & \text{otherwise} \end{cases}$

    (b) $x(t) = \begin{cases} t & 0 < t \le T \\ 0 & \text{otherwise} \end{cases}$, $h(t) = \begin{cases} 1 & 0 < t \le 2T \\ 0 & \text{otherwise} \end{cases}$

    (c) $x(t) = u(t-1), h(t) = e^{-3t}u(t)$

2.47. Compute the convolution sum $y[n] = x[n] * h[n]$ of the following pairs of sequences:

(a) $x[n] = u[n], h[n] = 2^n u[-n]$

(b) $x[n] = u[n] - u[n - N], h[n] = \alpha^n u[n], 0 < \alpha < 1$

(c) $x[n] = (\frac{1}{2})^n u[n], h[n] = \delta[n] - \frac{1}{2}\delta[n - 1]$

2.48. Show that if $y(t) = x(t) * h(t)$, then

$$y'(t) = x'(t) * h(t) = x(t) * h'(t)$$

2.49. Show that

$$x(t) * \delta'(t) = x'(t)$$

2.50. Let $y[n] = x[n] * h[n]$. Then show that

$$x[n - n_1] * h[n - n_2] = y[n - n_1 - n_2]$$

2.51. Show that

$$x_1[n] \otimes x_2[n] = \sum_{k=n_0}^{n_0+N-1} x_1[k]x_2[n - k]$$

for an arbitrary starting point n_0.

2.52. The step response $s(t)$ of a continuous-time LTI system is given by

$$s(t) = [\cos \omega_0 t]u(t)$$

Find the impulse response $h(t)$ of the system.

2.53. The system shown in Fig. 2-31 is formed by connection two systems *in parallel*. The impulse responses of the systems are given by

$$h_1(t) = e^{-2t} u(t) \qquad \text{and} \qquad h_2(t) = 2e^{-t} u(t)$$

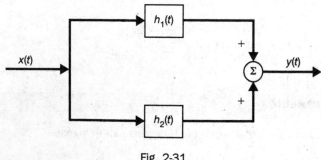

Fig. 2-31

(a)   Find the impulse response $h(t)$ of the overall system.

(b)   Is the overall system stable?

2.54. Consider an integrator whose input $x(t)$ and output $y(t)$ are related by

$$y(t) = \int_{-\infty}^{t} x(\tau)\, d\tau$$

(a) Find the impulse response $h(t)$ of the integrator.

(b) Is the integrator stable?

2.55. Consider a discrete-time LTI system with impulse response $h[n]$ given by

$$h[n] = \delta[n-1]$$

Is this system memoryless?

2.56. The impulse response of a discrete-time LTI system is given by

$$h[n] = \left(\frac{1}{2}\right)^{n} u[n]$$

Let $y[n]$ be the output of the system with the input

$$x[n] = 2\delta[n] + \delta[n-3]$$

Find $y[1]$ and $y[4]$.

2.57. Consider a discrete-time LTI system with impulse response $h[n]$ given by

$$h[n] = \left(-\frac{1}{2}\right)^{n} u[n-1]$$

(a) Is the system causal?

(b) Is the system stable?

2.58. Consider the *RLC* circuit shown in Fig. 2-32. Find the differential equation relating the output current $y(t)$ and the input voltage $x(t)$.

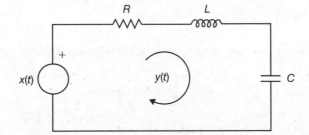

Fig. 2-32

2.59. Consider the *RL* circuit shown in Fig. 2-33.

(a) Find the differential equation relating the output voltage $y(t)$ across R and the the input voltage $x(t)$.

(b) Find the impulse response $h(t)$ of the circuit.

(c) Find the step response $s(t)$ of the circuit.

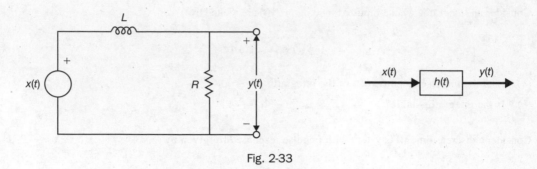

Fig. 2-33

2.60. Consider the system in Prob. 2.20. Find the output $y(t)$ if $x(t) = e^{-at}u(t)$ and $y(0) = 0$.

2.61. Is the system described by the differential equation

$$\frac{dy(t)}{dt} + 5y(t) + 2 = x(t)$$

linear?

2.62. Write the input-output equation for the system shown in Fig. 2-34.

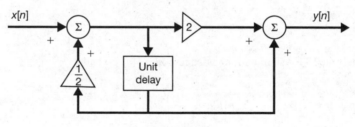

Fig. 2-34

2.63. Consider a discrete-time LTI system with impulse response

$$h[n] = \begin{cases} 1 & n = 0,1 \\ 0 & \text{otherwise} \end{cases}$$

Find the input-output relationship of the system.

2.64. Consider a discrete-time system whose input $x[n]$ and output $y[n]$ are related by

$$y[n] - \frac{1}{2}y[n-1] = x[n]$$

with $y[-1] = 0$. Find the output $y[n]$ for the following inputs:

(a) $x[n] = \left(\frac{1}{3}\right)^n u[n]$;

(b) $x[n] = \left(\frac{1}{2}\right)^n u[n]$

2.65. Consider the system in Prob. 2.42. Find the eigenfunction and the corresponding eigenvalue of the system.

ANSWERS TO SUPPLEMENTARY PROBLEMS

2.46. (*a*) $y(t) = \begin{cases} 2a - |t| & |t| < 2a \\ 0 & |t| \geq 2a \end{cases}$

(*b*) $y(t) = \begin{cases} 0 & t < 0 \\ \dfrac{1}{2}t^2 & 0 < t \leq T \\ \dfrac{1}{2}T^2 & T < t \leq 2T \\ -\dfrac{1}{2}t^2 + 2T - \dfrac{5}{2}T^2 & 2T < t \leq 3T \\ 0 & 3T < t \end{cases}$

(*c*) $\dfrac{1}{3}(1 - e^{-3(t-1)})u(t-1)$

2.47. (*a*) $y[n] = \begin{cases} 2^{1-n} & n \leq 0 \\ 2 & n > 0 \end{cases}$

(*b*) $y[n] = \begin{cases} 0 & n < 0 \\ \dfrac{1 - \alpha^{n+1}}{1 - \alpha} & n \leq 0 \leq N - 1 \\ \alpha^{n-N+1}\left(\dfrac{1 - \alpha^N}{1 - \alpha}\right) & N - 1 < n \end{cases}$

(*c*) $y[n] = \delta[n]$

2.48. *Hint:* Differentiate Eqs. (2.6) and (2.10) with respect to *t*.

2.49. *Hint:* Use the result from Prob. 2.48 and Eq. (2.58).

2.50. *Hint:* See Prob. 2.3.

2.51. *Hint:* See Probs. 2.31 and 2.8.

2.52. $h(t) = \delta(t) - \omega_0[\sin \omega_0 t]u(t)$

2.53. (*a*) $h(t) = (e^{-2t} + 2e^{-t})u(t)$

(*b*) Yes

2.54. (*a*) $h(t) = u(t)$

(*b*) No

2.55. No, the system has memory.

2.56. $y[1] = 1$ and $y[4] = \dfrac{5}{8}$

2.57. (*a*) Yes; (*b*) Yes

2.58. $\dfrac{d^2 y(t)}{dt^2} + \dfrac{R}{L}\dfrac{dy(t)}{dt} + \dfrac{1}{LC}y(t) = \dfrac{1}{L}\dfrac{dx(t)}{dt}$

2.59. (a) $\dfrac{dy(t)}{dt} + \dfrac{R}{L} y(t) = \dfrac{R}{L} x(t)$

(b) $h(t) = \dfrac{R}{L} e^{-(R/L)t} u(t)$

(c) $s(t) = [1 - e^{-(R/L)t}] u(t)$

2.60. $te^{-at} u(t)$

2.61. No, it is nonlinear.

2.62. $2y[n] - y[n-1] = 4x[n] + 2x[n-1]$

2.63. $y[n] = x[n] + x[n-1]$

2.64. (a) $y[n] = 6\left[\left(\dfrac{1}{2}\right)^{n+1} - \left(\dfrac{1}{3}\right)^{n+1} \right] u[n]$

(b) $y[n] = (n+1)\left(\dfrac{1}{2}\right)^{n} u[n]$

2.65. $z^{n}, \lambda = \dfrac{z}{z-a}$

CHAPTER 3

Laplace Transform and Continuous-Time LTI Systems

3.1 Introduction

A basic result from Chap. 2 is that the response of an LTI system is given by convolution of the input and the impulse response of the system. In this chapter and the following one we present an alternative representation for signals and LTI systems. In this chapter, the Laplace transform is introduced to represent continuous-time signals in the s-domain (s is a complex variable), and the concept of the system function for a continuous-time LTI system is described. Many useful insights into the properties of continuous-time LTI systems, as well as the study of many problems involving LTI systems, can be provided by application of the Laplace transform technique.

3.2 The Laplace Transform

In Sec. 2.4 we saw that for a continuous-time LTI system with impulse response $h(t)$, the output $y(t)$ of the system to the complex exponential input of the form e^{st} is

$$y(t) = T\{e^{st}\} = H(s)e^{st} \tag{3.1}$$

where

$$H(s) = \int_{-\infty}^{\infty} h(t)e^{-st}dt \tag{3.2}$$

A. Definition:

The function $H(s)$ in Eq. (3.2) is referred to as the Laplace transform of $h(t)$. For a general continuous-time signal $x(t)$, the Laplace transform $X(s)$ is defined as

$$X(s) = \int_{-\infty}^{\infty} x(t)e^{-st}dt \tag{3.3}$$

The variable s is generally complex-valued and is expressed as

$$s = \sigma + j\omega \tag{3.4}$$

The Laplace transform defined in Eq. (3.3) is often called the *bilateral* (or *two-sided*) Laplace transform in contrast to the *unilateral* (or *one-sided*) Laplace transform, which is defined as

$$X_I(s) = \int_{0^-}^{\infty} x(t)e^{-st}dt \tag{3.5}$$

where $0^- = \lim_{\varepsilon \to 0}(0 - \varepsilon)$. Clearly the bilateral and unilateral transforms are equivalent only if $x(t) = 0$ for $t < 0$. The unilateral Laplace transform is discussed in Sec. 3.8. We will omit the word "bilateral" except where it is needed to avoid ambiguity.

Equation (3.3) is sometimes considered an operator that transforms a signal $x(t)$ into a function $X(s)$ symbolically represented by

$$X(s) = \mathscr{L}\{x(t)\} \tag{3.6}$$

and the signal $x(t)$ and its Laplace transform $X(s)$ are said to form a Laplace transform pair denoted as

$$x(t) \leftrightarrow X(s) \tag{3.7}$$

B.   The Region of Convergence:

The range of values of the complex variables s for which the Laplace transform converges is called the *region of convergence* (ROC). To illustrate the Laplace transform and the associated ROC, let us consider some examples.

EXAMPLE 3.1   Consider the signal

$$x(t) = e^{-at}u(t) \qquad a \text{ real} \tag{3.8}$$

Then by Eq. (3.3) the Laplace transform of $x(t)$ is

$$X(s) = \int_{-\infty}^{\infty} e^{-at} u(t)e^{-st} dt = \int_{0^+}^{\infty} e^{-(s+a)t} dt$$

$$= -\frac{1}{s+a} e^{-(s+a)t} \bigg|_{0^+}^{\infty} = \frac{1}{s+a} \qquad \text{Re}(s) > -a \tag{3.9}$$

because $\lim_{t \to \infty} e^{-(s+a)t} = 0$ only if $\text{Re}(s + a) > 0$ or $\text{Re}(s) > -a$.

Thus, the ROC for this example is specified in Eq. (3.9) as $\text{Re}(s) > -a$ and is displayed in the complex plane as shown in Fig. 3-1 by the shaded area to the right of the line $\text{Re}(s) = -a$. In Laplace transform applications, the complex plane is commonly referred to as the *s*-plane. The horizontal and vertical axes are sometimes referred to as the σ-axis and the $j\omega$-axis, respectively.

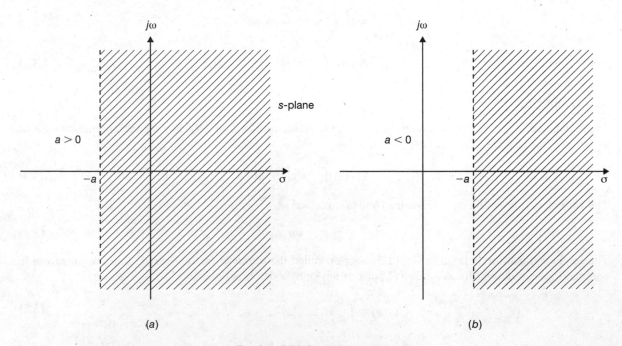

Fig. 3-1 ROC for Example 3.1.

EXAMPLE 3.2   Consider the signal

$$x(t) = -e^{-at}u(-t) \qquad a \text{ real} \tag{3.10}$$

Its Laplace transform $X(s)$ is given by (Prob. 3.1)

$$X(s) = \frac{1}{s+a} \qquad \text{Re}(s) < -a \tag{3.11}$$

Thus, the ROC for this example is specified in Eq. (3.11) as $\text{Re}(s) < -a$ and is displayed in the complex plane as shown in Fig. 3-2 by the shaded area to the left of the line $\text{Re}(s) = -a$. Comparing Eqs. (3.9) and (3.11), we see that the algebraic expressions for $X(s)$ for these two different signals are identical except for the ROCs. Therefore, in order for the Laplace transform to be unique for each signal $x(t)$, *the ROC must be specified as part of the transform.*

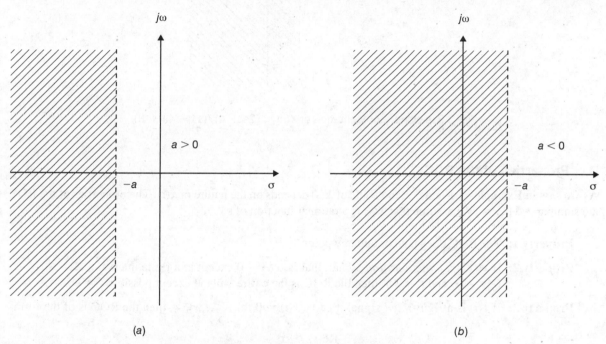

Fig. 3-2 ROC for Example 3.2.

C. Poles and Zeros of $X(s)$:

Usually, $X(s)$ will be a rational function in s; that is,

$$X(s) = \frac{a_0 s^m + a_1 s^{m-1} + \cdots + a_m}{b_0 s^n + b_1 s^{n-1} + \cdots + b_n} = \frac{a_0}{b_0} \frac{(s-z_1)\cdots(s-z_m)}{(s-p_1)\cdots(s-p_n)} \tag{3.12}$$

The coefficients a_k and b_k are real constants, and m and n are positive integers. The $X(s)$ is called a *proper* rational function if $n > m$, and an *improper* rational function if $n \leq m$. The roots of the numerator polynomial, z_k, are called the *zeros* of $X(s)$ because $X(s) = 0$ for those values of s. Similarly, the roots of the denominator polynomial, p_k, are called the *poles* of $X(s)$ because $X(s)$ is infinite for those values of s. Therefore, the poles of $X(s)$ lie outside the ROC since $X(s)$ does not converge at the poles, by definition. The zeros, on the other hand, may lie inside or outside the ROC. Except for a scale factor a_0/b_0, $X(s)$ can be completely specified by its zeros and poles. Thus, a very compact representation of $X(s)$ in the s-plane is to show the locations of poles and zeros in addition to the ROC.

Traditionally, an "×" is used to indicate each pole location and an "o" is used to indicate each zero. This is illustrated in Fig. 3-3 for $X(s)$ given by

$$X(s) = \frac{2s+4}{s^2+4s+3} = 2\frac{s+2}{(s+1)(s+3)} \qquad \text{Re}(s) > -1$$

Note that $X(s)$ has one zero at $s = -2$ and two poles at $s = -1$ and $s = -3$ with scale factor 2.

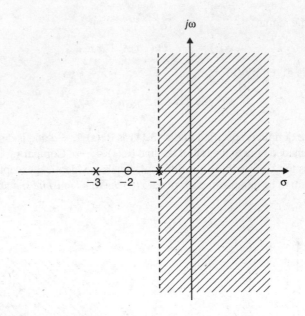

Fig. 3-3 s-plane representation of $X(s) = (2s + 4)/(s^2 + 4s + 3)$.

D. Properties of the ROC:

As we saw in Examples 3.1 and 3.2, the ROC of $X(s)$ depends on the nature of $x(t)$. The properties of the ROC are summarized below. We assume that $X(s)$ is a rational function of s.

Property 1: The ROC does not contain any poles.

Property 2: If $x(t)$ is a *finite-duration* signal, that is, $x(t) = 0$ except in a finite interval $t_1 \le t \le t_2$ $(-\infty < t_1$ and $t_2 < \infty)$, then the ROC is the entire s-plane except possibly $s = 0$ or $s = \infty$.

Property 3: If $x(t)$ is a *right-sided* signal, that is, $x(t) = 0$ for $t < t_1 < \infty$, then the ROC is of the form

$$\text{Re}(s) > \sigma_{\max}$$

where $\sigma_{\max}$ equals the maximum real part of any of the poles of $X(s)$. Thus, the ROC is a half-plane to the right of the vertical line $\text{Re}(s) = \sigma_{\max}$ in the s-plane and thus to the right of all of the poles of $X(s)$.

Property 4: If $x(t)$ is a *left-sided* signal, that is, $x(t) = 0$ for $t > t_2 > -\infty$, then the ROC is of the form

$$\text{Re}(s) < \sigma_{\min}$$

where $\sigma_{\min}$ equals the minimum real part of any of the poles of $X(s)$. Thus, the ROC is a half-plane to the left of the vertical line $\text{Re}(s) = \sigma_{\min}$ in the s-plane and thus to the left of all of the poles of $X(s)$.

Property 5: If $x(t)$ is a *two-sided* signal, that is, $x(t)$ is an infinite-duration signal that is neither right-sided nor left-sided, then the ROC is of the form

$$\sigma_1 < \text{Re}(s) < \sigma_2$$

where σ_1 and σ_2 are the real parts of the two poles of $X(s)$. Thus, the ROC is a vertical strip in the s-plane between the vertical lines $\text{Re}(s) = \sigma_1$ and $\text{Re}(s) = \sigma_2$.

Note that Property 1 follows immediately from the definition of poles; that is, $X(s)$ is infinite at a pole. For verification of the other properties see Probs. 3.2 to 3.7.

3.3 Laplace Transforms of Some Common Signals

A. Unit Impulse Function $\delta(t)$:

Using Eqs. (3.3) and (1.20), we obtain

$$\mathcal{L}[\delta(t)] = \int_{-\infty}^{\infty} \delta(t)e^{-st}dt = 1 \qquad \text{all } s \tag{3.13}$$

B. Unit Step Function $u(t)$:

$$\mathcal{L}[u(t)] = \int_{-\infty}^{\infty} u(t)e^{-st}dt = \int_{0^+}^{\infty} e^{-st}dt$$

$$= -\frac{1}{s}e^{-st}\Big|_{0^+}^{\infty} = \frac{1}{s} \qquad \text{Re}(s) > 0 \tag{3.14}$$

where $0^+ = \lim_{\varepsilon \to 0}(0 + \varepsilon)$.

C. Laplace Transform Pairs for Common Signals:

The Laplace transforms of some common signals are tabulated in Table 3-1. Instead of having to reevaluate the transform of a given signal, we can simply refer to such a table and read out the desired transform.

TABLE 3-1 Some Laplace Transforms Pairs

$x(t)$	$X(s)$	ROC
$\delta(t)$	1	All s
$u(t)$	$\dfrac{1}{s}$	$\text{Re}(s) > 0$
$-u(-t)$	$\dfrac{1}{s}$	$\text{Re}(s) < 0$
$tu(t)$	$\dfrac{1}{s^2}$	$\text{Re}(s) > 0$
$t^k u(t)$	$\dfrac{k!}{s^{k+1}}$	$\text{Re}(s) > 0$
$e^{-at}u(t)$	$\dfrac{1}{s+a}$	$\text{Re}(s) > -\text{Re}(a)$
$-e^{-at}u(-t)$	$\dfrac{1}{s+a}$	$\text{Re}(s) < -\text{Re}(a)$
$te^{-at}u(t)$	$\dfrac{1}{(s+a)^2}$	$\text{Re}(s) > -\text{Re}(a)$
$-te^{-at}u(-t)$	$\dfrac{1}{(s+a)^2}$	$\text{Re}(s) < -\text{Re}(a)$
$\cos \omega_0 tu(t)$	$\dfrac{s}{s^2+\omega_0^2}$	$\text{Re}(s) > 0$
$\sin \omega_0 tu(t)$	$\dfrac{\omega_0}{s^2+\omega_0^2}$	$\text{Re}(s) > 0$
$e^{-at}\cos \omega_0 tu(t)$	$\dfrac{s+a}{(s+a)^2+\omega_0^2}$	$\text{Re}(s) > -\text{Re}(a)$
$e^{-at}\sin \omega_0 tu(t)$	$\dfrac{\omega_0}{(s+a)^2+\omega_0^2}$	$\text{Re}(s) > -\text{Re}(a)$

3.4   Properties of the Laplace Transform

Basic properties of the Laplace transform are presented in the following. Verification of these properties is given in Probs. 3.8 to 3.16.

A.   Linearity:

If

$$x_1(t) \leftrightarrow X_1(s) \qquad \text{ROC} = R_1$$
$$x_2(t) \leftrightarrow X_2(s) \qquad \text{ROC} = R_2$$

Then $a_1 x_1(t) + a_2 x_2(t) \leftrightarrow a_1 X_1(s) + a_2 X_2(s) \qquad R' \supset R_1 \cap R_2$ (3.15)

The set notation $A \supset B$ means that set A contains set B, while $A \cap B$ denotes the intersection of sets A and B, that is, the set containing all elements in both A and B. Thus, Eq. (3.15) indicates that the ROC of the resultant Laplace transform is at least as large as the region in common between R_1 and R_2. Usually we have simply $R' = R_1 \cap R_2$. This is illustrated in Fig. 3-4.

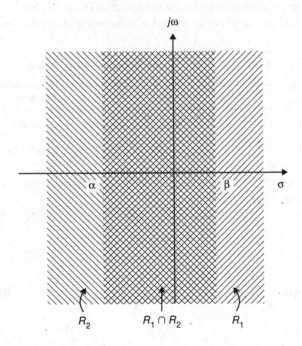

Fig. 3-4 ROC of $a_1 X_1(s) + a_2 X_2(s)$.

B.   Time Shifting:

If

$$x(t) \leftrightarrow X(s) \qquad \text{ROC} = R$$

then $x(t - t_0) \leftrightarrow e^{-s t_0} X(s) \qquad R' = R$ (3.16)

Equation (3.16) indicates that the ROCs before and after the time-shift operation are the same.

C.   Shifting in the *s*-Domain:

If

$$x(t) \leftrightarrow X(s) \qquad \text{ROC} = R$$

then $e^{s_0 t} x(t) \leftrightarrow X(s - s_0) \qquad R' = R + \text{Re}(s_0)$ (3.17)

Equation (3.17) indicates that the ROC associated with $X(s-s_0)$ is that of $X(s)$ shifted by $\mathrm{Re}(s_0)$. This is illustrated in Fig. 3-5.

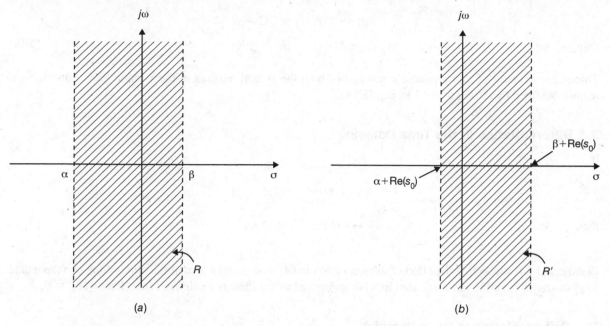

Fig. 3-5 Effect on the ROC of shifting in the s-domain. (a) ROC of $X(s)$; (b) ROC of $X(s - s_0)$.

D. Time Scaling:

If

$$x(t) \leftrightarrow X(s) \qquad \mathrm{ROC} = R$$

then

$$x(at) \leftrightarrow \frac{1}{|a|} X\left(\frac{s}{a}\right) \qquad R' = aR \tag{3.18}$$

Equation (3.18) indicates that scaling the time variable t by the factor a causes an inverse scaling of the variable s by $1/a$ as well as an amplitude scaling of $X(s/a)$ by $1/|a|$. The corresponding effect on the ROC is illustrated in Fig. 3-6.

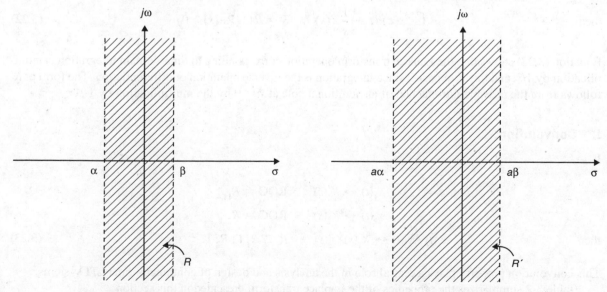

Fig. 3-6 Effect on the ROC of time scaling. (a) ROC of $X(s)$; (b) ROC of $X(s/a)$.

E. Time Reversal:

If

$$x(t) \leftrightarrow X(s) \qquad \text{ROC} = R$$

then
$$x(-t) \leftrightarrow X(-s) \qquad R' = -R \tag{3.19}$$

Thus, time reversal of $x(t)$ produces a reversal of both the σ- and $j\omega$-axes in the s-plane. Equation (3.19) is readily obtained by setting $a = -1$ in Eq. (3.18).

F. Differentiation in the Time Domain:

If

$$x(t) \leftrightarrow X(s) \qquad \text{ROC} = R$$

then
$$\frac{dx(t)}{dt} \leftrightarrow sX(s) \qquad R' \supset R \tag{3.20}$$

Equation (3.20) shows that the effect of differentiation in the time domain is multiplication of the corresponding Laplace transform by s. The associated ROC is unchanged unless there is a pole-zero cancellation at $s = 0$.

G. Differentiation in the s-Domain:

If

$$x(t) \leftrightarrow X(s) \qquad \text{ROC} = R$$

then
$$-tx(t) \leftrightarrow \frac{dX(s)}{ds} \qquad R' = R \tag{3.21}$$

H. Integration in the Time Domain:

If

$$x(t) \leftrightarrow X(s) \qquad \text{ROC} = R$$

then
$$\int_{-\infty}^{t} x(\tau)d\tau \leftrightarrow \frac{1}{s}X(s) \qquad R' = R \cap \{\text{Re}(s) > 0\} \tag{3.22}$$

Equation (3.22) shows that the Laplace transform operation corresponding to time-domain integration is multiplication by $1/s$, and this is expected since integration is the inverse operation of differentiation. The form of R' follows from the possible introduction of an additional pole at $s = 0$ by the multiplication by $1/s$.

I. Convolution:

If

$$x_1(t) \leftrightarrow X_1(s) \qquad \text{ROC} = R_1$$
$$x_2(t) \leftrightarrow X_2(s) \qquad \text{ROC} = R_2$$

then
$$x_1(t) * x_2(t) \leftrightarrow X_1(s)X_2(s) \qquad R' \supset R_1 \cap R_2 \tag{3.23}$$

This convolution property plays a central role in the analysis and design of continuous-time LTI systems.

Table 3-2 summarizes the properties of the Laplace transform presented in this section.

TABLE 3-2 Properties of the Laplace Transform

PROPERTY	SIGNAL	TRANSFORM	ROC
	$x(t)$	$X(s)$	R
	$x_1(t)$	$X_1(s)$	R_1
	$x_2(t)$	$X_2(s)$	R_2
Linearity	$a_1 x_1(t) + a_2 x_2(t)$	$a_1 X_1(s) + a_2 X_2(s)$	$R' \supset R_1 \cap R_2$
Time shifting	$x(t - t_0)$	$e^{-s t_0} X(s)$	$R' = R$
Shifting in s	$e^{s_0 t} x(t)$	$X(s - s_0)$	$R' = R + \mathrm{Re}(s_0)$
Time scaling	$x(at)$	$\dfrac{1}{\|a\|} X(a)$	$R' = aR$
Time reversal	$x(-t)$	$X(-s)$	$R' = -R$
Differentiation in t	$\dfrac{dx(t)}{dt}$	$sX(s)$	$R' \supset R$
Differentiation in s	$-tx(t)$	$\dfrac{dX(s)}{ds}$	$R' = R$
Integration	$\displaystyle\int_{-\infty}^{t} x(\tau)\,d\tau$	$\dfrac{1}{s} X(s)$	$R' \supset R \cap \{\mathrm{Re}(s) > 0\}$
Convolution	$x_1(t) * x_2(t)$	$X_1(s)\,X_2(s)$	$R' \supset R_1 \cap R_2$

3.5 The Inverse Laplace Transform

Inversion of the Laplace transform to find the signal $x(t)$ from its Laplace transform $X(s)$ is called the *inverse Laplace transform*, symbolically denoted as

$$x(t) = \mathscr{L}^{-1}\{X(s)\} \tag{3.24}$$

A. Inversion Formula:

There is a procedure that is applicable to all classes of transform functions that involves the evaluation of a line integral in complex s-plane; that is,

$$x(t) = \frac{1}{2\pi j} \int_{c-j\infty}^{c+j\infty} X(s) e^{st}\, ds \tag{3.25}$$

In this integral, the real c is to be selected such that if the ROC of $X(s)$ is $\sigma_1 < \mathrm{Re}(s) < \sigma_2$, then $\sigma_1 < c < \sigma_2$. The evaluation of this inverse Laplace transform integral requires understanding of complex variable theory.

B. Use of Tables of Laplace Transform Pairs:

In the second method for the inversion of $X(s)$, we attempt to express $X(s)$ as a sum

$$X(s) = X_1(s) + \cdots + X_n(s) \tag{3.26}$$

where $X_1(s), \ldots, X_n(s)$ are functions with known inverse transforms $x_1(t), \ldots, x_n(t)$. From the linearity property (3.15) it follows that

$$x(t) = x_1(t) + \cdots + x_n(t) \tag{3.27}$$

C. Partial-Fraction Expansion:

If $X(s)$ is a rational function, that is, of the form

$$X(s) = \frac{N(s)}{D(s)} = k \frac{(s-z_1)\cdots(s-z_m)}{(s-p_1)\cdots(s-p_n)} \tag{3.28}$$

a simple technique based on partial-fraction expansion can be used for the inversion of $X(s)$.

(*a*)   When $X(s)$ is a proper rational function, that is, when $m < n$:

1. Simple Pole Case:

If all poles of $X(s)$, that is, all zeros of $D(s)$, are simple (or distinct), then $X(s)$ can be written as

$$X(s) = \frac{c_1}{s-p_1} + \cdots + \frac{c_n}{s-p_n} \tag{3.29}$$

where coefficients c_k are given by

$$c_k = (s-p_k)X(s)\big|_{s=p_k} \tag{3.30}$$

2. Multiple Pole Case:

If $D(s)$ has multiple roots, that is, if it contains factors of the form $(s-p_i)^r$, we say that p_i is the *multiple pole of $X(s)$ with multiplicity r*. Then the expansion of $X(s)$ will consist of terms of the form

$$\frac{\lambda_1}{s-p_i} + \frac{\lambda_2}{(s-p_i)^2} + \cdots + \frac{\lambda_r}{(s-p_i)^r} \tag{3.31}$$

where

$$\lambda_{r-k} = \frac{1}{k!} \frac{d^k}{ds^k} \left[(s-p_i)^r X(s) \right]\big|_{s=p_i} \tag{3.32}$$

(*b*) When $X(s)$ is an improper rational function, that is, when $m \geq n$:

If $m \geq n$, by long division we can write $X(s)$ in the form

$$X(s) = \frac{N(s)}{D(s)} = Q(s) + \frac{R(s)}{D(s)} \tag{3.33}$$

where $N(s)$ and $D(s)$ are the numerator and denominator polynomials in s, respectively, of $X(s)$, the quotient $Q(s)$ is a polynomial in s with degree $m-n$, and the remainder $R(s)$ is a polynomial in s with degree strictly less than n. The inverse Laplace transform of $X(s)$ can then be computed by determining the inverse Laplace transform of $Q(s)$ and the inverse Laplace transform of $R(s)/D(s)$. Since $R(s)/D(s)$ is proper, the inverse Laplace transform of $R(s)/D(s)$ can be computed by first expanding into partial fractions as given above. The inverse Laplace transform of $Q(s)$ can be computed by using the transform pair

$$\frac{d^k \delta(t)}{dt^k} \leftrightarrow s^k \qquad k = 1, 2, 3, \ldots \tag{3.34}$$

3.6 The System Function

A. The System Function:

In Sec. 2.2 we showed that the output $y(t)$ of a continuous-time LTI system equals the convolution of the input $x(t)$ with the impulse response $h(t)$; that is,

$$y(t) = x(t) * h(t) \tag{3.35}$$

Applying the convolution property (3.23), we obtain

$$Y(s) = X(s)H(s) \tag{3.36}$$

where $Y(s), X(s)$, and $H(s)$ are the Laplace transforms of $y(t), x(t)$, and $h(t)$, respectively. Equation (3.36) can be expressed as

$$H(s) = \frac{Y(s)}{X(s)} \tag{3.37}$$

The Laplace transform $H(s)$ of $h(t)$ is referred to as the *system function* (or the *transfer function*) of the system. By Eq. (3.37), the system function $H(s)$ can also be defined as the ratio of the Laplace transforms of the output $y(t)$ and the input $x(t)$. The system function $H(s)$ completely characterizes the system because the impulse response $h(t)$ completely characterizes the system. Fig. 3-7 illustrates the relationship of Eqs. (3.35) and (3.36).

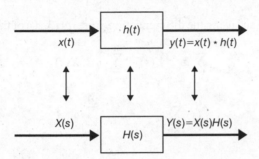

Fig. 3-7 Impulse response and system function.

B. Characterization of LTI Systems:

Many properties of continuous-time LTI systems can be closely associated with the characteristics of $H(s)$ in the s-plane and in particular with the pole locations and the ROC.

1. Causality:

For a causal continuous-time LTI system, we have

$$h(t) = 0 \qquad t < 0$$

Since $h(t)$ is a right-sided signal, the corresponding requirement on $H(s)$ is that the ROC of $H(s)$ must be of the form

$$\text{Re}(s) > \sigma_{max}$$

That is, the ROC is the region in the s-plane to the right of all of the system poles. Similarly, if the system is anticausal, then

$$h(t) = 0 \qquad t > 0$$

and $h(t)$ is left-sided. Thus, the ROC of $H(s)$ must be of the form

$$\text{Re}(s) < \sigma_{min}$$

That is, the ROC is the region in the s-plane to the left of all of the system poles.

2. Stability:

In Sec. 2.3 we stated that a continuous-time LTI system is BIBO stable if and only if [Eq. (2.21)]

$$\int_{-\infty}^{\infty} |h(t)| \, dt < \infty$$

The corresponding requirement on $H(s)$ is that the ROC of $H(s)$ contains the $j\omega$-axis (that is, $s = j\omega$) (Prob. 3.26).

3. Causal and Stable Systems:

If the system is both causal and stable, then all the poles of $H(s)$ must lie in the left half of the s-plane; that is, they all have negative real parts because the ROC is of the form $\text{Re}(s) > \sigma_{max}$, and since the $j\omega$ axis is included in the ROC, we must have $\sigma_{max} < 0$.

C. System Function for LTI Systems Described by Linear Constant-Coefficient Differential Equations:

In Sec. 2.5 we considered a continuous-time LTI system for which input $x(t)$ and output $y(t)$ satisfy the general linear constant-coefficient differential equation of the form

$$\sum_{k=0}^{N} a_k \frac{d^k y(t)}{dt^k} = \sum_{k=0}^{M} b_k \frac{d^k x(t)}{dt^k} \tag{3.38}$$

Applying the Laplace transform and using the differentiation property (3.20) of the Laplace transform, we obtain

$$\sum_{k=0}^{N} a_k s^k Y(s) = \sum_{k=0}^{M} b_k s^k X(s)$$

or

$$Y(s) \sum_{k=0}^{N} a_k s^k = X(s) \sum_{k=0}^{M} b_k s^k \tag{3.39}$$

Thus,

$$H(s) = \frac{Y(s)}{X(s)} = \frac{\sum_{k=0}^{M} b_k s^k}{\sum_{k=0}^{N} a_k s^k} \tag{3.40}$$

Hence, $H(s)$ is always rational. Note that the ROC of $H(s)$ is not specified by Eq. (3.40) but must be inferred with additional requirements on the system such as the causality or the stability.

D. Systems Interconnection:

For two LTI systems [with $h_1(t)$ and $h_2(t)$, respectively] in cascade [Fig. 3-8(a)], the overall impulse response $h(t)$ is given by [Eq. (2.81), Prob. 2.14]

$$h(t) = h_1(t) * h_2(t)$$

Thus, the corresponding system functions are related by the product

$$H(s) = H_1(s)H_2(s) \tag{3.41}$$

This relationship is illustrated in Fig. 3-8(b).

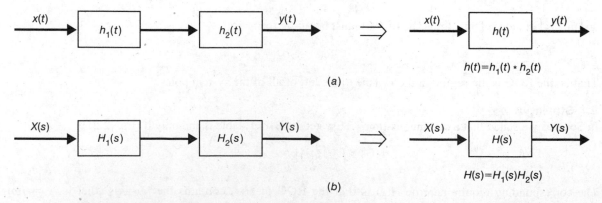

Fig. 3-8 Two systems in cascade. (a) Time-domain representation; (b) s-domain representation.

Similarly, the impulse response of a parallel combination of two LTI systems [Fig. 3-9(a)] is given by (Prob. 2.53)

$$h(t) = h_1(t) + h_2(t)$$

Thus,

$$H(s) = H_1(s) + H_2(s) \tag{3.42}$$

This relationship is illustrated in Fig. 3-9(b).

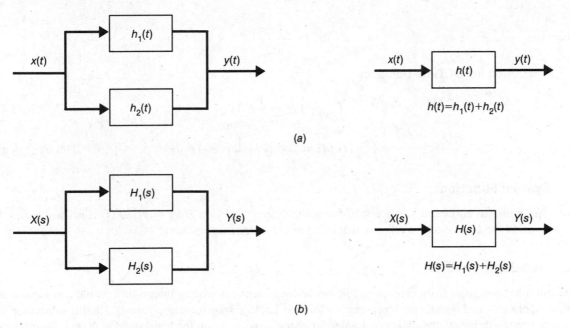

(a)

(b)

Fig. 3-9 Two systems in parallel. (a) Time-domain representation; (b) s-domain representation.

3.7 The Unilateral Laplace Transform

A. Definitions:

The *unilateral* (or *one-sided*) Laplace transform $X_I(s)$ of a signal $x(t)$ is defined as [Eq. (3.5)]

$$X_I(s) = \int_{0^-}^{\infty} x(t)s^{-st}\,dt \tag{3.43}$$

The lower limit of integration is chosen to be 0^- (rather than 0 or 0^+) to permit $x(t)$ to include $\delta(t)$ or its derivatives. Thus, we note immediately that the integration from 0^- to 0^+ is zero except when there is an impulse function or its derivative at the origin. The unilateral Laplace transform ignores $x(t)$ for $t < 0$. Since $x(t)$ in Eq. (3.43) is a right-sided signal, the ROC of $X_I(s)$ is always of the form $\text{Re}(s) > \sigma_{max}$, that is, a right half-plane in the s-plane.

B. Basic Properties:

Most of the properties of the unilateral Laplace transform are the same as for the bilateral transform. The unilateral Laplace transform is useful for calculating the response of a causal system to a causal input when the system is described by a linear constant-coefficient differential equation with nonzero initial conditions. The basic properties of the unilateral Laplace transform that are useful in this application are the time-differentiation and time-integration properties which are different from those of the bilateral transform. They are presented in the following.

1.   Differentiation in the Time Domain:

$$\frac{dx(t)}{dt} \leftrightarrow sX_I(s) - x(0^-) \tag{3.44}$$

provided that $\lim_{t \to \infty} x(t)e^{-st} = 0$. Repeated application of this property yields

$$\frac{d^2 x(t)}{dt^2} \leftrightarrow s^2 X_I(s) - sx(0^-) - x'(0^-) \tag{3.45}$$

$$\frac{d^n x(t)}{dt^n} \leftrightarrow s^n X_I(s) - s^{n-1} x(0^-) - s^{n-2} x'(0^-) - \cdots - x^{(n-1)}(0^-) \tag{3.46}$$

where

$$x^{(r)}(0^-) = \frac{d^r x(t)}{dt^r} \Big|_{t=0^-}$$

2.   Integration in the Time Domain:

$$\int_{0^-}^{t} x(\tau)\, d\tau \leftrightarrow \frac{1}{s} X_I(s) \tag{3.47}$$

$$\int_{-\infty}^{t} x(\tau)\, d\tau \leftrightarrow \frac{1}{s} X_I(s) + \frac{1}{s} \int_{-\infty}^{0^-} x(\tau)\, d\tau \tag{3.48}$$

C.   System Function:

Note that with the unilateral Laplace transform, the system function $H(s) = Y(s)/X(s)$ is defined under the condition that the LTI system is relaxed, that is, all initial conditions are zero.

D.   Transform Circuits:

The solution for signals in an electric circuit can be found without writing integrodifferential equations if the circuit operations and signals are represented with their Laplace transform equivalents. [In this subsection the Laplace transform means the unilateral Laplace transform and we drop the subscript I in $X_I(s)$.] We refer to a circuit produced from these equivalents as a *transform circuit*. In order to use this technique, we require the Laplace transform models for individual circuit elements. These models are developed in the following discussion and are shown in Fig. 3-10. Applications of this transform model technique to electric circuits problems are illustrated in Probs. 3.40 to 3.42.

1.   Signal Sources:

$$v(t) \leftrightarrow V(s) \qquad i(t) \leftrightarrow I(s)$$

where $v(t)$ and $i(t)$ are the voltage and current source signals, respectively.

2.   Resistance R:

$$v(t) = Ri(t) \leftrightarrow V(s) = RI(s) \tag{3.49}$$

3.   Inductance L:

$$v(t) = L\frac{di(t)}{dt} \leftrightarrow V(s) = sLI(s) - Li(0^-) \tag{3.50}$$

The second model of the inductance L in Fig. 3-10 is obtained by rewriting Eq. (3.50) as

$$i(t) \leftrightarrow I(s) = \frac{1}{sL} V(s) + \frac{1}{s} i(0^-) \tag{3.51}$$

4. Capacitance C:

$$i(t) = C\frac{dv(t)}{dt} \leftrightarrow I(s) = sCV(s) - Cv(0^-) \tag{3.52}$$

The second model of the capacitance C in Fig. 3-10 is obtained by rewriting Eq. (3.52) as

$$v(t) \leftrightarrow V(s) = \frac{1}{sC}I(s) + \frac{1}{s}v(0^-) \tag{3.53}$$

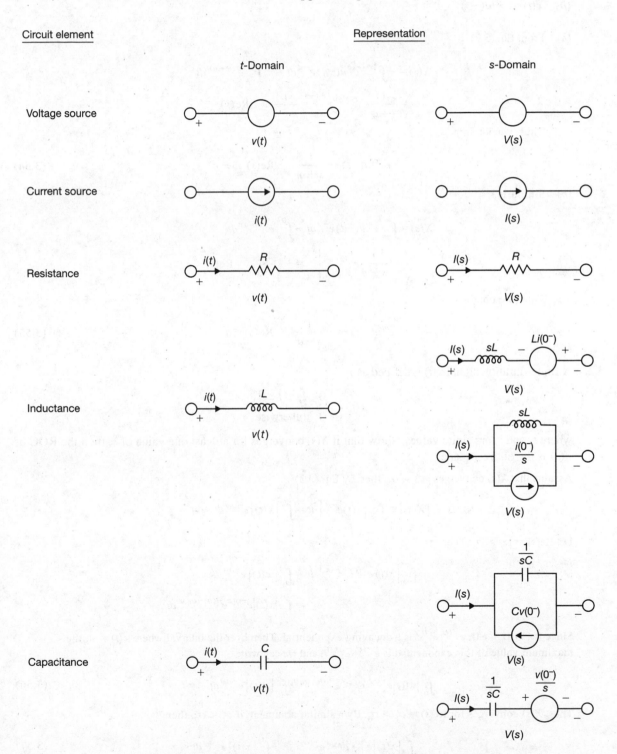

Fig. 3-10 Representation of Laplace transform circuit-element models.

Laplace Transform

3.1. Find the Laplace transform of

  (a) $x(t) = -e^{-at}u(-t)$

  (b) $x(t) = e^{at}u(-t)$

  (a) From Eq. (3.3)

$$X(s) = -\int_{-\infty}^{\infty} e^{-at}u(-t)e^{-st}\,dt = -\int_{-\infty}^{0^-} e^{-(s+a)t}\,dt$$

$$= \frac{1}{s+a}e^{-(s+a)t}\Big|_{-\infty}^{0^-} = \frac{1}{s+a} \qquad \text{Re}(s) < -a$$

Thus, we obtain

$$-e^{-at}u(-t) \leftrightarrow \frac{1}{s+a} \qquad \text{Re}(s) < -a \tag{3.54}$$

  (b) Similarly,

$$X(s) = \int_{-\infty}^{\infty} e^{at}u(-t)e^{-st}\,dt = \int_{-\infty}^{0^-} e^{-(s-a)t}\,dt$$

$$= -\frac{1}{s-a}e^{-(s-a)t}\Big|_{-\infty}^{0^-} = -\frac{1}{s-a} \qquad \text{Re}(s) < a$$

Thus, we obtain

$$e^{at}u(-t) \leftrightarrow -\frac{1}{s-a} \qquad \text{Re}(s) < a \tag{3.55}$$

3.2. A finite-duration signal $x(t)$ is defined as

$$x(t)\begin{cases} \neq 0 & t_1 \le t \le t_2 \\ = 0 & \text{otherwise} \end{cases}$$

where t_1 and t_2 are finite values. Show that if $X(s)$ converges for at least one value of s, then the ROC of $X(s)$ is the entire s-plane.

Assume that $X(s)$ converges at $s = \sigma_0$; then by Eq. (3.3)

$$|X(s)| \le \int_{-\infty}^{\infty} \left| x(t)e^{-st} \right| dt = \int_{t_1}^{t_2} |x(t)| e^{-\sigma_0 t}\,dt < \infty$$

Let $\text{Re}(s) = \sigma_1 > \sigma_0$. Then

$$\int_{-\infty}^{\infty} \left| x(t)e^{-(\sigma_1 + j\omega)t} \right| dt = \int_{t_1}^{t_2} |x(t)| e^{-\sigma_1 t}\,dt$$

$$= \int_{t_1}^{t_2} |x(t)| e^{-\sigma_0 t}e^{-(\sigma_1 - \sigma_0)t}\,dt$$

Since $(\sigma_1 - \sigma_0) > 0$, $e^{-(\sigma_1 - \sigma_0)t}$ is a decaying exponential. Then over the interval where $x(t) \neq 0$, the maximum value of this exponential is $e^{-(\sigma_1 - \sigma_0)t_1}$, and we can write

$$\int_{t_1}^{t_2} |x(t)| e^{-\sigma_1 t}\,dt < e^{-(\sigma_1 - \sigma_0)t_1} \int_{t_1}^{t_2} |x(t)| e^{-\sigma_0 t}\,dt < \infty \tag{3.56}$$

Thus, $X(s)$ converges for $\text{Re}(s) = \sigma_1 > \sigma_0$. By a similar argument, if $\sigma_1 < \sigma_0$, then

$$\int_{t_1}^{t_2} |x(t)| e^{-\sigma_1 t}\,dt < e^{-(\sigma_1 - \sigma_0)t_2} \int_{t_1}^{t_2} |x(t)| e^{-\sigma_0 t}\,dt < \infty \tag{3.57}$$

and again $X(s)$ converges for $\text{Re}(s) = \sigma_1 < \sigma_0$. Thus, the ROC of $X(s)$ includes the entire s-plane.

3.3. Let

$$x(t) = \begin{cases} e^{-at} & 0 \le t \le T \\ 0 & \text{otherwise} \end{cases}$$

Find the Laplace transform of $x(t)$.

By Eq. (3.3)

$$X(s) = \int_0^T e^{-at} e^{-st}\, dt = \int_0^T e^{-(s+a)t}\, dt$$

$$= -\frac{1}{s+a} e^{-(s+a)t} \Big|_0^T = \frac{1}{s+a}[1 - e^{-(s+a)T}] \tag{3.58}$$

Since $x(t)$ is a finite-duration signal, the ROC of $X(s)$ is the entire s-plane. Note that from Eq. (3.58) it appears that $X(s)$ does not converge at $s = -a$. But this is not the case. Setting $s = -a$ in the integral in Eq. (3.58), we have

$$X(-a) = \int_0^T e^{-(a+a)t}\, dt = \int_0^T dt = T$$

The same result can be obtained by applying L'Hospital's rule to Eq. (3.58).

3.4. Show that if $x(t)$ is a right-sided signal and $X(s)$ converges for some value of s, then the ROC of $X(s)$ is of the form

$$\text{Re}(s) > \sigma_{max}$$

where σ_{max} equals the maximum real part of any of the poles of $X(s)$.

Consider a right-sided signal $x(t)$ so that

$$x(t) = 0 \qquad t < t_1$$

and $X(s)$ converges for $\text{Re}(s) = \sigma_0$. Then

$$|X(s)| \le \int_{-\infty}^{\infty} |x(t) e^{-st}|\, dt = \int_{-\infty}^{\infty} |x(t)| e^{-\sigma_0 t}\, dt$$

$$= \int_{t_1}^{\infty} |x(t)| e^{-\sigma_0 t}\, dt < \infty$$

Let $\text{Re}(s) = \sigma_1 > \sigma_0$. Then

$$\int_{t_1}^{\infty} |x(t)| e^{-\sigma_1 t}\, dt = \int_{t_1}^{\infty} |x(t)| e^{-\sigma_0 t} e^{-(\sigma_1 - \sigma_0)t}\, dt$$

$$< e^{-(\sigma_1 - \sigma_0)t_1} \int_{t_1}^{\infty} |x(t)| e^{-\sigma_0 t}\, dt < \infty$$

Thus, $X(s)$ converges for $\text{Re}(s) = \sigma_1$ and the ROC of $X(s)$ is of the form $\text{Re}(s) > \sigma_0$. Since the ROC of $X(s)$ cannot include any poles of $X(s)$, we conclude that it is of the form

$$\text{Re}(s) > \sigma_{max}$$

where σ_{max} equals the maximum real part of any of the poles of $X(s)$.

3.5. Find the Laplace transform $X(s)$ and sketch the pole-zero plot with the ROC for the following signals $x(t)$:

(a) $x(t) = e^{-2t}u(t) + e^{-3t}u(t)$

(b) $x(t) = e^{-3t}u(t) + e^{2t}u(-t)$

(c) $x(t) = e^{2t}u(t) + e^{-3t}u(-t)$

(*a*)   From Table 3-1

$$e^{-2t}u(t) \leftrightarrow \frac{1}{s+2} \qquad \text{Re}(s) > -2 \tag{3.59}$$

$$e^{-3t}u(t) \leftrightarrow \frac{1}{s+3} \qquad \text{Re}(s) > -3 \tag{3.60}$$

We see that the ROCs in Eqs. (3.59) and (3.60) overlap, and thus,

$$X(s) = \frac{1}{s+2} + \frac{1}{s+3} = \frac{2\left(s+\frac{5}{2}\right)}{(s+2)(s+3)} \qquad \text{Re}(s) > -2 \tag{3.61}$$

From Eq. (3.61) we see that $X(s)$ has one zero at $s = -\frac{5}{2}$ and two poles at $s = -2$ and $s = -3$ and that the ROC is $\text{Re}(s) > -2$, as sketched in Fig. 3-11(*a*).

(*b*)   From Table 3-1

$$e^{-3t}u(t) \leftrightarrow \frac{1}{s+3} \qquad \text{Re}(s) > -3 \tag{3.62}$$

$$e^{2t}u(-t) \leftrightarrow -\frac{1}{s-2} \qquad \text{Re}(s) < 2 \tag{3.63}$$

We see that the ROCs in Eqs. (3.62) and (3.63) overlap, and thus,

$$X(s) = \frac{1}{s+3} - \frac{1}{s-2} = \frac{-5}{(s-2)(s+3)} \qquad -3 < \text{Re}(s) < 2 \tag{3.64}$$

From Eq. (3.64) we see that $X(s)$ has no zeros and two poles at $s = 2$ and $s = -3$ and that the ROC is $-3 < \text{Re}(s) < 2$, as sketched in Fig. 3-11(*b*).

(*c*)   From Table 3-1

$$e^{2t}u(t) \leftrightarrow \frac{1}{s-2} \qquad \text{Re}(s) > 2 \tag{3.65}$$

$$e^{-3t}u(-t) \leftrightarrow -\frac{1}{s+3} \qquad \text{Re}(s) < -3 \tag{3.66}$$

We see that the ROCs in Eqs. (3.65) and (3.66) do not overlap and that there is no common ROC; thus, $x(t)$ has no transform $X(s)$.

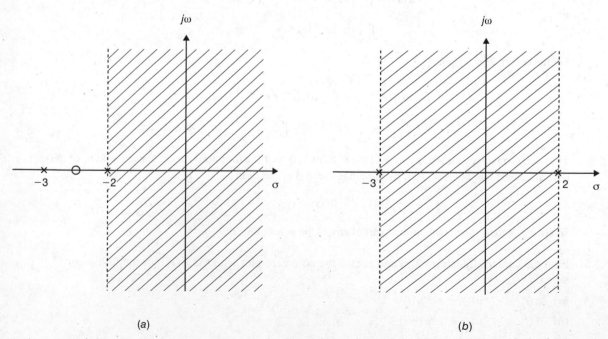

(a)   (b)

Fig. 3-11

3.6. Let

$$x(t) = e^{-a|t|}$$

Find $X(s)$ and sketch the zero-pole plot and the ROC for $a > 0$ and $a < 0$.

The signal $x(t)$ is sketched in Figs. 3-12 (*a*) and (*b*) for both $a > 0$ and $a < 0$. Since $x(t)$ is a two-sided signal, we can express it as

$$x(t) = e^{-at}u(t) + e^{at}u(-t) \tag{3.67}$$

Note that $x(t)$ is continuous at $t = 0$ and $x(0^-) = x(0) = x(0^+) = 1$. From Table 3-1

$$e^{-at}u(t) \leftrightarrow \frac{1}{s+a} \qquad \mathrm{Re}(s) > -a \tag{3.68}$$

$$e^{at}u(-t) \leftrightarrow -\frac{1}{s-a} \qquad \mathrm{Re}(s) < a \tag{3.69}$$

If $a > 0$, we see that the ROCs in Eqs. (3.68) and (3.69) overlap, and thus,

$$X(s) = \frac{1}{s+a} - \frac{1}{s-a} = \frac{-2a}{s^2 - a^2} \qquad -a < \mathrm{Re}(s) < a \tag{3.70}$$

From Eq. (3.70) we see that $X(s)$ has no zeros and two poles at $s = a$ and $s = -a$ and that the ROC is $-a < \mathrm{Re}(s) < a$, as sketched in Fig. 3-12(*c*). If $a < 0$, we see that the ROCs in Eqs. (3.68) and (3.69) do not overlap and that there is no common ROC; thus, $x(t)$ has no transform $X(s)$.

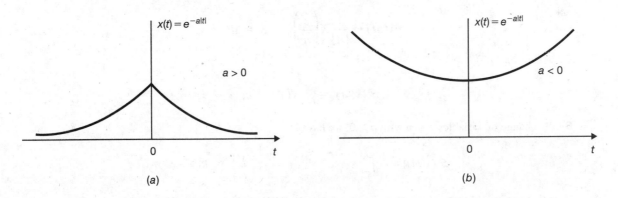

(*a*)   (*b*)

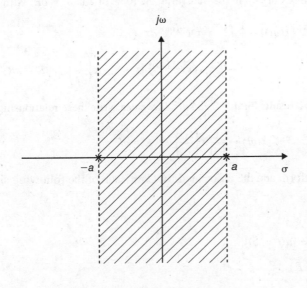

(*c*)

Fig. 3-12

Properties of the Laplace Transform

3.7. Verify the time-shifting property (3.16); that is,

$$x(t - t_0) \leftrightarrow e^{-st_0} X(s) \qquad R' = R$$

By definition (3.3)

$$\mathcal{L}\{x(t - t_0)\} = \int_{-\infty}^{\infty} x(t - t_0) e^{-st} dt$$

By the change of variables $\tau = t - t_0$ we obtain

$$\mathcal{L}\{x(t - t_0)\} = \int_{-\infty}^{\infty} x(\tau) e^{-s(\tau + t_0)} d\tau$$

$$= e^{-st_0} \int_{-\infty}^{\infty} x(\tau) e^{-s\tau} d\tau = e^{-st_0} X(s)$$

with the same ROC as for $X(s)$ itself. Hence,

$$x(t - t_0) \leftrightarrow e^{-st_0} X(s) \qquad R' = R$$

where R and R' are the ROCs before and after the time-shift operation.

3.8. Verify the time-scaling property (3.18); that is,

$$x(at) \leftrightarrow \frac{1}{|a|} X\left(\frac{s}{a}\right) \qquad R' = aR$$

By definition (3.3)

$$\mathcal{L}\{x(at)\} = \int_{-\infty}^{\infty} x(at) e^{-st} dt$$

By the change of variables $\tau = at$ with $a > 0$, we have

$$\mathcal{L}\{x(at)\} = \frac{1}{a} \int_{-\infty}^{\infty} x(\tau) e^{-(s/a)\tau} d\tau = \frac{1}{a} X\left(\frac{s}{a}\right) \qquad R' = aR$$

Note that because of the scaling s/a in the transform, the ROC of $X(s/a)$ is aR. With $a < 0$, we have

$$\mathcal{L}\{x(at)\} = \frac{1}{a} \int_{\infty}^{-\infty} x(\tau) e^{-(s/a)\tau} d\tau$$

$$= -\frac{1}{a} \int_{-\infty}^{\infty} x(\tau) e^{-(s/a)\tau} d\tau = -\frac{1}{a} X\left(\frac{s}{a}\right) \qquad R' = aR$$

Thus, combining the two results for $a > 0$ and $a < 0$, we can write these relationships as

$$x(at) \leftrightarrow \frac{1}{|a|} X\left(\frac{s}{a}\right) \qquad R' = aR$$

3.9. Find the Laplace transform and the associated ROC for each of the following signals:

(a) $x(t) = \delta(t - t_0)$

(b) $x(t) = u(t - t_0)$

(c) $x(t) = e^{-2t} [u(t) - u(t - 5)]$

(d) $x(t) = \displaystyle\sum_{k=0}^{\infty} \delta(t - kT)$

(e) $x(t) = \delta(at + b)$, a, b real constants

(a)   Using Eqs. (3.13) and (3.16), we obtain

$$\delta(t - t_0) \leftrightarrow e^{-st_0} \qquad \text{all } s \tag{3.71}$$

(b)   Using Eqs. (3.14) and (3.16), we obtain

$$u(t - t_0) \leftrightarrow \frac{e^{-st_0}}{s} \qquad \text{Re}(s) > 0 \tag{3.72}$$

(c)   Rewriting $x(t)$ as

$$x(t) = e^{-2t}[u(t) - u(t - 5)] = e^{-2t}u(t) - e^{-2t}u(t - 5)$$
$$= e^{-2t}u(t) - e^{-10}e^{-2(t-5)}u(t - 5)$$

Then, from Table 3-1 and using Eq. (3.16), we obtain

$$X(s) = \frac{1}{s+2} - e^{-10}e^{-5s}\frac{1}{s+2} = \frac{1}{s+2}(1 - e^{-5(s+2)}) \qquad \text{Re}(s) > -2$$

(d)   Using Eqs. (3.71) and (1.99), we obtain

$$X(s) = \sum_{k=0}^{\infty} e^{-skT} = \sum_{k=0}^{\infty} (e^{-sT})^k = \frac{1}{1 - e^{-sT}} \qquad \text{Re}(s) > 0 \tag{3.73}$$

(e)   Let

$$f(t) = \delta(at)$$

Then from Eqs. (3.13) and (3.18) we have

$$f(t) = \delta(at) \leftrightarrow F(s) = \frac{1}{|a|} \qquad \text{all } s \tag{3.74}$$

Now

$$x(t) = \delta(at + b) = \delta\left[a\left(t + \frac{b}{a}\right)\right] = f\left(t + \frac{b}{a}\right)$$

Using Eqs. (3.16) and (3.74), we obtain

$$X(s) = e^{sb/a}F(s) = \frac{1}{|a|}e^{sb/a} \qquad \text{all } s \tag{3.75}$$

3.10.   Verify the time differentiation property (3.20); that is,

$$\frac{dx(t)}{dt} \leftrightarrow sX(s) \qquad R' \supset R$$

From Eq. (3.24) the inverse Laplace transform is given by

$$x(t) = \frac{1}{2\pi j}\int_{c-j\infty}^{c+j\infty} X(s)e^{st}ds \tag{3.76}$$

Differentiating both sides of the above expression with respect to t, we obtain

$$\frac{dx(t)}{dt} = \frac{1}{2\pi j}\int_{c-j\infty}^{c+j\infty} sX(s)e^{st}ds \tag{3.77}$$

Comparing Eq. (3.77) with Eq. (3.76), we conclude that $dx(t)/dt$ is the inverse Laplace transform of $sX(s)$. Thus,

$$\frac{dx(t)}{dt} \leftrightarrow sX(s) \qquad R' \supset R$$

Note that the associated ROC is unchanged unless a pole-zero cancellation exists at $s = 0$.

3.11. Verify the differentiation in s property (3.21); that is,

$$-tx(t) \leftrightarrow \frac{dX(s)}{ds} \qquad R' = R$$

From definition (3.3)

$$X(s) = \int_{-\infty}^{\infty} x(t)e^{-st}dt$$

Differentiating both sides of the above expression with respect to s, we have

$$\frac{dX(s)}{ds} = \int_{-\infty}^{\infty} (-t)x(t)e^{-st}dt = \int_{-\infty}^{\infty} [-tx(t)]e^{-st}dt$$

Thus, we conclude that

$$-tx(t) \leftrightarrow \frac{dX(s)}{ds} \qquad R' = R$$

3.12. Verify the integration property (3.22); that is,

$$\int_{-\infty}^{t} x(\tau)d\tau \leftrightarrow \frac{1}{s}X(s) \qquad R' = R \cap \{\text{Re}(s) > 0\}$$

Let

$$f(t) = \int_{-\infty}^{t} x(\tau)d\tau \leftrightarrow F(s)$$

Then

$$x(t) = \frac{df(t)}{dt}$$

Applying the differentiation property (3.20), we obtain

$$X(s) = sF(s)$$

Thus,

$$F(s) = \frac{1}{s}X(s) \qquad R' = R \cap \{\text{Re}(s) > 0\}$$

The form of the ROC R' follows from the possible introduction of an additional pole at $s = 0$ by the multiplying by $1/s$.

3.13. Using the various Laplace transform properties, derive the Laplace transforms of the following signals from the Laplace transform of $u(t)$.

(a) $\delta(t)$ (b) $\delta'(t)$

(c) $tu(t)$ (d) $e^{-at}u(t)$

(e) $te^{-at}u(t)$ (f) $\cos \omega_0 t\, u(t)$

(g) $e^{-at} \cos \omega_0 t u(t)$

(a) From Eq. (3.14) we have

$$u(t) \leftrightarrow \frac{1}{s} \qquad \text{for } \text{Re}(s) > 0$$

From Eq. (1.30) we have

$$\delta(t) = \frac{du(t)}{dt}$$

Thus, using the time-differentiation property (3.20), we obtain

$$\delta(t) \leftrightarrow s\frac{1}{s} = 1 \qquad \text{all } s$$

(b)   Again applying the time-differentiation property (3.20) to the result from part (a), we obtain

$$\delta'(t) \leftrightarrow s \qquad \text{all } s \tag{3.78}$$

(c)   Using the differentiation in s property (3.21), we obtain

$$tu(t) \leftrightarrow -\frac{d}{ds}\left(\frac{1}{s}\right) = \frac{1}{s^2} \qquad \text{Re}(s) > 0 \tag{3.79}$$

(d)   Using the shifting in the s-domain property (3.17), we have

$$e^{-at}u(t) \leftrightarrow \frac{1}{s+a} \qquad \text{Re}(s) > -a$$

(e)   From the result from part (c) and using the differentiation in s property (3.21), we obtain

$$te^{-at}u(t) \leftrightarrow -\frac{d}{ds}\left(\frac{1}{s+a}\right) = \frac{1}{(s+a)^2} \qquad \text{Re}(s) > -a \tag{3.80}$$

(f)   From Euler's formula we can write

$$\cos \omega_0 \, tu(t) = \frac{1}{2}(e^{j\omega_0 t} + e^{-j\omega_0 t})u(t) = \frac{1}{2}e^{j\omega_0 t}u(t) + \frac{1}{2}e^{-j\omega_0 t}u(t)$$

Using the linearity property (3.15) and the shifting in the s-domain property (3.17), we obtain

$$\cos \omega_0 \, tu(t) \leftrightarrow \frac{1}{2}\frac{1}{s - j\omega_0} + \frac{1}{2}\frac{1}{s + j\omega_0} = \frac{s}{s^2 + \omega_0^2} \qquad \text{Re}(s) > 0 \tag{3.81}$$

(g)   Applying the shifting in the s-domain property (3.17) to the result from part (f), we obtain

$$e^{-at}\cos \omega_0 \, tu(t) \leftrightarrow \frac{s + a}{(s+a)^2 + \omega_0^2} \qquad \text{Re}(s) > -a \tag{3.82}$$

3.14. Verify the convolution property (3.23); that is,

$$x_1(t) * x_2(t) \leftrightarrow X_1(s)\, X_2(s) \qquad R' \supset R_1 \cap R_2$$

Let

$$y(t) = x_1(t) * x_2(t) = \int_{-\infty}^{\infty} x_1(\tau)x_2(t - \tau)\, d\tau$$

Then, by definition (3.3)

$$Y(s) = \int_{-\infty}^{\infty}\left[\int_{-\infty}^{\infty} x_1(\tau)x_2(t - \tau)\, d\tau\right]e^{-st}\, dt$$

$$= \int_{-\infty}^{\infty} x_1(\tau)\left[\int_{-\infty}^{\infty} x_2(t - \tau)e^{-st}\, dt\right]d\tau$$

Noting that the bracketed term in the last expression is the Laplace transform of the shifted signal $x_2(t - \tau)$, by Eq. (3.16) we have

$$Y(s) = \int_{-\infty}^{\infty} x_1(\tau)e^{-s\tau}X_2(s)\, d\tau$$

$$= \left[\int_{-\infty}^{\infty} x_1(\tau)e^{-s\tau}d\tau\right]X_2(s) = X_1(s)\, X_2(s)$$

with an ROC that contains the intersection of the ROC of $X_1(s)$ and $X_2(s)$. If a zero of one transform cancels a pole of the other, the ROC of $Y(s)$ may be larger. Thus, we conclude that

$$x_1(t) * x_2(t) \leftrightarrow X_1(s)\, X_2(s) \qquad R' \supset R_1 \cap R_2$$

3.15 Using the convolution property (3.23), verify Eq. (3.22); that is,

$$\int_{-\infty}^{t} x(\tau)\,d\tau \leftrightarrow \frac{1}{s}X(s) \qquad R' = R \cap \{\operatorname{Re}(s)>0\}$$

We can write [Eq. (2.60), Prob. 2.2]

$$\int_{-\infty}^{t} x(\tau)\,d\tau = x(t)*u(t) \tag{3.83}$$

From Eq. (3.14)

$$u(t) \leftrightarrow \frac{1}{s} \qquad \operatorname{Re}(s)>0$$

and thus, from the convolution property (3.23) we obtain

$$x(t)*u(t) \leftrightarrow \frac{1}{s}X(s)$$

with the ROC that includes the intersection of the ROC of $X(s)$ and the ROC of the Laplace transform of $u(t)$. Thus,

$$\int_{-\infty}^{t} x(\tau)d\tau \leftrightarrow \frac{1}{s}X(s) \qquad R' = R \cap \{\operatorname{Re}(s)>0\}$$

Inverse Laplace Transform

3.16. Find the inverse Laplace transform of the following $X(s)$:

(a) $X(s)=\dfrac{1}{s+1},\operatorname{Re}(s)>-1$

(b) $X(s)=\dfrac{1}{s+1},\operatorname{Re}(s)<-1$

(c) $X(s)=\dfrac{s}{s^2+4},\operatorname{Re}(s)>0$

(d) $X(s)=\dfrac{s+1}{(s+1)^2+4},\operatorname{Re}(s)>-1$

(a) From Table 3-1 we obtain

$$x(t) = e^{-t}u(t)$$

(b) From Table 3-1 we obtain

$$x(t) = -e^{-t}u(-t)$$

(c) From Table 3-1 we obtain

$$x(t) = \cos 2t\,u(t)$$

(d) From Table 3-1 we obtain

$$s(t) = e^{-t}\cos 2t\,u(t)$$

3.17. Find the inverse Laplace transform of the following $X(s)$:

(a) $X(s)=\dfrac{2s+4}{s^2+4s+3},\operatorname{Re}(s)>-1$

(b) $X(s)=\dfrac{2s+4}{s^2+4s+3},\operatorname{Re}(s)<-3$

(c) $X(s)=\dfrac{2s+4}{s^2+4s+3},-3<\operatorname{Re}(s)<-1$

Expanding by partial fractions, we have

$$X(s) = \frac{2s+4}{s^2+4s+3} = 2\frac{s+2}{(s+1)(s+3)} = \frac{c_1}{s+1} + \frac{c_2}{s+3}$$

Using Eq. (3.30), we obtain

$$c_1 = (s+1)X(s)\big|_{s=-1} = 2\frac{s+2}{s+3}\bigg|_{s=-1} = 1$$

$$c_2 = (s+3)X(s)\big|_{s=-3} = 2\frac{s+2}{s+1}\bigg|_{s=-3} = 1$$

Hence,

$$X(s) = \frac{1}{s+1} + \frac{1}{s+3}$$

(a) The ROC of $X(s)$ is $\text{Re}(s) > -1$. Thus, $x(t)$ is a right-sided signal and from Table 3-1 we obtain

$$x(t) = e^{-t}u(t) + e^{-3t}u(t) = (e^{-t} + e^{-3t})u(t)$$

(b) The ROC of $X(s)$ is $\text{Re}(s) < -3$. Thus, $x(t)$ is a left-sided signal and from Table 3-1 we obtain

$$x(t) = -e^{-t}u(-t) - e^{-3t}u(-t) = -(e^{-t} + e^{-3t})u(-t)$$

(c) The ROC of $X(s)$ is $-3 < \text{Re}(s) < -1$. Thus, $x(t)$ is a double-sided signal and from Table 3-1 we obtain

$$x(t) = -e^{-t}u(-t) + e^{-3t}u(t)$$

3.18. Find the inverse Laplace transform of

$$X(s) = \frac{5s+13}{s(s^2+4s+13)} \qquad \text{Re}(s) > 0$$

We can write

$$s^2 + 4s + 13 = (s+2)^2 + 9 = (s+2-j3)(s+2+j3)$$

Then

$$X(s) = \frac{5s+13}{s(s^2+4s+13)} = \frac{5s+13}{s(s+2-j3)(s+2+j3)}$$

$$= \frac{c_1}{s} + \frac{c_2}{s-(-2+j3)} + \frac{c_3}{s-(-2-j3)}$$

where

$$c_1 = sX(s)\big|_{s=0} = \frac{5s+13}{s^2+4s+13}\bigg|_{s=0} = 1$$

$$c_2 = (s+2-j3)X(s)\big|_{s=-2+j3} = \frac{5s+13}{s(s+2+j3)}\bigg|_{s=-2+j3} = -\frac{1}{2}(1+j)$$

$$c_3 = (s+2+j3)X(s)\big|_{s=-2-j3} = \frac{5s+13}{s(s+2-j3)}\bigg|_{s=-2-j3} = -\frac{1}{2}(1-j)$$

Thus,

$$X(s) = \frac{1}{s} - \frac{1}{2}(1+j)\frac{1}{s-(-2+j3)} - \frac{1}{2}(1-j)\frac{1}{s-(-2-j3)}$$

The ROC of $X(s)$ is $\text{Re}(s) > 0$. Thus, $x(t)$ is a right-sided signal and from Table 3-1 we obtain

$$x(t) = u(t) - \frac{1}{2}(1+j)e^{(-2+j3)t}u(t) - \frac{1}{2}(1-j)e^{(-2-j3)t}u(t)$$

Inserting the identity

$$e^{(-2 \pm j3)t} = e^{-2t}e^{\pm j3t} = e^{-2t}(\cos 3t \pm j \sin 3t)$$

into the above expression, after simple computations we obtain

$$x(t) = u(t) - e^{-2t}(\cos 3t - \sin 3t)\, u(t)$$
$$= [1 - e^{-2t}(\cos 3t - \sin 3t)]\, u(t)$$

Alternate Solution:

We can write $X(s)$ as

$$X(s) = \frac{5s + 13}{s(s^2 + 4s + 13)} = \frac{c_1}{s} + \frac{c_2 s + c_3}{s^2 + 4s + 13}$$

As before, by Eq. (3.30) we obtain

$$c_1 = sX(s)\big|_{s=0} = \frac{5s + 13}{s^2 + 4s + 13}\bigg|_{s=0} = 1$$

Then we have

$$\frac{c_2 s + c_3}{s^2 + 4s + 13} = \frac{5s + 13}{s(s^2 + 4s + 13)} - \frac{1}{s} = \frac{-s + 1}{s^2 + 4s + 13}$$

Thus,

$$X(s) = \frac{1}{s} - \frac{s - 1}{s^2 + 4s + 13} = \frac{1}{s} - \frac{s + 2 - 3}{(s+2)^2 + 9}$$
$$= \frac{1}{s} - \frac{s + 2}{(s+2)^2 + 3^2} + \frac{3}{(s+2)^2 + 3^2}$$

Then from Table 3-1 we obtain

$$x(t) = u(t) - e^{-2t}\cos 3t\, u(t) + e^{-2t}\sin 3t\, u(t)$$
$$= [1 - e^{-2t}(\cos 3t - \sin 3t)]u(t)$$

3.19. Find the inverse Laplace transform of

$$X(s) = \frac{s^2 + 2s + 5}{(s+3)(s+5)^2} \qquad \mathrm{Re}(s) > -3$$

We see that $X(s)$ has one simple pole at $s = -3$ and one multiple pole at $s = -5$ with multiplicity 2. Then by Eqs. (3.29) and (3.31) we have

$$X(s) = \frac{c_1}{s+3} + \frac{\lambda_1}{s+5} + \frac{\lambda_2}{(s+5)^2} \tag{3.84}$$

By Eqs. (3.30) and (3.32) we have

$$c_1 = (s+3)X(s)\big|_{s=-3} = \frac{s^2 + 2s + 5}{(s+5)^2}\bigg|_{s=-3} = 2$$

$$\lambda_2 = (s+5)^2 X(s)\big|_{s=-5} = \frac{s^2 + 2s + 5}{s+3}\bigg|_{s=-5} = -10$$

$$\lambda_1 = \frac{d}{ds}\big[(s+5)^2 X(s)\big]\bigg|_{s=-5} = \frac{d}{ds}\left[\frac{s^2 + 2s + 5}{s+3}\right]\bigg|_{s=-5}$$

$$= \frac{s^2 + 6s + 1}{(s+3)^2}\bigg|_{s=-5} = -1$$

Hence,

$$X(s) = \frac{2}{s+3} - \frac{1}{s+5} - \frac{10}{(s+5)^2}$$

The ROC of $X(s)$ is $\text{Re}(s) > -3$. Thus, $x(t)$ is a right-sided signal and from Table 3-1 we obtain

$$x(t) = 2e^{-3t}u(t) - e^{-5t}u(t) - 10te^{-5t}u(t)$$
$$= [2e^{-3t} - e^{-5t} - 10te^{-5t}]\, u(t)$$

Note that there is a simpler way of finding λ_1 without resorting to differentiation. This is shown as follows: First find c_1 and λ_2 according to the regular procedure. Then substituting the values of c_1 and λ_2 into Eq. (3.84), we obtain

$$\frac{s^2 + 2s + 5}{(s+3)(s+5)^2} = \frac{2}{s+3} + \frac{\lambda_1}{s+5} - \frac{10}{(s+5)^2}$$

Setting $s = 0$ on both sides of the above expression, we have

$$\frac{5}{75} = \frac{2}{3} + \frac{\lambda_1}{5} - \frac{10}{25}$$

from which we obtain $\lambda_1 = -1$.

3.20. Find the inverse Laplace transform of the following $X(s)$:

(a) $X(s) = \dfrac{2s+1}{s+2}$, $\text{Re}(s) > -2$

(b) $X(s) = \dfrac{s^2 + 6s + 7}{s^2 + 3s + 2}$, $\text{Re}(s) > -1$

(c) $X(s) = \dfrac{s^3 + 2s^2 + 6}{s^2 + 3s}$, $\text{Re}(s) > 0$

(a) $X(s) = \dfrac{2s+1}{s+2} = \dfrac{2(s+2) - 3}{s+2} = 2 - \dfrac{3}{s+2}$

Since the ROC of $X(s)$ is $\text{Re}(s) > -2$, $x(t)$ is a right-sided signal and from Table 3-1 we obtain

$$x(t) = 2\delta(t) - 3e^{-2t}u(t)$$

(b) Performing long division, we have

$$X(s) = \frac{s^2 + 6s + 7}{s^2 + 3s + 2} = 1 + \frac{3s+5}{s^2 + 3s + 2} = 1 + \frac{3s+5}{(s+1)(s+2)}$$

Let

$$X_1(s) = \frac{3s+5}{(s+1)(s+2)} = \frac{c_1}{s+1} + \frac{c_2}{s+2}$$

where

$$c_1 = (s+1)X_1(s)\big|_{s=-1} = \frac{3s+5}{s+2}\bigg|_{s=-1} = 2$$

$$c_2 = (s+2)X_1(s)\big|_{s=-2} = \frac{3s+5}{s+1}\bigg|_{s=-2} = 1$$

Hence,

$$X(s) = 1 + \frac{2}{s+1} + \frac{1}{s+2}$$

The ROC of $X(s)$ is $\text{Re}(s) > -1$. Thus, $x(t)$ is a right-sided signal and from Table 3-1 we obtain

$$x(t) = \delta(t) + (2e^{-t} + e^{-2t})u(t)$$

(c)   Proceeding similarly, we obtain

$$X(s) = \frac{s^3 + 2s^2 + 6}{s^2 + 3s} = s - 1 + \frac{3s + 6}{s(s + 3)}$$

Let

$$X_1(s) = \frac{3s + 6}{s(s + 3)} = \frac{c_1}{s} + \frac{c_2}{s + 3}$$

where

$$c_1 = sX_1(s)\Big|_{s=0} = \frac{3s + 6}{s + 3}\Big|_{s=0} = 2$$

$$c_2 = (s + 3)X_1(s)\Big|_{s=-3} = \frac{3s + 6}{s}\Big|_{s=-3} = 1$$

Hence,

$$X(s) = s - 1 + \frac{2}{s} + \frac{1}{s + 3}$$

The ROC of $X(s)$ is $\text{Re}(s) > 0$. Thus, $x(t)$ is a right-sided signal and from Table 3-1 and Eq. (3.78) we obtain

$$x(t) = \delta'(t) - \delta(t) + (2 + e^{-3t})u(t)$$

Note that all $X(s)$ in this problem are improper fractions and that $x(t)$ contains $\delta(t)$ or its derivatives.

3.21. Find the inverse Laplace transform of

$$X(s) = \frac{2 + 2se^{-2s} + 4e^{-4s}}{s^2 + 4s + 3} \qquad \text{Re}(s) > -1$$

We see that $X(s)$ is a sum

$$X(s) = X_1(s) + X_2(s)\,e^{-2s} + X_3(s)e^{-4s}$$

where

$$X_1(s) = \frac{2}{s^2 + 4s + 3} \qquad X_2(s) = \frac{2s}{s^2 + 4s + 3} \qquad X_3(s) = \frac{4}{s^2 + 4s + 3}$$

If

$$x_1(t) \leftrightarrow X_1(s) \qquad x_2(t) \leftrightarrow X_2(s) \qquad x_3(t) \leftrightarrow X_3(s)$$

then by the linearity property (3.15) and the time-shifting property (3.16) we obtain

$$x(t) = x_1(t) + x_2(t - 2) + x_3(t - 4) \tag{3.85}$$

Next, using partial-fraction expansions and from Table 3-1, we obtain

$$X_1(s) = \frac{1}{s + 1} - \frac{1}{s + 3} \leftrightarrow x_1(t) = (e^{-t} - e^{-3t})u(t)$$

$$X_2(s) = \frac{-1}{s + 1} + \frac{3}{s + 3} \leftrightarrow x_2(t) = (-e^{-t} + 3e^{-3t})u(t)$$

$$X_3(s) = \frac{2}{s + 1} - \frac{2}{s + 3} \leftrightarrow x_3(t) = 2(e^{-t} - e^{-3t})u(t)$$

Thus, by Eq. (3.85) we have

$$x(t) = (e^{-t} - e^{-3t})u(t) + [-e^{-(t-2)} + 3e^{-3(t-2)}]u(t-2)$$
$$+ 2[e^{-(t-4)} - e^{-3(t-4)}]u(t-4)$$

3.22. Using the differentiation in s property (3.21), find the inverse Laplace transform of

$$X(s) = \frac{1}{(s+a)^2} \qquad \text{Re}(s) > -a$$

We have

$$-\frac{d}{ds}\left(\frac{1}{s+a}\right) = \frac{1}{(s+a)^2}$$

and from Eq. (3.9) we have

$$e^{-at}u(t) \leftrightarrow \frac{1}{s+a} \qquad \text{Re}(s) > -a$$

Thus, using the differentiation in s property (3.21), we obtain

$$x(t) = te^{-at}u(t)$$

System Function

3.23 Find the system function $H(s)$ and the impulse response $h(t)$ of the RC circuit in Fig. 1-32 (Prob. 1.32).

(a) Let

$$x(t) = v_s(t) \qquad y(t) = v_c(t)$$

In this case, the RC circuit is described by [Eq. (1.105)]

$$\frac{dy(t)}{dt} + \frac{1}{RC}y(t) = \frac{1}{RC}x(t)$$

Taking the Laplace transform of the above equation, we obtain

$$sY(s) + \frac{1}{RC}Y(s) = \frac{1}{RC}X(s)$$

or

$$\left(s + \frac{1}{RC}\right)Y(s) = \frac{1}{RC}X(s)$$

Hence, by Eq. (3.37) the system function $H(s)$ is

$$H(s) = \frac{Y(s)}{X(s)} = \frac{1/RC}{s + 1/RC} = \frac{1}{RC}\frac{1}{s + 1/RC}$$

Since the system is causal, taking the inverse Laplace transform of $H(s)$, the impulse response $h(t)$ is

$$h(t) = \mathcal{L}^{-1}\{H(s)\} = \frac{1}{RC}e^{-t/RC}u(t)$$

(b) Let

$$x(t) = v_s(t) \qquad y(t) = i(t)$$

In this case, the RC circuit is described by [Eq. (1.107)]

$$\frac{dy(t)}{dt} + \frac{1}{RC}y(t) = \frac{1}{R}\frac{dx(t)}{dt}$$

Taking the Laplace transform of the above equation, we have

$$sY(s) + \frac{1}{RC}Y(s) = \frac{1}{R}sX(s)$$

or

$$\left(s + \frac{1}{RC}\right)Y(s) = \frac{1}{R}sX(s)$$

Hence, the system function $H(s)$ is

$$H(s) = \frac{Y(s)}{X(s)} = \frac{s/R}{s + 1/RC} = \frac{1}{R}\frac{s}{s + 1/RC}$$

In this case, the system function $H(s)$ is an improper fraction and can be rewritten as

$$H(s) = \frac{1}{R}\frac{s + 1/RC - 1/RC}{s + 1/RC} = \frac{1}{R} - \frac{1}{R^2 C}\frac{1}{s + 1/RC}$$

Since the system is causal, taking the inverse Laplace transform of $H(s)$, the impulse response $h(t)$ is

$$h(t) = \mathcal{L}^{-1}\{H(s)\} = \frac{1}{R}\delta(t) - \frac{1}{R^2 C}e^{-t/RC}u(t)$$

Note that we obtained different system functions depending on the different sets of input and output.

3.24. Using the Laplace transform, redo Prob. 2.5.

From Prob. 2.5 we have

$$h(t) = e^{-\alpha t}u(t) \qquad x(t) = e^{\alpha t}u(-t) \qquad \alpha > 0$$

Using Table 3-1, we have

$$H(s) = \frac{1}{s + \alpha} \qquad \text{Re}(s) > -\alpha$$

$$X(s) = -\frac{1}{s - \alpha} \qquad \text{Re}(s) < \alpha$$

Thus,

$$Y(s) = X(s)H(s) = -\frac{1}{(s + \alpha)(s - \alpha)} = -\frac{1}{s^2 - \alpha^2} \qquad -\alpha < \text{Re}(s) < \alpha$$

and from Table 3-1 (or Prob. 3.6) the output is

$$y(t) = \frac{1}{2\alpha}e^{-\alpha|t|}$$

which is the same as Eq. (2.67).

3.25. The output $y(t)$ of a continuous-time LTI system is found to be $2e^{-3t}u(t)$ when the input $x(t)$ is $u(t)$.

(a)   Find the impulse response $h(t)$ of the system.

(b)   Find the output $y(t)$ when the input $x(t)$ is $e^{-t}u(t)$.

(a) $x(t) = u(t), y(t) = 2e^{-3t}u(t)$

Taking the Laplace transforms of $x(t)$ and $y(t)$, we obtain

$$X(s) = \frac{1}{s} \qquad \text{Re}(s) > 0$$

$$Y(s) = \frac{2}{s + 3} \qquad \text{Re}(s) > -3$$

Hence, the system function $H(s)$ is

$$H(s) = \frac{Y(s)}{X(s)} = \frac{2s}{s+3} \qquad \text{Re}(s) > -3$$

Rewriting $H(s)$ as

$$H(s) = \frac{2s}{s+3} = \frac{2(s+3)-6}{s+3} = 2 - \frac{6}{s+3} \qquad \text{Re}(s) > -3$$

and taking the inverse Laplace transform of $H(s)$, we have

$$h(t) = 2\delta(t) - 6e^{-3t}u(t)$$

Note that $h(t)$ is equal to the derivative of $2e^{-3t}u(t)$, which is the step response $s(t)$ of the system [see Eq. (2.13)].

(b)
$$x(t) = e^{-t}u(t) \leftrightarrow \frac{1}{s+1} \qquad \text{Re}(s) > -1$$

Thus,

$$Y(s) = X(s)\,H(s) = \frac{2s}{(s+1)(s+3)} \qquad \text{Re}(s) > -1$$

Using partial-fraction expansions, we get

$$Y(s) = -\frac{1}{s+1} + \frac{3}{s+3}$$

Taking the inverse Laplace transform of $Y(s)$, we obtain

$$y(t) = (-e^{-t} + 3e^{-3t})\,u(t)$$

3.26. If a continuous-time LTI system is BIBO stable, then show that the ROC of its system function $H(s)$ must contain the imaginary axis; that is, $s = j\omega$.

A continuous-time LTI system is BIBO stable if and only if its impulse response $h(t)$ is absolutely integrable, that is [Eq. (2.21)],

$$\int_{-\infty}^{\infty} |h(t)|\,dt < \infty$$

By Eq. (3.3)

$$H(s) = \int_{-\infty}^{\infty} h(t)e^{-st}\,dt$$

Let $s = j\omega$. Then

$$|H(j\omega)| = \left| \int_{-\infty}^{\infty} h(t)e^{-j\omega t}\,dt \right| \le \int_{-\infty}^{\infty} \left| h(t)e^{-j\omega t} \right| dt = \int_{-\infty}^{\infty} |h(t)|\,dt < \infty$$

Therefore, we see that if the system is stable, then $H(s)$ converges for $s = j\omega$. That is, for a stable continuous-time LTI system, the ROC of $H(s)$ must contain the imaginary axis $s = j\omega$.

3.27 Using the Laplace transfer, redo Prob. 2.14

(a) Using Eqs. (3.36) and (3.41), we have

$$Y(s) = X(s)H_1(s)H_2(s) = X(s)H(s)$$

where $H(s) = H_1(s)H_2(s)$ is the system function of the overall system. Now from Table 3-1 we have

$$h_1(t) = e^{-2t}u(t) \leftrightarrow H_1(s) = \frac{1}{s+2} \qquad \text{Re}(s) > -2$$

$$h_2(t) = 2e^{-t}u(t) \leftrightarrow H_2(s) = \frac{2}{s+1} \qquad \text{Re}(s) > -1$$

Hence,

$$H(s) = H_1(s)H_2(s) = \frac{2}{(s+1)(s+2)} = \frac{2}{s+1} - \frac{2}{s+2} \qquad \text{Re}(s) > -1$$

Taking the inverse Laplace transfer of $H(s)$, we get

$$h(t) = 2(e^{-t} - e^{-2t})u(t)$$

(b)   Since the ROC of $H(s)$, $\text{Re}(s) > -1$, contains the $j\omega$-axis, the overall system is stable.

3.28. Using the Laplace transform, redo Prob. 2.23.

The system is described by

$$\frac{dy(t)}{dt} + ay(t) = x(t)$$

Taking the Laplace transform of the above equation, we obtain

$$sY(s) + aY(s) = X(s) \qquad \text{or} \qquad (s+a)Y(s) = X(s)$$

Hence, the system function $H(s)$ is

$$H(s) = \frac{Y(s)}{X(s)} = \frac{1}{s+a}$$

Assuming the system is causal and taking the inverse Laplace transform of $H(s)$, the impulse response $h(t)$ is

$$h(t) = e^{-at}u(t)$$

which is the same as Eq. (2.124).

3.29. Using the Laplace transform, redo Prob. 2.25.

The system is described by

$$y'(t) + 2y(t) = x(t) + x'(t)$$

Taking the Laplace transform of the above equation, we get

$$sY(s) + 2Y(s) = X(s) + sX(s)$$

or
$$(s+2)Y(s) = (s+1)X(s)$$

Hence, the system function $H(s)$ is

$$H(s) = \frac{Y(s)}{X(s)} = \frac{s+1}{s+2} = \frac{s+2-1}{s+2} = 1 - \frac{1}{s+2}$$

Assuming the system is causal and taking the inverse Laplace transform of $H(s)$, the impulse response $h(t)$ is

$$h(t) = \delta(t) - e^{-2t}u(t)$$

3.30. Consider a continuous-time LTI system for which the input $x(t)$ and output $y(t)$ are related by

$$y''(t) + y'(t) - 2y(t) = x(t) \tag{3.86}$$

(a)   Find the system function $H(s)$.

(b)   Determine the impulse response $h(t)$ for each of the following three cases: (i) the system is causal, (ii) the system is stable, (iii) the system is neither causal nor stable.

(a)   Taking the Laplace transform of Eq. (3.86), we have

$$s^2Y(s) + sY(s) - 2Y(s) = X(s)$$

or
$$(s^2 + s - 2)Y(s) = X(s)$$

Hence, the system function $H(s)$ is

$$H(s) = \frac{Y(s)}{X(s)} = \frac{1}{s^2 + s - 2} = \frac{1}{(s+2)(s-1)}$$

(b) Using partial-fraction expansions, we get

$$H(s) = \frac{1}{(s+2)(s-2)} = -\frac{1}{3}\frac{1}{s+2} + \frac{1}{3}\frac{1}{s-1}$$

(i) If the system is causal, then $h(t)$ is causal (that is, a right-sided signal) and the ROC of $H(s)$ is $\mathrm{Re}(s) > 1$. Then from Table 3-1 we get

$$h(t) = -\frac{1}{3}(e^{-2t} - e^{t})u(t)$$

(ii) If the system is stable, then the ROC of $H(s)$ must contain the $j\omega$-axis. Consequently the ROC of $H(s)$ is $-2 < \mathrm{Re}(s) < 1$. Thus, $h(t)$ is two-sided and from Table 3-1 we get

$$h(t) = -\frac{1}{3}e^{-2t}u(t) - \frac{1}{3}e^{t}u(-t)$$

(iii) If the system is neither causal nor stable, then the ROC of $H(s)$ is $\mathrm{Re}(s) < -2$. Then $h(t)$ is noncausal (that is, a left-sided signal) and from Table 3-1 we get

$$h(t) = \frac{1}{3}e^{-2t}u(-t) - \frac{1}{3}e^{t}u(-t)$$

3.31. The feedback interconnection of two causal subsystems with system functions $F(s)$ and $G(s)$ is depicted in Fig. 3-13. Find the overall system function $H(s)$ for this feedback system.

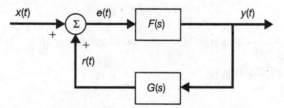

Fig. 3-13 Feedback system.

Let
$$x(t) \leftrightarrow X(s) \qquad y(t) \leftrightarrow Y(s) \qquad r(t) \leftrightarrow R(s) \qquad e(t) \leftrightarrow E(s)$$

Then,

$$Y(s) = E(s)F(s) \tag{3.87}$$
$$R(s) = Y(s)G(s) \tag{3.88}$$

Since

$$e(t) = x(t) + r(t)$$

we have

$$E(s) = X(s) + R(s) \tag{3.89}$$

Substituting Eq. (3.88) into Eq. (3.89) and then substituting the result into Eq. (3.87), we obtain

$$Y(s) = [X(s) + Y(s)G(s)]F(s)$$

or

$$[1 - F(s)G(s)]\, Y(s) = F(s)X(s)$$

Thus, the overall system function is

$$H(s) = \frac{Y(s)}{X(s)} = \frac{F(s)}{1 - F(s)G(s)} \tag{3.90}$$

Unilateral Laplace Transform

3.32. Verify Eqs. (3.44) and (3.45); that is,

(a) $\dfrac{dx(t)}{dt} \leftrightarrow sX_I(s) - x(0^-)$

(b) $\dfrac{d^2x(t)}{dt^2} \leftrightarrow s^2X_I(s) - sx(0^-) - x'(0^-)$

(a) Using Eq. (3.43) and integrating by parts, we obtain

$$\mathscr{L}_I\left\{\frac{dx(t)}{dt}\right\} = \int_{0^-}^{\infty} \frac{dx(t)}{dt} e^{-st} \, dt$$
$$= x(t)e^{-st}\Big|_{0^-}^{\infty} + s\int_{0^-}^{\infty} x(t)e^{-st} \, dt$$
$$= -x(0^-) + sX_I(s) \qquad \text{Re}(s) > 0$$

Thus, we have

$$\frac{dx(t)}{dt} \leftrightarrow sX_I(s) - x(0^-)$$

(b) Applying the above property to signal $x'(t) = dx(t)/dt$, we obtain

$$\frac{d^2x(t)}{dt^2} = \frac{d}{dt}\frac{dx(t)}{dt} \leftrightarrow s[sX_I(s) - x(0^-)] - x'(0^-)$$
$$= s^2X_I(s) - sx(0^-) - x'(0^-)$$

Note that Eq. (3.46) can be obtained by continued application of the above procedure.

3.33. Verify Eqs. (3.47) and (3.48); that is,

(a) $\displaystyle\int_{0^-}^{t} x(\tau)d\tau \leftrightarrow \frac{1}{s}X_I(s)$

(b) $\displaystyle\int_{-\infty}^{t} x(\tau)d\tau \leftrightarrow \frac{1}{s}X_I(s) + \frac{1}{s}\int_{-\infty}^{0^-} x(\tau)d\tau$

(a) Let

$$g(t) = \int_{0^-}^{t} x(\tau)d\tau$$

Then

$$\frac{dg(t)}{dt} = x(t) \qquad \text{and} \qquad g(0^-) = 0$$

Now if

$$g(t) \leftrightarrow G_I(s)$$

then by Eq. (3.44)

$$X_I(s) = sG_I(s) - g(0^-) = sG_I(s)$$

Thus,

$$G_I(s) = \frac{1}{s}X_I(s)$$

or

$$\int_{0^-}^{t} x(\tau)d\tau \leftrightarrow \frac{1}{s}X_I(s)$$

(b) We can write

$$\int_{-\infty}^{t} x(\tau)\,d\tau = \int_{-\infty}^{0^-} x(\tau)\,d\tau + \int_{0^-}^{t} x(\tau)\,d\tau$$

Note that the first term on the right-hand side is a constant. Thus, taking the unilateral Laplace transform of the above equation and using Eq. (3.47), we get

$$\int_{-\infty}^{t} x(\tau)\,d\tau \leftrightarrow \frac{1}{s}X_I(s) + \frac{1}{s}\int_{-\infty}^{0^-} x(\tau)\,d\tau$$

3.34. (*a*) Show that the bilateral Laplace transform of $x(t)$ can be computed from two unilateral Laplace transforms.

(*b*) Using the result obtained in part (*a*), find the bilateral Laplace transform of $e^{-2|t|}$.

(*a*) The bilateral Laplace transform of $x(t)$ defined in Eq. (3.3) can be expressed as

$$X(s) = \int_{-\infty}^{\infty} x(t)e^{-st}dt = \int_{-\infty}^{0^-} x(t)e^{-st}dt + \int_{0^-}^{\infty} x(t)e^{-st}dt$$

$$= \int_{0^-}^{\infty} x(-t)e^{st}dt + \int_{0^-}^{\infty} x(t)e^{-st}dt \tag{3.91}$$

Now

$$\int_{0^-}^{\infty} x(t)e^{-st}dt = X_I(s) \qquad \mathrm{Re}(s) > \sigma^+ \tag{3.92}$$

Next, let

$$\mathcal{L}_I\{x(-t)\} = X_I^-(s) = \int_{0^-}^{\infty} x(-t)e^{-st}dt \qquad \mathrm{Re}(s) > \sigma^- \tag{3.93}$$

Then

$$\int_{0^-}^{\infty} x(-t)e^{st}dt = \int_{0^-}^{\infty} x(-t)e^{-(-s)t}dt = X_I^-(-s) \qquad \mathrm{Re}(s) < \sigma^- \tag{3.94}$$

Thus, substituting Eqs. (3.92) and (3.94) into Eq. (3.91), we obtain

$$X(s) = X_I(s) + X_I^-(-s) \qquad \sigma^+ < \mathrm{Re}(s) < \sigma^- \tag{3.95}$$

(*b*) $$x(t) = e^{-2|t|}$$

(1) $x(t) = e^{-2t}$ for $t > 0$, which gives

$$\mathcal{L}_I\{x(t)\} = X_I(s) = \frac{1}{s+2} \qquad \mathrm{Re}(s) > -2$$

(2) $x(t) = e^{2t}$ for $t < 0$. Then $x(-t) = e^{-2t}$ for $t > 0$, which gives

$$\mathcal{L}_I\{x(-t)\} = X_I^-(s) = \frac{1}{s+2} \qquad \mathrm{Re}(s) > -2$$

Thus,

$$X_I^-(-s) = \frac{1}{-s+2} = -\frac{1}{s-2} \qquad \mathrm{Re}(s) < 2$$

(3) According to Eq. (3.95), we have

$$X(s) = X_I(s) + X_I^-(-s) = \frac{1}{s+2} - \frac{1}{s-2}$$

$$= -\frac{4}{s^2-4} \qquad -2 < \mathrm{Re}(s) < 2 \tag{3.96}$$

which is equal to Eq. (3.70), with $a = 2$, in Prob. 3.6.

3.35. Show that

(*a*) $$x(0^+) = \lim_{s\to\infty} sX_I(s) \tag{3.97}$$

(*b*) $$\lim_{t\to\infty} x(t) = \lim_{s\to 0} sX_I(s) \tag{3.98}$$

Equation (3.97) is called the *initial value theorem*, while Eq. (3.98) is called the *final value theorem* for the unilateral Laplace transform.

(a) Using Eq. (3.44), we have

$$sX_I(s) - x(0^-) = \int_{0^-}^{\infty} \frac{dx(t)}{dt} e^{-st} dt$$

$$= \int_{0^-}^{0^+} \frac{dx(t)}{dt} e^{-st} dt + \int_{0^+}^{\infty} \frac{dx(t)}{dt} e^{-st} dt$$

$$= x(t) \Big|_{0^-}^{0^+} + \int_{0^+}^{\infty} \frac{dx(t)}{dt} e^{-st} dt$$

$$= x(0^+) - x(0^-) + \int_{0^+}^{\infty} \frac{dx(t)}{dt} e^{-st} dt$$

Thus,

$$sX_I(s) = x(0^+) = \int_{0^+}^{\infty} \frac{dx(t)}{dt} e^{-st} dt$$

and

$$\lim_{s \to \infty} sX_I(s) = x(0^+) + \lim_{s \to \infty} \int_{0^+}^{\infty} \frac{dx(t)}{dt} e^{-st} dt$$

$$= x(0^+) + \int_{0^+}^{\infty} \frac{dx(t)}{dt} \left(\lim_{s \to \infty} e^{-st} \right) dt = x(0^+)$$

since $\lim_{s \to \infty} e^{-st} = 0$.

(b) Again using Eq. (3.44), we have

$$\lim_{s \to 0} [sX_I(s) - x(0^-)] = \lim_{s \to 0} \int_{0^-}^{\infty} \frac{dx(t)}{dt} e^{-st} dt$$

$$= \int_{0^-}^{\infty} \frac{dx(t)}{dt} \left(\lim_{s \to 0} e^{-st} \right) dt$$

$$= \int_{0^-}^{\infty} \frac{dx(t)}{dt} dt = x(t) \Big|_{0^-}^{\infty}$$

$$= \lim_{t \to \infty} x(t) - x(0^-)$$

Since

$$\lim_{s \to 0} [sX_I(s) - x(0^-)] = \lim_{s \to 0} [sX_I(s)] - x(0^-)$$

we conclude that

$$\lim_{t \to \infty} x(t) = \lim_{s \to 0} sX_I(s)$$

3.36. The unilateral Laplace transform is sometimes defined as

$$\mathcal{L}_+\{x(t)\} = X_I^+(s) = \int_{0^+}^{\infty} x(t) e^{-st} dt \qquad (3.99)$$

with 0^+ as the lower limit. (This definition is sometimes referred to as the 0^+ definition.)

(a) Show that

$$\mathcal{L}_+ \left\{ \frac{dx(t)}{dt} \right\} = sX_I^+(s) - x(0^+) \qquad \text{Re}(s) > 0 \qquad (3.100)$$

(b) Show that

$$\mathcal{L}_+\{u(t)\} = \frac{1}{s} \qquad (3.101)$$

$$\mathcal{L}_+\{\delta(t)\} = 0 \qquad (3.102)$$

(a) Let $x(t)$ have unilateral Laplace transform $X_I^+(s)$. Using Eq. (3.99) and integrating by parts, we obtain

$$\mathcal{L}_+ \left\{ \frac{dx(t)}{dt} \right\} = \int_{0^+}^{\infty} \frac{dx(t)}{dt} e^{-st} dt$$

$$= x(t) e^{-st} \Big|_{0^+}^{\infty} + s \int_{0^+}^{\infty} x(t) e^{-st} dt$$

$$= -x(0^+) + sX_I^+(s) \qquad \text{Re}(s) > 0$$

Thus, we have

$$\frac{dx(t)}{dt} \leftrightarrow sX_I^+(s) - x(0^+)$$

(b) By definition (3.99)

$$\mathcal{L}_+\{u(t)\} = \int_{0^+}^{\infty} u(t)e^{-st}dt = \int_{0^+}^{\infty} e^{-st}dt$$
$$= -\frac{1}{s}e^{-st}\Big|_{0^+}^{\infty} = \frac{1}{s} \qquad \text{Re}(s) > 0$$

From Eq. (1.30) we have

$$\delta(t) = \frac{du(t)}{dt} \tag{3.103}$$

Taking the 0^+ unilateral Laplace transform of Eq. (3.103) and using Eq. (3.100), we obtain

$$\mathcal{L}_+\{\delta(t)\} = s\frac{1}{s} - u(0^+) = 1 - 1 = 0$$

This is consistent with Eq. (1.21); that is,

$$\mathcal{L}_+\{\delta(t)\} = \int_{0^+}^{\infty} \delta(t)e^{-st}dt = 0$$

Note that taking the 0^- unilateral Laplace transform of Eq. (3.103) and using Eq. (3.44), we obtain

$$\mathcal{L}_-\{\delta(t)\} = s\frac{1}{s} - u(0^-) = 1 - 0 = 1$$

Application of Unilateral Laplace Transform

3.37. Using the unilateral Laplace transform, redo Prob. 2.20.

The system is described by

$$y'(t) + ay(t) = x(t) \tag{3.104}$$

with $y(0) = y_0$ and $x(t) = Ke^{-bt}u(t)$.

Assume that $y(0) = y(0^-)$. Let

$$y(t) \leftrightarrow Y_I(s)$$

Then from Eq. (3.44)

$$y'(t) \leftrightarrow sY_I(s) - y(0^-) = sY_I(s) - y_0$$

From Table 3-1 we have

$$x(t) \leftrightarrow X_I(s) = \frac{K}{s+b} \qquad \text{Re}(s) > -b$$

Taking the unilateral Laplace transform of Eq. (3.104), we obtain

$$[sY_I(s) - y_0] + aY_I(s) = \frac{K}{s+b}$$

or

$$(s+a)Y_I(s) = y_0 + \frac{K}{s+b}$$

Thus,

$$Y_I(s) = \frac{y_0}{s+a} + \frac{K}{(s+a)(s+b)}$$

Using partial-fraction expansions, we obtain

$$Y_I(s) = \frac{y_0}{s+a} + \frac{K}{a-b}\left(\frac{1}{s+b} - \frac{1}{s+a}\right)$$

Taking the inverse Laplace transform of $Y_I(s)$, we obtain

$$y(t) = \left[y_0 e^{-at} + \frac{K}{a-b}(e^{-bt} - e^{-at})\right]u(t)$$

which is the same as Eq. (2.107). Noting that $y(0^+) = y(0) = y(0^-) = y_0$, we write $y(t)$ as

$$y(t) = y_0 e^{-at} + \frac{K}{a-b}(e^{-bt} - e^{-at}) \qquad t \geq 0$$

3.38. Solve the second-order linear differential equation

$$y''(t) + 5y'(t) + 6y(t) = x(t) \qquad\qquad (3.105)$$

with the initial conditions $y(0) = 2$, $y'(0) = 1$, and $x(t) = e^{-t}u(t)$.

Assume that $y(0) = y(0^-)$ and $y'(0) = y'(0^-)$. Let

$$y(t) \leftrightarrow Y_I(s)$$

Then from Eqs. (3.44) and (3.45)

$$y'(t) \leftrightarrow sY_I(s) - y(0^-) = sY_I(s) - 2$$
$$y''(t) \leftrightarrow s^2 Y_I(s) - sy(0^-) - y'(0^-) = s^2 Y_I(s) - 2s - 1$$

From Table 3-1 we have

$$x(t) \leftrightarrow X_I(s) = \frac{1}{s+1}$$

Taking the unilateral Laplace transform of Eq. (3.105), we obtain

$$[s^2 Y_I(s) - 2s - 1] + 5[sY_I(s) - 2] + 6Y_I(s) = \frac{1}{s+1}$$

or

$$(s^2 + 5s + 6)Y_I(s) = \frac{1}{s+1} + 2s + 11 = \frac{2s^2 + 13s + 12}{s+1}$$

Thus,

$$Y_I(s) = \frac{2s^2 + 13s + 12}{(s+1)(s^2 + 5s + 6)} = \frac{2s^2 + 13s + 12}{(s+1)(s+2)(s+3)}$$

Using partial-fraction expansions, we obtain

$$Y_I(s) = \frac{1}{2}\frac{1}{s+1} + 6\frac{1}{s+2} - \frac{9}{2}\frac{1}{s+3}$$

Taking the inverse Laplace transform of $Y_I(s)$, we have

$$y(t) = \left(\frac{1}{2}e^{-t} + 6e^{-2t} - \frac{9}{2}e^{-3t}\right)u(t)$$

Notice that $y(0^+) = 2 = y(0)$ and $y'(0^+) = 1 = y'(0)$; and we can write $y(t)$ as

$$y(t) = \frac{1}{2}e^{-t} + 6e^{-2t} - \frac{9}{2}e^{-3t} \qquad t \geq 0$$

3.39. Consider the *RC* circuit shown in Fig. 3-14(*a*). The switch is closed at $t = 0$. Assume that there is an initial voltage on the capacitor and $v_c(0^-) = v_0$.

(*a*) Find the current $i(t)$.

(*b*) Find the voltage across the capacitor $v_c(t)$.

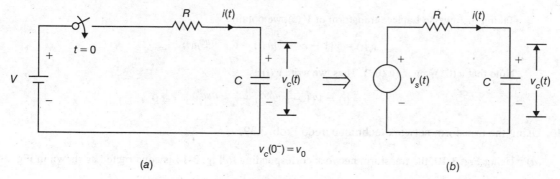

$$v_c(0^-) = v_0$$

(*a*)　　　　　　　　　　　　　　　　　　　　　(*b*)

Fig. 3-14 *RC* circuit.

(*a*) With the switching action, the circuit shown in Fig. 3-14(*a*) can be represented by the circuit shown in Fig. 3-14(*b*) with $v_s(t) = Vu(t)$. When the current $i(t)$ is the output and the input is $v_s(t)$, the differential equation governing the circuit is

$$Ri(t) + \frac{1}{C}\int_{-\infty}^{t} i(\tau)\, d\tau = v_s(t) \tag{3.106}$$

Taking the unilateral Laplace transform of Eq. (3.106) and using Eq. (3.48), we obtain

$$RI(s) + \frac{1}{C}\left[\frac{1}{s}I(s) + \frac{1}{s}\int_{-\infty}^{0^-} i(\tau)\, d\tau\right] = \frac{V}{s} \tag{3.107}$$

where　　　　　$I(s) = \mathcal{L}_I\{i(t)\}$

Now　　　　　　$v_c(t) = \frac{1}{C}\int_{-\infty}^{t} i(\tau)\, d\tau$

and　　　　　　$v_c(0^-) = \frac{1}{C}\int_{-\infty}^{0^-} i(\tau)\, d\tau = v_0$

Hence, Eq. (3.107) reduces to

$$\left(R + \frac{1}{Cs}\right)I(s) + \frac{v_0}{s} = \frac{V}{s}$$

Solving for $I(s)$, we obtain

$$I(s) = \frac{V - v_0}{s}\frac{1}{R + 1/Cs} = \frac{V - v_0}{R}\frac{1}{s + 1/RC}$$

Taking the inverse Laplace transform of $I(s)$, we get

$$i(t) = \frac{V - v_0}{R}e^{-t/RC}u(t)$$

(*b*) When $v_c(t)$ is the output and the input is $v_s(t)$, the differential equation governing the circuit is

$$\frac{dv_c(t)}{dt} + \frac{1}{RC}v_c(t) = \frac{1}{RC}v_s(t) \tag{3.108}$$

Taking the unilateral Laplace transform of Eq. (3.108) and using Eq. (3.44), we obtain

$$sV_c(s) - v_c(0^-) + \frac{1}{RC}V_c(s) = \frac{1}{RC}\frac{V}{s}$$

or　　　　　　$\left(s + \frac{1}{RC}\right)V_c(s) = \frac{1}{RC}\frac{V}{s} + v_0$

Solving for $V_c(s)$, we have

$$V_c(s) = \frac{V}{RC} \frac{1}{s(s+1/RC)} + \frac{v_0}{s+1/RC}$$

$$= V\left(\frac{1}{s} - \frac{1}{s+1/RC}\right) + \frac{v_0}{s+1/RC}$$

Taking the inverse Laplace transform of $V_c(s)$, we obtain

$$v_c(t) = V[1 - e^{-t/RC}]u(t) + v_0 e^{-t/RC}u(t)$$

Note that $v_c(0^+) = v_0 = v_c(0^-)$. Thus, we write $v_c(t)$ as

$$v_c(t) = V(1 - e^{-t/RC}) + v_0 e^{-t/RC} \qquad t \geq 0$$

3.40. Using the transform network technique, redo Prob. 3.39.

(a) Using Fig. 3-10, the transform network corresponding to Fig. 3-14 is constructed as shown in Fig. 3-15.

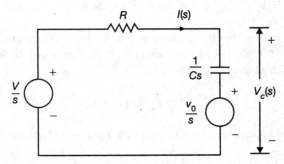

Fig. 3-15 Transform circuit.

Writing the voltage law for the loop, we get

$$\left(R + \frac{1}{Cs}\right)I(s) + \frac{v_0}{s} = \frac{V}{s}$$

Solving for $I(s)$, we have

$$I(s) = \frac{V - v_0}{s} \frac{1}{R+1/Cs} = \frac{V - v_0}{R} \frac{1}{s+1/RC}$$

Taking the inverse Laplace transform of $I(s)$, we obtain

$$i(t) = \frac{V - v_0}{R} e^{-t/RC}u(t)$$

(b) From Fig. 3.15 we have

$$V_c(s) = \frac{1}{Cs}I(s) + \frac{v_0}{s}$$

Substituting $I(s)$ obtained in part (a) into the above equation, we get

$$V_c(s) = \frac{V - v_0}{RC} \frac{1}{s(s+1/RC)} + \frac{v_0}{s}$$

$$= (V - v_0)\left(\frac{1}{s} - \frac{1}{s+1/RC}\right) + \frac{v_0}{s}$$

$$= V\left(\frac{1}{s} - \frac{1}{s+1/RC}\right) + \frac{v_0}{s+1/RC}$$

Taking the inverse Laplace transform of $V_c(s)$, we have

$$v_c(t) = V(1 - e^{-t/RC})u(t) + v_0 e^{-t/RC}u(t)$$

3.41. In the circuit in Fig. 3-16(a) the switch is in the closed position for a long time before it is opened at $t = 0$. Find the inductor current $i(t)$ for $t \geq 0$.

When the switch is in the closed position for a long time, the capacitor voltage is charged to 10 V and there is no current flowing in the capacitor. The inductor behaves as a short circuit, and the inductor current is $\frac{10}{5} = 2$ A.

Thus, when the switch is open, we have $i(0^-) = 2$ and $v_c(0^-) = 10$; the input voltage is 10 V, and therefore it can be represented as $10u(t)$. Next, using Fig. 3-10, we construct the transform circuit as shown in Fig. 3-16(b).

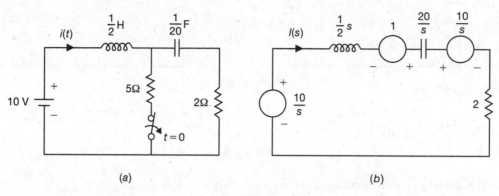

(a) (b)

Fig. 3-16

From Fig. 3-16(b) the loop equation can be written as

$$\frac{1}{2}sI(s) - 1 + 2I(s) + \frac{20}{s}I(s) + \frac{10}{s} = \frac{10}{s}$$

or

$$\left(\frac{1}{2}s + 2 + \frac{20}{s}\right)I(s) = 1$$

Hence,

$$I(s) = \frac{1}{\frac{1}{2}s + 2 + 20/s} = \frac{2s}{s^2 + 4s + 40}$$

$$= \frac{2(s+2) - 4}{(s+2)^2 + 6^2} = 2\frac{(s+2)}{(s+2)^2 + 6^2} - \frac{2}{3}\frac{6}{(s+2)^2 + 6^2}$$

Taking the inverse Laplace transform of $I(s)$, we obtain

$$i(t) = e^{-2t}\left(2\cos 6t - \frac{2}{3}\sin 6t\right)u(t)$$

Note that $i(0^+) = 2 = i(0^-)$; that is, there is no discontinuity in the inductor current before and after the switch is opened. Thus, we have

$$i(t) = e^{-2t}\left(2\cos 6t - \frac{2}{3}\sin 6t\right) \qquad t \geq 0$$

3.42. Consider the circuit shown in Fig. 3-17(a). The two switches are closed simultaneously at $t = 0$. The voltages on capacitors C_1 and C_2 before the switches are closed are 1 and 2 V, respectively.

(a) Find the currents $i_1(t)$ and $i_2(t)$.

(b) Find the voltages across the capacitors at $t = 0^+$.

(a) From the given initial conditions, we have

$$v_{C_1}(0^-) = 1 \text{ V} \qquad \text{and} \qquad v_{C_2}(0^-) = 2 \text{ V}$$

Thus, using Fig. 3-10, we construct a transform circuit as shown in Fig. 3-17(b). From

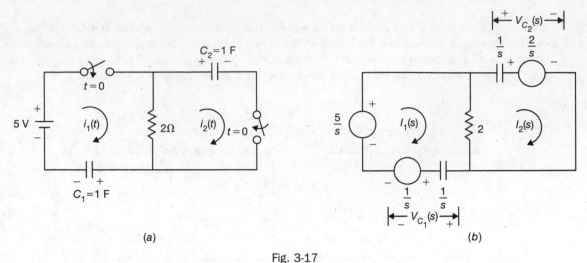

(a)   (b)

Fig. 3-17

Fig. 3-17(b) the loop equations can be written directly as

$$\left(2 + \frac{1}{s}\right)I_1(s) - 2I_2(s) = \frac{4}{s}$$

$$-2I_1(s) + \left(2 + \frac{1}{s}\right)I_2(s) = -\frac{2}{s}$$

Solving for $I_1(s)$ and $I_2(s)$ yields

$$I_1(s) = \frac{s+1}{s+\dfrac{1}{4}} = \frac{s + \dfrac{1}{4} + \dfrac{3}{4}}{s + \dfrac{1}{4}} = 1 + \frac{3}{4}\frac{1}{s + \dfrac{1}{4}}$$

$$I_2(s) = \frac{s - \dfrac{1}{2}}{s + \dfrac{1}{4}} = \frac{s + \dfrac{1}{4} - \dfrac{3}{4}}{s + \dfrac{1}{4}} = 1 - \frac{3}{4}\frac{1}{s + \dfrac{1}{4}}$$

Taking the inverse Laplace transforms of $I_1(s)$ and $I_2(s)$, we get

$$i_1(t) = \delta(t) + \frac{3}{4}e^{-t/4}u(t)$$

$$i_2(t) = \delta(t) - \frac{3}{4}e^{-t/4}u(t)$$

(b)   From Fig. 3-17(b) we have

$$V_{C_1}(s) = \frac{1}{s}I_1(s) + \frac{1}{s}$$

$$V_{C_2}(s) = \frac{1}{s}I_2(s) + \frac{2}{s}$$

Substituting $I_1(s)$ and $I_2(s)$ obtained in part (a) into the above expressions, we get

$$V_{C_1}(s) = \frac{1}{s}\frac{s+1}{s+\dfrac{1}{4}} + \frac{1}{s}$$

$$V_{C_2}(s) = \frac{1}{s}\frac{s - \dfrac{1}{2}}{s + \dfrac{1}{4}} + \frac{2}{s}$$

Then, using the initial value theorem (3.97), we have

$$v_{C_1}(0^+) = \lim_{s \to \infty} sV_{C_1}(s) = \lim_{s \to \infty} \frac{s+1}{s+\dfrac{1}{4}} + 1 = 1 + 1 = 2\,\text{V}$$

$$v_{C_2}(0^+) = \lim_{s \to \infty} sV_{C_2}(s) = \lim_{s \to \infty} \frac{s-\dfrac{1}{2}}{s+\dfrac{1}{4}} + 2 = 1 + 2 = 3\,\text{V}$$

Note that $v_{C_1}(0^+) \neq v_{C_1}(0^-)$ and $v_{C_2}(0^+) \neq v_{C_2}(0^-)$. This is due to the existence of a capacitor loop in the circuit resulting in a sudden change in voltage across the capacitors. This step change in voltages will result in impulses in $i_1(t)$ and $i_2(t)$. Circuits having a capacitor loop or an inductor star connection are known as *degenerative circuits*.

SUPPLEMENTARY PROBLEMS

3.43. Find the Laplace transform of the following $x(t)$:

    (a) $x(t) = \sin \omega_0 t\, u(t)$

    (b) $x(t) = \cos(\omega_0 t + \phi)\, u(t)$

    (c) $x(t) = e^{-at}u(t) - e^{at}u(-t)$

    (d) $x(t) = 1$

    (e) $x(t) = \operatorname{sgn} t$

3.44. Find the Laplace transform of $x(t)$ given by

$$x(t) = \begin{cases} 1 & t_1 \leq t \leq t_2 \\ 0 & \text{otherwise} \end{cases}$$

3.45. Show that if $x(t)$ is a left-sided signal and $X(s)$ converges for some value of s, then the ROC of $X(s)$ is of the form

$$\operatorname{Re}(s) < \sigma_{\min}$$

where $\sigma_{\min}$ equals the minimum real part of any of the poles of $X(s)$.

3.46. Verify Eq. (3.21); that is,

$$-tx(t) \leftrightarrow \frac{dX(s)}{ds} \qquad R' = R$$

3.47. Show the following properties for the Laplace transform:

    (a) If $x(t)$ is even, then $X(-s) = X(s)$; that is, $X(s)$ is also even.

    (b) If $x(t)$ is odd, then $X(-s) = -X(s)$; that is, $X(s)$ is also odd.

    (c) If $x(t)$ is odd, then there is a zero in $X(s)$ at $s = 0$.

3.48. Find the Laplace transform of

$$x(t) = (e^{-t}\cos 2t - 5e^{-2t})u(t) + \tfrac{1}{2}e^{2t}u(-t)$$

3.49. Find the inverse Laplace transform of the following $X(s)$:

(a) $X(s) = \dfrac{1}{s(s+1)^2}, \operatorname{Re}(s) > -1$

(b) $X(s) = \dfrac{1}{s(s+1)^2}, -1 < \operatorname{Re}(s) < 0$

(c) $X(s) = \dfrac{1}{s(s+1)^2}, \operatorname{Re}(s) < -1$

(d) $X(s) = \dfrac{s+1}{s^2 + 4s + 13}, \operatorname{Re}(s) > -2$

(e) $X(s) = \dfrac{s}{(s^2 + 4)^2}, \operatorname{Re}(s) > 0$

(f) $X(s) = \dfrac{s}{s^3 + 2s^2 + 9s + 18}, \operatorname{Re}(s) > -2$

3.50. Using the Laplace transform, redo Prob. 2.46.

3.51. Using the Laplace transform, show that

(a) $x(t) * \delta(t) = x(t)$

(b) $x(t) * \delta'(t) = x'(t)$

3.52. Using the Laplace transform, redo Prob. 2.54.

3.53. Find the output $y(t)$ of the continuous-time LTI system with

$$h(t) = e^{-2t}u(t)$$

for the each of the following inputs:

(a) $x(t) = e^{-t}u(t)$

(b) $x(t) = e^{-t}u(-t)$

3.54. The step response of an continuous-time LTI system is given by $(1 - e^{-t})\,u(t)$. For a certain unknown input $x(t)$, the output $y(t)$ is observed to be $(2 - 3e^{-t} + e^{-3t})u(t)$. Find the input $x(t)$.

3.55. Determine the overall system function $H(s)$ for the system shown in Fig. 3-18.

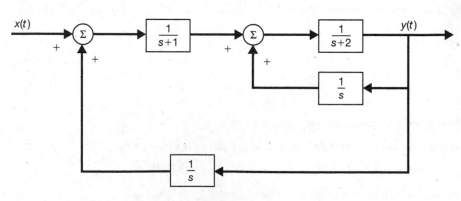

Fig. 3-18

3.56. If $x(t)$ is a periodic function with fundamental period T, find the unilateral Laplace transform of $x(t)$.

3.57. Find the unilateral Laplace transforms of the periodic signals shown in Fig. 3-19.

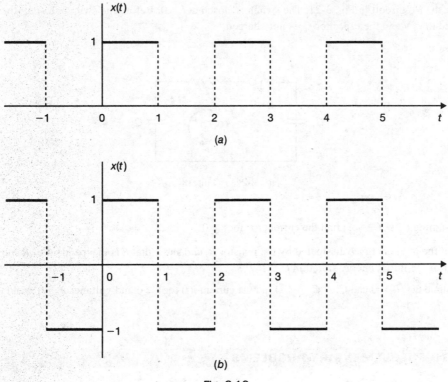

(a)

(b)

Fig. 3-19

3.58. Using the unilateral Laplace transform, find the solution of

$$y''(t) - y'(t) - 6y(t) = e^t$$

with the initial conditions $y(0) = 1$ and $y'(0) = 0$ for $t \geq 0$.

3.59. Using the unilateral Laplace transform, solve the following simultaneous differential equations:

$$y'(t) + y(t) + x'(t) + x(t) = 1$$

$$y'(t) - y(t) - 2x(t) = 0$$

with $x(0) = 0$ and $y(0) = 1$ for $t \geq 0$.

3.60. Using the unilateral Laplace transform, solve the following integral equations:

(a) $y(t) = 1 + a\int_0^t y(\tau)\, d\tau, t \geq 0$

(b) $y(t) = e^t \left[1 + \int_0^t e^{-\tau} y(\tau)\, d\tau \right], t \geq 0$

3.61. Consider the *RC* circuit in Fig. 3-20. The switch is closed at $t = 0$. The capacitor voltage before the switch closing is v_0. Find the capacitor voltage for $t \geq 0$.

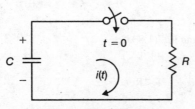

Fig. 3-20 *RC* circuit.

3.62. Consider the *RC* circuit in Fig. 3-21. The switch is closed at $t = 0$. Before the switch closing, the capacitor C_1 is charged to v_0 V and the capacitor C_2 is not charged.

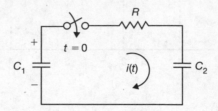

Fig. 3-21 *RC* circuit.

(a)   Assuming $C_1 = C_2 = C$, find the current $i(t)$ for $t \geq 0$.

(b)   Find the total energy E dissipated by the resistor R, and show that E is independent of R and is equal to half of the initial energy stored in C_1.

(c)   Assume that $R = 0$ and $C_1 = C_2 = C$. Find the current $i(t)$ for $t \geq 0$ and voltages $v_{C_1}(0^+)$ and $v_{C_2}(0^+)$.

ANSWERS TO SUPPLEMENTARY PROBLEMS

3.43. (a) $X(s) = \dfrac{\omega_0}{s^2 + \omega_0^2}, \mathrm{Re}(s) > 0$

(b) $X(s) = \dfrac{s \cos \phi - \omega_0 \sin \phi}{s^2 + \omega_0^2}, \mathrm{Re}(s) > 0$

(c)   If $a > 0$, $X(s) = \dfrac{2s}{s^2 - a^2}$, $-a < \mathrm{Re}(s) < a$. If $a < 0$, $X(s)$ does not exist since $X(s)$ does not have an ROC.

(d)   *Hint:* $x(t) = u(t) + u(-t)$

$X(s)$ does not exist since $X(s)$ does not have an ROC.

(e)   *Hint:* $x(t) = u(t) - u(-t)$

$X(s)$ does not exist since $X(s)$ does not have an ROC.

3.44. $X(s) = \dfrac{1}{s}[e^{-st_1} - e^{-st_2}]$, all s

3.45. *Hint:*   Proceed in a manner similar to Prob. 3.4.

3.46. *Hint:*   Differentiate both sides of Eq. (3.3) with respect to s.

3.47. *Hint:*

(a)   Use Eqs. (1.2) and (3.17).

(b)   Use Eqs. (1.3) and (3.17).

(c)   Use the result from part (b) and Eq. (1.83a).

3.48. $X(s) = \dfrac{s+1}{(s+1)^2 + 4} - \dfrac{5}{s+2} - \dfrac{1}{2} \dfrac{1}{s-2}, -1 < \mathrm{Re}(s) < 2$

3.49. (a) $x(t) = (1 - e^{-t} - te^{-t})u(t)$

    (b) $x(t) = -u(-t) - (1 + t)e^{-t}u(t)$

    (c) $x(t) = (-1 + e^{-t} + te^{-t})u(-t)$

    (d) $x(t) = e^{-2t}\left(\cos 3t - \dfrac{1}{3}\sin 3t\right)u(t)$

    (e) $x(t) = \dfrac{1}{4}t\sin 2t\, u(t)$

    (f) $x(t) = \left(-\dfrac{2}{13}e^{-2t} + \dfrac{2}{13}\cos 3t + \dfrac{3}{13}\sin 3t\right)u(t)$

3.50. *Hint:* Use Eq. (3.21) and Table 3-1.

3.51. *Hint:*

    (a) Use Eq. (3.21) and Table 3-1.

    (b) Use Eqs. (3.18) and (3.21) and Table 3-1.

3.52. *Hint:*

    (a) Find the system function $H(s)$ by Eq. (3.32) and take the inverse Laplace transform of $H(s)$.

    (b) Find the ROC of $H(s)$ and show that it does not contain the $j\omega$-axis.

3.53. (a) $y(t) = (e^{-t} - e^{-2t})u(t)$

    (b) $y(t) = e^{-t}u(-t) + e^{-2t}u(t)$

3.54. $x(t) = 2(1 - e^{-3t})u(t)$

3.55. *Hint:* Use the result from Prob. 3.31 to simplify the block diagram.

$$H(s) = \frac{2}{s^3 + 3s^2 + s - 2}$$

3.56. $X(s) = \dfrac{1}{1 - e^{-sT}}\displaystyle\int_{0^-}^{T} x(t)e^{-st}dt,\ \mathrm{Re}(s) > 0$

3.57. (a) $\dfrac{1}{s(1 + e^{-s})},\ \mathrm{Re}(s) > 0;$ (b) $\dfrac{1 - e^{-s}}{s(1 + e^{-s})},\ \mathrm{Re}(s) > 0$

3.58. $y(t) = -\dfrac{1}{6}e^{t} + \dfrac{2}{3}e^{-2t} + \dfrac{1}{2}e^{3t},\ t \geq 0$

3.59. $x(t) = e^{-t} - 1,\ y(t) = 2 - e^{-t},\ t \geq 0$

3.60. (a) $y(t) = e^{at},\ t \geq 0;$ (b) $y(t) = e^{2t},\ t \geq 0$

3.61. $v_c(t) = v_0 e^{-t/RC},\ t \geq 0$

3.62. (a) $i(t) = (v_0/R)e^{-2t/RC},\ t \geq 0$

    (b) $E = \dfrac{1}{4}v_0^2 C$

    (c) $i(t) = \dfrac{1}{2}v_0 C\,\delta(t),\ v_{C_1}(0^+) = v_0/2 \neq v_{C_1}(0^-) = v_0,\ v_{C_2}(0^+) = v_0/2 \neq v_{C_2}(0^-) = 0$

CHAPTER 4

The *z*-Transform and Discrete-Time LTI Systems

4.1 Introduction

In Chap. 3 we introduced the Laplace transform. In this chapter we present the z-transform, which is the discrete-time counterpart of the Laplace transform. The z-transform is introduced to represent discrete-time signals (or sequences) in the z-domain (z is a complex variable), and the concept of the system function for a discrete-time LTI system will be described. The Laplace transform converts integrodifferential equations into algebraic equations. In a similar manner, the z-transform converts difference equations into algebraic equations, thereby simplifying the analysis of discrete-time systems.

The properties of the z-transform closely parallel those of the Laplace transform. However, we will see some important distinctions between the z-transform and the Laplace transform.

4.2 The *z*-Transform

In Sec. 2.8 we saw that for a discrete-time LTI system with impulse response $h[n]$, the output $y[n]$ of the system to the complex exponential input of the form z^n is

$$y[n] = \mathbf{T}\{z^n\} = H(z)z^n \tag{4.1}$$

where

$$H(z) = \sum_{n=-\infty}^{\infty} h[n]z^{-n} \tag{4.2}$$

A. Definition:

The function $H(z)$ in Eq. (4.2) is referred to as the z-transform of $h[n]$. For a general discrete-time signal $x[n]$, the z-transform $X(z)$ is defined as

$$X(z) = \sum_{n=-\infty}^{\infty} x[n]z^{-n} \tag{4.3}$$

The variable z is generally complex-valued and is expressed in polar form as

$$z = re^{j\Omega} \tag{4.4}$$

where r is the magnitude of z and Ω is the angle of z. The z-transform defined in Eq. (4.3) is often called the *bilateral* (or *two-sided*) z-transform in contrast to the *unilateral* (or *one-sided*) z-transform, which is defined as

$$X_I(z) = \sum_{n=0}^{\infty} x[n]z^{-n} \tag{4.5}$$

Clearly the bilateral and unilateral z-transforms are equivalent only if $x[n] = 0$ for $n < 0$. The unilateral z-transform is discussed in Sec. 4.8. We will omit the word "bilateral" except where it is needed to avoid ambiguity.

As in the case of the Laplace transform, Eq. (4.3) is sometimes considered an operator that transforms a sequence $x[n]$ into a function $X(z)$, symbolically represented by

$$X(z) = \mathfrak{Z}\{x[n]\} \tag{4.6}$$

The $x[n]$ and $X(z)$ are said to form a z-transform pair denoted as

$$x[n] \leftrightarrow X(z) \tag{4.7}$$

B. The Region of Convergence:

As in the case of the Laplace transform, the range of values of the complex variable z for which the z-transform converges is called the region of convergence. To illustrate the z-transform and the associated ROC let us consider some examples.

EXAMPLE 4.1 Consider the sequence

$$x[n] = a^n u[n] \qquad a \text{ real} \tag{4.8}$$

Then by Eq. (4.3) the z-transform of $x[n]$ is

$$X(z) = \sum_{n=-\infty}^{\infty} a^n u[n]z^{-n} = \sum_{n=0}^{\infty} (az^{-1})^n$$

For the convergence of $X(z)$ we require that

$$\sum_{n=0}^{\infty} \left| az^{-1} \right|^n < \infty$$

Thus, the ROC is the range of values of z for which $\left| az^{-1} \right| < 1$ or, equivalently, $|z| > |a|$. Then

$$X(z) = \sum_{n=0}^{\infty} (az^{-1})^n = \frac{1}{1 - az^{-1}} \qquad |z| > |a| \tag{4.9}$$

Alternatively, by multiplying the numerator and denominator of Eq. (4.9) by z, we may write $X(z)$ as

$$X(z) = \frac{z}{z - a} \qquad |z| > |a| \tag{4.10}$$

Both forms of $X(z)$ in Eqs. (4.9) and (4.10) are useful depending upon the application. From Eq. (4.10) we see that $X(z)$ is a rational function of z. Consequently, just as with rational Laplace transforms, it can be characterized by its zeros (the roots of the numerator polynomial) and its poles (the roots of the denominator polynomial). From Eq. (4.10) we see that there is one zero at $z = 0$ and one pole at $z = a$. The ROC and the pole-zero plot for this example are shown in Fig. 4-1. In z-transform applications, the complex plane is commonly referred to as the z-plane.

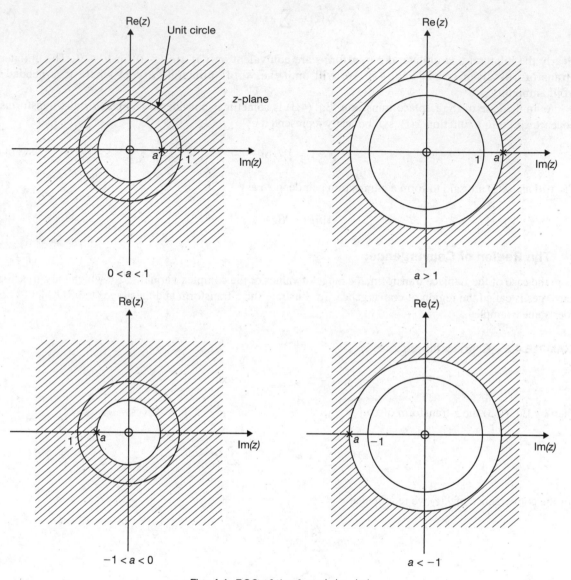

Fig. 4-1 ROC of the form $|z| > |a|$.

EXAMPLE 4.2   Consider the sequence

$$x[n] = -a^n u[-n-1] \tag{4.11}$$

Its z-transform $X(z)$ is given by (Prob. 4.1)

$$X(z) = \frac{1}{1 - az^{-1}} \qquad |z| < |a| \tag{4.12}$$

Again, as before, $X(z)$ may be written as

$$X(z) = \frac{z}{z - a} \qquad |z| < |a| \tag{4.13}$$

Thus, the ROC and the pole-zero plot for this example are shown in Fig. 4-2. Comparing Eqs. (4.9) and (4.12) [or Eqs. (4.10) and (4.13)], we see that the algebraic expressions of $X(z)$ for two different sequences are identical except for the ROCs. Thus, as in the Laplace transform, specification of the z-transform requires both the algebraic expression and the ROC.

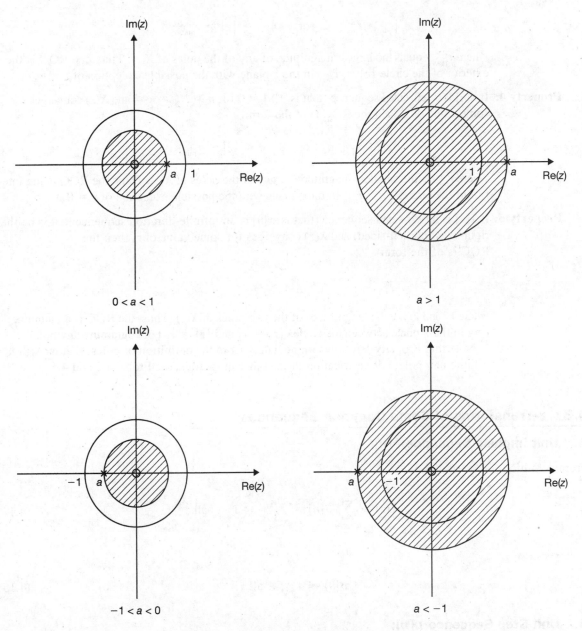

Fig. 4-2 ROC of the form $|z| < |a|$.

C. Properties of the ROC:

As we saw in Examples 4.1 and 4.2, the ROC of $X(z)$ depends on the nature of $x[n]$. The properties of the ROC are summarized below. We assume that $X(z)$ is a rational function of z.

Property 1: The ROC does not contain any poles.

Property 2: If $x[n]$ is a finite sequence (that is, $x[n] = 0$ except in a finite interval $N_1 \leq n \leq N_2$, where N_1 and N_2 are finite) and $X(z)$ converges for some value of z, then the ROC is the entire z-plane except possibly $z = 0$ or $z = \infty$.

Property 3: If $x[n]$ is a right-sided sequence (that is, $x[n] = 0$ for $n < N_1 < \infty$) and $X(z)$ converges for some value of z, then the ROC is of the form

$$|z| > r_{max} \qquad \text{or} \qquad \infty > |z| > r_{max}$$

where r_{max} equals the largest magnitude of any of the poles of $X(z)$. Thus, the ROC is the exterior of the circle $|z| = r_{max}$ in the z-plane with the possible exception of $z = \infty$.

Property 4: If $x[n]$ is a left-sided sequence (that is, $x[n] = 0$ for $n > N_2 > -\infty$) and $X(z)$ converges for some value of z, then the ROC is of the form

$$|z| < r_{min} \qquad \text{or} \qquad 0 < |z| < r_{min}$$

where r_{min} is the smallest magnitude of any of the poles of $X(z)$. Thus, the ROC is the interior of the circle $|z| = r_{min}$ in the z-plane with the possible exception of $z = 0$.

Property 5: If $x[n]$ is a two-sided sequence (that is, $x[n]$ is an infinite-duration sequence that is neither right-sided nor left-sided) and $X(z)$ converges for some value of z, then the ROC is of the form

$$r_1 < |z| < r_2$$

where r_1 and r_2 are the magnitudes of the two poles of $X(z)$. Thus, the ROC is an annular ring in the z-plane between the circles $|z| = r_1$ and $|z| = r_2$ not containing any poles.

Note that Property 1 follows immediately from the definition of poles; that is, $X(z)$ is infinite at a pole. For verification of the other properties, see Probs. 4.2 and 4.5.

4.3 *z*-Transforms of Some Common Sequences

A. Unit Impulse Sequence $\delta[n]$:

From definitions (1.45) and (4.3)

$$X(z) = \sum_{n=-\infty}^{\infty} \delta[n] z^{-n} = z^{-0} = 1 \qquad \text{all } z \tag{4.14}$$

Thus,

$$\delta[n] \leftrightarrow 1 \qquad \text{all } z \tag{4.15}$$

B. Unit Step Sequence $u[n]$:

Setting $a = 1$ in Eqs. (4.8) to (4.10), we obtain

$$u[n] \leftrightarrow \frac{1}{1 - z^{-1}} = \frac{z}{z-1} \qquad |z| > 1 \tag{4.16}$$

C. *z*-Transform Pairs:

The z-transforms of some common sequences are tabulated in Table 4-1.

TABLE 4-1 Some Common z-Transform Pairs

$x[n]$	$X(z)$	ROC
$\delta[n]$	1	All z
$u[n]$	$\dfrac{1}{1-z^{-1}}, \dfrac{z}{z-1}$	$|z| > 1$
$-u[-n-1]$	$\dfrac{1}{1-z^{-1}}, \dfrac{z}{z-1}$	$|z| < 1$
$\delta[n-m]$	z^{-m}	All z except 0 if $(m > 0)$ or ∞ if $(m < 0)$
$a^n u[n]$	$\dfrac{1}{1-az^{-1}}, \dfrac{z}{z-a}$	$|z| > |a|$
$-a^n u[-n-1]$	$\dfrac{1}{1-az^{-1}}, \dfrac{z}{z-a}$	$|z| < |a|$
$na^n u[n]$	$\dfrac{az^{-1}}{(1-az^{-1})^2}, \dfrac{az}{(z-a)^2}$	$|z| > |a|$
$-na^n u[-n-1]$	$\dfrac{az^{-1}}{(1-az^{-1})^2}, \dfrac{az}{(z-a)^2}$	$|z| < |a|$
$(n+1)a^n u[n]$	$\dfrac{1}{(1-az^{-1})^2}, \left[\dfrac{z}{z-a}\right]^2$	$|z| > |a|$
$(\cos \Omega_0 n)u[n]$	$\dfrac{z^2-(\cos \Omega_0)z}{z^2-(2\cos \Omega_0)z+1}$	$|z| > 1$
$(\sin \Omega_0 n)u[n]$	$\dfrac{(\sin \Omega_0)z}{z^2-(2\cos \Omega_0)z+1}$	$|z| > 1$
$(r^n \cos \Omega_0 n)u[n]$	$\dfrac{z^2-(r\cos \Omega_0)z}{z^2-(2r\cos \Omega_0)z+r^2}$	$|z| > r$
$(r^n \sin \Omega_0 n)u[n]$	$\dfrac{(r\sin \Omega_0)z}{z^2-(2r\cos \Omega_0)z+r^2}$	$|z| > r$
$\begin{cases} a^n & 0 \le n \le N-1 \\ 0 & \text{otherwise} \end{cases}$	$\dfrac{1-a^N z^{-N}}{1-az^{-1}}$	$|z| > 0$

4.4 Properties of the z-Transform

Basic properties of the z-transform are presented in the following discussion. Verification of these properties is given in Probs. 4.8 to 4.14.

A. Linearity:

If

$$x_1[n] \leftrightarrow X_1(z) \qquad \text{ROC} = R_1$$
$$x_2[n] \leftrightarrow X_2(z) \qquad \text{ROC} = R_2$$

then

$$a_1 x_1[n] + a_2 x_2[n] \leftrightarrow a_1 X_1(z) + a_2 X_2(z) \qquad R' \supset R_1 \cap R_2 \tag{4.17}$$

where a_1 and a_2 are arbitrary constants.

B. Time Shifting:

If

$$x[n] \leftrightarrow X(z) \qquad \text{ROC} = R$$

then

$$x[n - n_0] \leftrightarrow z^{-n_0} X(z) \qquad R' = R \cap \{0 < |z| < \infty\} \tag{4.18}$$

Special Cases:

$$x[n - 1] \leftrightarrow z^{-1} X(z) \qquad R' = R \cap \{0 < |z|\} \tag{4.19}$$
$$x[n + 1] \leftrightarrow z X(z) \qquad R' = R \cap \{|z| < \infty\} \tag{4.20}$$

Because of these relationship [Eqs. (4.19) and (4.20)], z^{-1} is often called the *unit-delay operator* and z is called the *unit-advance operator*. Note that in the Laplace transform the operators $s^{-1} = 1/s$ and s correspond to time-domain integration and differentiation, respectively [Eqs. (3.22) and (3.20)].

C. Multiplication by z_0^n:

If

$$x[n] \leftrightarrow X(z) \qquad \text{ROC} = R$$

then

$$z_0^n x[n] \leftrightarrow X\left(\frac{z}{z_0}\right) \qquad R' = |z_0| R \tag{4.21}$$

In particular, a pole (or zero) at $z = z_k$ in $X(z)$ moves to $z = z_0 z_k$ after multiplication by z_0^n and the ROC expands or contracts by the factor $|z_0|$.

Special Case:

$$e^{j\Omega_0 n} x[n] \leftrightarrow X(e^{-j\Omega_0} z) \qquad R' = R \tag{4.22}$$

In this special case, all poles and zeros are simply rotated by the angle Ω_0 and the ROC is unchanged.

D. Time Reversal:

If

$$x[n] \leftrightarrow X(z) \qquad \text{ROC} = R$$

then

$$x[-n] \leftrightarrow X\left(\frac{1}{z}\right) \qquad R' = \frac{1}{R} \tag{4.23}$$

Therefore, a pole (or zero) in $X(z)$ at $z = z_k$ moves to $1/z_k$ after time reversal. The relationship $R' = 1/R$ indicates the inversion of R, reflecting the fact that a right-sided sequence becomes left-sided if time-reversed, and vice versa.

E. Multiplication by *n* (or Differentiation in *z*):

If

$$x[n] \leftrightarrow X(z) \qquad \text{ROC} = R$$

then

$$nx[n] \leftrightarrow -z\frac{dX(z)}{dz} \qquad R' = R \tag{4.24}$$

F. Accumulation:

If

$$x[n] \leftrightarrow X(z) \qquad \text{ROC} = R$$

then

$$\sum_{k=-\infty}^{n} x[k] \leftrightarrow \frac{1}{1-z^{-1}}X(z) = \frac{z}{z-1}X(z) \qquad R' \supset R \cap \{|z|>1\} \tag{4.25}$$

Note that $\sum_{k=-\infty}^{n} x[k]$ is the discrete-time counterpart to integration in the time domain and is called the *accumulation*. The comparable Laplace transform operator for integration is $1/s$.

G. Convolution:

If

$$x_1[n] \leftrightarrow X_1(z) \qquad \text{ROC} = R_1$$
$$x_2[n] \leftrightarrow X_2(z) \qquad \text{ROC} = R_2$$

then

$$x_1[n] * x_2[n] \leftrightarrow X_1(z)X_2(z) \qquad R' \supset R_1 \cap R_2 \tag{4.26}$$

This relationship plays a central role in the analysis and design of discrete-time LTI systems, in analogy with the continuous-time case.

H. Summary of Some *z*-transform Properties:

For convenient reference, the properties of the z-transform presented above are summarized in Table 4-2.

TABLE 4-2.   Some Properties of the z-Transform

PROPERTY	SEQUENCE	TRANSFORM	ROC
	$x[n]$	$X(z)$	R
	$x_1[n]$	$X_1(z)$	R_1
	$x_2[n]$	$X_2(z)$	R_2
Linearity	$a_1 x_1[n] + a_2 x_2[n]$	$a_1 X_1(z) + a_2 X_2(z)$	$R' \supset R_1 \cap R_2$
Time shifting	$x[n-n_0]$	$z^{-n_0} X(z)$	$R' \supset R \cap \{0 < |z| < \infty\}$
Multiplication by z_0^n	$z_0^n x[n]$	$X\left(\dfrac{z}{z_0}\right)$	$R' = |z_0| R$
Multiplication by $e^{j\Omega_0 n}$	$e^{j\Omega_0 n} x[n]$	$X(e^{-j\Omega_0} z)$	$R' = R$
Time reversal	$x[-n]$	$X\left(\dfrac{1}{z}\right)$	$R' = \dfrac{1}{R}$
Multiplication by n	$n x[n]$	$-z \dfrac{dX(z)}{dz}$	$R' = R$
Accumulation	$\displaystyle\sum_{k=-\infty}^{n} x[n]$	$\dfrac{1}{1 - z^{-1}} X(z)$	$R' \supset R \cap \{|z| > 1\}$
Convolution	$x_1[n] * x_2[n]$	$X_1(z) X_2(z)$	$R' \supset R_1 \cap R_2$

4.5   The Inverse z-Transform

Inversion of the z-transform to find the sequence $x[n]$ from its z-transform $X(z)$ is called the inverse z-transform, symbolically denoted as

$$x[n] = \mathcal{3}^{-1}\{X(z)\} \tag{4.27}$$

A.   Inversion Formula:

As in the case of the Laplace transform, there is a formal expression for the inverse z-transform in terms of an integration in the z-plane; that is,

$$x[n] = \frac{1}{2\pi j} \oint_C X(z) z^{n-1} \, dz \tag{4.28}$$

where C is a counterclockwise contour of integration enclosing the origin. Formal evaluation of Eq. (4.28) requires an understanding of complex variable theory.

B.   Use of Tables of z-Transform Pairs:

In the second method for the inversion of $X(z)$, we attempt to express $X(z)$ as a sum

$$X(z) = X_1(z) + \cdots + X_n(z) \tag{4.29}$$

where $X_1(z), \ldots, X_n(z)$ are functions with known inverse transforms $x_1[n], \ldots, x_n[n]$. From the linearity property (4.17) it follows that

$$x[n] = x_1[n] + \cdots + x_n[n] \tag{4.30}$$

C. Power Series Expansion:

The defining expression for the z-transform [Eq. (4.3)] is a power series where the sequence values $x[n]$ are the coefficients of z^{-n}. Thus, if $X(z)$ is given as a power series in the form

$$X[z] = \sum_{n=-\infty}^{\infty} x[n]z^{-n}$$

$$= \cdots + x[-2]z^2 + x[-1]z + x[0] + x[1]z^{-1} + x[2]z^{-2} + \cdots \tag{4.31}$$

we can determine any particular value of the sequence by finding the coefficient of the appropriate power of z^{-1}. This approach may not provide a closed-form solution but is very useful for a finite-length sequence where $X(z)$ may have no simpler form than a polynomial in z^{-1} (see Prob. 4.15). For rational z-transforms, a power series expansion can be obtained by long division as illustrated in Probs. 4.16 and 4.17.

D. Partial-Fraction Expansion:

As in the case of the inverse Laplace transform, the partial-fraction expansion method provides the most generally useful inverse z-transform, especially when $X(z)$ is a rational function of z. Let

$$X(z) = \frac{N(z)}{D(z)} = k\frac{(z - z_1)\cdots(z - z_m)}{(z - p_1)\cdots(z - p_n)} \tag{4.32}$$

Assuming $n \geq m$ and all poles p_k are simple, then

$$\frac{X(z)}{z} = \frac{c_0}{z} + \frac{c_1}{z - p_1} + \frac{c_2}{z - p_2} + \cdots + \frac{c_n}{z - p_n} = \frac{c_0}{z} + \sum_{k=1}^{n}\frac{c_k}{z - p_k} \tag{4.33}$$

where

$$c_0 = X(z)\big|_{z=0} \qquad c_k = (z - p_k)\frac{X(z)}{z}\bigg|_{z=p_k} \tag{4.34}$$

Hence, we obtain

$$X(z) = c_0 + c_1\frac{z}{z - p_1} + \cdots + c_n\frac{z}{z - p_n} = c_0 + \sum_{k=1}^{n} c_k\frac{z}{z - p_k} \tag{4.35}$$

Inferring the ROC for each term in Eq. (4.35) from the overall ROC of $X(z)$ and using Table 4-1, we can then invert each term, producing thereby the overall inverse z-transform (see Probs. 4.19 to 4.23).

If $m > n$ in Eq. (4.32), then a polynomial of z must be added to the right-hand side of Eq. (4.35), the order of which is $(m - n)$. Thus for $m > n$, the complete partial-fraction expansion would have the form

$$X(z) = \sum_{q=0}^{m-n} b_q z^q + \sum_{k=1}^{n} c_k\frac{z}{z - p_k} \tag{4.36}$$

If $X(z)$ has multiple-order poles, say, p_i is the multiple pole with multiplicity r, then the expansion of $X(z)/z$ will consist of terms of the form

$$\frac{\lambda_1}{z - p_i} + \frac{\lambda_2}{(z - p_i)^2} + \cdots + \frac{\lambda_r}{(z - p_i)^r} \tag{4.37}$$

where

$$\lambda_{r-k} = \frac{1}{k!}\frac{d^k}{dz^k}\left[(z - p_i)^r\frac{X(z)}{z}\right]\bigg|_{z=p_i} \tag{4.38}$$

4.6   The System Function of Discrete-Time LTI Systems

A.   The System Function:

In Sec. 2.6 we showed that the output $y[n]$ of a discrete-time LTI system equals the convolution of the input $x[n]$ with the impulse response $h[n]$; that is [Eq. (2.35)],

$$y[n] = x[n] * h[n] \tag{4.39}$$

Applying the convolution property (4.26) of the z-transform, we obtain

$$Y(z) = X(z)H(z) \tag{4.40}$$

where $Y(z)$, $X(z)$, and $H(z)$ are the z-transforms of $y[n]$, $x[n]$, and $h[n]$, respectively. Equation (4.40) can be expressed as

$$H(z) = \frac{Y(z)}{X(z)} \tag{4.41}$$

The z-transform $H(z)$ of $h[n]$ is referred to as the *system function* (or the *transfer function*) of the system. By Eq. (4.41) the system function $H(z)$ can also be defined as the ratio of the z-transforms of the output $y[n]$ and the input $x[n]$. The system function $H(z)$ completely characterizes the system. Fig. 4-3 illustrates the relationship of Eqs. (4.39) and (4.40).

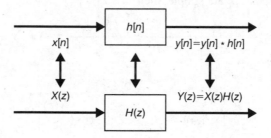

Fig. 4-3 Impulse response and system function.

B.   Characterization of Discrete-Time LTI Systems:

Many properties of discrete-time LTI systems can be closely associated with the characteristics of $H(z)$ in the z-plane and in particular with the pole locations and the ROC.

1.   Causality:

For a causal discrete-time LTI system, we have [Eq. (2.44)]

$$h[n] = 0 \qquad n < 0$$

since $h[n]$ is a right-sided signal, the corresponding requirement on $H(z)$ is that the ROC of $H(z)$ must be of the form

$$|z| > r_{max}$$

That is, the ROC is the exterior of a circle containing all of the poles of $H(z)$ in the z-plane. Similarly, if the system is anticausal, that is,

$$h[n] = 0 \qquad n \geq 0$$

then $h[n]$ is left-sided and the ROC of $H(z)$ must be of the form

$$|z| < r_{min}$$

That is, the ROC is the interior of a circle containing no poles of $H(z)$ in the z-plane.

2. Stability:

In Sec. 2.7 we stated that a discrete-time LTI system is BIBO stable if and only if [Eq. (2.49)]

$$\sum_{n=-\infty}^{\infty} |h[n]| < \infty$$

The corresponding requirement on $H(z)$ is that the ROC of $H(z)$ contains the unit circle (that is, $|z| = 1$). (See Prob. 4.30.)

3. Causal and Stable Systems:

If the system is both causal and stable, then all of the poles of $H(z)$ must lie inside the unit circle of the z-plane because the ROC is of the form $|z| > r_{max}$, and since the unit circle is included in the ROC, we must have $r_{max} < 1$.

C. System Function for LTI Systems Described by Linear Constant-Coefficient Difference Equations:

In Sec. 2.9 we considered a discrete-time LTI system for which input $x[n]$ and output $y[n]$ satisfy the general linear constant-coefficient difference equation of the form

$$\sum_{k=0}^{N} a_k y[n-k] = \sum_{k=0}^{M} b_k x[n-k] \qquad (4.42)$$

Applying the z-transform and using the time-shift property (4.18) and the linearity property (4.17) of the z-transform, we obtain

$$\sum_{k=0}^{N} a_k z^{-k} Y(z) = \sum_{k=0}^{M} b_k z^{-k} X(z)$$

or

$$Y(z) \sum_{k=0}^{N} a_k z^{-k} = X(z) \sum_{k=0}^{M} b_k z^{-k} \qquad (4.43)$$

Thus,

$$H(z) = \frac{Y(z)}{X(z)} = \frac{\displaystyle\sum_{k=0}^{M} b_k z^{-k}}{\displaystyle\sum_{k=0}^{N} a_k z^{-k}} \qquad (4.44)$$

Hence, $H(z)$ is always rational. Note that the ROC of $H(z)$ is not specified by Eq. (4.44) but must be inferred with additional requirements on the system such as the causality or the stability.

D. Systems Interconnection:

For two LTI systems (with $h_1[n]$ and $h_2[n]$, respectively) in cascade, the overall impulse response $h[n]$ is given by

$$h[n] = h_1[n] * h_2[n] \qquad (4.45)$$

Thus, the corresponding system functions are related by the product

$$H(z) = H_1(z)H_2(z) \qquad R \supset R_1 \cap R_2 \qquad (4.46)$$

Similarly, the impulse response of a parallel combination of two LTI systems is given by

$$h[n] = h_1[n] + h_2[n] \qquad (4.47)$$

and

$$H(z) = H_1(z) + H_2(z) \qquad R \supset R_1 \cap R_2 \tag{4.48}$$

4.7   The Unilateral z-Transform

A.   Definition:

The *unilateral* (or *one-sided*) z-transform $X_I(z)$ of a sequence $x[n]$ is defined as [Eq. (4.5)]

$$X_I(z) = \sum_{n=0}^{\infty} x[n] z^{-n} \tag{4.49}$$

and differs from the bilateral transform in that the summation is carried over only $n \geq 0$. Thus, the unilateral z-transform of $x[n]$ can be thought of as the bilateral transform of $x[n]u[n]$. Since $x[n]u[n]$ is a right-sided sequence, the ROC of $X_I(z)$ is always outside a circle in the z-plane.

B.   Basic Properties:

Most of the properties of the unilateral z-transform are the same as for the bilateral z-transform. The unilateral z-transform is useful for calculating the response of a causal system to a causal input when the system is described by a linear constant-coefficient difference equation with nonzero initial conditions. The basic property of the unilateral z-transform that is useful in this application is the following time-shifting property which is different from that of the bilateral transform.

Time-Shifting Property:

If $x[n] \leftrightarrow X_I(z)$, then for $m \geq 0$,

$$x[n - m] \leftrightarrow z^{-m} X_I(z) + z^{-m+1} x[-1] + z^{-m+2} x[-2] + \cdots + x[-m] \tag{4.50}$$

$$x[n + m] \leftrightarrow z^m X_I(z) - z^m x[0] - z^{m-1} x[1] - \cdots - z x[m - 1] \tag{4.51}$$

The proofs of Eqs. (4.50) and (4.51) are given in Prob. 4.36.

D.   System Function:

Similar to the case of the continuous-time LTI system, with the unilateral z-transform, the system function $H(z) = Y(z)/X(z)$ is defined under the condition that the system is relaxed; that is, all initial conditions are zero.

SOLVED PROBLEMS

The z-Transform

4.1.   Find the z-transform of

(a) $x[n] = -a^n u[-n - 1]$

(b) $x[n] = a^{-n} u[-n - 1]$

(a)   From Eq. (4.3)

$$X(z) = -\sum_{n=-\infty}^{\infty} a^n u[-n-1] z^{-n} = -\sum_{n=-\infty}^{-1} a^n z^{-n}$$

$$= -\sum_{n=1}^{\infty} (a^{-1}z)^n = 1 - \sum_{n=0}^{\infty} (a^{-1}z)^n$$

By Eq. (1.91)

$$\sum_{n=0}^{\infty} (a^{-1}z)^n = \frac{1}{1-a^{-1}z} \quad \text{if} \left| a^{-1}z \right| < 1 \text{ or } |z| < |a|$$

Thus,

$$X(z) = 1 - \frac{1}{1-a^{-1}z} = \frac{-a^{-1}z}{1-a^{-1}z} = \frac{z}{z-a} = \frac{1}{1-az^{-1}} \quad |z| < |a| \tag{4.52}$$

(*b*) Similarly,

$$X(z) = \sum_{n=-\infty}^{\infty} a^{-n}u[-n-1]z^{-n} = \sum_{n=-\infty}^{-1} (az)^{-n}$$

$$= \sum_{n=1}^{\infty} (az)^n = \sum_{n=0}^{\infty} (az)^n - 1$$

Again by Eq. (1.91)

$$\sum_{n=0}^{\infty} (az)^n = \frac{1}{1-az} \quad \text{if} \left| az \right| < 1 \text{ or } |z| < \frac{1}{|a|}$$

Thus,

$$X(z) = \frac{1}{1-az} - 1 = \frac{az}{1-az} = -\frac{z}{z-1/a} \quad |z| < \frac{1}{|a|} \tag{4.53}$$

4.2. A finite sequence $x[n]$ is defined as

$$x[n] \begin{cases} \neq 0 & N_1 \leq n \leq N_2 \\ = 0 & \text{otherwise} \end{cases}$$

where N_1 and N_2 are finite. Show that the ROC of $X(z)$ is the entire z-plane except possibly $z = 0$ or $z = \infty$.

From Eq. (4.3)

$$X(z) = \sum_{n=N_1}^{N_2} x[n]z^{-n} \tag{4.54}$$

For z not equal to zero or infinity, each term in Eq. (4.54) will be finite and thus $X(z)$ will converge. If $N_1 < 0$ and $N_2 > 0$, then Eq. (4.54) includes terms with both positive powers of z and negative powers of z. As $|z| \to 0$, terms with negative powers of z become unbounded, and as $|z| \to \infty$, terms with positive powers of z become unbounded. Hence, the ROC is the entire z-plane except for $z = 0$ and $z = \infty$. If $N_1 \geq 0$, Eq. (4.54) contains only negative powers of z, and hence the ROC includes $z = \infty$. If $N_2 \leq 0$, Eq. (4.54) contains only positive powers of z, and hence the ROC includes $z = 0$.

4.3. A finite sequence $x[n]$ is defined as

$$x[n] = \{5, 3, -2, 0, 4, -3\}$$
$$\uparrow$$

Find $X(z)$ and its ROC.

From Eq. (4.3) and given $x[n]$ we have

$$X(z) = \sum_{n=-\infty}^{\infty} x[n]z^{-n} = \sum_{n=-2}^{3} x[n]z^{-n}$$

$$= x[-2]z^2 + x[-1]z + x[0] + x[1]z^{-1} + x[2]z^{-2} + x[3]z^{-3}$$

$$= 5z^2 + 3z - 2 + 4z^{-2} - 3z^{-3}.$$

For z not equal to zero or infinity, each term in $X(z)$ will be finite and consequently $X(z)$ will converge. Note that $X(z)$ includes both positive powers of z and negative powers of z. Thus, from the result of Prob. 4.2 we conclude that the ROC of $X(z)$ is $0 < |z| < \infty$.

4.4. Consider the sequence

$$x[n] = \begin{cases} a^n & 0 \le n \le N-1, a > 0 \\ 0 & \text{otherwise} \end{cases}$$

Find $X(z)$ and plot the poles and zeros of $X(z)$.

By Eq. (4.3) and using Eq. (1.90), we get

$$X(z) = \sum_{n=0}^{N-1} a^n z^{-n} = \sum_{n=0}^{N-1} (az^{-1})^n = \frac{1 - (az^{-1})^N}{1 - az^{-1}} = \frac{1}{z^{N-1}} \frac{z^N - a^N}{z - a} \tag{4.55}$$

From Eq. (4.55) we see that there is a pole of $(N-1)$th order at $z = 0$ and a pole at $z = a$. Since $x[n]$ is a finite sequence and is zero for $n < 0$, the ROC is $|z| > 0$. The N roots of the numerator polynomial are at

$$z_k = ae^{j(2\pi k/N)} \qquad k = 0, 1, \ldots, N-1 \tag{4.56}$$

The root at $k = 0$ cancels the pole at $z = a$. The remaining zeros of $X(z)$ are at

$$z_k = ae^{j(2\pi k/N)} \qquad k = 1, \ldots, N-1 \tag{4.57}$$

The pole-zero plot is shown in Fig. 4-4 with $N = 8$.

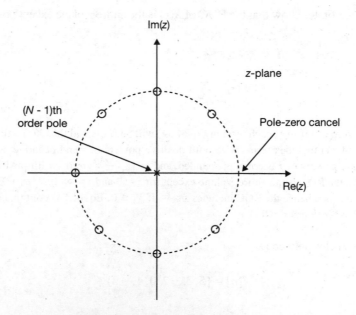

Fig. 4-4 Pole-zero plot with $N = 8$.

4.5. Show that if $x[n]$ is a right-sided sequence and $X(z)$ converges for some value of z, then the ROC of $X(z)$ is of the form

$$|z| > r_{max} \qquad \text{or} \qquad \infty > |z| > r_{max}$$

where r_{max} is the maximum magnitude of any of the poles of $X(z)$.

Consider a right-sided sequence $x[n]$ so that

$$x[n] = 0 \qquad n < N_1$$

and $X(z)$ converges for $|z| = r_0$. Then from Eq. (4.3)

$$|X(z)| \le \sum_{n=-\infty}^{\infty} |x[n]| r_0^{-n} = \sum_{n=N_1}^{\infty} |x[n]| r_0^{-n} < \infty$$

Now if $r_1 > r_0$, then

$$\sum_{n=N_1}^{\infty} |x[n]| r_1^{-n} = \sum_{n=N_1}^{\infty} |x[n]| \left(r_0 \frac{r_1}{r_0} \right)^{-n} = \sum_{n=N_1}^{\infty} |x[n]| r_0^{-n} \left(\frac{r_1}{r_0} \right)^{-n}$$

$$\le \left(\frac{r_1}{r_0} \right)^{-N_1} \sum_{n=N_1}^{\infty} |x[n]| r_0^{-n} < \infty$$

since $(r_1/r_0)^{-n}$ is a decaying sequence. Thus, $X(z)$ converges for $r = r_1$ and the ROC of $X(z)$ is of the form

$$|z| > r_0$$

Since the ROC of $X(z)$ cannot contain the poles of $X(z)$, we conclude that the ROC of $X(z)$ is of the form

$$|z| > r_{max}$$

where r_{max} is the maximum magnitude of any of the poles of $X(z)$.

If $N_1 < 0$, then

$$X(z) = \sum_{n=N_1}^{\infty} x[n] z^{-n} = x[N_1] z^{-N_1} + \cdots + x[-1]z + \sum_{n=0}^{\infty} x[n] z^{-n}$$

That is, $X(z)$ contains the positive powers of z and becomes unbounded at $z = \infty$. In this case the ROC is of the form

$$\infty > |z| > r_{max}$$

From the above result we can tell that a sequence $x[n]$ is causal (not just right-sided) from the ROC of $X(z)$ if $z = \infty$ is included. Note that this is not the case for the Laplace transform.

4.6. Find the z-transform $X(z)$ and sketch the pole-zero plot with the ROC for each of the following sequences:

(a) $\quad x[n] = \left(\dfrac{1}{2} \right)^n u[n] + \left(\dfrac{1}{3} \right)^n u[n]$

(b) $\quad x[n] = \left(\dfrac{1}{3} \right)^n u[n] + \left(\dfrac{1}{2} \right)^n u[-n-1]$

(c) $\quad x[n] = \left(\dfrac{1}{2} \right)^n u[n] + \left(\dfrac{1}{3} \right)^n u[-n-1]$

(a)   From Table 4-1

$$\left(\frac{1}{2}\right)^n u[n] \leftrightarrow \frac{z}{z - \frac{1}{2}} \qquad |z| > \frac{1}{2} \tag{4.58}$$

$$\left(\frac{1}{3}\right)^n u[n] \leftrightarrow \frac{z}{z - \frac{1}{3}} \qquad |z| > \frac{1}{3} \tag{4.59}$$

We see that the ROCs in Eqs. (4.58) and (4.59) overlap, and thus,

$$X(z) = \frac{z}{z - \frac{1}{2}} + \frac{z}{z - \frac{1}{3}} = \frac{2z\left(z - \frac{5}{12}\right)}{\left(s - \frac{1}{2}\right)\left(z - \frac{1}{3}\right)} \qquad |z| > \frac{1}{2} \tag{4.60}$$

From Eq. (4.60) we see that $X(z)$ has two zeros at $z = 0$ and $z = \frac{5}{12}$ and two poles at $z = \frac{1}{2}$ and $z = \frac{1}{3}$ and that the ROC is $|z| > \frac{1}{2}$, as sketched in Fig. 4-5(a).

(b)   From Table 4-1

$$\left(\frac{1}{3}\right)^n u[n] \leftrightarrow \frac{z}{z - \frac{1}{3}} \qquad |z| > \frac{1}{3} \tag{4.61}$$

$$\left(\frac{1}{2}\right)^n u[-n-1] \leftrightarrow -\frac{z}{z - \frac{1}{2}} \qquad |z| < \frac{1}{2} \tag{4.62}$$

We see that the ROCs in Eqs. (4.61) and (4.62) overlap, and thus

$$X(z) = \frac{z}{z - \frac{1}{3}} - \frac{z}{z - \frac{1}{2}} = -\frac{1}{6} \frac{z}{\left(z - \frac{1}{2}\right)\left(z - \frac{1}{3}\right)} \qquad \frac{1}{3} < |z| < \frac{1}{2} \tag{4.63}$$

From Eq. (4.63) we see that $X(z)$ has one zero at $z = 0$ and two poles at $z = \frac{1}{2}$ and $z = \frac{1}{3}$ and that the ROC is $\frac{1}{3} < |z| < \frac{1}{2}$, as sketched in Fig. 4-5(b).

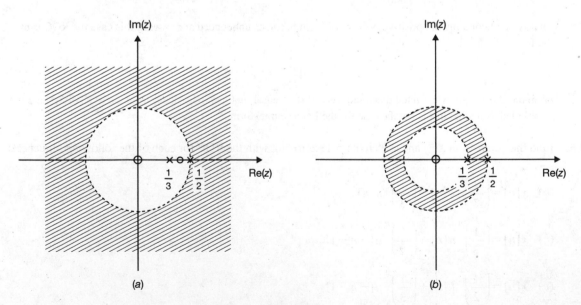

(a)

(b)

Fig. 4-5

(c)   From Table 4-1

$$\left(\frac{1}{2}\right)^n u[n] \leftrightarrow \frac{z}{z-\frac{1}{2}} \quad |z| > \frac{1}{2} \tag{4.64}$$

$$\left(\frac{1}{3}\right)^n u[-n-1] \leftrightarrow -\frac{z}{z-\frac{1}{3}} \quad |z| < \frac{1}{3} \tag{4.65}$$

We see that the ROCs in Eqs. (4.64) and (4.65) do not overlap and that there is no common ROC, and thus $x[n]$ will not have $X(z)$.

4.7.   Let

$$x[n] = a^{|n|} \qquad a > 0 \tag{4.66}$$

(a)   Sketch $x[n]$ for $a < 1$ and $a > 1$.

(b)   Find $X(z)$ and sketch the zero-pole plot and the ROC for $a < 1$ and $a > 1$.

(a)   The sequence $x[n]$ is sketched in Figs. 4-6(a) and (b) for both $a < 1$ and $a > 1$.

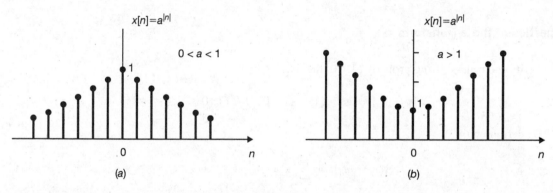

Fig. 4-6

(b)   Since $x[n]$ is a two-sided sequence, we can express it as

$$x[n] = a^n u[n] + a^{-n} u[-n-1] \tag{4.67}$$

From Table 4-1

$$a^n u[n] \leftrightarrow \frac{z}{z-a} \quad |z| > a \tag{4.68}$$

$$a^{-n} u[-n-1] \leftrightarrow -\frac{z}{z-1/a} \quad |z| < \frac{1}{a} \tag{4.69}$$

If $a < 1$, we see that the ROCs in Eqs. (4.68) and (4.69) overlap, and thus,

$$X(z) = \frac{z}{z-a} - \frac{z}{z-1/a} = \frac{a^2-1}{a} \frac{z}{(z-a)(z-1/a)} \quad a < |z| < \frac{1}{a} \tag{4.70}$$

From Eq. (4.70) we see that $X(z)$ has one zero at the origin and two poles at $z = a$ and $z = 1/a$ and that the ROC is $a < |z| < 1/a$, as sketched in Fig. 4-7. If $a > 1$, we see that the ROCs in Eqs. (4.68) and (4.69) do not overlap and that there is no common ROC, and thus $x[n]$ will not have $X(z)$.

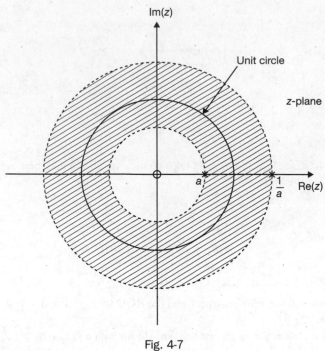

Fig. 4-7

Properties of the *z*-Transform

4.8. Verify the time-shifting property (4.18); that is,

$$x[n - n_0] \leftrightarrow z^{-n_0} X(z) \qquad R' \supset R \cap \{0 < |z| < \infty\}$$

By definition (4.3)

$$\mathfrak{Z}\{x[n - n_0]\} = \sum_{n=-\infty}^{\infty} x[n - n_0] z^{-n}$$

By the change of variables $m = n - n_0$, we obtain

$$\mathfrak{Z}\{x[n - n_0]\} = \sum_{m=-\infty}^{\infty} x[m] z^{-(m+n_0)}$$

$$= z^{-n_0} \sum_{m=-\infty}^{\infty} x[m] z^{-m} = z^{-n_0} X(z)$$

Because of the multiplication by z^{-n_0}, for $n_0 > 0$, additional poles are introduced at $z = 0$ and will be deleted at $z = \infty$. Similarly, if $n_0 < 0$, additional zeros are introduced at $z = 0$ and will be deleted at $z = \infty$. Therefore, the points $z = 0$ and $z = \infty$ can be either added to or deleted from the ROC by time shifting. Thus, we have

$$x[n - n_0] \leftrightarrow z^{-n_0} X(z) \qquad R' \supset R \cap \{0 < |z| < \infty\}$$

where R and R' are the ROCs before and after the time-shift operation.

4.9. Verify Eq. (4.21); that is,

$$z_0^n x[n] \leftrightarrow X\left(\frac{z}{z_0}\right) \qquad R' = |z_0| R$$

By definition (4.3)

$$3\{z_0^n x[n]\} = \sum_{n=-\infty}^{\infty} (z_0^n x[n]) z^{-n} = \sum_{n=-\infty}^{\infty} x[n] \left(\frac{z}{z_0}\right)^{-n} = X\left(\frac{z}{z_0}\right)$$

A pole (or zero) at $z = z_k$ in $X(z)$ moves to $z = z_0 z_k$, and the ROC expands or contracts by the factor $|z_0|$. Thus, we have

$$z_0^n x[n] \leftrightarrow X\left(\frac{z}{z_0}\right) \qquad R' = |z_0| R$$

4.10. Find the z-transform and the associated ROC for each of the following sequences:

(a) $x[n] = \delta[n - n_0]$ 　　　　　　　(b) $x[n] = u[n - n_0]$

(c) $x[n] = a^{n+1} u[n + 1]$ 　　　　　(d) $x[n] = u[-n]$

(e) $x[n] = a^{-n} u[-n]$

(a) From Eq. (4.15)

$$\delta[n] \leftrightarrow 1 \qquad \text{all } z$$

Applying the time-shifting property (4.18), we obtain

$$\delta[n - n_0] \leftrightarrow z^{-n_0} \qquad \begin{array}{l} 0 < |z|, n_0 > 0 \\[6pt] |z| < \infty, n_0 < 0 \end{array} \tag{4.71}$$

(b) From Eq. (4.16)

$$u[n] \leftrightarrow \frac{z}{z-1} \qquad |z| > 1$$

Again by the time-shifting property (4.18) we obtain

$$u[n - n_0] \leftrightarrow z^{-n_0} \frac{z}{z-1} = \frac{z^{-(n_0-1)}}{z-1} \qquad 1 < |z| < \infty \tag{4.72}$$

(c) From Eqs. (4.8) and (4.10)

$$a^n u[n] \leftrightarrow \frac{z}{z-a} \qquad |z| > |a|$$

By Eq. (4.20) we obtain

$$a^{n+1} u[n+1] \leftrightarrow z \frac{z}{z-a} = \frac{z^2}{z-a} \qquad |a| < |z| < \infty \tag{4.73}$$

(d) From Eq. (4.16)

$$u[n] \leftrightarrow \frac{z}{z-1} \qquad |z| > 1$$

By the time-reversal property (4.23) we obtain

$$u[-n] \leftrightarrow \frac{1/z}{1/z-1} = \frac{1}{1-z} \qquad |z| < 1 \tag{4.74}$$

(e) From Eqs. (4.8) and (4.10)

$$a^n u[n] \leftrightarrow \frac{z}{z-a} \qquad |z| > |a|$$

Again by the time-reversal property (4.23) we obtain

$$a^{-n}u[-n] \leftrightarrow \frac{1/z}{1/z - a} = \frac{1}{1 - az} \qquad |z| < \frac{1}{|a|} \qquad (4.75)$$

4.11. Verify the multiplication by n (or differentiation in z) property (4.24); that is,

$$nx[n] \leftrightarrow -z\frac{dX(z)}{dz} \qquad R' = R$$

From definition (4.3)

$$X(z) = \sum_{n=-\infty}^{\infty} x[n]z^{-n}$$

Differentiating both sides with respect to z, we have

$$\frac{dX(z)}{dz} = \sum_{n=-\infty}^{\infty} -nx[n]z^{-n-1}$$

and

$$-z\frac{dX(z)}{dz} = \sum_{n=-\infty}^{\infty} \{nx[n]\}z^{-n} = \mathfrak{Z}\{nx[n]\}$$

Thus, we conclude that

$$nx[n] \leftrightarrow -z\frac{dX(z)}{dz} \qquad R' = R$$

4.12. Find the z-transform of each of the following sequences:

(a) $x[n] = na^n u[n]$

(b) $x[n] = na^{n-1} u[n]$

(a)   From Eqs. (4.8) and (4.10)

$$a^n u[n] \leftrightarrow \frac{z}{z - a} \qquad |z| > |a| \qquad (4.76)$$

Using the multiplication by n property (4.24), we get

$$na^n u[n] \leftrightarrow -z\frac{d}{dz}\left(\frac{z}{z - a}\right) = \frac{az}{(z - a)^2} \qquad |z| > |a| \qquad (4.77)$$

(b)   Differentiating Eq. (4.76) with respect to a, we have

$$na^{n-1} u[n] \leftrightarrow \frac{d}{da}\left(\frac{z}{z - a}\right) = \frac{z}{(z - a)^2} \qquad |z| > |a| \qquad (4.78)$$

Note that dividing both sides of Eq. (4.77) by a, we obtain Eq. (4.78).

4.13. Verify the convolution property (4.26); that is,

$$x_1[n] * x_2[n] \leftrightarrow X_1(z)X_2(z) \qquad R' \supset R_1 \cap R_2$$

By definition (2.35)

$$y[n] = x_1[n] * x_2[n] = \sum_{k=-\infty}^{\infty} x_1[k]x_2[n - k]$$

Thus, by definition (4.3)

$$Y[z] = \sum_{n=-\infty}^{\infty} \left(\sum_{k=-\infty}^{\infty} x_1[k] x_2[n-k] \right) z^{-n} = \sum_{k=-\infty}^{\infty} x_1[k] \left(\sum_{n=-\infty}^{\infty} x_2[n-k] z^{-n} \right)$$

Noting that the term in parentheses in the last expression is the z-transform of the shifted signal $x_2[n-k]$, then by the time-shifting property (4.18), we have

$$Y[z] = \sum_{k=-\infty}^{\infty} x_1[k] \left[z^{-k} X_2(z) \right] = \left(\sum_{n=-\infty}^{\infty} x_1[k] z^{-k} \right) X_2(z) = X_1(z) X_2(z)$$

with an ROC that contains the intersection of the ROC of $X_1(z)$ and $X_2(z)$. If a zero of one transform cancels a pole of the other, the ROC of $Y(z)$ may be larger. Thus, we conclude that

$$x_1[n] * x_2[n] \leftrightarrow X_1(z) X_2(z) \qquad R' \supset R_1 \cap R_2$$

4.14. Verify the accumulation property (4.25); that is,

$$\sum_{k=-\infty}^{n} x[k] \leftrightarrow \frac{1}{1-z^{-1}} X(z) = \frac{z}{z-1} X(z) \qquad R' \supset R \cap \{|z| > 1\}$$

From Eq. (2.40) we have

$$y[n] = \sum_{k=-\infty}^{n} x[k] = x[n] * u[n]$$

Thus, using Eq. (4.16) and the convolution property (4.26), we obtain

$$Y(z) = X(z) \left(\frac{1}{1-z^{-1}} \right) = X(z) \left(\frac{z}{z-1} \right)$$

with the ROC that includes the intersection of the ROC of $X(z)$ and the ROC of the z-transform of $u[n]$. Thus,

$$\sum_{k=-\infty}^{n} x[k] \leftrightarrow \frac{1}{1-z^{-1}} X(z) = \frac{z}{z-1} X(z) \qquad R' \supset R \cap \{|z| > 1\}$$

Inverse z-Transform

4.15. Find the inverse z-transform of

$$X(z) = z^2 \left(1 - \frac{1}{2} z^{-1} \right) (1 - z^{-1})(1 + 2z^{-1}) \qquad 0 < |z| < \infty \tag{4.79}$$

Multiplying out the factors of Eq. (4.79), we can express $X(z)$ as

$$X(z) = z^2 + \frac{1}{2} z - \frac{5}{2} + z^{-1}$$

Then, by definition (4.3),

$$X(z) = x[-2]z^2 + x[-1]z + x[0] + x[1]z^{-1}$$

and we get

$$X[n] = \left\{ ..., 0, 1, \frac{1}{2}, -\frac{5}{2}, 1, 0, ... \right\}$$
$$\uparrow$$

4.16. Using the power series expansion technique, find the inverse z-transform of the following $X(z)$:

(a) $X(z) = \dfrac{1}{1 - az^{-1}}, \qquad |z| > |a|$

(b) $X(z) = \dfrac{1}{1 - az^{-1}}, \qquad |z| < |a|$

(a) Since the ROC is $|z| > |a|$, that is, the exterior of a circle, $x[n]$ is a right-sided sequence. Thus, we must divide to obtain a series in the power of z^{-1}. Carrying out the long division, we obtain

$$
1 - az^{-1} \overline{\smash{\big)}\,1} \begin{array}{l} 1 + az^{-1} + a^2z^{-2} + \cdots \\[4pt] \end{array}
$$

$$
\begin{array}{r}
1 - az^{-1} \\ \hline
az^{-1} \\
az^{-1} - a^2z^{-2} \\ \hline
a^2z^{-2} \\
\vdots
\end{array}
$$

Thus,

$$
X(z) = \frac{1}{1 - az^{-1}} = 1 + az^{-1} + a^2z^{-2} + \cdots + a^kz^{-k} + \cdots
$$

and so by definition (4.3) we have

$$
x[n] = 0 \qquad n < 0
$$
$$
x[0] = 1 \qquad x[1] = a \qquad x[2] = a^2 \qquad \cdots \qquad x[k] = a^k \qquad \cdots
$$

Thus, we obtain

$$
x[n] = a^n u[n]
$$

(b) Since the ROC is $|z| < |a|$, that is, the interior of a circle, $x[n]$ is a left-sided sequence. Thus, we must divide so as to obtain a series in the power of z as follows. Multiplying both the numerator and denominator of $X(z)$ by z, we have

$$
X(z) = \frac{z}{z - a}
$$

and carrying out the long division, we obtain

$$
-a + z \overline{\smash{\big)}\,z} \begin{array}{l} -a^{-1}z - a^{-2}z^2 - a^{-3}z^3 - \cdots \\[4pt] \end{array}
$$

$$
\begin{array}{r}
z - a^{-1}z^2 \\ \hline
a^{-1}z^2 \\
a^{-1}z^2 - a^{-2}z^3 \\ \hline
a^{-2}z^3 \\
\vdots
\end{array}
$$

Thus,

$$
X(z) = \frac{1}{1 - az^{-1}} = -a^{-1}z - a^{-2}z^2 - a^{-3}z^3 - \cdots - a^{-k}z^k - \cdots
$$

and so by definition (4.3) we have

$$
x[n] = 0 \qquad n \geq 0
$$
$$
x[-1] = -a^{-1} \qquad x[-2] = -a^{-2} \qquad x[-3] = -a^{-3} \qquad \cdots \qquad x[-k] = -a^{-k} \qquad \cdots
$$

Thus, we get

$$x[n] = -a^n u[-n-1]$$

4.17. Find the inverse z-transform of the following $X(z)$:

(a) $X(z) = \log\left(\dfrac{1}{1 - az^{-1}}\right), |z| > |a|$

(b) $X(z) = \log\left(\dfrac{1}{1 - a^{-1}z}\right), |z| < |a|$

(a) The power series expansion for $\log(1 - r)$ is given by

$$\log(1 - r) = -\sum_{n=1}^{\infty} \frac{1}{n} r^n \qquad |r| < 1 \tag{4.80}$$

Now

$$X(z) = \log\left(\frac{1}{1 - az^{-1}}\right) = -\log(1 - az^{-1}) \qquad |z| > |a|$$

Since the ROC is $|z| > |a|$, that is, $|az^{-1}| < 1$, by Eq. (4.80), $X(z)$ has the power series expansion

$$X(z) = \sum_{n=1}^{\infty} \frac{1}{n}(az^{-1})^n = \sum_{n=1}^{\infty} \frac{1}{n} a^n z^{-n}$$

from which we can indentify $x[n]$ as

$$x[n] = \begin{cases} (1/n)a^n & n \geq 1 \\ 0 & n \leq 0 \end{cases}$$

or

$$x[n] = \frac{1}{n} a^n u[n-1] \tag{4.81}$$

(b)

$$X(z) = \log\left(\frac{1}{1 - a^{-1}z}\right) = -\log(1 - a^{-1}z) \qquad |z| < |a|$$

Since the ROC is $|z| < |a|$, that is, $|a^{-1}z| < 1$, by Eq. (4.80), $X(z)$ has the power series expansion

$$X(z) = \sum_{n=1}^{\infty} \frac{1}{n}(a^{-1}z)^n = \sum_{n=-1}^{-\infty} -\frac{1}{n}(a^{-1}z)^{-n} = \sum_{n=-1}^{-\infty} -\frac{1}{n} a^n z^{-n}$$

from which we can identify $x[n]$ as

$$x[n] = \begin{cases} 0 & n \geq 0 \\ -(1/n)a^n & n \leq -1 \end{cases}$$

or

$$x[n] = -\frac{1}{n} a^n u[-n-1] \tag{4.82}$$

4.18. Using the power series expansion technique, find the inverse z-transform of the following $X(z)$:

(a) $X(z) = \dfrac{z}{2z^2 - 3z + 1} \qquad |z| < \dfrac{1}{2}$

(b) $X(z) = \dfrac{z}{2z^2 - 3z + 1} \qquad |z| > 1$

(a) Since the ROC is $|z| < \frac{1}{2}$, $x[n]$ is a left-sided sequence. Thus, we must divide to obtain a series in power of z. Carrying out the long division, we obtain

$$
1 - 3z + 2z^2 \overline{\smash{\big)}\, z} \quad \begin{array}{l} z + 3z^2 + 7z^3 + 15z^4 + \cdots \end{array}
$$

$$
\begin{array}{l}
\underline{z - 3z^2 + 2z^3} \\
\quad\ 3z^2 - 2z^3 \\
\quad\ \underline{3z^2 - 9z^3 + 6z^4} \\
\qquad\quad 7z^3 - 6z^4 \\
\qquad\quad \underline{7z^3 - 21z^4 + 14z^5} \\
\qquad\qquad\quad 15z^4 \cdots
\end{array}
$$

Thus,

$$
X(z) = \cdots + 15z^4 + 7z^3 + 3z^2 + z
$$

and so by definition (4.3) we obtain

$$
x[n] = \{\dots, 15, 7, 3, 1, 0\}
$$
$$
\qquad\qquad\qquad\qquad\uparrow
$$

(b) Since the ROC is $|z| > 1$, $x[n]$ is a right-sided sequence. Thus, we must divide so as to obtain a series in power of z^{-1} as follows:

$$
2z^2 - 3z + 1 \overline{\smash{\big)}\, z} \quad \begin{array}{l} \frac{1}{2}z^{-1} + \frac{3}{4}z^{-2} + \frac{7}{8}z^{-3} + \cdots \end{array}
$$

$$
\begin{array}{l}
\underline{z - \dfrac{3}{2} - \dfrac{1}{2}z^{-1}} \\
\quad \dfrac{3}{2} - \dfrac{1}{2}z^{-1} \\
\quad \underline{\dfrac{3}{2} - \dfrac{9}{4}z^{-1} + \dfrac{3}{4}z^{-2}} \\
\qquad \dfrac{7}{4}z^{-1} - \dfrac{3}{4}z^{-2} \\
\qquad\qquad \vdots
\end{array}
$$

Thus,

$$
X(z) = \frac{1}{2}z^{-1} + \frac{3}{4}z^{-2} + \frac{7}{8}z^{-3} + \cdots
$$

and so by definition (4.3) we obtain

$$
x[n] = \left\{0, \frac{1}{2}, \frac{3}{4}, \frac{7}{8}, \dots\right\}
$$

4.19. Using partial-fraction expansion, redo Prob. 4.18.

(a)
$$
X(z) = \frac{z}{2z^2 - 3z + 1} = \frac{z}{2(z-1)\left(z - \dfrac{1}{2}\right)} \qquad |z| < \frac{1}{2}
$$

Using partial-fraction expansion, we have

$$
\frac{X(z)}{z} = \frac{1}{2z^2 - 3z + 1} = \frac{1}{2(z-1)\left(z - \dfrac{1}{2}\right)} = \frac{c_1}{z-1} + \frac{c_2}{z - \dfrac{1}{2}}
$$

where
$$c_1 = \frac{1}{2\left(z - \frac{1}{2}\right)}\Bigg|_{z=1} = 1 \qquad c_2 = \frac{1}{2(z-1)}\Bigg|_{z=1/2} = -1$$

and we get
$$X(z) = \frac{z}{z-1} - \frac{z}{z - \frac{1}{2}} \qquad |z| < \frac{1}{2}$$

Since the ROC of $X(z)$ is $|z| < \frac{1}{2}$, $x[n]$ is a left-sided sequence, and from Table 4-1 we get

$$x[n] = -u[-n-1] + \left(\frac{1}{2}\right)^n u[-n-1] = \left[\left(\frac{1}{2}\right)^n - 1\right] u[-n-1]$$

which gives

$$x[n] = \{\ldots, 15, 7, 3, 1, 0\}$$
$$\uparrow$$

(b)
$$X(z) = \frac{z}{z-1} - \frac{z}{z - \frac{1}{2}} \qquad |z| > 1$$

Since the ROC of $X(z)$ is $|z| > 1$, $x[n]$ is a right-sided sequence, and from Table 4-1 we get

$$x[n] = u[n] - \left(\frac{1}{2}\right)^n u[n] = \left[1 - \left(\frac{1}{2}\right)^n\right] u[n]$$

which gives

$$x[n] = \left\{0, \frac{1}{2}, \frac{3}{4}, \frac{7}{8}, \ldots\right\}$$

4.20. Find the inverse z-transform of

$$X(z) = \frac{z}{z(z-1)(z-2)^2} \qquad |z| > 2$$

Using partial-fraction expansion, we have

$$\frac{X(z)}{z} = \frac{1}{(z-1)(z-2)^2} = \frac{c_1}{z-1} + \frac{\lambda_1}{z-2} + \frac{\lambda_2}{(z-2)^2} \tag{4.83}$$

where
$$c_1 = \frac{1}{(z-2)^2}\Bigg|_{z=1} = 1 \qquad \lambda_2 = \frac{1}{z-1}\Bigg|_{z=2} = 1$$

Substituting these values into Eq. (4.83), we have

$$\frac{1}{(z-1)(z-2)^2} = \frac{1}{z-1} + \frac{\lambda_1}{z-2} + \frac{1}{(z-2)^2}$$

Setting $z = 0$ in the above expression, we have

$$-\frac{1}{4} = -1 - \frac{\lambda_1}{2} + \frac{1}{4} \rightarrow \lambda_1 = -1$$

Thus,

$$X(z) = \frac{z}{z-1} - \frac{z}{z-2} + \frac{z}{(z-2)^2} \qquad |z| > 2$$

Since the ROC is $|z| > 2$, $x[n]$ is a right-sided sequence, and from Table 4-1 we get

$$x[n] = (1 - 2^n + n2^{n-1})u[n]$$

4.21. Find the inverse z-transform of

$$X(z) = \frac{2z^3 - 5z^2 + z + 3}{(z-1)(z-2)} \qquad |z| < 1$$

$$X(z) = \frac{2z^3 - 5z^2 + z + 3}{(z-1)(z-2)} = \frac{2z^3 - 5z^2 + z + 3}{z^2 - 3z + 2}$$

Note that $X(z)$ is an improper rational function; thus, by long division, we have

$$X(z) = 2z + 1 + \frac{1}{z^2 - 3z + 2} = 2z + 1 + \frac{1}{(z-1)(z-2)}$$

Let

$$X_1(z) = \frac{1}{(z-1)(z-2)}$$

Then

$$\frac{X_1(z)}{z} = \frac{1}{z(z-1)(z-2)} = \frac{c_1}{z} + \frac{c_2}{z-1} + \frac{c_3}{z-2}$$

where

$$c_1 = \frac{1}{(z-1)(z-2)}\bigg|_{z=0} = \frac{1}{2} \qquad c_2 = \frac{1}{z(z-2)}\bigg|_{z=1} = -1$$

$$c_3 = \frac{1}{z(z-1)}\bigg|_{z=2} = \frac{1}{2}$$

Thus,

$$X_1(z) = \frac{1}{2} - \frac{z}{z-1} + \frac{1}{2}\frac{z}{z-2}$$

and

$$X(z) = 2z + \frac{3}{2} - \frac{z}{z-1} + \frac{1}{2}\frac{z}{z-2} \qquad |z| < 1$$

Since the ROC of $X(z)$ is $|z| < 1$, $x[n]$ is a left-sided sequence, and from Table 4-1 we get

$$x[n] = 2\delta[n+1] + \frac{3}{2}\delta[n] + u[-n-1] - \frac{1}{2}2^n u[-n-1]$$

$$= 2\delta[n+1] + \frac{3}{2}\delta[n] + (1 - 2^{n-1})u[-n-1]$$

4.22. Find the inverse z-transform of

$$X(z) = \frac{3}{z-2} \qquad |z| > 2$$

$X(z)$ can be rewritten as

$$X(z) = \frac{3}{z-2} = 3z^{-1}\left(\frac{z}{z-2}\right) \qquad |z| > 2$$

Since the ROC is $|z| > 2$, $x[n]$ is a right-sided sequence, and from Table 4-1 we have

$$2^n u[n] \leftrightarrow \frac{z}{z-2}$$

Using the time-shifting property (4.18), we have

$$2^{n-1} u[n-1] \leftrightarrow z^{-1}\left(\frac{z}{z-2}\right) = \frac{1}{z-2}$$

Thus, we conclude that

$$x[n] = 3(2)^{n-1} u[n-1]$$

4.23. Find the inverse z-transform of

$$X(z) = \frac{2 + z^{-2} + 3z^{-4}}{z^2 + 4z + 3} \qquad |z| > 0$$

We see that $X(z)$ can be written as

$$X(z) = (2z^{-1} + z^{-3} + 3z^{-5}) X_1(z)$$

where

$$X_1(z) = \frac{z}{z^2 + 4z + 3}$$

Thus, if

$$x_1[n] \leftrightarrow X_1(z)$$

then by the linearity property (4.17) and the time-shifting property (4.18), we get

$$x[n] = 2x_1[n-1] + x_1[n-3] + 3x_1[n-5] \tag{4.84}$$

Now

$$\frac{X_1(z)}{z} = \frac{1}{z^2 + 4z + 3} = \frac{1}{(z+1)(z+3)} = \frac{c_1}{z+1} + \frac{c_2}{z+3}$$

Where

$$c_1 = \left.\frac{1}{z+3}\right|_{z=-1} = \frac{1}{2} \qquad c_2 = \left.\frac{1}{z+1}\right|_{z=-3} = -\frac{1}{2}$$

Then

$$X_1(z) = \frac{1}{2}\frac{z}{z+1} - \frac{1}{2}\frac{z}{z+3} \qquad |z| > 0$$

Since the ROC of $X_1(z)$ is $|z| > 0$, $x_1[n]$ is a right-sided sequence, and from Table 4-1 we get

$$x_1[n] = \frac{1}{2}[(-1)^n - (-3)^n] u[n]$$

Thus, from Eq. (4.84) we get

$$x[n] = [(-1)^{n-1} - (-3)^{n-1}] u[n-1] + \frac{1}{2}[(-1)^{n-3} - (-3)^{n-3}] u[n-3]$$

$$+ \frac{3}{2}[(-1)^{n-5} - (-3)^{n-5}] u[n-5]$$

4.24. Find the inverse z-transform of

$$X(z) = \frac{1}{(1 - az^{-1})^2} \qquad |z| > |a|$$

$$X(z) = \frac{1}{(1 - az^{-1})^2} = \frac{z^2}{(z-a)^2} \qquad |z| > |a| \tag{4.85}$$

From Eq. (4.78) (Prob. 4.12)

$$na^{n-1}u[n] \leftrightarrow \frac{z}{(z-a)^2} \qquad |z|>|a| \tag{4.86}$$

Now, from Eq. (4.85)

$$X(z) = z\left[\frac{z}{(z-a)^2}\right] \qquad |z|>|a|$$

and applying the time-shifting property (4.20) to Eq. (4.86), we get

$$x[n] = (n+1)\,a^n u[n+1] = (n+1)a^n u[n] \tag{4.87}$$

since $x[-1] = 0$ at $n = -1$.

System Function

4.25. Using the z-transform, redo Prob. 2.28.

From Prob. 2.28, $x[n]$ and $h[n]$ are given by

$$x[n] = u[n] \qquad h[n] = \alpha^n u[n] \qquad 0 < \alpha < 1$$

From Table 4-1

$$x[n] = u[n] \leftrightarrow X(z) = \frac{z}{z-1} \qquad |z|>|1|$$

$$h[n] = \alpha^n u[n] \leftrightarrow H(z) = \frac{z}{z-\alpha} \qquad |z|>|\alpha|$$

Then, by Eq. (4.40)

$$Y(z) = X(z)H(z) = \frac{z^2}{(z-1)(z-\alpha)} \qquad |z|>1$$

Using partial-fraction expansion, we have

$$\frac{Y(z)}{z} = \frac{z}{(z-1)(z-\alpha)} = \frac{c_1}{z-1} + \frac{c_2}{z-\alpha}$$

where

$$c_1 = \frac{z}{z-\alpha}\bigg|_{z=1} = \frac{1}{1-\alpha} \qquad c_2 = \frac{z}{z-1}\bigg|_{z=\alpha} = -\frac{\alpha}{1-\alpha}$$

Thus,

$$Y(z) = \frac{1}{1-\alpha}\frac{z}{z-1} - \frac{\alpha}{1-\alpha}\frac{z}{z-\alpha} \qquad |z|>1$$

Taking the inverse z-transform of $Y(z)$, we get

$$y[n] = \frac{1}{1-\alpha}u[n] - \frac{\alpha}{1-\alpha}\alpha^n u[n] = \left(\frac{1-\alpha^{n-1}}{1-\alpha}\right)u[n]$$

which is the same as Eq. (2.134).

4.26. Using the z-transform, redo Prob. 2.29.

(a) From Prob. 2.29(a), $x[n]$ and $h[n]$ are given by

$$x[n] = \alpha^n u[n] \qquad h[n] = \beta^n u[n]$$

From Table 4-1

$$x[n] = \alpha^n u[n] \leftrightarrow X(z) = \frac{z}{z - \alpha} \qquad |z| > |\alpha|$$

$$h[n] = \beta^n u[n] \leftrightarrow H(z) = \frac{z}{z - \beta} \qquad |z| > |\beta|$$

Then

$$Y(z) = X(z)H(z) = \frac{z^2}{(z - \alpha)(z - \beta)} \qquad |z| > \max(\alpha, \beta)$$

Using partial-fraction expansion, we have

$$\frac{Y(z)}{z} = \frac{z}{(z - \alpha)(z - \beta)} = \frac{c_1}{z - \alpha} + \frac{c_2}{z - \beta}$$

where

$$c_1 = \frac{z}{z - \beta}\bigg|_{z=\alpha} = \frac{\alpha}{\alpha - \beta} \qquad c_2 = \frac{z}{z - \alpha}\bigg|_{z=\beta} = -\frac{\beta}{\alpha - \beta}$$

Thus,

$$Y(z) = \frac{\alpha}{\alpha - \beta}\frac{z}{z - \alpha} - \frac{\beta}{\alpha - \beta}\frac{z}{z - \beta} \qquad |z| > \max(\alpha, \beta)$$

and

$$y[n] = \left[\frac{\alpha}{\alpha - \beta}\alpha^n - \frac{\beta}{\alpha - \beta}\beta^n\right]u[n] = \left(\frac{\alpha^{n+1} - \beta^{n+1}}{\alpha - \beta}\right)u[n]$$

which is the same as Eq. (2.135). When $\alpha = \beta$,

$$Y(z) = \frac{z^2}{(z - \alpha)^2} \qquad |z| > \alpha$$

Using partial-fraction expansion, we have

$$\frac{Y(z)}{z} = \frac{z}{(z - \alpha)^2} = \frac{\lambda_1}{z - \alpha} + \frac{\lambda_2}{(z - \alpha)^2}$$

where

$$\lambda_2 = z\big|_{z=\alpha} = \alpha$$

and

$$\frac{z}{(z - \alpha)^2} = \frac{\lambda_1}{z - \alpha} + \frac{\alpha}{(z - \alpha)^2}$$

Setting $z = 0$ in the above expression, we have

$$0 = -\frac{\lambda_1}{\alpha} + \frac{1}{\alpha} \rightarrow \lambda_1 = 1$$

Thus,

$$Y(z) = \frac{z}{z - \alpha} + \frac{\alpha z}{(z - \alpha)^2} \qquad |z| > \alpha$$

and from Table 4-1 we get

$$y[n] = (\alpha^n + n\alpha^n)u[n] = \alpha^n(1 + n)u[n]$$

Thus, we obtain the same results as Eq. (2.135).

(b)   From Prob. 2.29(b), $x[n]$ and $h[n]$ are given by

$$x[n] = \alpha^n u[n] \qquad h[n] = \alpha^{-n} u[-n] \qquad 0 < \alpha < 1$$

From Table 4-1 and Eq. (4.75)

$$x[n] = \alpha^n u[n] \leftrightarrow X(z) = \frac{z}{z - \alpha} \qquad |z| > |\alpha|$$

$$h[n] = \alpha^{-n} u[-n] \leftrightarrow H(z) = \frac{1}{1 - \alpha z} = -\frac{1}{\alpha(z - 1/\alpha)} \qquad |z| < \frac{1}{|\alpha|}$$

Then
$$Y(z) = X(z)H(z) = -\frac{1}{\alpha}\frac{z}{(z-\alpha)(z-1/\alpha)} \qquad \alpha < |z| < \frac{1}{\alpha}$$

Using partial-fraction expansion, we have

$$\frac{Y(z)}{z} = -\frac{1}{\alpha}\frac{1}{(z-\alpha)(z-1/\alpha)} = -\frac{1}{\alpha}\left(\frac{c_1}{z-\alpha} + \frac{c_2}{z-1/\alpha}\right)$$

where
$$c_1 = \frac{1}{z-1/\alpha}\bigg|_{z=\alpha} = -\frac{\alpha}{1-\alpha^2} \qquad c_2 = \frac{1}{z-\alpha}\bigg|_{z=1/\alpha} = \frac{\alpha}{1-\alpha^2}$$

Thus,

$$Y(z) = \frac{1}{1-\alpha^2}\frac{z}{z-\alpha} - \frac{1}{1-\alpha^2}\frac{z}{z-1/\alpha} \qquad \alpha < |z| < \frac{1}{\alpha}$$

and from Table 4-1 we obtain

$$y[n] = \frac{1}{1-\alpha^2}\alpha^n u[n] - \frac{1}{1-\alpha^2}\left\{-\left(\frac{1}{\alpha}\right)^n u[-n-1]\right\}$$

$$= \frac{1}{1-\alpha^2}\alpha^n u[n] + \frac{1}{1-\alpha^2}\alpha^{-n} u[-n-1] = \frac{1}{1-\alpha^2}\alpha^{|n|}$$

which is the same as Eq. (2.137).

4.27. Using the z-transform, redo Prob. 2.30.

From Fig. 2-23 and definition (4.3)

$$x[n] = \{1,1,1,1\} \leftrightarrow X(z) = 1 + z^{-1} + z^{-2}z^{-3}$$
$$h[n] = \{1,1,1\} \leftrightarrow H(z) = 1 + z^{-1} + z^{-2}$$

Thus, by the convolution property (4.26)

$$Y(z) = X(z)H(z) = (1 + z^{-1} + z^{-2} + z^{-3})(1 + z^{-1} + z^{-2})$$
$$= 1 + 2z^{-1} + 3z^{-2} + 3z^{-3} + 2z^{-4} + z^{-5}$$

Hence,

$$h[n] = \{1,2,3,3,2,1\}$$

which is the same result obtained in Prob. 2.30.

4.28. Using the z-transform, redo Prob. 2.32.

Let $x[n]$ and $y[n]$ be the input and output of the system. Then

$$x[n] = u[n] \quad \leftrightarrow X(z) = \frac{z}{z-1} \qquad |z| > 1$$

$$y[n] = \alpha^n u[n] \leftrightarrow Y(z) = \frac{z}{z-\alpha} \qquad |z| > |\alpha|$$

Then, by Eq. (4.41)

$$H(z) = \frac{Y(z)}{X(z)} = \frac{z-1}{z-\alpha} \qquad |z| > \alpha$$

Using partial-fraction expansion, we have

$$\frac{H(z)}{z} = \frac{z-1}{z(z-\alpha)} = \frac{c_1}{z} + \frac{c_2}{z-\alpha}$$

where

$$c_1 = \frac{z-1}{z-\alpha}\bigg|_{z=0} = \frac{1}{\alpha} \qquad c_2 = \frac{z-1}{z}\bigg|_{z=\alpha} = \frac{\alpha-1}{\alpha} = -\frac{1-\alpha}{\alpha}$$

Thus,

$$H(z) = \frac{1}{\alpha} - \frac{1-\alpha}{\alpha}\frac{z}{z-\alpha} \qquad |z| > \alpha$$

Taking the inverse z-transform of $H(z)$, we obtain

$$h[n] = \frac{1}{\alpha}\delta[n] - \frac{1-\alpha}{\alpha}\alpha^n u[n]$$

When $n = 0$,

$$h[0] = \frac{1}{\alpha} - \frac{1-\alpha}{\alpha} = 1$$

Then

$$h[n] = \begin{cases} 1 & n=0 \\ -(1-\alpha)\alpha^{n-1} & n \geq 1 \end{cases}$$

Thus, $h[n]$ can be rewritten as

$$h[n] = \delta[n] - (1-\alpha)\,\alpha^{n-1}u[n-1]$$

which is the same result obtained in Prob. 2.32.

4.29. The output $y[n]$ of a discrete-time LTI system is found to be $2(\frac{1}{3})^n u[n]$ when the input $x[n]$ is $u[n]$.

  (a) Find the impulse response $h[n]$ of the system.

  (b) Find the output $y[n]$ when the input $x[n]$ is $(\frac{1}{2})^n u[n]$.

  (a)

$$x[n] = u[n] \leftrightarrow X(z) = \frac{z}{z-1} \qquad |z| > 1$$

$$y[n] = 2\left(\frac{1}{3}\right)^n u[n] \leftrightarrow Y(z) = \frac{2z}{z-\frac{1}{3}} \qquad |z| > \frac{1}{3}$$

Hence, the system function $H(z)$ is

$$H(z) = \frac{Y(z)}{X(z)} = \frac{2(z-1)}{z-\frac{1}{3}} \qquad |z| > \frac{1}{3}$$

Using partial-fraction expansion, we have

$$\frac{H(z)}{z} = \frac{2(z-1)}{z\left(z-\frac{1}{3}\right)} = \frac{c_1}{z} + \frac{c_2}{z-\frac{1}{3}}$$

where

$$c_1 = \frac{2(z-1)}{z-\frac{1}{3}}\bigg|_{z=0} = 6 \qquad c_2 = \frac{2(z-1)}{z}\bigg|_{z=1/3} = -4$$

Thus,

$$H[z] = 6 - 4\frac{z}{z - \frac{1}{3}} \qquad |z| > \frac{1}{3}$$

Taking the inverse z-transform of $H(z)$, we obtain

$$h[n] = 6\delta[n] - 4\left(\frac{1}{3}\right)^n u[n]$$

(b)

$$x[n] = \left(\frac{1}{2}\right)^n u[n] \leftrightarrow X(z) = \frac{z}{z - \frac{1}{2}} \qquad |z| > \frac{1}{2}$$

Then,

$$Y(z) = X(z)H(z) = \frac{2z(z-1)}{\left(z - \frac{1}{2}\right)\left(z - \frac{1}{3}\right)} \qquad |z| > \frac{1}{2}$$

Again by partial-fraction expansion we have

$$\frac{Y(z)}{z} = \frac{2(z-1)}{\left(z - \frac{1}{2}\right)\left(z - \frac{1}{3}\right)} = \frac{c_1}{z - \frac{1}{2}} + \frac{c_2}{z - \frac{1}{3}}$$

where

$$c_1 = \frac{2(z-1)}{z - \frac{1}{3}}\Bigg|_{z=1/2} = -6 \qquad c_2 = \frac{2(z-1)}{z - \frac{1}{2}}\Bigg|_{z=1/3} = 8$$

Thus,

$$Y(z) = -6\frac{z}{z - \frac{1}{2}} + 8\frac{z}{z - \frac{1}{3}} \qquad |z| > \frac{1}{2}$$

Taking the inverse z-transform of $Y(z)$, we obtain

$$y[n] = \left[-6\left(\frac{1}{2}\right)^n + 8\left(\frac{1}{3}\right)^n\right]u[n]$$

4.30. If a discrete-time LTI system is BIBO stable, show that the ROC of its system function $H(z)$ must contain the unit circle; that is, $|z| = 1$.

A discrete-time LTI system is BIBO stable if and only if its impulse response $h[n]$ is absolutely summable, that is [Eq. (2.49)],

$$\sum_{n=-\infty}^{\infty} |h[n]| < \infty$$

Now

$$H(z) = \sum_{n=-\infty}^{\infty} h[n]z^{-n}$$

Let $z = e^{j\Omega}$ so that $|z| = |e^{j\Omega}| = 1$. Then

$$\left|H(e^{j\Omega})\right| = \left|\sum_{n=-\infty}^{\infty} h[n]e^{-j\Omega n}\right|$$

$$\leq \sum_{n=-\infty}^{\infty} \left|h[n]e^{-j\Omega n}\right| = \sum_{n=-\infty}^{\infty} |h[n]| < \infty$$

Therefore, we see that if the system is stable, then $H(z)$ converges for $z = e^{j\Omega}$. That is, for a stable discrete-time LTI system, the ROC of $H(z)$ must contain the unit circle $|z| = 1$.

4.31. Using the z-transform, redo Prob. 2.38.

(a) From Prob. 2.38 the impulse response of the system is

$$h[n] = \alpha^n u[n]$$

Then

$$H(z) = \frac{z}{z - \alpha} \qquad |z| > |\alpha|$$

Since the ROC of $H(z)$ is $|z| > |\alpha|$, $z = \infty$ is included. Thus, by the result from Prob. 4.5 we conclude that $h[n]$ is a causal sequence. Thus, the system is causal.

(b) If $|\alpha| > 1$, the ROC of $H(z)$ does not contain the unit circle $|z| = 1$, and hence the system will not be stable. If $|\alpha| < 1$, the ROC of $H(z)$ contains the unit circle $|z| = 1$, and hence the system will be stable.

4.32. A causal discrete-time LTI system is described by

$$y[n] - \frac{3}{4}y[n-1] + \frac{1}{8}y[n-2] = x[n] \tag{4.88}$$

where $x[n]$ and $y[n]$ are the input and output of the system, respectively.

(a) Determine the system function $H(z)$.

(b) Find the impulse response $h[n]$ of the system.

(c) Find the step response $s[n]$ of the system.

(a) Taking the z-transform of Eq. (4.88), we obtain

$$Y(z) - \frac{3}{4}z^{-1}Y(z) + \frac{1}{8}z^{-2}Y(z) = X(z)$$

or

$$\left(1 - \frac{3}{4}z^{-1} + \frac{1}{8}z^{-2}\right)Y(z) = X(z)$$

Thus,

$$H(z) = \frac{Y(z)}{X(z)} = \frac{1}{1 - \frac{3}{4}z^{-1} + \frac{1}{8}z^{-2}} = \frac{z^2}{z^2 - \frac{3}{4}z + \frac{1}{8}}$$

$$= \frac{z^2}{\left(z - \frac{1}{2}\right)\left(z - \frac{1}{4}\right)} \qquad |z| > \frac{1}{2}$$

(b) Using partial-fraction expansion, we have

$$\frac{H(z)}{z} = \frac{z}{\left(z - \frac{1}{2}\right)\left(z - \frac{1}{4}\right)} = \frac{c_1}{z - \frac{1}{2}} + \frac{c_2}{z - \frac{1}{4}}$$

where

$$c_1 = \frac{z}{z - \frac{1}{4}}\bigg|_{z=1/2} = 2 \qquad c_2 = \frac{z}{z - \frac{1}{2}}\bigg|_{z=1/4} = -1$$

Thus,

$$H(z) = 2\frac{z}{z - \frac{1}{2}} - \frac{z}{z - \frac{1}{4}} \qquad |z| > \frac{1}{2}$$

Taking the inverse z-transform of $H(z)$, we get

$$h[n] = \left[2\left(\frac{1}{2}\right)^n - \left(\frac{1}{4}\right)^n \right] u[n]$$

(c)

$$x[n] = u[n] \leftrightarrow X(z) = \frac{z}{z-1} \qquad |z| > 1$$

Then

$$Y(z) = X(z)H(z) = \frac{z^3}{(z-1)\left(z-\frac{1}{2}\right)\left(z-\frac{1}{4}\right)} \qquad |z| > 1$$

Again using partial-fraction expansion, we have

$$\frac{Y(z)}{z} = \frac{z^2}{(z-1)\left(z-\frac{1}{2}\right)\left(z-\frac{1}{4}\right)} = \frac{c_1}{z-1} + \frac{c_2}{z-\frac{1}{2}} + \frac{c_3}{z-\frac{1}{4}}$$

where

$$c_1 = \left. \frac{z^2}{\left(z-\frac{1}{2}\right)\left(z-\frac{1}{4}\right)} \right|_{z=1} = \frac{8}{3} \qquad c_2 = \left. \frac{z^2}{(z-1)\left(z-\frac{1}{4}\right)} \right|_{z=1/2} = -2$$

$$c_3 = \left. \frac{z^2}{(z-1)\left(z-\frac{1}{2}\right)} \right|_{z=1/4} = \frac{1}{3}$$

Thus,

$$Y(z) = \frac{8}{3}\frac{z}{z-1} - 2\frac{z}{z-\frac{1}{2}} + \frac{1}{3}\frac{z}{z-\frac{1}{4}} \qquad |z| > 1$$

Taking the inverse z-transformation of $Y(z)$, we obtain

$$y[n] = s[n] = \left[\frac{8}{3} - 2\left(\frac{1}{2}\right)^n + \frac{1}{3}\left(\frac{1}{4}\right)^n \right] u[n]$$

4.33. Using the z-transform, redo Prob. 2.41.

As in Prob. 2.41, from Fig. 2-30 we see that

$$q[n] = 2q[n-1] + x[n]$$
$$y[n] = q[n] + 3q[n-1]$$

Taking the z-transform of the above equations, we get

$$Q(z) = 2z^{-1} Q(z) + X(z)$$
$$Y(z) = Q(z) + 3z^{-1} Q(z)$$

Rearranging, we get

$$(1 - 2z^{-1})Q(z) = X(z)$$
$$(1 + 3z^{-1})Q(z) = Y(z)$$

from which we obtain

$$H(z) = \frac{Y(z)}{X(z)} = \frac{1 + 3z^{-1}}{1 - 2z^{-1}} \tag{4.89}$$

Rewriting Eq. (4.89), we have

$$(1 - 2z^{-1})Y(z) = (1 + 3z^{-1})X(z)$$

or

$$Y(z) - 2z^{-1}Y(z) = X(z) + 3z^{-1}X(z) \tag{4.90}$$

Taking the inverse z-transform of Eq. (4.90) and using the time-shifting property (4.18), we obtain

$$y[n] - 2y[n - 1] = x[n] + 3x[n - 1]$$

which is the same as Eq. (2.148).

4.34. Consider the discrete-time system shown in Fig. 4-8. For what values of k is the system BIBO stable?

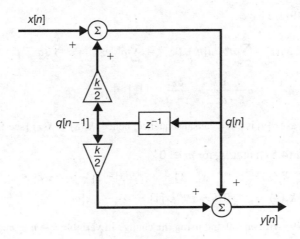

Fig. 4-8

From Fig. 4-8 we see that

$$q[n] = x[n] + \frac{k}{2}q[n - 1]$$

$$y[n] = q[n] + \frac{k}{3}q[n - 1]$$

Taking the z-transform of the above equations, we obtain

$$Q(z) = X(z) + \frac{k}{2}z^{-1}Q(z)$$

$$Y(z) = Q(z) + \frac{k}{3}z^{-1}Q(z)$$

Rearranging, we have

$$\left(1 - \frac{k}{2}z^{-1}\right)Q(z) = X(z)$$

$$\left(1 + \frac{k}{3}z^{-1}\right)Q(z) = Y(z)$$

from which we obtain

$$H(z) = \frac{Y(z)}{X(z)} = \frac{1 + (k/3)z^{-1}}{1 - (k/2)z^{-1}} = \frac{z + k/3}{z - k/2} \qquad |z| > \left|\frac{k}{2}\right|$$

which shows that the system has one zero at $z = -k/3$ and one pole at $z = k/2$ and that the ROC is $|z| > |k/2|$. Thus, as shown in Prob. 4.30, the system will be BIBO stable if the ROC contains the unit circle, $|z| = 1$. Hence, the system is stable only if $|k| < 2$.

Unilateral z-Transform

4.35. Find the unilateral z-transform of the following $x[n]$:

(a) $x[n] = a^n u[n]$

(b) $x[n] = a^{n+1} u[n + 1]$

(a)   Since $x[n] = 0$ for $n < 0$, $X_I(z) = X(z)$, and from Example 4.1 we have

$$X_I(z) = \frac{1}{1 - az^{-1}} = \frac{z}{z - a} \qquad |z| > |a| \tag{4.91}$$

(b)   By definition (4.49) we have

$$X_I(z) = \sum_{n=0}^{\infty} a^{n+1} u[n+1] z^{-n} = \sum_{n=0}^{\infty} a^{n+1} z^{-n} = a \sum_{n=0}^{\infty} (az^{-1})^n$$

$$= a \frac{1}{1 - az^{-1}} = \frac{az}{z - a} \qquad |z| > |a| \tag{4.92}$$

Note that in this case $x[n]$ is not a causal sequence; hence, $X_I(z) \neq X(z)$ [see Eq. (4.73) in Prob. 4.10].

4.36. Verify Eqs. (4.50) and (4.51); that is, for $m \geq 0$,

(a) $x[n - m] \leftrightarrow z^{-m} X_I(z) + z^{-m+1} x[-1] + z^{-m+2} x[-2] + \cdots + x[-m]$

(b) $x[n + m] \leftrightarrow z^m X_I(z) - z^m x[0] - z^{m-1} x[1] - \cdots - z x[m - 1]$

(a)   By definition (4.49) with $m \geq 0$ and using the change in variable $k = n - m$, we have

$$\mathfrak{Z}_I\{x[n - m]\} = \sum_{n=0}^{\infty} x[n - m] z^{-n} = \sum_{k=-m}^{\infty} x[k] z^{-(m+k)}$$

$$= z^{-m} \left\{ \sum_{k=0}^{\infty} x[k] z^{-k} + \sum_{k=-1}^{-m} x[k] z^{-k} \right\}$$

$$= z^{-m} \{ X_I(z) + x[-1] z + x[-2] z^2 + \cdots + x[-m] z^m \}$$

$$= z^{-m} X_I(z) + z^{-m+1} x[-1] + z^{-m+2} x[-2] + \cdots + x[-m]$$

(b)   With $m \geq 0$

$$\mathfrak{Z}_I\{x[n + m]\} = \sum_{n=0}^{\infty} x[n + m] z^{-n} = \sum_{k=m}^{\infty} x[k] z^{-(k-m)}$$

$$= z^m \left\{ \sum_{k=0}^{\infty} x[k] z^{-k} - \sum_{k=0}^{m-1} x[k] z^{-k} \right\}$$

$$= z^m \{ X_I(z) - (x[0] + x[1] z^{-1} + \cdots + x[m - 1] z^{-(m-1)}) \}$$

$$= z^m X_I(z) - z^m x[0] - z^{m-1} x[1] - \cdots - z x[m - 1]$$

4.37. Using the unilateral z-transform, redo Prob. 2.42.

The system is described by

$$y[n] - ay[n - 1] = x[n] \tag{4.93}$$

with $y[-1] = y_{-1}$ and $x[n] = Kb^n u[n]$. Let

$$y[n] \leftrightarrow Y_I(z)$$

Then from Eq. (4.50)

$$y[n-1] \leftrightarrow z^{-1}Y_I(z) + y[-1] = z^{-1}Y_I(z) + y_{-1}$$

From Table 4-1 we have

$$x[n] \leftrightarrow X_I(z) = K\frac{z}{z-b} \qquad |z| > |b|$$

Taking the unilateral z-transform of Eq. (4.93), we obtain

$$Y_I(z) - a\{z^{-1}Y_I(z) + y_{-1}\} = K\frac{z}{z-b}$$

or

$$(1 - az^{-1})Y_I(z) = ay_{-1} + K\frac{z}{z-b}$$

or

$$\left(\frac{z-a}{z}\right)Y_I(z) = ay_{-1} + K\frac{z}{z-b}$$

Thus,

$$Y_I(z) = ay_{-1}\frac{z}{z-a} + K\frac{z^2}{(z-a)(z-b)}$$

Using partial-fraction expansion, we obtain

$$Y_I(z) = ay_{-1}\frac{z}{z-a} + \frac{K}{b-a}\left(b\frac{z}{z-b} - a\frac{z}{z-a}\right)$$

Taking the inverse z-transform of $Y_I(z)$, we get

$$y[n] = ay_{-1}a^n u[n] + K\frac{b}{b-a}b^n u[n] - K\frac{a}{b-a}a^n u[n]$$

$$= \left(y_{-1}a^{n+1} + K\frac{b^{n+1} - a^{n+1}}{b-a}\right)u[n]$$

which is the same as Eq. (2.158).

4.38. For each of the following difference equations and associated input and initial conditions, determine the output $y[n]$:

(a) $y[n] - \frac{1}{2}y[n-1] = x[n]$, with $x[n] = (\frac{1}{3})^n$, $y[-1] = 1$

(b) $3y[n] - 4y[n-1] + y[n-2] = x[n]$, with $x[n] = (\frac{1}{2})^n$, $y[-1] = 1$, $y[-2] = 2$

(a)

$$x[n] \leftrightarrow X_I(z) = \frac{z}{z - \frac{1}{3}} \qquad |z| > \left|\frac{1}{3}\right|$$

Taking the unilateral z-transform of the given difference equation, we get

$$Y_I(z) - \frac{1}{2}\{z^{-1}Y_I(z) + y[-1]\} = X_I(z)$$

Substituting $y[-1] = 1$ and $X_I(z)$ into the above expression, we get

$$\left(1 - \frac{1}{2}z^{-1}\right)Y_I(z) = \frac{1}{2} + \frac{z}{z - \frac{1}{3}}$$

or

$$\left(\frac{z - \frac{1}{2}}{z}\right)Y_I(z) = \frac{1}{2} + \frac{z}{z - \frac{1}{3}}$$

Thus,

$$Y_I(z) = \frac{1}{2}\frac{z}{z-\frac{1}{2}} + \frac{z^2}{\left(z-\frac{1}{2}\right)\left(z-\frac{1}{3}\right)} = \frac{7}{2}\frac{z}{z-\frac{1}{2}} - 2\frac{z}{z-\frac{1}{3}}$$

Hence,

$$y[n] = 7\left(\frac{1}{2}\right)^{n+1} - 2\left(\frac{1}{3}\right)^n \quad n \geq -1$$

(b)

$$x[n] \leftrightarrow X_I(z) = \frac{z}{z-\frac{1}{2}} \quad |z| > \left|\frac{1}{2}\right|$$

Taking the unilateral z-transform of the given difference equation, we obtain

$$3Y_I(z) - 4\{z^{-1}Y_I(z) + y[-1]\} + \{z^{-2}Y_I(z) + z^{-1}y[-1] + y[-2]\} = X_I(z)$$

Substituting $y[-1] = 1$, $y[-2] = 2$, and $X_I(z)$ into the above expression, we get

$$(3 - 4z^{-1} + z^{-2})Y_I(z) = 2 - z^{-1} + \frac{z}{z-\frac{1}{2}}$$

or

$$\frac{3(z-1)\left(z-\frac{1}{3}\right)}{z^2}Y_I(z) = \frac{3z^2 - 2z + \frac{1}{2}}{z\left(z-\frac{1}{2}\right)}$$

Thus,

$$Y_I(z) = \frac{z\left(3z^2 - 2z + \frac{1}{2}\right)}{3(z-1)\left(z-\frac{1}{2}\right)\left(z-\frac{1}{3}\right)}$$

$$= \frac{3}{2}\frac{z}{z-1} - \frac{z}{z-\frac{1}{2}} + \frac{1}{2}\frac{z}{z-\frac{1}{3}}$$

Hence,

$$y[n] = \frac{3}{2} - \left(\frac{1}{2}\right)^n + \frac{1}{2}\left(\frac{1}{3}\right)^n \quad n \geq -2$$

4.39. Let $x[n]$ be a causal sequence and

$$x[n] \leftrightarrow X(z)$$

Show that

$$x[0] = \lim_{z \to \infty} X(z) \quad\quad (4.94)$$

Equation (4.94) is called the *initial value theorem* for the z-transform.

Since $x[n] = 0$ for $n < 0$, we have

$$X[z] = \sum_{n=0}^{\infty} x[n]z^{-n} = x[0] + x[1]z^{-1} + x[2]z^{-2} + \cdots$$

As $z \to \infty$, $z^{-n} \to 0$ for $n > 0$. Thus, we get

$$\lim_{z \to \infty} X(z) = x[0]$$

4.40. Let $x[n]$ be a causal sequence and

$$x[n] \leftrightarrow X(z)$$

Show that if $X(z)$ is a rational function with all its poles strictly inside the unit circle except possibly for a first-order pole at $z = 1$, then

$$\lim_{N \to \infty} x[N] = \lim_{z \to 1} (1 - z^{-1}) X(z) \qquad (4.95)$$

Equation (4.95) is called the *final value theorem* for the z-transform.

From the time-shifting property (4.19) we have

$$\mathcal{3}\{x[n] - x[n-1]\} = (1 - z^{-1}) X(z) \qquad (4.96)$$

The left-hand side of Eq. (4.96) can be written as

$$\sum_{n=0}^{\infty} \{x[n] - x[n-1]\} z^{-n} = \lim_{N \to \infty} \sum_{n=0}^{N} \{x[n] - x[n-1]\} z^{-n}$$

If we now let $z \to 1$, then from Eq. (4.96) we have

$$\lim_{z \to 1} (1 - z^{-1}) X(z) = \lim_{N \to \infty} \sum_{n=0}^{N} \{x[n] - x[n-1]\} = \lim_{N \to \infty} x[N]$$

SUPPLEMENTARY PROBLEMS

4.41. Find the z-transform of the following $x[n]$:

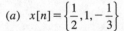

(a) $x[n] = \left\{ \dfrac{1}{2}, 1, -\dfrac{1}{3} \right\}$

(b) $x[n] = 2\delta[n+2] - 3\delta[n-2]$

(c) $x[n] = 3\left(-\dfrac{1}{2}\right)^n u[n] - 2(3)^n u[-n-1]$

(d) $x[n] = 3\left(\dfrac{1}{2}\right)^n u[n] - 2\left(\dfrac{1}{4}\right)^n u[-n-1]$

4.42. Show that if $x[n]$ is a left-sided sequence and $X(z)$ converges from some value of z, then the ROC of $X(z)$ is of the form

$$|z| < r_{\min} \qquad \text{or} \qquad 0 < |z| < r_{\min}$$

where $r_{\min}$ is the smallest magnitude of any of the poles of $X(z)$.

4.43. Given

$$X(z) = \frac{z(z-4)}{(z-1)(z-2)(z-3)}$$

(a) State all the possible regions of convergence.

(b) For which ROC is $X(z)$ the z-transform of a causal sequence?

4.44. Verify the time-reversal property (4.23); that is,

$$x[-n] \leftrightarrow X\left(\frac{1}{z}\right) \qquad R' = \frac{1}{R}$$

4.45. Show the following properties for the z-transform.

(a) If $x[n]$ is even, then $X(z^{-1}) = X(z)$.

(b) If $x[n]$ is odd, then $X(z^{-1}) = -X(z)$.

(c) If $x[n]$ is odd, then there is a zero in $X(z)$ at $z = 1$.

4.46. Consider the continuous-time signal

$$x(t) = e^{-\alpha t} \qquad t \geq 0$$

Let the sequence $x[n]$ be obtained by uniform sampling of $x(t)$ such that $x[n] = x(nT_s)$, where T_s is the sampling interval. Find the z-transform of $x[n]$.

4.47. Derive the following transform pairs:

$$(\cos \Omega_0 n)u[n] \leftrightarrow \frac{z^2 - (\cos \Omega_0)z}{z^2 - (2\cos \Omega_0)z + 1} \qquad |z| > 1$$

$$(\sin \Omega_0 n)u[n] \leftrightarrow \frac{(\sin \Omega_0)z}{z^2 - (2\cos \Omega_0)z + 1} \qquad |z| > 1$$

4.48. Find the z-transforms of the following $x[n]$:

(a) $x[n] = (n-3)u[n-3]$

(b) $x[n] = (n-3)u[n]$

(c) $x[n] = u[n] - u[n-3]$

(d) $x[n] = n\{u[n] - u[n-3]\}$

4.49. Using the relation

$$a^n u[n] \leftrightarrow \frac{z}{z-a} \qquad |z| > |a|$$

find the z-transform of the following $x[n]$:

(a) $x[n] = na^{n-1}u[n]$

(b) $x[n] = n(n-1)a^{n-2}u[n]$

(c) $x[n] = n(n-1)\cdots(n-k+1)a^{n-k}u[n]$

4.50. Using the z-transform, verify Eqs. (2.130) and (2.131) in Prob. 2.27; that is,

(a) $x[n] * \delta[n] = x[n]$

(b) $x[n] * \delta[n - n_0] = x[n - n_0]$

4.51. Using the z-transform, redo Prob. 2.47.

4.52. Find the inverse z-transform of

$$X(z) = e^{a/z} \qquad |z| > 0$$

4.53. Using the method of long division, find the inverse z-transform of the following $X(z)$:

(a) $X(z) = \dfrac{z}{(z-1)(z-2)}, \quad |z| < 1$

(b) $X(z) = \dfrac{z}{(z-1)(z-2)}, \quad 1 < |z| < 2$

(c) $X(z) = \dfrac{z}{(z-1)(z-2)}, \quad |z| > 2$

4.54. Using the method of partial-fraction expansion, redo Prob. 4.53.

4.55. Consider the system shown in Fig. 4-9. Find the system function $H(z)$ and its impulse response $h[n]$.

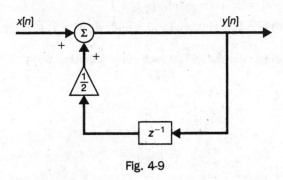

Fig. 4-9

4.56. Consider the system shown in Fig. 4-10.

(a) Find the system function $H(z)$.

(b) Find the difference equation relating the output $y[n]$ and input $x[n]$.

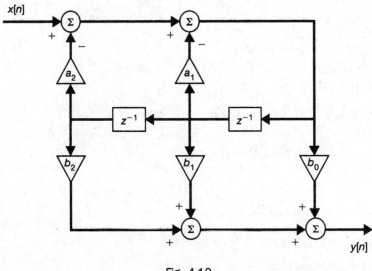

Fig. 4-10

4.57. Consider a discrete-time LTI system whose system function $H(z)$ is given by

$$H(z) = \frac{z}{z - \dfrac{1}{2}} \quad |z| > \frac{1}{2}$$

(a) Find the step response $s[n]$.

(b) Find the output $y[n]$ to the input $x[n] = nu[n]$.

4.58. Consider a causal discrete-time system whose output $y[n]$ and input $x[n]$ are related by

$$y[n] - \frac{5}{6}y[n-1] + \frac{1}{6}y[n-2] = x[n]$$

(a) Find its system function $H(z)$.

(b) Find its impulse response $h[n]$.

4.59. Using the unilateral z-transform, solve the following difference equations with the given initial conditions:

(a) $y[n] - 3y[n-1] = x[n]$, with $x[n] = 4u[n]$, $y[-1] = 1$

(b) $y[n] - 5y[n-1] + 6y[n-2] = x[n]$, with $x[n] = u[n]$, $y[-1] = 3$, $y[-2] = 2$

4.60. Determine the initial and final values of $x[n]$ for each of the following $X(z)$:

(a) $X(z) = \dfrac{2z\left(z - \dfrac{5}{12}\right)}{\left(z - \dfrac{1}{2}\right)\left(z - \dfrac{1}{3}\right)}$, $|z| > \dfrac{1}{2}$

(b) $X(z) = \dfrac{z}{2z^2 - 3z + 1}$, $|z| > 1$

ANSWERS TO SUPPLEMENTARY PROBLEMS

4.41. (a) $X(z) = \dfrac{1}{2} + z^{-1} - \dfrac{1}{3}z^{-2}$, $0 < |z|$

(b) $X(z) = 2z^3 - 3z^{-3}$, $0 < |z| < \infty$

(c) $X(z) = \dfrac{z(5z - 8)}{\left(z + \dfrac{1}{2}\right)(z - 3)}$, $\dfrac{1}{2} < |z| < 3$

(d) $X(z)$ does not exist.

4.42. *Hint:* Proceed in a manner similar to Prob. 4.5.

4.43. (a) $0 < |z| < 1, 1 < |z| < 2, 2 < |z| < 3, |z| > 3$

(b) $|z| > 3$

4.44. *Hint:* Change n to $-n$ in definition (4.3).

4.45. *Hint:* (a) Use Eqs. (1.2) and (4.23).

(b) Use Eqs. (1.3) and (4.23).

(c) Use the result from part (b).

4.46. $X(z) = \dfrac{1}{1 - e^{-\alpha T_s}z^{-1}}$

4.47. *Hint:* Use Euler's formulas.

$$\cos \Omega_0 n = \frac{1}{2}(e^{j\Omega_0 n} + e^{-j\Omega_0 n}) \qquad \sin \Omega_0 n = \frac{1}{2j}(e^{j\Omega_0 n} - e^{-j\Omega_0 n})$$

and use Eqs. (4.8) and (4.10) with $a = e^{\pm j\Omega_0}$.

4.48. (a) $\dfrac{z^{-2}}{(z-1)^2}$, $|z|>1$

    (b) $\dfrac{-3z^2+4z}{(z-1)^2}$, $|z|>1$

    (c) $\dfrac{z-z^{-2}}{z-1}$, $|z|>1$

    (d) $\dfrac{z-4z^{-2}+3z^{-3}}{(z-1)^2}$, $|z|>1$

4.49. *Hint:* Differentiate both sides of the given relation consecutively with respect to a.

    (a) $\dfrac{z}{(z-a)^2}$, $|z|>|a|$

    (b) $\dfrac{2z}{(z-a)^3}$, $|z|>|a|$

    (c) $\dfrac{k!z}{(z-a)^{k+1}}$, $|z|>|a|$

4.50. *Hint:* Use Eq. (4.26) of the z-transform and transform pairs 1 and 4 from Table 4-1.

4.51. *Hint:* Use Eq. (4.26) and Table 4-1.

4.52. *Hint:* Use the power series expansion of the exponential function e^r.

$$x[n]=\frac{a^n}{n!}u[n]$$

4.53. (a) $x[n]=\left\{\ldots,\dfrac{7}{8},\dfrac{3}{4},\dfrac{1}{2},0\right\}$
                       $\uparrow$

    (b) $x[n]=\left\{\ldots,-\dfrac{1}{8},-\dfrac{1}{4},-\dfrac{1}{2},-1,-1,-1,\ldots\right\}$
                               $\uparrow$

    (c) $x[n]=\{0,1,3,7,15,\ldots\}$

4.54. (a) $x[n]=(1-2^n)u[-n-1]$

    (b) $x[n]=-u[n]-2^n u[-n-1]$

    (c) $x[n]=(-1+2^n)u[n]$

4.55. $H(z)=\dfrac{1}{1-\dfrac{1}{2}z^{-1}}$, $h[n]=\left(\dfrac{1}{2}\right)^n u[n]$

4.56. (a) $H(z)=\dfrac{b_0+b_1z^{-1}+b_2z^{-2}}{1+a_1z^{-1}+a_2z^{-2}}$

    (b) $y[n]+a_1y[n-1]+a_2y[n-2]=b_0x[n]+b_1x[n-1]+b_2x[n-2]$

4.57. (a) $s[n] = \left[2 - \left(\frac{1}{2}\right)^n\right]u[n]$

(b) $y[n] = 2\left[\left(\frac{1}{2}\right)^n + n - 1\right]u[n]$

4.58. (a) $H(z) = \dfrac{z^2}{\left(z - \frac{1}{2}\right)\left(z - \frac{1}{3}\right)}, \ |z| > \frac{1}{2}$

(b) $h[n] = \left[3\left(\frac{1}{2}\right)^n - 2\left(\frac{1}{3}\right)^n\right]u[n]$

4.59. (a) $y[n] = -2 + 9(3)^n, \ n \geq -1$

(b) $y[n] = \dfrac{1}{2} + 8(2)^n - \dfrac{9}{2}(3)^n, \ n \geq -2$

4.60. (a) $x[0] = 2, \ x[\infty] = 0$

(b) $x[0] = 0, \ x[\infty] = 1$

CHAPTER 5

Fourier Analysis of Continuous-Time Signals and Systems

5.1 Introduction

In previous chapters we introduced the Laplace transform and the z-transform to convert time-domain signals into the complex s-domain and z-domain representations that are, for many purposes, more convenient to analyze and process. In addition, greater insights into the nature and properties of many signals and systems are provided by these transformations. In this chapter and the following one, we shall introduce other transformations known as Fourier series and Fourier transform which convert time-domain signals into frequency-domain (or *spectral*) representations. In addition to providing spectral representations of signals, Fourier analysis is also essential for describing certain types of systems and their properties in the frequency domain. In this chapter we shall introduce Fourier analysis in the context of continuous-time signals and systems.

5.2 Fourier Series Representation of Periodic Signals

A. Periodic Signals:

In Chap. 1 we defined a continuous-time signal $x(t)$ to be periodic if there is a positive nonzero value of T for which

$$x(t + T) = x(t) \qquad \text{all } t \tag{5.1}$$

The fundamental period T_0 of $x(t)$ is the smallest positive value of T for which Eq. (5.1) is satisfied, and $1/T_0 = f_0$ is referred to as the *fundamental frequency*.

Two basic examples of periodic signals are the real sinusoidal signal

$$x(t) = \cos(\omega_0 t + \phi) \tag{5.2}$$

and the complex exponential signal

$$x(t) = e^{j\omega_0 t} \tag{5.3}$$

where $\omega_0 = 2\pi/T_0 = 2\pi f_0$ is called the *fundamental angular frequency*.

B.   Complex Exponential Fourier Series Representation:

The complex exponential Fourier series representation of a periodic signal $x(t)$ with fundamental period T_0 is given by

$$x(t) = \sum_{k=-\infty}^{\infty} c_k e^{jk\omega_0 t} \qquad \omega_0 = \frac{2\pi}{T_0} \tag{5.4}$$

where c_k are known as the *complex Fourier coefficients* and are given by

$$c_k = \frac{1}{T_0} \int_{T_0} x(t) e^{-jk\omega_0 t} \, dt \tag{5.5}$$

where $\int_{T_0}$ denotes the integral over any one period and 0 to T_0 or $-T_0/2$ to $T_0/2$ is commonly used for the integration. Setting $k = 0$ in Eq. (5.5), we have

$$c_0 = \frac{1}{T_0} \int_{T_0} x(t) \, dt \tag{5.6}$$

which indicates that c_0 equals the average value of $x(t)$ over a period.

When $x(t)$ is real, then from Eq. (5.5) it follows that

$$c_{-k} = c_k^* \tag{5.7}$$

where the asterisk indicates the complex conjugate.

C.   Trigonometric Fourier Series:

The trigonometric Fourier series representation of a periodic signal $x(t)$ with fundamental period T_0 is given by

$$x(t) = \frac{a_0}{2} + \sum_{k=1}^{\infty} (a_k \cos k\omega_0 t + b_k \sin k\omega_0 t) \qquad \omega_0 = \frac{2\pi}{T_0} \tag{5.8}$$

where a_k and b_k are the Fourier coefficients given by

$$a_k = \frac{2}{T_0} \int_{T_0} x(t) \cos k\omega_0 t \, dt \tag{5.9a}$$

$$b_k = \frac{2}{T_0} \int_{T_0} x(t) \sin k\omega_0 t \, dt \tag{5.9b}$$

The coefficients a_k and b_k and the complex Fourier coefficients c_k are related by (Prob. 5.3)

$$\frac{a_0}{2} = c_0 \qquad a_k = c_k + c_{-k} \qquad b_k = j(c_k - c_{-k}) \tag{5.10}$$

From Eq. (5.10) we obtain

$$c_k = \frac{1}{2}(a_k - jb_k) \qquad c_{-k} = \frac{1}{2}(a_k + jb_k) \tag{5.11}$$

When $x(t)$ is real, then a_k and b_k are real and by Eq. (5.10) we have

$$a_k = 2\,\text{Re}[c_k] \qquad b_k = -2\,\text{Im}[c_k] \tag{5.12}$$

Even and Odd Signals:

If a periodic signal $x(t)$ is even, then $b_k = 0$ and its Fourier series (5.8) contains only cosine terms:

$$x(t) = \frac{a_0}{2} + \sum_{k=1}^{\infty} a_k \cos k\omega_0 t \qquad \omega_0 = \frac{2\pi}{T_0} \tag{5.13}$$

If $x(t)$ is odd, then $a_k = 0$ and its Fourier series contains only sine terms:

$$x(t) = \sum_{k=1}^{\infty} b_k \sin k\omega_0 t \qquad \omega_0 = \frac{2\pi}{T_0} \tag{5.14}$$

D. Harmonic Form Fourier Series:

Another form of the Fourier series representation of a real periodic signal $x(t)$ with fundamental period T_0 is

$$x(t) = C_0 + \sum_{k=1}^{\infty} C_k \cos(k\omega_0 t - \theta_k) \qquad \omega_0 = \frac{2\pi}{T_0} \tag{5.15}$$

Equation (5.15) can be derived from Eq. (5.8) and is known as the *harmonic form* Fourier series of $x(t)$. The term C_0 is known as the *dc component*, and the term $C_k \cos(k\omega_0 t - \theta_k)$ is referred to as the *kth harmonic component* of $x(t)$. The first harmonic component $C_1 \cos(\omega_0 t - \theta_1)$ is commonly called the *fundamental component* because it has the same fundamental period as $x(t)$. The coefficients C_k and the angles θ_k are called the *harmonic amplitudes* and *phase angles*, respectively, and they are related to the Fourier coefficients a_k and b_k by

$$C_0 = \frac{a_0}{2} \qquad C_k = \sqrt{a_k^2 + b_k^2} \qquad \theta_k = \tan^{-1} \frac{b_k}{a_k} \tag{5.16}$$

For a real periodic signal $x(t)$, the Fourier series in terms of complex exponentials as given in Eq. (5.4) is mathematically equivalent to either of the two forms in Eqs. (5.8) and (5.15). Although the latter two are common forms for Fourier series, the complex form in Eq. (5.4) is more general and usually more convenient, and we will use that form almost exclusively.

E. Convergence of Fourier Series:

It is known that a periodic signal $x(t)$ has a Fourier series representation if it satisfies the following Dirichlet conditions:

1. $x(t)$ is absolutely integrable over any period; that is,

$$\int_{T_0} |x(t)| \, dt < \infty \tag{5.17}$$

2. $x(t)$ has a finite number of maxima and minima within any finite interval of t.
3. $x(t)$ has a finite number of discontinuities within any finite interval of t, and each of these discontinuities is finite.

Note that the Dirichlet conditions are sufficient but not necessary conditions for the Fourier series representation (Prob. 5.8).

F. Amplitude and Phase Spectra of a Periodic Signal:

Let the complex Fourier coefficients c_k in Eq. (5.4) be expressed as

$$c_k = |c_k| \, e^{j\phi_k} \tag{5.18}$$

A plot of $|c_k|$ versus the angular frequency ω is called the *amplitude spectrum* of the periodic signal $x(t)$, and a plot of ϕ_k versus ω is called the *phase spectrum* of $x(t)$. Since the index k assumes only integers, the amplitude and phase spectra are not continuous curves but appear only at the discrete frequencies $k\omega_0$. They are therefore referred to as *discrete frequency spectra* or *line spectra*.

For a real periodic signal $x(t)$ we have $c_{-k} = c_k^*$. Thus,

$$|c_{-k}| = |c_k| \qquad \phi_{-k} = -\phi_k \tag{5.19}$$

Hence, the amplitude spectrum is an even function of ω, and the phase spectrum is an odd function of ω for a real periodic signal.

G.   Power Content of a Periodic Signal:

In Chap. 1 (Prob. 1.18) we introduced the average power of a periodic signal $x(t)$ over any period as

$$P = \frac{1}{T_0} \int_{T_0} |x(t)|^2 \, dt \tag{5.20}$$

If $x(t)$ is represented by the complex exponential Fourier series in Eq. (5.4), then it can be shown that (Prob. 5.14)

$$\frac{1}{T_0} \int_{T_0} |x(t)|^2 \, dt = \sum_{k=-\infty}^{\infty} |c_k|^2 \tag{5.21}$$

Equation (5.21) is called *Parseval's identity* (or *Parseval's theorem*) for the Fourier series.

5.3   The Fourier Transform

A.   From Fourier Series to Fourier Transform:

Let $x(t)$ be a nonperiodic signal of finite duration; that is,

$$x(t) = 0 \qquad |t| > T_1$$

Such a signal is shown in Fig. 5-1(a). Let $x_{T_0}(t)$ be a periodic signal formed by repeating $x(t)$ with fundamental period T_0 as shown in Fig. 5-1(b). If we let $T_0 \to \infty$, we have

$$\lim_{T_0 \to \infty} x_{T_0}(t) = x(t) \tag{5.22}$$

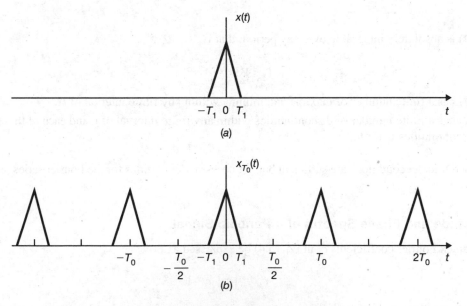

Fig. 5-1 (a) Nonperiodic signal $x(t)$; (b) periodic signal formed by periodic extension of $x(t)$.

The complex exponential Fourier series of $x_{T_0}(t)$ is given by

$$x_{T_0}(t) = \sum_{k=-\infty}^{\infty} C_k \, e^{jk\omega_0 t} \qquad \omega_0 = \frac{2\pi}{T_0} \tag{5.23}$$

where
$$c_k = \frac{1}{T_0} \int_{-T_0/2}^{T_0/2} x_{T_0}(t) e^{-jk\omega_0 t} \, dt \tag{5.24a}$$

Since $x_{T_0}(t) = x(t)$ for $|t| < T_0/2$ and also since $x(t) = 0$ outside this interval, Eq. (5.24a) can be rewritten as

$$c_k = \frac{1}{T_0} \int_{-T_0/2}^{T_0/2} x(t) e^{-jk\omega_0 t} \, dt = \frac{1}{T_0} \int_{-\infty}^{\infty} x(t) e^{-jk\omega_0 t} \, dt \tag{5.24b}$$

Let us define $X(\omega)$ as

$$X(\omega) = \int_{-\infty}^{\infty} x(t) e^{-j\omega t} \, dt \tag{5.25}$$

Then from Eq. (5.24b) the complex Fourier coefficients c_k can be expressed as

$$c_k = \frac{1}{T_0} X(k\omega_0) \tag{5.26}$$

Substituting Eq. (5.26) into Eq. (5.23), we have

$$x_{T_0}(t) = \sum_{k=-\infty}^{\infty} \frac{1}{T_0} X(k\omega_0) e^{jk\omega_0 t}$$

or
$$x_{T_0}(t) = \frac{1}{2\pi} \sum_{k=-\infty}^{\infty} X(k\omega_0) e^{jk\omega_0 t} \omega_0 \tag{5.27}$$

As $T_0 \to \infty$, $\omega_0 = 2\pi/T_0$ becomes infinitesimal ($\omega_0 \to 0$). Thus, let $\omega_0 = \Delta\omega$. Then Eq. (5.27) becomes

$$x_{T_0}(t)\big|_{T_0 \to \infty} \to \frac{1}{2\pi} \sum_{k=-\infty}^{\infty} X(k\Delta\omega) e^{jk\Delta\omega t} \Delta\omega \tag{5.28}$$

Therefore,

$$x(t) = \lim_{T_0 \to \infty} x_{T_0}(t) = \lim_{\Delta\omega \to 0} \frac{1}{2\pi} \sum_{k=-\infty}^{\infty} X(k\Delta\omega) e^{jk\Delta\omega t} \Delta\omega \tag{5.29}$$

The sum on the right-hand side of Eq. (5.29) can be viewed as the area under the function $X(\omega) e^{j\omega t}$, as shown in Fig. 5-2. Therefore, we obtain

$$x(t) = \frac{1}{2\pi} \int_{-\infty}^{\infty} X(\omega) e^{j\omega t} \, d\omega \tag{5.30}$$

which is the Fourier representation of a nonperiodic $x(t)$.

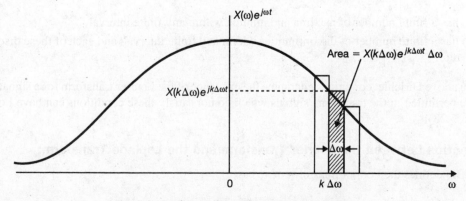

Fig. 5-2 Graphical interpretation of Eq. (5.29).

B. Fourier Transform Pair:

The function $X(\omega)$ defined by Eq. (5.25) is called the *Fourier transform* of $x(t)$, and Eq. (5.30) defines the *inverse Fourier transform of* $X(\omega)$. Symbolically they are denoted by

$$X(\omega) = \mathscr{F}\{x(t)\} = \int_{-\infty}^{\infty} x(t) e^{j\omega t} \, dt \tag{5.31}$$

$$x(t) = \mathscr{F}^{-1}\{X(\omega)\} = \frac{1}{2\pi}\int_{-\infty}^{\infty} X(\omega) e^{j\omega t} \, d\omega \tag{5.32}$$

and we say that $x(t)$ and $X(\omega)$ form a Fourier transform pair denoted by

$$x(t) \leftrightarrow X(\omega) \tag{5.33}$$

C. Fourier Spectra:

The Fourier transform $X(\omega)$ of $x(t)$ is, in general, complex, and it can be expressed as

$$X(\omega) = |X(\omega)| \, e^{j\phi(\omega)} \tag{5.34}$$

By analogy with the terminology used for the complex Fourier coefficients of a periodic signal $x(t)$, the Fourier transform $X(\omega)$ of a nonperiodic signal $x(t)$ is the frequency-domain specification of $x(t)$ and is referred to as the *spectrum* (or *Fourier spectrum*) of $x(t)$. The quantity $|X(\omega)|$ is called the *magnitude spectrum* of $x(t)$, and $\phi(\omega)$ is called the *phase spectrum* of $x(t)$.

   If $x(t)$ is a real signal, then from Eq. (5.31) we get

$$X(-\omega) = \int_{-\infty}^{\infty} x(t) e^{j\omega t} \, dt \tag{5.35}$$

Then it follows that

$$X(-\omega) = X^*(\omega) \tag{5.36a}$$

and $\qquad\qquad\qquad |X(-\omega)| = |X(\omega)| \qquad \phi(-\omega) = -\phi(\omega) \tag{5.36b}$

Hence, as in the case of periodic signals, the amplitude spectrum $|X(\omega)|$ is an even function and the phase spectrum $\phi(\omega)$ is an odd function of ω.

D. Convergence of Fourier Transforms:

Just as in the case of periodic signals, the sufficient conditions for the convergence of $X(\omega)$ are the following (again referred to as the Dirichlet conditions):

   1. $x(t)$ is absolutely integrable; that is,

$$\int_{-\infty}^{\infty} |x(t)| \, dt < \infty \tag{5.37}$$

   2. $x(t)$ has a finite number of maxima and minima within any finite interval.
   3. $x(t)$ has a finite number of discontinuities within any finite interval, and each of these discontinuities is finite.

Although the above Dirichlet conditions guarantee the existence of the Fourier transform for a signal, if impulse functions are permitted in the transform, signals which do not satisfy these conditions can have Fourier transforms (Prob. 5.23).

E. Connection between the Fourier Transform and the Laplace Transform:

Equation (5.31) defines the Fourier transform of $x(t)$ as

$$X(\omega) = \int_{-\infty}^{\infty} x(t) e^{-j\omega t} \, dt \tag{5.38}$$

The bilateral Laplace transform of $x(t)$, as defined in Eq. (4.3), is given by

$$X(s) = \int_{-\infty}^{\infty} x(t)e^{-st}\,dt \tag{5.39}$$

Comparing Eqs. (5.38) and (5.39), we see that the Fourier transform is a special case of the Laplace transform in which $s = j\omega$; that is,

$$X(s)|_{s=j\omega} = \mathcal{F}\{x(t)\} \tag{5.40}$$

Setting $s = \sigma + j\omega$ in Eq. (5.39), we have

$$X(\sigma + j\omega) = \int_{-\infty}^{\infty} x(t)e^{-(\sigma+j\omega)t}\,dt = \int_{-\infty}^{\infty} [x(t)e^{-\sigma t}]e^{-j\omega t}\,dt$$

or

$$X(\sigma + j\omega) = \mathcal{F}\{x(t)e^{-\sigma t}\} \tag{5.41}$$

which indicates that the bilateral Laplace transform of $x(t)$ can be interpreted as the Fourier transform of $x(t)\,e^{-\sigma t}$.

Since the Laplace transform may be considered a generalization of the Fourier transform in which the frequency is generalized from $j\omega$ to $s = \sigma + j\omega$, the complex variable s is often referred to as the *complex frequency*.

Note that since the integral in Eq. (5.39) is denoted by $X(s)$, the integral in Eq. (5.38) may be denoted as $X(j\omega)$. Thus, in the remainder of this book both $X(\omega)$ and $X(j\omega)$ mean the same thing whenever we connect the Fourier transform with the Laplace transform. Because the Fourier transform is the Laplace transform with $s = j\omega$, it should not be assumed automatically that the Fourier transform of a signal $x(t)$ is the Laplace transform with s replaced by $j\omega$. If $x(t)$ is absolutely integrable, that is, if $x(t)$ satisfies condition (5.37), the Fourier transform of $x(t)$ can be obtained from the Laplace transform of $x(t)$ with $s = j\omega$. This is not generally true of signals which are not absolutely integrable. The following examples illustrate the above statements.

EXAMPLE 5.1   Consider the unit impulse function $\delta(t)$.

From Eq. (3.13) the Laplace transform of $\delta(t)$ is

$$\mathcal{L}\{\delta(t)\} = 1 \qquad \text{all } s \tag{5.42}$$

By definitions (5.31) and (1.20) the Fourier transform of $\delta(t)$ is

$$\mathcal{F}\{\delta(t)\} = \int_{-\infty}^{\infty} \delta(t)e^{-j\omega t}\,dt = 1 \tag{5.43}$$

Thus, the Laplace transform and the Fourier transform of $\delta(t)$ are the same.

EXAMPLE 5.2   Consider the exponential signal

$$x(t) = e^{-at}u(t) \qquad a > 0$$

From Eq. (3.8) the Laplace transform of $x(t)$ is given by

$$\mathcal{L}\{x(t)\} = X(s) = \frac{1}{s+a} \qquad \text{Re}(s) > -a \tag{5.44}$$

By definition (5.31) the Fourier transform of $x(t)$ is

$$\mathcal{F}\{x(t)\} = X(\omega) = \int_{-\infty}^{\infty} e^{-at}\,u(t)\,e^{-j\omega t}\,dt$$

$$= \int_{0^+}^{\infty} e^{-(a+j\omega)t}\,dt = \frac{1}{a+j\omega} \tag{5.45}$$

Thus, comparing Eqs. (5.44) and (5.45), we have

$$X(\omega) = X(s)|_{s=j\omega} \tag{5.46}$$

Note that $x(t)$ is absolutely integrable.

EXAMPLE 5.3   Consider the unit step function $u(t)$.

From Eq. (3.14) the Laplace transform of $u(t)$ is

$$\mathscr{L}\{u(t)\} = \frac{1}{s} \qquad \mathrm{Re}(s) > 0 \tag{5.47}$$

The Fourier transform of $u(t)$ is given by (Prob. 5.30)

$$\mathscr{F}\{u(t)\} = \pi\delta(\omega) + \frac{1}{j\omega} \tag{5.48}$$

Thus, the Fourier transform of $u(t)$ cannot be obtained from its Laplace transform. Note that the unit step function $u(t)$ is not absolutely integrable.

5.4   Properties of the Continuous-Time Fourier Transform

Basic properties of the Fourier transform are presented in the following. Many of these properties are similar to those of the Laplace transform (see Sec. 3.4).

A.   Linearity:

$$a_1 x_1(t) + a_2 x_2(t) \leftrightarrow a_1 X_1(\omega) + a_2 X_2(\omega) \tag{5.49}$$

B.   Time Shifting:

$$x(t - t_0) \leftrightarrow e^{-j\omega t_0} X(\omega) \tag{5.50}$$

Equation (5.50) shows that the effect of a shift in the time domain is simply to add a linear term $-\omega t_0$ to the original phase spectrum $\theta(\omega)$. This is known as a *linear phase shift* of the Fourier transform $X(\omega)$.

C.   Frequency Shifting:

$$e^{j\omega_0 t} x(t) \leftrightarrow X(\omega - \omega_0) \tag{5.51}$$

The multiplication of $x(t)$ by a complex exponential signal $e^{j\omega_0 t}$ is sometimes called *complex modulation*. Thus, Eq. (5.51) shows that complex modulation in the time domain corresponds to a shift of $X(\omega)$ in the frequency domain. Note that the frequency-shifting property Eq. (5.51) is the dual of the time-shifting property Eq. (5.50).

D.   Time Scaling:

$$x(at) \leftrightarrow \frac{1}{|a|} X\left(\frac{\omega}{a}\right) \tag{5.52}$$

where a is a real constant. This property follows directly from the definition of the Fourier transform. Equation (5.52) indicates that scaling the time variable t by the factor a causes an inverse scaling of the frequency variable ω by $1/a$, as well as an amplitude scaling of $X(\omega/a)$ by $1/|a|$. Thus, the scaling property (5.52) implies that time compression of a signal $(a > 1)$ results in its spectral expansion and that time expansion of the signal $(a < 1)$ results in its spectral compression.

E.   Time Reversal:

$$x(-t) \leftrightarrow X(-\omega) \tag{5.53}$$

Thus, time reversal of $x(t)$ produces a like reversal of the frequency axis for $X(\omega)$. Equation (5.53) is readily obtained by setting $a = -1$ in Eq. (5.52).

F.   Duality (or Symmetry):

$$X(t) \leftrightarrow 2\pi x(-\omega) \tag{5.54}$$

The duality property of the Fourier transform has significant implications. This property allows us to obtain both of these dual Fourier transform pairs from one evaluation of Eq. (5.31) (Probs. 5.20 and 5.22).

G. Differentiation in the Time Domain:

$$\frac{dx(t)}{dt} \leftrightarrow j\omega X(\omega) \tag{5.55}$$

Equation (5.55) shows that the effect of differentiation in the time domain is the multiplication of $X(\omega)$ by $j\omega$ in the frequency domain (Prob. 5.28).

H. Differentiation in the Frequency Domain:

$$(-jt)x(t) \leftrightarrow \frac{dX(\omega)}{d\omega} \tag{5.56}$$

Equation (5.56) is the dual property of Eq. (5.55).

I. Integration in the Time Domain:

$$\int_{-\infty}^{t} x(\tau)\, d\tau \leftrightarrow \pi X(0)\, \delta(\omega) + \frac{1}{j\omega}\, X(\omega) \tag{5.57}$$

Since integration is the inverse of differentiation, Eq. (5.57) shows that the frequency domain operation corresponding to time-domain integration is multiplication by $1/j\omega$, but an additional term is needed to account for a possible dc component in the integrator output. Hence, unless $X(0) = 0$, a dc component is produced by the integrator (Prob. 5.33).

J. Convolution:

$$x_1(t) * x_2(t) \leftrightarrow X_1(\omega)\, X_2(\omega) \tag{5.58}$$

Equation (5.58) is referred to as the *time convolution theorem*, and it states that convolution in the time domain becomes multiplication in the frequency domain (Prob. 5.31). As in the case of the Laplace transform, this convolution property plays an important role in the study of continuous-time LTI systems (Sec. 5.5) and also forms the basis for our discussion of filtering (Sec. 5.6).

K. Multiplication:

$$x_1(t)x_2(t) \leftrightarrow \frac{1}{2\pi} X_1(\omega) * X_2(\omega) \tag{5.59}$$

The multiplication property (5.59) is the dual property of Eq. (5.58) and is often referred to as the *frequency convolution theorem*. Thus, multiplication in the time domain becomes convolution in the frequency domain (Prob. 5.35).

L. Additional Properties:

If $x(t)$ is real, let

$$x(t) = x_e(t) + x_o(t) \tag{5.60}$$

where $x_e(t)$ and $x_o(t)$ are the even and odd components of $x(t)$, respectively. Let

$$x(t) \leftrightarrow X(\omega) = A(\omega) + jB(\omega)$$

Then

$$X(-\omega) = X^*(\omega) \tag{5.61a}$$

$$x_e(t) \leftrightarrow \text{Re}\{X(\omega)\} = A(\omega) \tag{5.61b}$$

$$x_o(t) \leftrightarrow j\, \text{Im}\{X(\omega)\} = jB(\omega) \tag{5.61c}$$

Equation (5.61a) is the necessary and sufficient condition for $x(t)$ to be real (Prob. 5.39). Equations (5.61b) and (5.61c) show that the Fourier transform of an even signal is a real function of ω and that the Fourier transform of an odd signal is a pure imaginary function of ω.

M. Parseval's Relations:

$$\int_{-\infty}^{\infty} x_1(\lambda)X_2(\lambda)\,d\lambda = \int_{-\infty}^{\infty} X_1(\lambda)x_2(\lambda)\,d\lambda \tag{5.62}$$

$$\int_{-\infty}^{\infty} x_1(t)x_2(t)\,dt = \frac{1}{2\pi}\int_{-\infty}^{\infty} X_1(\omega)X_2(-\omega)\,d\omega \tag{5.63}$$

$$\int_{-\infty}^{\infty} |x(t)|^2\,dt = \frac{1}{2\pi}\int_{-\infty}^{\infty} |X(\omega)|^2\,d\omega \tag{5.64}$$

Equation (5.64) is called *Parseval's identity* (or *Parseval's theorem*) for the Fourier transform. Note that the quantity on the left-hand side of Eq. (5.64) is the normalized energy content E of $x(t)$ [Eq. (1.14)]. Parseval's identity says that this energy content E can be computed by integrating $|X(\omega)|^2$ over all frequencies ω. For this reason, $|X(\omega)|^2$ is often referred to as the *energy-density spectrum* of $x(t)$, and Eq. (5.64) is also known as the *energy theorem*.

Table 5-1 contains a summary of the properties of the Fourier transform presented in this section. Some common signals and their Fourier transforms are given in Table 5-2.

TABLE 5-1   Properties of the Fourier Transform

PROPERTY	SIGNAL	FOURIER TRANSFORM
	$x(t)$	$X(\omega)$
	$x_1(t)$	$X_1(\omega)$
	$x_2(t)$	$X_2(\omega)$
Linearity	$a_1 x_1(t) + a_2 x_2(t)$	$a_1 X_1(\omega) + a_2 X_2(\omega)$
Time shifting	$x(t - t_0)$	$e^{-j\omega t_0}\,X(\omega)$
Frequency shifting	$e^{j\omega_0 t}\,x(t)$	$X(\omega - \omega_0)$
Time scaling	$x(at)$	$\dfrac{1}{|a|}X\!\left(\dfrac{\omega}{a}\right)$
Time reversal	$x(-t)$	$X(-\omega)$
Duality	$X(t)$	$2\pi x(-\omega)$
Time differentiation	$\dfrac{dx(t)}{dt}$	$j\omega\,X(\omega)$
Frequency differentiation	$(-jt)x(t)$	$\dfrac{dX(\omega)}{d\omega}$
Integration	$\displaystyle\int_{-\infty}^{t} x(\tau)\,d\tau$	$\pi X(0)\,\delta(\omega) + \dfrac{1}{j\omega}X(\omega)$
Convolution	$x_1(t) * x_2(t)$	$X_1(\omega)X_2(\omega)$
Multiplication	$x_1(t)x_2(t)$	$\dfrac{1}{2\pi}X_1(\omega) * X_2(\omega)$
Real signal	$x(t) = x_e(t) + x_o(t)$	$X(\omega) = A(\omega) + jB(\omega)$
		$X(-\omega) = X^*(\omega)$
Even component	$x_e(t)$	$\operatorname{Re}\{X(\omega)\} = A(\omega)$
Odd component	$x_o(t)$	$j\operatorname{Im}\{X(\omega)\} = jB(\omega)$
Parseval's relations		

$$\int_{-\infty}^{\infty} x_1(\lambda)X_2(\lambda)\,d\lambda = \int_{-\infty}^{\infty} X_1(\lambda)x_2(\lambda)\,d\lambda$$

$$\int_{-\infty}^{\infty} x_1(t)x_2(t)\,dt = \frac{1}{2\pi}\int_{-\infty}^{\infty} X_1(\omega)X_2(-\omega)\,d\omega$$

$$\int_{-\infty}^{\infty} |x(t)|^2\,dt = \frac{1}{2\pi}\int_{-\infty}^{\infty} |X(\omega)|^2\,d\omega$$

TABLE 5.2 Common Fourier Transforms Pairs

$x(t)$	$X(\omega)$
$\delta(t)$	1
$\delta(t - t_0)$	$e^{-j\omega t_0}$
1	$2\pi\delta(\omega)$
$e^{j\omega_0 t}$	$2\pi\delta(\omega - \omega_0)$
$\cos\omega_0 t$	$\pi[\delta(\omega - \omega_0) + \delta(\omega + \omega_0)]$
$\sin\omega_0 t$	$-j\pi[\delta(\omega - \omega_0) - \delta(\omega + \omega_0)]$
$u(t)$	$\pi\delta(\omega) + \dfrac{1}{j\omega}$
$u(-t)$	$\pi\delta(\omega) - \dfrac{1}{j\omega}$
$e^{-at}u(t), a > 0$	$\dfrac{1}{j\omega + a}$
$te^{-at}u(t), a > 0$	$\dfrac{1}{(j\omega + a)^2}$
$e^{-a\vert t\vert}, a > 0$	$\dfrac{2a}{a^2 + \omega^2}$
$\dfrac{1}{a^2 + t^2}$	$e^{-a\vert\omega\vert}$
$e^{-at^2}, a > 0$	$\sqrt{\dfrac{\pi}{a}}\, e^{-\omega^2/4a}$
$p_a(t) = \begin{cases} 1 & \vert t\vert < a \\ 0 & \vert t\vert > a \end{cases}$	$2a\dfrac{\sin\omega a}{\omega a}$
$\dfrac{\sin at}{\pi t}$	$p_a(\omega) = \begin{cases} 1 & \vert\omega\vert < a \\ 0 & \vert\omega\vert > a \end{cases}$
$\operatorname{sgn} t$	$\dfrac{2}{j\omega}$
$\displaystyle\sum_{k=-\infty}^{\infty} \delta(t - kT)$	$\omega_0\displaystyle\sum_{k=-\infty}^{\infty} \delta(\omega - k\omega_0), \omega_0 = \dfrac{2\pi}{T}$

5.5 The Frequency Response of Continuous-Time LTI Systems

A. Frequency Response:

In Sec. 2.2 we showed that the output $y(t)$ of a continuous-time LTI system equals the convolution of the input $x(t)$ with the impulse response $h(t)$; that is,

$$y(t) = x(t) * h(t) \tag{5.65}$$

Applying the convolution property (5.58), we obtain

$$Y(\omega) = X(\omega)H(\omega) \tag{5.66}$$

where $Y(\omega)$, $X(\omega)$, and $H(\omega)$ are the Fourier transforms of $y(t)$, $x(t)$, and $h(t)$, respectively. From Eq. (5.66) we have

$$H(\omega) = \frac{Y(\omega)}{X(\omega)} \tag{5.67}$$

The function $H(\omega)$ is called the *frequency response* of the system. Relationships represented by Eqs. (5.65) and (5.66) are depicted in Fig. 5-3. Let

$$H(\omega) = |H(\omega)|e^{j\theta_H(\omega)} \tag{5.68}$$

Then $|H(\omega)|$ is called the *magnitude response* of the system, and $\theta_H(\omega)$ the *phase response* of the system.

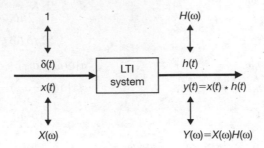

Fig. 5-3 Relationships between inputs and outputs in an LTI system.

Consider the complex exponential signal

$$x(t) = e^{j\omega_0 t} \tag{5.69}$$

with Fourier transform (Prob. 5.23)

$$X(\omega) = 2\pi\delta(\omega - \omega_0) \tag{5.70}$$

Then from Eqs. (5.66) and (1.26) we have

$$Y(\omega) = 2\pi H(\omega_0)\,\delta(\omega - \omega_0) \tag{5.71}$$

Taking the inverse Fourier transform of $Y(\omega)$, we obtain

$$y(t) = H(\omega_0)\,e^{j\omega_0 t} \tag{5.72}$$

which indicates that the complex exponential signal $e^{j\omega_0 t}$ is an eigenfunction of the LTI system with corresponding eigenvalue $H(\omega_0)$, as previously observed in Chap. 2 (Sec. 2.4 and Prob. 2.17]. Furthermore, by the linearity property (5.49), if the input $x(t)$ is periodic with the Fourier series

$$x(t) = \sum_{k=-\infty}^{\infty} c_k\,e^{jk\omega_0 t} \tag{5.73}$$

then the corresponding output $y(t)$ is also periodic with the Fourier series

$$y(t) = \sum_{k=-\infty}^{\infty} c_k H(k\omega_0)e^{jk\omega_0 t} \tag{5.74}$$

If $x(t)$ is not periodic, then from Eq. (5.30)

$$x(t) = \frac{1}{2\pi}\int_{-\infty}^{\infty} X(\omega)\,e^{j\omega t}\,d\omega \tag{5.75}$$

and using Eq. (5.66), the corresponding output $y(t)$ can be expressed as

$$y(t) = \frac{1}{2\pi}\int_{-\infty}^{\infty} H(\omega)X(\omega)\,e^{j\omega t}\,d\omega \tag{5.76}$$

Thus, the behavior of a continuous-time LTI system in the frequency domain is completely characterized by its frequency response $H(\omega)$. Let

$$X(\omega) = |X(\omega)|e^{j\theta_X(\omega)} \qquad Y(\omega) = |Y(\omega)|e^{j\theta_Y(\omega)} \tag{5.77}$$

Then from Eq. (5.66) we have

$$|Y(\omega)| = |X(\omega)||H(\omega)| \tag{5.78a}$$

$$\theta_Y(\omega) = \theta_X(\omega) + \theta_H(\omega) \tag{5.78b}$$

Hence, the magnitude spectrum $|X(\omega)|$ of the input is multiplied by the magnitude response $|H(\omega)|$ of the system to determine the magnitude spectrum $|Y(\omega)|$ of the output, and the phase response $\theta_H(\omega)$ is added to the phase spectrum $\theta_X(\omega)$ of the input to produce the phase spectrum $\theta_Y(\omega)$ of the output. The magnitude response $|H(\omega)|$ is sometimes referred to as the *gain* of the system.

B. Distortionless Transmission:

For distortionless transmission through an LTI system we require that the exact input signal shape be reproduced at the output, although its amplitude may be different and it may be delayed in time. Therefore, if $x(t)$ is the input signal, the required output is

$$y(t) = Kx(t - t_d) \tag{5.79}$$

where t_d is the *time delay* and K (> 0) is a *gain constant*. This is illustrated in Figs. 5-4(a) and (b). Taking the Fourier transform of both sides of Eq. (5.79), we get

$$Y(\omega) = Ke^{-j\omega t_d} X(\omega) \tag{5.80}$$

Thus, from Eq. (5.66) we see that for distortionless transmission, the system must have

$$H(\omega) = |H(\omega)|e^{j\theta_H(\omega)} = Ke^{-j\omega t_d} \tag{5.81}$$

Thus,

$$|H(\omega)| = K \tag{5.82a}$$

$$\theta_H(\omega) = -j\omega t_d \tag{5.82b}$$

That is, the amplitude of $H(\omega)$ must be constant over the entire frequency range, and the phase of $H(\omega)$ must be linear with the frequency. This is illustrated in Figs. 5-4(c) and (d).

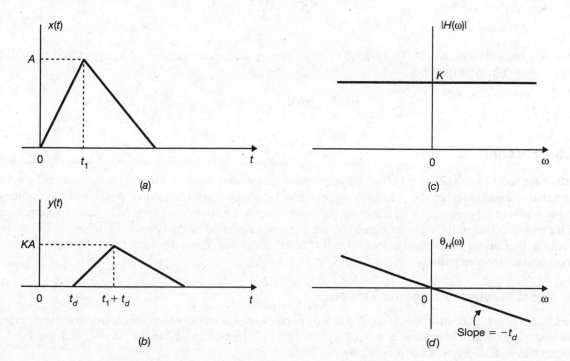

Fig. 5-4 Distortionless transmission.

Amplitude Distortion and Phase Distortion:

When the amplitude spectrum $|H(\omega)|$ of the system is not constant within the frequency band of interest, the frequency components of the input signal are transmitted with a different amount of gain or attenuation. This effect is called *amplitude distortion*. When the phase spectrum $\theta_H(\omega)$ of the system is not linear with the frequency, the output signal has a different waveform than the input signal because of different delays in passing through the system for different frequency components of the input signal. This form of distortion is called *phase distortion*.

C. LTI Systems Characterized by Differential Equations:

As discussed in Sec. 2.5, many continuous-time LTI systems of practical interest are described by linear constant-coefficient differential equations of the form

$$\sum_{k=0}^{N} a_k \frac{d^k y(t)}{dt^k} = \sum_{k=0}^{M} b_k \frac{d^k x(t)}{dt^k} \tag{5.83}$$

with $M \le N$. Taking the Fourier transform of both sides of Eq. (5.83) and using the linearity property (5.49) and the time-differentiation property (5.55), we have

$$\sum_{k=0}^{N} a_k (j\omega)^k Y(\omega) = \sum_{k=0}^{M} b_k (j\omega)^k X(\omega)$$

or

$$Y(\omega) \sum_{k=0}^{N} a_k (j\omega)^k = X(\omega) \sum_{k=0}^{M} b_k (j\omega)^k \tag{5.84}$$

Thus, from Eq. (5.67)

$$H(\omega) = \frac{Y(\omega)}{X(\omega)} = \frac{\displaystyle\sum_{k=0}^{M} b_k (j\omega)^k}{\displaystyle\sum_{k=0}^{N} a_k (j\omega)^k} \tag{5.85}$$

which is a rational function of ω. The result (5.85) is the same as the Laplace transform counterpart $H(s) = Y(s)/X(s)$ with $s = j\omega$ [Eq. (3.40)]; that is,

$$H(\omega) = H(s)\big|_{s=j\omega} = H(j\omega)$$

5.6 Filtering

One of the most basic operations in any signal processing system is *filtering*. Filtering is the process by which the relative amplitudes of the frequency components in a signal are changed or perhaps some frequency components are suppressed. As we saw in the preceding section, for continuous-time LTI systems, the spectrum of the output is that of the input multiplied by the frequency response of the system. Therefore, an LTI system acts as a filter on the input signal. Here the word "filter" is used to denote a system that exhibits some sort of frequency-selective behavior.

A. Ideal Frequency-Selective Filters:

An *ideal* frequency-selective filter is one that exactly passes signals at one set of frequencies and completely rejects the rest. The band of frequencies passed by the filter is referred to as the *pass band*, and the band of frequencies rejected by the filter is called the *stop band*.

The most common types of ideal frequency-selective filters are the following.

1. Ideal Low-Pass Filter:

An ideal low-pass filter (LPF) is specified by

$$|H(\omega)| = \begin{cases} 1 & |\omega| < \omega_c \\ 0 & |\omega| > \omega_c \end{cases} \tag{5.86}$$

which is shown in Fig. 5-5(a). The frequency ω_c is called the *cutoff* frequency.

2. Ideal High-Pass Filter:

An ideal high-pass filter (HPF) is specified by

$$|H(\omega)| = \begin{cases} 0 & |\omega| < \omega_c \\ 1 & |\omega| > \omega_c \end{cases} \tag{5.87}$$

which is shown in Fig. 5-5(b).

3. Ideal Bandpass Filter:

An ideal bandpass filter (BPF) is specified by

$$|H(\omega)| = \begin{cases} 1 & \omega_1 < |\omega| < \omega_2 \\ 0 & \text{otherwise} \end{cases} \tag{5.88}$$

which is shown in Fig. 5-5(c).

4. Ideal Bandstop Filter:

An ideal bandstop filter (BSF) is specified by

$$|H(\omega)| = \begin{cases} 0 & \omega_1 < |\omega| < \omega_2 \\ 1 & \text{otherwise} \end{cases} \tag{5.89}$$

which is shown in Fig. 5-5(d).

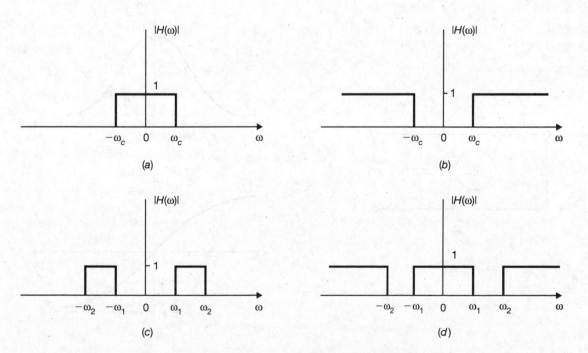

Fig. 5-5 Magnitude responses of ideal frequency-selective filters.

In the above discussion, we said nothing regarding the phase response of the filters. To avoid phase distortion in the filtering process, a filter should have a linear phase characteristic over the pass band of the filter; that is [Eq. (5.82b)],

$$\theta_H(\omega) = -\omega t_d \tag{5.90}$$

where t_d is a constant.

Note that all ideal frequency-selective filters are noncausal systems.

B. Nonideal Frequency-Selective Filters:

As an example of a simple continuous-time causal frequency-selective filter, we consider the *RC* filter shown in Fig. 5-6(*a*). The output $y(t)$ and the input $x(t)$ are related by (Prob. 1.32)

$$RC\frac{dy(t)}{dt} + y(t) = x(t)$$

Taking the Fourier transforms of both sides of the above equation, the frequency response $H(\omega)$ of the *RC* filter is given by

$$H(\omega) = \frac{Y(\omega)}{X(\omega)} = \frac{1}{1 + j\omega RC} = \frac{1}{1 + j\omega/\omega_0} \tag{5.91}$$

where $\omega_0 = 1/RC$. Thus, the amplitude response $|H(\omega)|$ and phase response $\theta_H(\omega)$ are given by

$$|H(\omega)| = \frac{1}{|1 + j\omega/\omega_0|} = \frac{1}{\left[1 + (\omega/\omega_0)^2\right]^{1/2}} \tag{5.92}$$

$$\theta_H(\omega) = -\tan^{-1}\frac{\omega}{\omega_0} \tag{5.93}$$

which are plotted in Fig. 5-6(*b*). From Fig. 5-6(*b*) we see that the *RC* network in Fig. 5-6(*a*) performs as a low-pass filter.

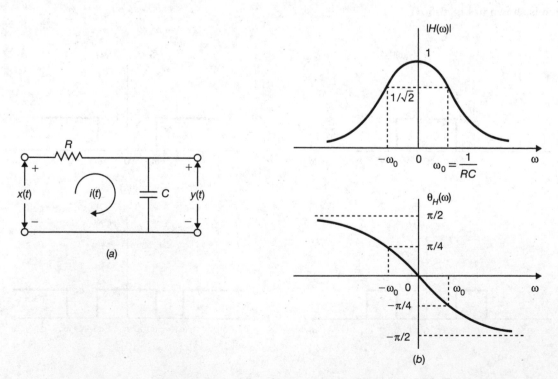

(a)

(b)

Fig. 5-6 *RC* filter and its frequency response.

5.7 Bandwidth

A. Filter (or System) Bandwidth:

One important concept in system analysis is the *bandwidth* of an LTI system. There are many different definitions of system bandwidth.

1. Absolute Bandwidth:

The bandwidth W_B of an ideal low-pass filter equals its cutoff frequency; that is, $W_B = \omega_c$ [Fig. 5-5(a)]. In this case W_B is called the *absolute bandwidth*. The absolute bandwidth of an ideal bandpass filter is given by $W_B = \omega_2 - \omega_1$ [Fig. 5-5(c)]. A bandpass filter is called *narrowband* if $W_B \ll \omega_0$, where $\omega_0 = \frac{1}{2}(\omega_1 + \omega_2)$ is the center frequency of the filter. No bandwidth is defined for a high-pass or a bandstop filter.

2. 3-dB (or Half-Power) Bandwidth:

For causal or practical filters, a common definition of filter (or system) bandwidth is the 3-dB bandwidth $W_{3\,dB}$. In the case of a low-pass filter, such as the *RC* filter described by Eq. (5.92) or in Fig. 5-6(b), $W_{3\,dB}$ is defined as the positive frequency at which the amplitude spectrum $|H(\omega)|$ drops to a value equal to $|H(0)|/\sqrt{2}$, as illustrated in Fig. 5-7(a). Note that $|H(0)|$ is the peak value of $H(\omega)$ for the low-pass *RC* filter. The 3-dB bandwidth is also known as the *half-power* bandwidth because a voltage or current attenuation of 3 dB is equivalent to a power attenuation by a factor of 2. In the case of a bandpass filter, $W_{3\,dB}$ is defined as the difference between the frequencies at which $|H(\omega)|$ drops to a value equal to $1/\sqrt{2}$ times the peak value $|H(\omega_m)|$ as illustrated in Fig. 5-7(b). This definition of $W_{3\,dB}$ is useful for systems with unimodal amplitude response (in the positive frequency range) and is a widely accepted criterion for measuring a system's bandwidth, but it may become ambiguous and nonunique with systems having multiple peak amplitude responses.

Note that each of the preceding bandwidth definitions is defined along the positive frequency axis only and always defines positive frequency, or one-sided, bandwidth only.

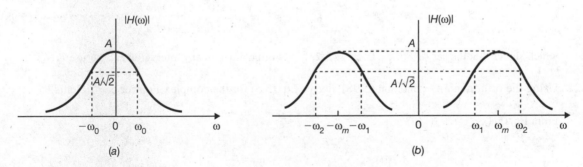

Fig. 5-7 Filter bandwidth.

B. Signal Bandwidth:

The *bandwidth* of a signal can be defined as the range of positive frequencies in which "most" of the energy or power lies. This definition is rather ambiguous and is subject to various conventions (Probs. 5.57 and 5.76).

3-dB Bandwidth:

The bandwidth of a signal $x(t)$ can also be defined on a similar basis as a filter bandwidth such as the 3-dB bandwidth, using the magnitude spectrum $|X(\omega)|$ of the signal. Indeed, if we replace $|H(\omega)|$ by $|X(\omega)|$ in Figs. 5-5(a) to (c), we have frequency-domain plots of *low-pass*, *high-pass*, and *bandpass* signals.

Band-Limited Signal:

A signal $x(t)$ is called a *band-limited* signal if

$$|X(\omega)| = 0 \qquad |\omega| > \omega_M \tag{5.94}$$

Thus, for a band-limited signal, it is natural to define ω_M as the bandwidth.

SOLVED PROBLEMS

Fourier Series

5.1. We call a set of signals $\{\Psi_n(t)\}$ *orthogonal* on an interval (a, b) if any two signals $\Psi_m(t)$ and $\Psi_k(t)$ in the set satisfy the condition

$$\int_a^b \Psi_m(t)\,\Psi_k^*(t)\,dt = \begin{cases} 0 & m \neq k \\ \alpha & m = k \end{cases} \tag{5.95}$$

where $*$ denotes the complex conjugate and $\alpha \neq 0$. Show that the set of complex exponentials $\{e^{jk\omega_0 t}: k = 0, \pm 1, \pm 2, \ldots\}$ is orthogonal on any interval over a period T_0, where $T_0 = 2\pi/\omega_0$.

For any t_0 we have

$$\int_{t_0}^{t_0+T_0} e^{jm\omega_0 t}\,dt = \frac{1}{jm\omega_0}\,e^{jm\omega_0 t}\bigg|_{t_0}^{t_0+T_0} = \frac{1}{jm\omega_0}\left(e^{jm\omega_0(t_0+T_0)} - e^{jm\omega_0 t_0}\right)$$

$$= \frac{1}{jm\omega_0}\,e^{jm\omega_0 t_0}\left(e^{jm2\pi} - 1\right) = 0 \qquad m \neq 0 \tag{5.96}$$

since $e^{jm2\pi} = 1$. When $m = 0$, we have $e^{jm\omega_0 t}\big|_{m=0} = 1$ and

$$\int_{t_0}^{t_0+T_0} e^{jm\omega_0 t}\,dt = \int_{t_0}^{t_0+T_0} dt = T_0 \tag{5.97}$$

Thus, from Eqs. (5.96) and (5.97) we conclude that

$$\int_{t_0}^{t_0+T_0} e^{jm\omega_0 t}\left(e^{jk\omega_0 t}\right)^*\,dt = \int_{t_0}^{t_0+T_0} e^{j(m-k)\omega_0 t}\,dt = \begin{cases} 0 & m \neq k \\ T_0 & m = k \end{cases} \tag{5.98}$$

which shows that the set $\{e^{jk\omega_0 t}: k = 0, \pm 1, \pm 2, \ldots\}$ is orthogonal on any interval over a period T_0.

5.2. Using the orthogonality condition (5.98), derive Eq. (5.5) for the complex Fourier coefficients.

From Eq. (5.4)

$$x(t) = \sum_{k=-\infty}^{\infty} c_k e^{jk\omega_0 t} \qquad \omega_0 = \frac{2\pi}{T_0}$$

Multiplying both sides of this equation by $e^{-jm\omega_0 t}$ and integrating the result from t_0 to $(t_0 + T_0)$, we obtain

$$\int_{t_0}^{t_0+T_0} x(t)e^{-jm\omega_0 t}\,dt = \int_{t_0}^{t_0+T_0}\left(\sum_{k=-\infty}^{\infty} c_k e^{jk\omega_0 t}\right)e^{-jm\omega_0 t}\,dt$$

$$= \sum_{k=-\infty}^{\infty} c_k \int_{t_0}^{t_0+T_0} e^{j(k-m)\omega_0 t}\,dt \tag{5.99}$$

Then by Eq. (5.98), Eq. (5.99) reduces to

$$\int_{t_0}^{t_0+T_0} x(t)e^{-jm\omega_0 t}\,dt = c_m T_0 \tag{5.100}$$

Changing index m to k, we obtain Eq. (5.5); that is,

$$c_k = \frac{1}{T_0}\int_{t_0}^{t_0+T_0} x(t)e^{-jk\omega_0 t}\,dt \tag{5.101}$$

We shall mostly use the following two special cases for Eq. (5.101): $t_0 = 0$ and $t_0 = -T_0/2$, respectively. That is,

$$c_k = \frac{1}{T_0} \int_0^{T_0} x(t) e^{-jk\omega_0 t} \, dt \qquad (5.102a)$$

$$c_k = \frac{1}{T_0} \int_{-T_0/2}^{T_0/2} x(t) e^{-jk\omega_0 t} \, dt \qquad (5.102b)$$

5.3. Derive the trigonometric Fourier series Eq. (5.8) from the complex exponential Fourier series Eq. (5.4).

Rearranging the summation in Eq. (5.4) as

$$x(t) = \sum_{k=-\infty}^{\infty} c_k e^{jk\omega_0 t} = c_0 + \sum_{k=1}^{\infty} (c_k e^{jk\omega_0 t} + c_{-k} e^{-jk\omega_0 t})$$

and using Euler's formulas

$$e^{\pm jk\omega_0 t} = \cos k\omega_0 t \pm j \sin k\omega_0 t$$

we have

$$x(t) = c_0 + \sum_{k=1}^{\infty} [(c_k + c_{-k}) \cos k\omega_0 t + j(c_k - c_{-k}) \sin k\omega_0 t] \qquad (5.103)$$

Setting

$$c_0 = \frac{a_0}{2} \qquad c_k + c_{-k} = a_k \qquad j(c_k - c_{-k}) = b_k \qquad (5.104)$$

Eq. (5.103) becomes

$$x(t) = \frac{a_0}{2} + \sum_{k=1}^{\infty} (a_k \cos k\omega_0 t + b_k \sin k\omega_0 t)$$

5.4. Determine the complex exponential Fourier series representation for each of the following signals:

(a) $x(t) = \cos \omega_0 t$

(b) $x(t) = \sin \omega_0 t$

(c) $x(t) = \cos\left(2t + \frac{\pi}{4}\right)$

(d) $x(t) = \cos 4t + \sin 6t$

(e) $x(t) = \sin^2 t$

(a) Rather than using Eq. (5.5) to evaluate the complex Fourier coefficients c_k using Euler's formula, we get

$$\cos \omega_0 t = \frac{1}{2}(e^{j\omega_0 t} + e^{-j\omega_0 t}) = \frac{1}{2} e^{-j\omega_0 t} + \frac{1}{2} e^{j\omega_0 t} = \sum_{k=-\infty}^{\infty} c_k e^{jk\omega_0 t}$$

Thus, the complex Fourier coefficients for $\cos \omega_0 t$ are

$$c_1 = \frac{1}{2} \qquad c_{-1} = \frac{1}{2} \qquad c_k = 0, |k| \neq 1$$

(b) In a similar fashion we have

$$\sin \omega_0 t = \frac{1}{2j}(e^{j\omega_0 t} - e^{-j\omega_0 t}) = -\frac{1}{2j} e^{-j\omega_0 t} + \frac{1}{2j} e^{j\omega_0 t} = \sum_{k=-\infty}^{\infty} c_k e^{jk\omega_0 t}$$

Thus, the complex Fourier coefficients for $\sin \omega_0 t$ are

$$c_1 = \frac{1}{2j} \qquad\qquad c_{-1} = -\frac{1}{2j} \qquad\qquad c_k = 0, |k| \neq 1$$

(c)   The fundamental angular frequency ω_0 of $x(t)$ is 2. Thus,

$$x(t) = \cos\left(2t + \frac{\pi}{4}\right) = \sum_{k=-\infty}^{\infty} c_k e^{jk\omega_0 t} = \sum_{k=-\infty}^{\infty} c_k e^{j2kt}$$

Now
$$x(t) = \cos\left(2t + \frac{\pi}{4}\right) = \frac{1}{2}\left(e^{j(2t+\pi/4)} + e^{-j(2t+\pi/4)}\right)$$

$$= \frac{1}{2} e^{-j\pi/4} e^{-j2t} + \frac{1}{2} e^{j\pi/4} e^{j2t} = \sum_{k=-\infty}^{\infty} c_k e^{j2kt}$$

Thus, the complex Fourier coefficients for $\cos(2t + \pi/4)$ are

$$c_1 = \frac{1}{2} e^{j\pi/4} = \frac{1}{2} \frac{1+j}{\sqrt{2}} = \frac{\sqrt{2}}{4}(1+j)$$

$$c_{-1} = \frac{1}{2} e^{-j\pi/4} = \frac{1}{2} \frac{1-j}{\sqrt{2}} = \frac{\sqrt{2}}{4}(1-j)$$

$$c_k = 0 \qquad\qquad |k| \neq 1$$

(d)   By the result from Prob. 1.14 the fundamental period T_0 of $x(t)$ is π and $\omega_0 = 2\pi/T_0 = 2$. Thus,

$$x(t) = \cos 4t + \sin 6t = \sum_{k=-\infty}^{\infty} c_k e^{jk\omega_0 t} = \sum_{k=-\infty}^{\infty} c_k e^{j2kt}$$

Again using Euler's formula, we have

$$x(t) = \cos 4t + \sin 6t = \frac{1}{2}\left(e^{j4t} + e^{-j4t}\right) + \frac{1}{2j}\left(e^{j6t} - e^{-j6t}\right)$$

$$= -\frac{1}{2j} e^{-j6t} + \frac{1}{2} e^{-j4t} + \frac{1}{2} e^{j4t} + \frac{1}{2j} e^{j6t} = \sum_{k=-\infty}^{\infty} c_k e^{j2kt}$$

Thus, the complex Fourier coefficients for $\cos 4t + \sin 6t$ are

$$c_{-3} = -\frac{1}{2j} \qquad c_{-2} = \frac{1}{2} \qquad c_2 = \frac{1}{2} \qquad c_3 = \frac{1}{2j}$$

and all other $c_k = 0$.

(e)   From Prob. 1.16(e) the fundamental period T_0 of $x(t)$ is π and $\omega_0 = 2\pi/T_0 = 2$. Thus,

$$x(t) = \sin^2 t = \sum_{k=-\infty}^{\infty} c_k e^{jk\omega_0 t} = \sum_{k=-\infty}^{\infty} c_k e^{j2kt}$$

Again using Euler's formula, we get

$$x(t) = \sin^2 t = \left(\frac{e^{jt} - e^{-jt}}{2j}\right)^2 = -\frac{1}{4}\left(e^{j2t} - 2 + e^{-j2t}\right)$$

$$= -\frac{1}{4} e^{-j2t} + \frac{1}{2} - \frac{1}{4} e^{j2t} = \sum_{k=-\infty}^{\infty} c_k e^{j2kt}$$

Thus, the complex Fourier coefficients for $\sin^2 t$ are

$$c_{-1} = -\frac{1}{4} \qquad c_0 = \frac{1}{2} \qquad c_1 = -\frac{1}{4}$$

and all other $c_k = 0$.

5.5. Consider the periodic square wave $x(t)$ shown in Fig. 5-8.

(a) Determine the complex exponential Fourier series of $x(t)$.

(b) Determine the trigonometric Fourier series of $x(t)$.

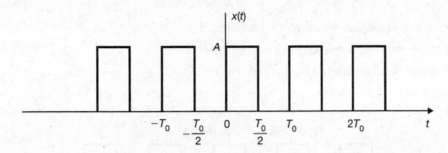

Fig. 5-8

(a) Let

$$x(t) = \sum_{k=-\infty}^{\infty} c_k e^{jk\omega_0 t} \qquad \omega_0 = \frac{2\pi}{T_0}$$

Using Eq. (5.102a), we have

$$c_k = \frac{1}{T_0} \int_0^{T_0} x(t) e^{-jk\omega_0 t} \, dt = \frac{1}{T_0} \int_0^{T_0/2} A\, e^{-jk\omega_0 t} \, dt$$

$$= \frac{A}{-jk\omega_0 T_0} e^{-jk\omega_0 t} \Big|_0^{T_0/2} = \frac{A}{-jk\omega_0 T_0} (e^{-jk\omega_0 T_0/2} - 1)$$

$$= \frac{A}{jk2\pi} (1 - e^{-jk\pi}) = \frac{A}{jk2\pi} \Big[1 - (-1)^k \Big]$$

since $\omega_0 T_0 = 2\pi$ and $e^{-jk\pi} = (-1)^k$. Thus,

$$c_k = 0 \qquad\qquad k = 2m \neq 0$$

$$c_k = \frac{A}{jk\pi} \qquad\qquad k = 2m + 1$$

$$c_0 = \frac{1}{T_0} \int_0^{T_0} x(t) \, dt = \frac{1}{T_0} \int_0^{T_0/2} A \, dt = \frac{A}{2}$$

Hence,

$$c_0 = \frac{A}{2} \qquad\qquad c_{2m} = 0 \qquad\qquad c_{2m+1} = \frac{A}{j(2m+1)\pi} \qquad (5.105)$$

and we obtain

$$x(t) = \frac{A}{2} + \frac{A}{j\pi} \sum_{m=-\infty}^{\infty} \frac{1}{2m+1} e^{j(2m+1)\omega_0 t} \qquad (5.106)$$

(*b*) From Eqs. (5.105), (5.10), and (5.12) we have

$$\frac{a_0}{2} = c_0 = \frac{A}{2} \qquad a_{2m} = b_{2m} = 0, m \neq 0$$

$$a_{2m+1} = 2\,\text{Re}[c_{2m+1}] = 0 \qquad b_{2m+1} = -2\,\text{Im}[c_{2m+1}] = \frac{2A}{(2m+1)\pi}$$

Substituting these values in Eq. (5.8), we get

$$x(t) = \frac{A}{2} + \frac{2A}{\pi} \sum_{m=0}^{\infty} \frac{1}{2m+1} \sin(2m+1)\omega_0 t$$

$$= \frac{A}{2} + \frac{2A}{\pi}\left(\sin\omega_0 t + \frac{1}{3}\sin 3\omega_0 t + \frac{1}{5}\sin 5\omega_0 t + \cdots\right) \qquad (5.107)$$

5.6. Consider the periodic square wave $x(t)$ shown in Fig. 5-9.

(*a*) Determine the complex exponential Fourier series of $x(t)$.

(*b*) Determine the trigonometric Fourier series of $x(t)$.

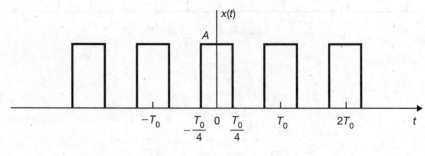

Fig. 5-9

(*a*) Let

$$x(t) = \sum_{k=-\infty}^{\infty} c_k\, e^{jk\omega_0 t} \qquad \omega_0 = \frac{2\pi}{T_0}$$

Using Eq. (5.102b), we have

$$c_k = \frac{1}{T_0}\int_{-T_0/2}^{T_0/2} x(t)\, e^{-jk\omega_0 t}\, dt = \frac{1}{T_0}\int_{-T_0/4}^{T_0/4} A\, e^{-jk\omega_0 t}\, dt$$

$$= \frac{A}{-jk\omega_0 T_0}(e^{-jk\omega_0 T_0/4} - e^{jk\omega_0 T_0/4})$$

$$= \frac{A}{-jk2\pi}(e^{-jk\pi/2} - e^{jk\pi/2}) = \frac{A}{k\pi}\sin\left(\frac{k\pi}{2}\right)$$

Thus,

$$c_k = 0 \qquad\qquad k = 2m \neq 0$$

$$c_k = (-1)^m\,\frac{A}{k\pi} \qquad k = 2m+1$$

$$c_0 = \frac{1}{T_0}\int_0^{T_0} x(t)\, dt = \frac{1}{T_0}\int_0^{T_0/2} A\, dt = \frac{A}{2}$$

Hence,

$$c_0 = \frac{A}{2} \qquad c_{2m} = 0, m \neq 0 \qquad c_{2m+1} = (-1)^m\,\frac{A}{(2m+1)\pi} \qquad (5.108)$$

and we obtain

$$x(t) = \frac{A}{2} + \frac{A}{\pi} \sum_{m=-\infty}^{\infty} \frac{(-1)^m}{2m+1} e^{j(2m+1)\omega_0 t} \qquad (5.109)$$

(b) From Eqs. (5.108), (5.10), and (5.12) we have

$$\frac{a_0}{2} = c_0 = \frac{A}{2} \qquad a_{2m} = 2\,\text{Re}[c_{2m}] = 0, m \neq 0$$

$$a_{2m+1} = 2\,\text{Re}[c_{2m+1}] = (-1)^m \frac{2A}{(2m+1)\pi} \qquad b_k = -2\,\text{Im}[c_k] = 0$$

Substituting these values into Eq. (5.8), we obtain

$$x(t) = \frac{A}{2} + \frac{2A}{\pi} \sum_{m=0}^{\infty} \frac{(-1)^m}{2m+1} \cos(2m+1)\omega_0 t$$

$$= \frac{A}{2} + \frac{2A}{\pi} \left(\cos \omega_0 t - \frac{1}{3} \cos 3\omega_0 t + \frac{1}{5} \cos 5\omega_0 t - \cdots \right) \qquad (5.110)$$

Note that $x(t)$ is even; thus, $x(t)$ contains only a dc term and cosine terms. Note also that $x(t)$ in Fig. 5-9 can be obtained by shifting $x(t)$ in Fig. 5-8 to the left by $T_0/4$.

5.7. Consider the periodic square wave $x(t)$ shown in Fig. 5-10.

(a) Determine the complex exponential Fourier series of $x(t)$.

(b) Determine the trigonometric Fourier series of $x(t)$.

Note that $x(t)$ can be expressed as

$$x(t) = x_1(t) - A$$

where $x_1(t)$ is shown in Fig. 5-11. Now comparing Fig. 5-11 and Fig. 5-8 in Prob. 5.5, we see that $x_1(t)$ is the same square wave of $x(t)$ in Fig. 5-8 except that A becomes $2A$.

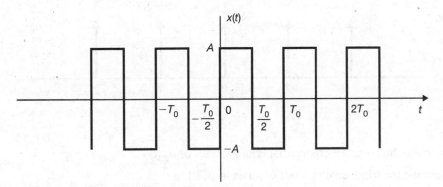

Fig. 5-10

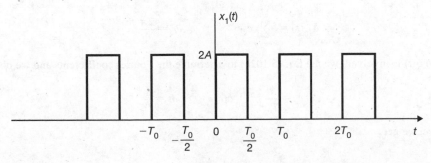

Fig. 5-11

(a) Replacing A by $2A$ in Eq. (5.106), we have

$$x_1(t) = A + \frac{2A}{j\pi} \sum_{m=-\infty}^{\infty} \frac{1}{2m+1} e^{j(2m+1)\omega_0 t}$$

Thus,

$$x(t) = x_1(t) - A = \frac{2A}{j\pi} \sum_{m=-\infty}^{\infty} \frac{1}{2m+1} e^{j(2m+1)\omega_0 t} \tag{5.111}$$

(b) Similarly, replacing A by $2A$ in Eq. (5.107), we have

$$x_1(t) = A + \frac{4A}{\pi} \sum_{m=0}^{\infty} \frac{1}{2m+1} \sin(2m+1)\omega_0 t$$

Thus,

$$x(t) = \frac{4A}{\pi} \sum_{m=0}^{\infty} \frac{1}{2m+1} \sin(2m+1)\omega_0 t$$

$$= \frac{4A}{\pi} \left(\sin \omega_0 t + \frac{1}{3} \sin 3\omega_0 t + \frac{1}{5} \sin 5\omega_0 t + \cdots \right) \tag{5.112}$$

Note that $x(t)$ is odd; thus, $x(t)$ contains only sine terms.

5.8. Consider the periodic impulse train $\delta_{T_0}(t)$ shown in Fig. 5-12 and defined by

$$\delta_{T_0}(t) = \sum_{k=-\infty}^{\infty} \delta(t - kT_0) \tag{5.113}$$

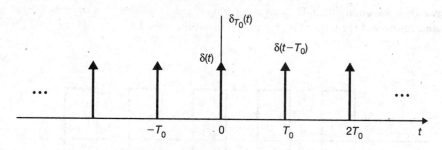

Fig. 5-12

(a) Determine the complex exponential Fourier series of $\delta_{T_0}(t)$.

(b) Determine the trigonometric Fourier series of $\delta_{T_0}(t)$.

(a) Let

$$\delta_{T_0}(t) = \sum_{k=-\infty}^{\infty} c_k e^{jk\omega_0 t} \qquad \omega_0 = \frac{2\pi}{T_0}$$

Since $\delta(t)$ is involved, we use Eq. (5.102b) to determine the Fourier coefficients and we obtain

$$c_k = \frac{1}{T_0} \int_{-T_0/2}^{T_0/2} \delta(t) e^{-jk\omega_0 t} \, dt = \frac{1}{T_0} \tag{5.114}$$

Hence, we get

$$\delta_{T_0}(t) = \sum_{k=-\infty}^{\infty} \delta(t - kT_0) = \frac{1}{T_0} \sum_{k=-\infty}^{\infty} e^{jk\omega_0 t} \qquad \omega_0 = \frac{2\pi}{T_0} \tag{5.115}$$

(b) Let

$$\delta_{T_0}(t) = \frac{a_0}{2} + \sum_{k=1}^{\infty} (a_k \cos k\omega_0 t + b_k \sin k\omega_0 t) \qquad \omega_0 = \frac{2\pi}{T_0}$$

Since $\delta_{T_0}(t)$ is even, $b_k = 0$, and by Eq. (5.9a), a_k are given by

$$a_k = \frac{2}{T_0} \int_{-T_0/2}^{T_0/2} \delta(t) \cos k\omega_0 t \, dt = \frac{2}{T_0} \tag{5.116}$$

Thus, we get

$$\delta_{T_0}(t) = \frac{1}{T_0} + \frac{2}{T_0} \sum_{k=1}^{\infty} \cos k\omega_0 t \qquad \omega_0 = \frac{2\pi}{T_0} \tag{5.117}$$

5.9. Consider the triangular wave $x(t)$ shown in Fig. 5-13(a). Using the differentiation technique, find (a) the complex exponential Fourier series of $x(t)$, and (b) the trigonometric Fourier series of $x(t)$.

The derivative $x'(t)$ of the triangular wave $x(t)$ is a square wave as shown in Fig. 5-13(b).

(a) Let

$$x(t) = \sum_{k=-\infty}^{\infty} c_k e^{jk\omega_0 t} \qquad \omega_0 = \frac{2\pi}{T_0} \tag{5.118}$$

Differentiating Eq. (5.118), we obtain

$$x'(t) = \sum_{k=-\infty}^{\infty} jk\omega_0 c_k e^{jk\omega_0 t} \tag{5.119}$$

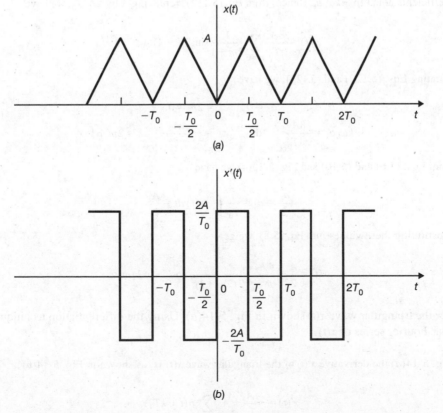

Fig. 5-13

Equation (5.119) shows that the complex Fourier coefficients of $x'(t)$ equal $jk\omega_0 c_k$. Thus, we can find c_k ($k \neq 0$) if the Fourier coefficients of $x'(t)$ are known. The term c_0 cannot be determined by Eq. (5.119) and must be evaluated directly in terms of $x(t)$ with Eq. (5.6). Comparing Fig. 5-13(b) and Fig. 5-10, we see that $x'(t)$ in Fig. 5-13(b) is the same as $x(t)$ in Fig. 5-10 with A replaced by $2A/T_0$. Hence, from Eq. (5.111), replacing A by $2A/T_0$, we have

$$x'(t) = \frac{4A}{j\pi T_0} \sum_{m=-\infty}^{\infty} \frac{1}{2m+1} e^{j(2m+1)\omega_0 t} \tag{5.120}$$

Equating Eqs. (5.119) and (5.120), we have

$$c_k = 0 \qquad\qquad k = 2m \neq 0$$

$$jk\omega_0 c_k = \frac{4A}{j\pi k T_0} \qquad \text{or} \qquad c_k = -\frac{2A}{\pi^2 k^2} \qquad k = 2m+1$$

From Fig. 5-13(a) and Eq. (5.6) we have

$$c_0 = \frac{1}{T_0} \int_0^{T_0} x(t)\, dt = \frac{A}{2}$$

Substituting these values into Eq. (5.118), we obtain

$$x(t) = \frac{A}{2} - \frac{2A}{\pi^2} \sum_{m=-\infty}^{\infty} \frac{1}{(2m+1)^2} e^{j(2m+1)\omega_0 t} \tag{5.121}$$

(b)   In a similar fashion, differentiating Eq. (5.8), we obtain

$$x'(t) = \sum_{k=1}^{\infty} k\omega_0 (b_k \cos k\omega_0 t - a_k \sin k\omega_0 t) \tag{5.122}$$

Equation (5.122) shows that the Fourier cosine coefficients of $x'(t)$ equal to $k\omega_0 b_k$ and that the sine coefficients equal to $-k\omega_0 a_k$. Hence, from Eq. (5.112), replacing A by $2A/T_0$, we have

$$x'(t) = \frac{8A}{\pi T_0} \sum_{m=0}^{\infty} \frac{1}{2m+1} \sin(2m+1)\omega_0 t \tag{5.123}$$

Equating Eqs. (5.122) and (5.123), we have

$$b_k = 0 \qquad a_k = 0 \qquad k = 2m \neq 0$$

$$-k\omega_0 a_k = \frac{8A}{\pi k T_0} \qquad \text{or} \qquad a_k = -\frac{4A}{\pi^2 k^2} \qquad k = 2m+1$$

From Eqs. (5.6) and (5.10) and Fig. 5-13(a) we have

$$\frac{a_0}{2} = c_0 = \frac{1}{T_0} \int_0^{T_0} x(t)\, dt = \frac{A}{2}$$

Substituting these values into Eq. (5.8), we get

$$x(t) = \frac{A}{2} - \frac{4A}{\pi^2} \sum_{m=0}^{\infty} \frac{1}{(2m+1)^2} \cos(2m+1)\,\omega_0 t \tag{5.124}$$

5.10. Consider the triangular wave $x(t)$ shown in Fig. 5-14(a). Using the differentiation technique, find the triangular Fourier series of $x(t)$.

From Fig. 5-14(a) the derivative $x'(t)$ of the triangular wave $x(t)$ is, as shown in Fig. 5-14(b),

$$x'(t) = -\frac{A}{T_0} + A \sum_{k=-\infty}^{\infty} \delta(t - kT_0) \tag{5.125}$$

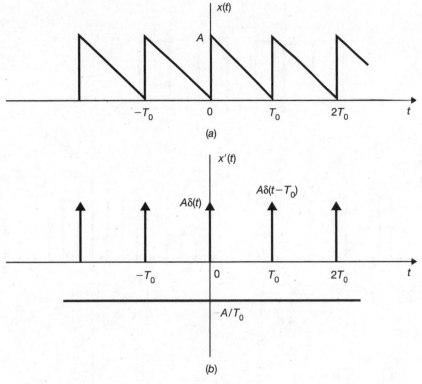

Fig. 5-14

Using Eq. (5.117), Eq. (5.125) becomes

$$x'(t) = \sum_{k=1}^{\infty} \frac{2A}{T_0} \cos k\omega_0 t \qquad \omega_0 = \frac{2\pi}{T_0} \tag{5.126}$$

Equating Eqs. (5.126) and (5.122), we have

$$a_k = 0, \ k \neq 0 \qquad k\omega_0 b_k = \frac{2A}{T_0} \qquad \text{or} \qquad b_k = \frac{A}{k\pi}$$

From Fig. 5-14(a) and Eq. (5.9a), we have

$$\frac{a_0}{2} = \frac{1}{T_0} \int_0^{T_0} x(t) \, dt = \frac{A}{2}$$

Thus, substituting these values into Eq. (5.8), we get

$$x(t) = \frac{A}{2} + \frac{A}{\pi} \sum_{k=1}^{\infty} \frac{1}{k} \sin k\omega_0 t \qquad \omega_0 = \frac{2\pi}{T_0} \tag{5.127}$$

5.11. Find and sketch the magnitude spectra for the periodic square pulse train signal $x(t)$ shown in Fig. 5-15(a) for (a) $d = T_0/4$, and (b) $d = T_0/8$.

Using Eq. (5.102a), we have

$$c_k = \frac{1}{T_0} \int_0^{T_0} x(t) e^{-jk\omega_0 t} \, dt = \frac{A}{T_0} \int_0^{d} e^{-jk\omega_0 t} \, dt$$

$$= \frac{A}{T_0} \frac{1}{-jk\omega_0} e^{-jk\omega_0 t} \Big|_0^{d} = \frac{A}{T_0} \frac{1}{jk\omega_0} (1 - e^{-jk\omega_0 d})$$

$$= \frac{A}{-jk\omega_0 T_0} e^{-jk\omega_0 d/2} (e^{jk\omega_0 d/2} - e^{-jk\omega_0 d/2})$$

$$= A \frac{d}{T_0} \frac{\sin(k\omega_0 d/2)}{k\omega_0 d/2} e^{-jk\omega_0 d/2} \tag{5.128}$$

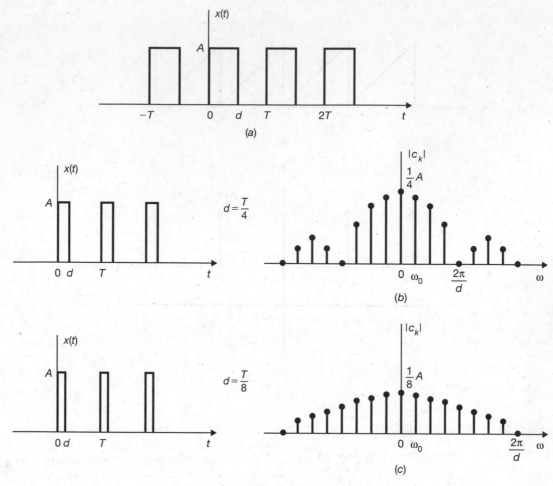

Fig. 5-15

Note that $c_k = 0$ whenever $k\omega_0 d/2 = m\pi$; that is,

$$n\omega_0 = \frac{m2\pi}{d} \qquad m = 0, \pm 1, \pm 2, \ldots$$

(a) $d = T_0/4$, $k\omega_0 d/2 = k\pi d/T_0 = k\pi/4$,

$$\left| c_k \right| = \frac{A}{4} \left| \frac{\sin(k\pi/4)}{k\pi/4} \right|$$

The magnitude spectrum for this case is shown in Fig. 5-15(b).

(b) $d = T_0/8$, $k\omega_0 d/2 = k\pi d/T_0 = k\pi/8$,

$$\left| c_k \right| = \frac{A}{8} \left| \frac{\sin(k\pi/8)}{k\pi/8} \right|$$

The magnitude spectrum for this case is shown in Fig. 5-15(c).

5.12. If $x_1(t)$ and $x_2(t)$ are periodic signals with fundamental period T_0 and their complex Fourier series expressions are

$$x_1(t) = \sum_{k=-\infty}^{\infty} d_k e^{jk\omega_0 t} \qquad x_2(t) = \sum_{k=-\infty}^{\infty} e_k e^{jk\omega_0 t} \qquad \omega_0 = \frac{2\pi}{T_0}$$

show that the signal $x(t) = x_1(t)x_2(t)$ is periodic with the same fundamental period T_0 and can be expressed as

$$x(t) = \sum_{k=-\infty}^{\infty} c_k e^{jk\omega_0 t} \qquad \omega_0 = \frac{2\pi}{\omega_0}$$

where c_k is given by

$$c_k = \sum_{m=-\infty}^{\infty} d_m e_{k-m} \qquad (5.129)$$

Now $\qquad x(t + T_0) = x_1(t + T_0)x_2(t + T_0) = x_1(t)x_2(t) = x(t)$

Thus, $x(t)$ is periodic with fundamental period T_0. Let

$$x(t) = \sum_{k=-\infty}^{\infty} c_k e^{jk\omega_0 t} \qquad \omega_0 = \frac{2\pi}{T_0}$$

Then
$$c_k = \frac{1}{T_0}\int_{-T_0/2}^{T_0/2} x(t)\,e^{-jk\omega_0 t}\,dt = \frac{1}{T_0}\int_{-T_0/2}^{T_0/2} x_1(t)x_2(t)\,e^{-jk\omega_0 t}\,dt$$

$$= \frac{1}{T_0}\int_{-T_0/2}^{T_0/2}\left(\sum_{m=-\infty}^{\infty} d_m e^{jm\omega_0 t}\right) x_2(t)\,e^{-jk\omega_0 t}\,dt$$

$$= \sum_{m=-\infty}^{\infty} d_m\left[\frac{1}{T_0}\int_{-T_0/2}^{T_0/2} x_2(t)\,e^{-j(k-m)\omega_0 t}\,dt\right] = \sum_{m=-\infty}^{\infty} d_m e_{k-m}$$

since
$$e_k = \frac{1}{T_0}\int_{-T_0/2}^{T_0/2} x_2(t)\,e^{-jk\omega_0 t}\,dt$$

and the term in brackets is equal to e_{k-m}.

5.13. Let $x_1(t)$ and $x_2(t)$ be the two periodic signals in Prob. 5.12. Show that

$$\frac{1}{T_0}\int_{-T_0/2}^{T_0/2} x_1(t)x_2(t)\,dt = \sum_{k=-\infty}^{\infty} d_k e_{-k} \qquad (5.130)$$

Equation (5.130) is known as *Parseval's relation* for periodic signals.

From Prob. 5.12 and Eq. (5.129) we have

$$c_k = \frac{1}{T_0}\int_{-T_0/2}^{T_0/2} x_1(t)x_2(t)\,e^{-jk\omega_0 t}\,dt = \sum_{m=-\infty}^{\infty} d_m e_{k-m}$$

Setting $k = 0$ in the above expression, we obtain

$$\frac{1}{T_0}\int_{-T_0/2}^{T_0/2} x_1(t)x_2(t)\,dt = \sum_{m=-\infty}^{\infty} d_m e_{-m} = \sum_{k=-\infty}^{\infty} d_k e_{-k}$$

5.14. Verify Parseval's identity (5.21) for the Fourier series; that is,

$$\frac{1}{T_0}\int_{T_0}|x(t)|^2\,dt = \sum_{k=-\infty}^{\infty}|c_k|^2$$

If
$$x(t) = \sum_{k=-\infty}^{\infty} c_k e^{jk\omega_0 t}$$

then
$$x^*(t) = \left(\sum_{k=-\infty}^{\infty} c_k e^{jk\omega_0 t} \right)^* = \sum_{k=-\infty}^{\infty} c_k^* e^{-jk\omega_0 t} = \sum_{k=-\infty}^{\infty} c_{-k}^* e^{jk\omega_0 t} \qquad (5.131)$$

where $*$ denotes the complex conjugate. Equation (5.131) indicates that if the Fourier coefficients of $x(t)$ are c_k, then the Fourier coefficients of $x^*(t)$ are c_{-k}^*. Setting $x_1(t) = x(t)$ and $x_2(t) = x^*(t)$ in Eq. (5.130), we have $d_k = c_k$ and $e_k = c_{-k}^*$ or $(e_{-k} = c_k^*)$, and we obtain

$$\frac{1}{T_0} \int_{-T_0/2}^{T_0/2} x(t)x^*(t)\,dt = \sum_{k=-\infty}^{\infty} c_k c_k^* \qquad (5.132)$$

or

$$\frac{1}{T_0} \int_{-T_0/2}^{T_0/2} |x(t)|^2\,dt = \sum_{k=-\infty}^{\infty} |c_k|^2$$

5.15. (a) The periodic convolution $f(t) = x_1(t) \otimes x_2(t)$ was defined in Prob. 2.8. If d_n and e_n are the complex Fourier coefficients of $x_1(t)$ and $x_2(t)$, respectively, then show that the complex Fourier coefficients c_k of $f(t)$ are given by

$$c_k = T_0 d_k e_k \qquad (5.133)$$

where T_0 is the fundamental period common to $x_1(t)$, $x_2(t)$, and $f(t)$.

(b) Find the complex exponential Fourier series of $f(t)$ defined in Prob. 2.8(c).

(a) From Eq. (2.70) (Prob. 2.8)

$$f(t) = x_1(t) \otimes x_2(t) = \int_0^{T_0} x_1(\tau)x_2(t - \tau)\,d\tau$$

Let
$$x_1(t) = \sum_{k=-\infty}^{\infty} d_k e^{jk\omega_0 t} \qquad x_2(t) = \sum_{k=-\infty}^{\infty} e_k e^{jk\omega_0 t}$$

Then
$$f(t) = \int_0^{T_0} x(\tau) \left(\sum_{k=-\infty}^{\infty} e_k e^{jk\omega_0(t-\tau)} \right) d\tau$$

$$= \sum_{k=-\infty}^{\infty} e_k e^{jk\omega_0 t} \int_0^{T_0} x(\tau) e^{-jk\omega_0 \tau}\,d\tau$$

Since
$$d_k = \frac{1}{T_0} \int_0^{T_0} x(\tau) e^{-jk\omega_0 \tau}\,d\tau$$

we get

$$f(t) = \sum_{k=-\infty}^{\infty} T_0 d_k e_k e^{jk\omega_0 t} \qquad (5.134)$$

which shows that the complex Fourier coefficients c_k of $f(t)$ equal $T_0 d_k e_k$.

(b) In Prob. 2.8(c), $x_1(t) = x_2(t) = x(t)$, as shown in Fig. 2-12, which is the same as Fig. 5-8 (Prob. 5.5). From Eq. (5.105) we have

$$d_0 = e_0 = \frac{A}{2} \qquad d_k = e_k = \begin{cases} 0 & k = 2m, m \neq 0 \\ A/jk\pi & k = 2m+1 \end{cases}$$

Thus, by Eq. (5.133) the complex Fourier coefficients c_k of $f(t)$ are

$$c_0 = T_0 d_0 e_0 = T_0 \frac{A^2}{4}$$

$$c_k = T_0 d_k e_k = \begin{cases} 0 & k = 2m, m \neq 0 \\ -T_0 A^2 / k^2 \pi^2 & k = 2m+1 \end{cases}$$

Note that in Prob. 2.8(c), $f(t) = x_1(t) \otimes x_2(t)$, shown in Fig. 2-13($b$), is proportional to $x(t)$, shown in Fig. 5-13(a). Thus, replacing A by $A^2 T_0/2$ in the result from Prob. 5.9, we get

$$c_0 = T_0 \frac{A^2}{4} \qquad c_k = \begin{cases} 0 & k = 2m, m \neq 0 \\ -T_0 A^2 / k^2 \pi^2 & k = 2m + 1 \end{cases}$$

which are the same results obtained by using Eq. (5.133).

Fourier Transform

5.16. (*a*) Verify the time-shifting property (5.50); that is,

$$x(t - t_0) \leftrightarrow e^{j\omega t_0} X(\omega)$$

By definition (5.31)

$$\mathscr{F}\{x(t - t_0)\} = \int_{-\infty}^{\infty} x(t - t_0) e^{-j\omega t} \, dt$$

By the change of variable $\tau = t - t_0$, we obtain

$$\mathscr{F}\{x(t - t_0)\} = \int_{-\infty}^{\infty} x(\tau) e^{-j\omega(\tau + t_0)} \, d\tau$$

$$= e^{-j\omega t_0} \int_{-\infty}^{\infty} x(\tau) e^{-j\omega \tau} \, d\tau = e^{-j\omega t_0} X(\omega)$$

Hence,

$$x(t - t_0) \leftrightarrow e^{-j\omega t_0} X(\omega)$$

5.17. Verify the frequency-shifting property (5.51); that is,

$$x(t) e^{j\omega_0 t} \leftrightarrow X(\omega - \omega_0)$$

By definition (5.31)

$$\mathscr{F}\{x(t) e^{j\omega_0 t}\} = \int_{-\infty}^{\infty} x(t) e^{j\omega_0 t} e^{-j\omega t} \, dt$$

$$= \int_{-\infty}^{\infty} x(t) e^{-j(\omega - \omega_0)t} \, dt = X(\omega - \omega_0)$$

Hence,

$$x(t) e^{j\omega_0 t} \leftrightarrow X(\omega - \omega_0)$$

5.18. Verify the duality property (5.54); that is,

$$X(t) \leftrightarrow 2\pi x(-\omega)$$

From the inverse Fourier transform definition (5.32), we have

$$\int_{-\infty}^{\infty} X(\omega) e^{j\omega t} \, d\omega = 2\pi x(t)$$

Changing t to $-t$, we obtain

$$\int_{-\infty}^{\infty} X(\omega) e^{-j\omega t} \, d\omega = 2\pi x(-t)$$

Now interchanging t and ω, we get

$$\int_{-\infty}^{\infty} X(t) e^{-j\omega t} \, dt = 2\pi x(-\omega)$$

Since
$$\mathscr{F}\{X(t)\} = \int_{-\infty}^{\infty} X(t)e^{-j\omega t}\, dt$$
we conclude that

$$X(t) \leftrightarrow 2\pi x(-\omega)$$

5.19. Find the Fourier transform of the rectangular pulse signal $x(t)$ [Fig. 5-16(a)] defined by

$$x(t) = p_a(t) = \begin{cases} 1 & |t| < a \\ 0 & |t| > a \end{cases} \tag{5.135}$$

By definition (5.31)

$$X(\omega) = \int_{-\infty}^{\infty} p_a(t)e^{-j\omega t}\, dt = \int_{-a}^{a} e^{-j\omega t}\, dt$$
$$= \frac{1}{j\omega}(e^{j\omega a} - e^{-j\omega a}) = 2\frac{\sin \omega a}{\omega} = 2a\frac{\sin \omega a}{\omega a}$$

Hence, we obtain

$$p_a(t) \leftrightarrow 2\frac{\sin \omega a}{\omega} = 2a\frac{\sin \omega a}{\omega a} \tag{5.136}$$

The Fourier transform $X(\omega)$ of $x(t)$ is sketched in Fig. 5-16(b).

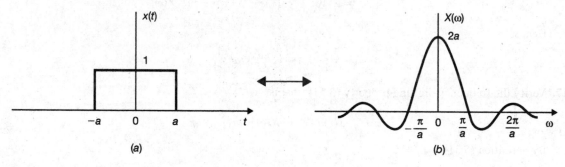

Fig. 5-16 Rectangular pulse and its Fourier transform.

5.20. Find the Fourier transform of the signal [Fig. 5-17(a)]

$$x(t) = \frac{\sin at}{\pi t}$$

From Eq. (5.136) we have

$$p_a(t) \leftrightarrow 2\frac{\sin \omega a}{\omega}$$

Now by the duality property (5.54), we have

$$2\frac{\sin at}{t} \leftrightarrow 2\pi p_a(-\omega)$$

Dividing both sides by 2π (and by the linearity property), we obtain

$$\frac{\sin at}{\pi t} \leftrightarrow p_a(-\omega) = p_a(\omega) \tag{5.137}$$

where $p_a(\omega)$ is defined by [see Eq. (5.135) and Fig. 5-17(b)]

$$p_a(\omega) = \begin{cases} 1 & |\omega| < a \\ 0 & |\omega| > a \end{cases}$$

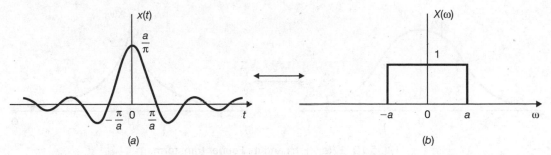

Fig. 5-17 $\sin at/\pi t$ and its Fourier transform.

5.21. Find the Fourier transform of the signal [Fig. 5-18(a)]

$$x(t) = e^{-a|t|} \qquad a > 0$$

Signal $x(t)$ can be rewritten as

$$x(t) = e^{-a|t|} = \begin{cases} e^{-at} & t > 0 \\ e^{at} & t < 0 \end{cases}$$

Then

$$X(\omega) = \int_{-\infty}^{0} e^{at} e^{-j\omega t} \, dt + \int_{0}^{\infty} e^{-at} e^{-j\omega t} \, dt$$

$$= \int_{-\infty}^{0} e^{(a-j\omega)t} \, dt + \int_{0}^{\infty} e^{-(a+j\omega)t} \, dt$$

$$= \frac{1}{a - j\omega} + \frac{1}{a + j\omega} = \frac{2a}{a^2 + \omega^2}$$

Hence, we get

$$e^{-a|t|} \leftrightarrow \frac{2a}{a^2 + \omega^2} \tag{5.138}$$

The Fourier transform $X(\omega)$ of $x(t)$ is shown in Fig. 5-18(b).

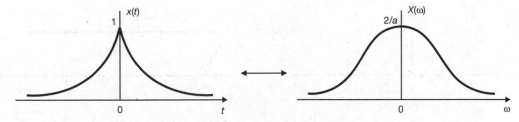

Fig. 5-18 $e^{-|a|t}$ and its Fourier transform.

5.22. Find the Fourier transform of the signal [Fig. 5-19(a)]

$$x(t) = \frac{1}{a^2 + t^2}$$

From Eq. (5.138) we have

$$e^{-a|t|} \leftrightarrow \frac{2a}{a^2 + \omega^2}$$

Now by the duality property (5.54) we have

$$\frac{2a}{a^2 + t^2} \leftrightarrow 2\pi e^{-a|-\omega|} = 2\pi e^{-a|\omega|}$$

Dividing both sides by $2a$, we obtain

$$\frac{1}{a^2 + t^2} \leftrightarrow \frac{\pi}{a} e^{-a|\omega|} \tag{5.139}$$

The Fourier transform $X(\omega)$ of $x(t)$ is shown in Fig. 5-19(b).

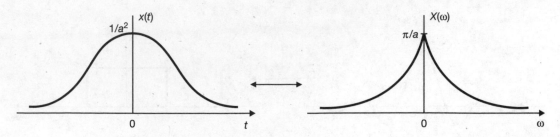

Fig. 5-19 $1/(a^2 + t^2)$ and its Fourier transform.

5.23. Find the Fourier transforms of the following signals:

(a) $x(t) = 1$ (b) $x(t) = e^{j\omega_0 t}$

(c) $x(t) = e^{-j\omega_0 t}$ (d) $x(t) = \cos \omega_0 t$

(e) $x(t) = \sin \omega_0 t$

(a)   By Eq. (5.43) we have

$$\delta(t) \leftrightarrow 1 \tag{5.140}$$

Thus, by the duality property (5.54) we get

$$1 \leftrightarrow 2\pi\delta(-\omega) = 2\pi\delta(\omega) \tag{5.141}$$

Figs. 5-20(a) and (b) illustrate the relationships in Eqs. (5.140) and (5.141), respectively.

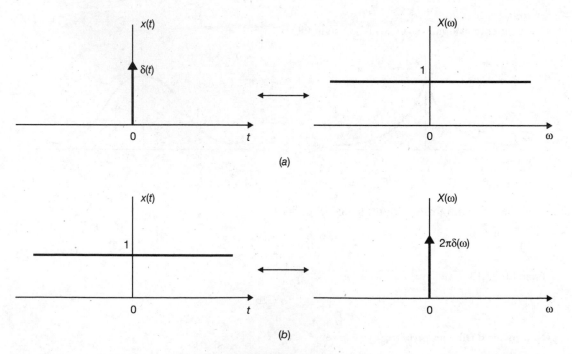

Fig. 5-20 (a) Unit impulse and its Fourier transform; (b) constant (dc) signal and its Fourier transform.

(b)   Applying the frequency-shifting property (5.51) to Eq. (5.141), we get

$$e^{j\omega_0 t} \leftrightarrow 2\pi\delta(\omega - \omega_0) \tag{5.142}$$

(c) From Eq. (5.142), it follows that

$$e^{-j\omega_0 t} \leftrightarrow 2\pi\delta(\omega + \omega_0) \tag{5.143}$$

(d) From Euler's formula we have

$$\cos \omega_0 t = \frac{1}{2}(e^{j\omega_0 t} + e^{j\omega_0 t})$$

Thus, using Eqs. (5.142) and (5.143) and the linearity property (5.49), we get

$$\cos \omega_0 t \leftrightarrow \pi[\delta(\omega - \omega_0) + \delta(\omega + \omega_0)] \tag{5.144}$$

Fig. 5-21 illustrates the relationship in Eq. (5.144).

(e) Similarly, we have

$$\sin \omega_0 t = \frac{1}{2j}(e^{j\omega_0 t} - e^{-j\omega_0 t})$$

and again using Eqs. (5.142) and (5.143), we get

$$\sin \omega_0 t \leftrightarrow -j\pi[\delta(\omega - \omega_0) - \delta(\omega + \omega_0)] \tag{5.145}$$

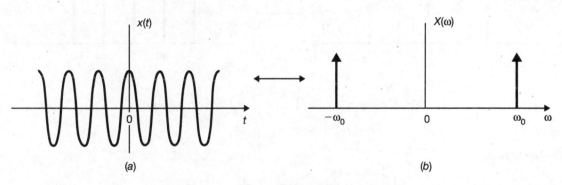

Fig. 5-21 Cosine signal and its Fourier transform.

5.24. Find the Fourier transform of a periodic signal $x(t)$ with period T_0.

We express $x(t)$ as

$$x(t) = \sum_{k=-\infty}^{\infty} c_k e^{jk\omega_0 t} \qquad \omega_0 = \frac{2\pi}{T_0}$$

Taking the Fourier transform of both sides and using Eq. (5.142) and the linearity property (5.49), we get

$$X(\omega) = 2\pi \sum_{k=-\infty}^{\infty} c_k \delta(\omega - k\omega_0) \tag{5.146}$$

which indicates that the Fourier transform of a periodic signal consists of a sequence of equidistant impulses located at the harmonic frequencies of the signal.

5.25. Find the Fourier transform of the periodic impulse train [Fig. 5-22(a)]

$$\delta_{T_0}(t) = \sum_{k=-\infty}^{\infty} \delta(t - kT_0)$$

From Eq. (5.115) in Prob. 5.8, the complex exponential Fourier series of $\delta_{T_0}(t)$ is given by

$$\delta_{T_0}(t) = \frac{1}{T_0} \sum_{k=-\infty}^{\infty} e^{jk\omega_0 t} \qquad \omega_0 = \frac{2\pi}{T_0}$$

Using Eq. (5.146), we get

$$\mathscr{F}[\delta_{T_0}(t)] = \frac{2\pi}{T_0} \sum_{k=-\infty}^{\infty} \delta(\omega - k\omega_0)$$

$$= \omega_0 \sum_{k=-\infty}^{\infty} \delta(\omega - k\omega_0) = \omega_0 \delta_{\omega_0}(\omega)$$

or
$$\sum_{k=-\infty}^{\infty} \delta(t - kT_0) \leftrightarrow \omega_0 \sum_{k=-\infty}^{\infty} \delta(\omega - k\omega_0) \tag{5.147}$$

Thus, the Fourier transform of a unit impulse train is also a similar impulse train [Fig. 5-22(b)].

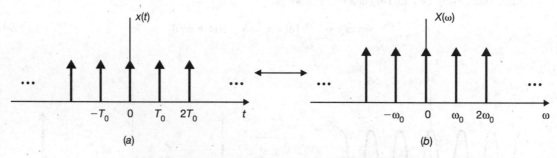

Fig. 5-22 Unit impulse train and its Fourier transform.

5.26. Show that

$$x(t) \cos \omega_0 t \leftrightarrow \frac{1}{2} X(\omega - \omega_0) + \frac{1}{2} X(\omega + \omega_0) \tag{5.148}$$

and
$$x(t) \sin \omega_0 t \leftrightarrow -j\left[\frac{1}{2} X(\omega - \omega_0) - \frac{1}{2} X(\omega + \omega_0)\right] \tag{5.149}$$

Equation (5.148) is known as the *modulation theorem*.

From Euler's formula we have

$$\cos \omega_0 t = \frac{1}{2}(e^{j\omega_0 t} + e^{-j\omega_0 t})$$

Then by the frequency-shifting property (5.51) and the linearity property (5.49), we obtain

$$\mathscr{F}[x(t) \cos \omega_0 t] = \mathscr{F}\left[\frac{1}{2} x(t) e^{j\omega_0 t} + \frac{1}{2} x(t) e^{-j\omega_0 t}\right]$$

$$= \frac{1}{2} X(\omega - \omega_0) + \frac{1}{2} X(\omega + \omega_0)$$

Hence,

$$x(t) \cos \omega_0 t \leftrightarrow \frac{1}{2} X(\omega - \omega_0) + \frac{1}{2} X(\omega + \omega_0)$$

In a similar manner we have

$$\sin \omega_0 t = \frac{1}{2j}(e^{j\omega_0 t} - e^{-j\omega_0 t})$$

and

$$\mathscr{F}[x(t)\sin \omega_0 t] = \mathscr{F}\left[\frac{1}{2j}x(t)e^{j\omega_0 t} - \frac{1}{2j}x(t)e^{-j\omega_0 t}\right]$$

$$= \frac{1}{2j}X(\omega - \omega_0) - \frac{1}{2j}X(\omega + \omega_0)$$

Hence,

$$x(t)\sin \omega_0 t \leftrightarrow -j\left[\frac{1}{2}X(\omega - \omega_0) - \frac{1}{2}X(\omega + \omega_0)\right]$$

5.27. The Fourier transform of a signal $x(t)$ is given by [Fig. 5-23(a)]

$$X(\omega) = \frac{1}{2}p_a(\omega - \omega_0) + \frac{1}{2}p_a(\omega + \omega_0)$$

Find and sketch $x(t)$.

From Eq. (5.137) and the modulation theorem (5.148), it follows that

$$x(t) = \frac{\sin at}{\pi t}\cos \omega_0 t$$

which is sketched in Fig. 5-23(b).

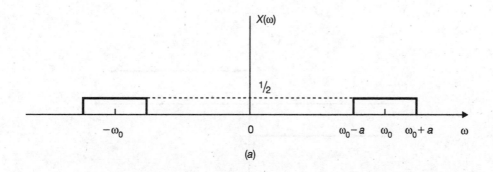

(a)

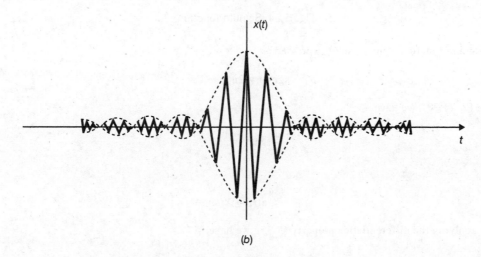

(b)

Fig. 5-23

5.28. Verify the differentiation property (5.55); that is,

$$\frac{dx(t)}{dt} \leftrightarrow j\omega X(\omega)$$

From Eq. (5.32) the inverse Fourier transform of $X(\omega)$ is

$$x(t) = \frac{1}{2\pi}\int_{-\infty}^{\infty} X(\omega)e^{j\omega t}\, d\omega \tag{5.150}$$

Then

$$\begin{aligned}
\frac{dx(t)}{dt} &= \frac{1}{2\pi}\frac{d}{dt}\left[\int_{-\infty}^{\infty} X(\omega)\,e^{j\omega t}\, d\omega\right] \\
&= \frac{1}{2\pi}\int_{-\infty}^{\infty} X(\omega)\frac{\partial}{\partial t}(e^{j\omega t})\, d\omega \\
&= \frac{1}{2\pi}\int_{-\infty}^{\infty} j\omega X(\omega)\,e^{j\omega t}\, d\omega
\end{aligned} \tag{5.151}$$

Comparing Eq. (5.151) with Eq. (5.150), we conclude that $dx(t)/dt$ is the inverse Fourier transform of $j\omega X(\omega)$. Thus,

$$\frac{dx(t)}{dt} \leftrightarrow j\omega X(\omega)$$

5.29. Find the Fourier transform of the *signum* function, sgn(t) (Fig. 5-24), which is defined as

$$\text{sgn}(t) = \begin{cases} 1 & t > 0 \\ -1 & t < 0 \end{cases} \tag{5.152}$$

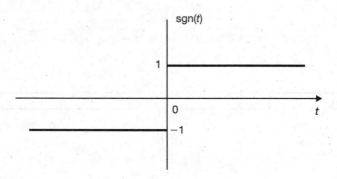

Fig. 5-24 Signum function.

The signum function, sgn(t), can be expressed as

$$\text{sgn}(t) = 2u(t) - 1$$

Using Eq. (1.30), we have

$$\frac{d}{dt}\text{sgn}(t) = 2\delta(t)$$

Let

$$\text{sgn}(t) \leftrightarrow X(\omega)$$

Then applying the differentiation property (5.55), we have

$$j\omega X(\omega) = \mathscr{F}[2\delta(t)] = 2 \rightarrow X(\omega) = \frac{2}{j\omega}$$

Hence,

$$\text{sgn}(t) \leftrightarrow \frac{2}{j\omega} \tag{5.153}$$

Note that sgn(t) is an odd function, and therefore its Fourier transform is a pure imaginary function of ω (Prob. 5.41).

5.30. Verify Eq. (5.48); that is,

$$u(t) \leftrightarrow \pi\delta(\omega) + \frac{1}{j\omega} \tag{5.154}$$

As shown in Fig. 5-25, $u(t)$ can be expressed as

$$u(t) = \frac{1}{2} + \frac{1}{2}\text{sgn}(t)$$

Note that $\frac{1}{2}$ is the even component of $u(t)$ and $\frac{1}{2}\text{sgn}(t)$ is the odd component of $u(t)$. Thus, by Eqs. (5.141) and (5.153) and the linearity property (5.49), we obtain

$$u(t) \leftrightarrow \pi\delta(\omega) + \frac{1}{j\omega}$$

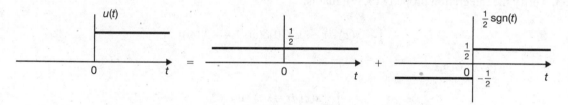

Fig. 5-25 Unit step function and its even and odd components.

5.31. Prove the time convolution theorem (5.58); that is,

$$x_1(t) * x_2(t) \leftrightarrow X_1(\omega) \, X_2(\omega)$$

By definitions (2.6) and (5.31), we have

$$\mathcal{F}[x_1(t) * x_2(t)] = \int_{-\infty}^{\infty}\left[\int_{-\infty}^{\infty} x_1(\tau) x_2(t-\tau)\, d\tau\right] e^{-j\omega t}\, dt$$

Changing the order of integration gives

$$\mathcal{F}[x_1(t) * x_2(t)] = \int_{-\infty}^{\infty} x_1(\tau)\left[\int_{-\infty}^{\infty} x_2(t-\tau)\, e^{-j\omega t}\, dt\right] d\tau$$

By the time-shifting property (5.50)

$$\int_{-\infty}^{\infty} x_2(t-\tau)\, e^{-j\omega t}\, dt = X_2(\omega)\, e^{-j\omega\tau}$$

Thus, we have

$$\mathcal{F}[x_1(t) * x_2(t)] = \int_{-\infty}^{\infty} x_1(\tau)\, X_2(\omega)\, e^{-j\omega\tau}\, d\tau$$

$$= \left[\int_{-\infty}^{\infty} x_1(\tau)\, e^{-j\omega\tau}\, d\tau\right] X_2(\omega) = X_1(\omega) X_2(\omega)$$

Hence,

$$x_1(t) * x_2(t) \leftrightarrow X_1(\omega) \, X_2(\omega)$$

5.32. Using the time convolution theorem (5.58), find the inverse Fourier transform of $X(\omega) = 1/(a + j\omega)^2$.

From Eq. (5.45) we have

$$e^{-at}u(t) \leftrightarrow \frac{1}{a + j\omega} \qquad\qquad (5.155)$$

Now

$$X(\omega) = \frac{1}{(a + j\omega)^2} = \left(\frac{1}{a + j\omega}\right)\left(\frac{1}{a + j\omega}\right)$$

Thus, by the time convolution theorem (5.58) we have

$$x(t) = e^{-at}u(t) * e^{-at}u(t)$$

$$= \int_{-\infty}^{\infty} e^{-a\tau}u(\tau)e^{-a(t-\tau)}u(t - \tau)\,d\tau$$

$$= e^{-at}\int_0^t d\tau = te^{-at}u(t)$$

Hence,

$$te^{-at}u(t) \leftrightarrow \frac{1}{(a + j\omega)^2} \qquad\qquad (5.156)$$

5.33. Verify the integration property (5.57); that is,

$$\int_{-\infty}^{t} x(\tau)\,d\tau \leftrightarrow \pi X(0)\delta(\omega) + \frac{1}{j\omega}X(\omega)$$

From Eq. (2.60) we have

$$\int_{-\infty}^{t} x(\tau)\,d\tau = x(t) * u(t)$$

Thus, by the time convolution theorem (5.58) and Eq. (5.154), we obtain

$$\mathcal{F}[x(t) * u(t)] = X(\omega)\left[\pi\delta(\omega) + \frac{1}{j\omega}\right] = \pi X(\omega)\,\delta(\omega) + \frac{1}{j\omega}X(\omega)$$

$$= \pi X(0)\delta(\omega) + \frac{1}{j\omega}X(\omega)$$

since $X(\omega)\delta(\omega) = X(0)\delta(\omega)$ by Eq. (1.25). Thus,

$$\left[\int_{-\infty}^{t} x(\tau)\,d\tau\right] \leftrightarrow \pi X(0)\delta(\omega) + \frac{1}{j\omega}X(\omega)$$

5.34. Using the integration property (5.57) and Eq. (1.31), find the Fourier transform of $u(t)$.

From Eq. (1.31) we have

$$u(t) = \int_{-\infty}^{t} \delta(\tau)\,d\tau$$

Now from Eq. (5.140) we have

$$\delta(t) \leftrightarrow 1$$

Setting $x(\tau) = \delta(\tau)$ in Eq. (5.57), we have

$$x(t) = \delta(t) \leftrightarrow X(\omega) = 1 \qquad \text{and} \qquad X(0) = 1$$

and

$$u(t) = \int_{-\infty}^{t} \delta(\tau)\,d\tau \leftrightarrow \pi\delta(\omega) + \frac{1}{j\omega}$$

5.35. Prove the frequency convolution theorem (5.59); that is,

$$x_1(t)x_2(t) \leftrightarrow \frac{1}{2\pi}X_1(\omega) * X_2(\omega)$$

By definitions (5.31) and (5.32) we have

$$\mathscr{F}[x_1(t)x_2(t)] = \int_{-\infty}^{\infty} x_1(t)x_2(t)\, e^{-j\omega t}\, dt$$

$$= \int_{-\infty}^{\infty} \left[\frac{1}{2\pi}\int_{-\infty}^{\infty} X_1(\lambda)\, e^{j\lambda t}\, d\lambda\right] x_2(t)e^{-j\omega t}\, dt$$

$$= \frac{1}{2\pi}\int_{-\infty}^{\infty} X_1(\lambda)\left[\int_{-\infty}^{\infty} x_2(t)\, e^{-j(\omega-\lambda)t}\, dt\right] d\lambda$$

$$= \frac{1}{2\pi}\int_{-\infty}^{\infty} X_1(\lambda)X_2(\omega - \lambda)\, d\lambda = \frac{1}{2\pi}X_1(\omega) * X_2(\omega)$$

Hence,

$$x_1(t)x_2(t) \leftrightarrow \frac{1}{2\pi}X_1(\omega) * X_2(\omega)$$

5.36. Using the frequency convolution theorem (5.59), derive the modulation theorem (5.148).

From Eq. (5.144) we have

$$\cos \omega_0 t \leftrightarrow \pi\delta(\omega - \omega_0) + \pi\delta(\omega + \omega_0)$$

By the frequency convolution theorem (5.59) we have

$$x(t)\cos \omega_0 t \leftrightarrow \frac{1}{2\pi}X(\omega) * \left[\pi\delta(\omega - \omega_0) + \pi\delta(\omega + \omega_0)\right]$$

$$= \frac{1}{2}X(\omega - \omega_0) + \frac{1}{2}X(\omega + \omega_0)$$

The last equality follows from Eq. (2.59).

5.37. Verify Parseval's relation (5.63); that is,

$$\int_{-\infty}^{\infty} x_1(t)x_2(t)\, dt = \frac{1}{2\pi}\int_{-\infty}^{\infty} X_1(\omega)X_2(-\omega)\, d\omega$$

From the frequency convolution theorem (5.59) we have

$$\mathscr{F}[x_1(t)x_2(t)] = \frac{1}{2\pi}\int_{-\infty}^{\infty} X_1(\lambda)X_2(\omega - \lambda)\, d\lambda$$

that is,

$$\int_{-\infty}^{\infty} [x_1(t)x_2(t)]\, e^{-j\omega t}\, dt = \frac{1}{2\pi}\int_{-\infty}^{\infty} X_1(\lambda)X_2(\omega - \lambda)\, d\lambda$$

Setting $\omega = 0$, we get

$$\int_{-\infty}^{\infty} x_1(t)x_2(t)\, dt = \frac{1}{2\pi}\int_{-\infty}^{\infty} X_1(\lambda)X_2(-\lambda)\, d\lambda$$

By changing the dummy variable of integration, we obtain

$$\int_{-\infty}^{\infty} x_1(t)x_2(t)\, dt = \frac{1}{2\pi}\int_{-\infty}^{\infty} X_1(\omega)X_2(-\omega)\, d\omega$$

5.38. Prove Parseval's identity [Eq. (5.64)] or Parseval's theorem for the Fourier transform; that is,

$$\int_{-\infty}^{\infty} |x(t)|^2\, dt = \frac{1}{2\pi} \int_{-\infty}^{\infty} |X(\omega)|^2\, d\omega$$

By definition (5.31) we have

$$\mathcal{F}\{x^*(t)\} = \int_{-\infty}^{\infty} x^*(t) e^{-j\omega t}\, dt$$

$$= \left[\int_{-\infty}^{\infty} x(t) e^{j\omega t}\, dt \right]^* = X^*(-\omega)$$

where $*$ denotes the complex conjugate. Thus,

$$x^*(t) \leftrightarrow X^*(-\omega) \tag{5.157}$$

Setting $x_1(t) = x(t)$ and $x_2(t) = x^*(t)$ in Parseval's relation (5.63), we get

$$\int_{-\infty}^{\infty} x(t) x^*(t)\, dt = \frac{1}{2\pi} \int_{-\infty}^{\infty} X(\omega) X^*(\omega)\, d\omega$$

or

$$\int_{-\infty}^{\infty} |x(t)|^2\, dt = \frac{1}{2\pi} \int_{-\infty}^{\infty} |X(\omega)|^2\, d\omega$$

5.39. Show that Eq. (5.61a); that is,

$$X^*(\omega) = X(-\omega)$$

is the necessary and sufficient condition for $x(t)$ to be real.

By definition (5.31)

$$X(\omega) = \int_{-\infty}^{\infty} x(t)\, e^{-j\omega t}\, dt$$

If $x(t)$ is real, then $x^*(t) = x(t)$ and

$$X^*(\omega) = \left[\int_{-\infty}^{\infty} x(t) e^{-j\omega t}\, dt \right]^* = \int_{-\infty}^{\infty} x^*(t) e^{j\omega t}\, dt$$

$$= \int_{-\infty}^{\infty} x(t) e^{j\omega t}\, dt = X(-\omega)$$

Thus, $X^*(\omega) = X(-\omega)$ is the necessary condition for $x(t)$ to be real. Next assume that $X^*(\omega) = X(-\omega)$. From the inverse Fourier transform definition (5.32)

$$x(t) = \frac{1}{2\pi} \int_{-\infty}^{\infty} X(\omega) e^{j\omega t}\, d\omega$$

Then

$$x^*(t) = \left[\frac{1}{2\pi} \int_{-\infty}^{\infty} X(\omega)\, e^{j\omega t}\, d\omega \right]^* = \frac{1}{2\pi} \int_{-\infty}^{\infty} X^*(\omega)\, e^{-j\omega t}\, d\omega$$

$$= \frac{1}{2\pi} \int_{-\infty}^{\infty} X(-\omega)\, e^{-j\omega t}\, d\omega = \frac{1}{2\pi} \int_{-\infty}^{\infty} X(\lambda)\, e^{j\lambda t} d\lambda = x(t)$$

which indicates that $x(t)$ is real. Thus, we conclude that

$$X^*(\omega) = X(-\omega)$$

is the necessary and sufficient condition for $x(t)$ to be real.

5.40. Find the Fourier transforms of the following signals:

     (a) $x(t) = u(-t)$

     (b) $x(t) = e^{at} u(-t), a > 0$

From Eq. (5.53) we have

$$x(-t) \leftrightarrow X(-\omega)$$

Thus, if $x(t)$ is real, then by Eq. (5.61a) we have

$$x(-t) \leftrightarrow X(-\omega) = X^*(\omega) \tag{5.158}$$

(*a*) From Eq. (5.154)

$$u(t) \leftrightarrow \pi\delta(\omega) + \frac{1}{j\omega}$$

Thus, by Eq. (5.158) we obtain

$$u(-t) \leftrightarrow \pi\delta(\omega) - \frac{1}{j\omega} \tag{5.159}$$

(*b*) From Eq. (5.155)

$$e^{-at}u(t) \leftrightarrow \frac{1}{a + j\omega}$$

Thus, by Eq. (5.158) we get

$$e^{at}u(-t) \leftrightarrow \frac{1}{a - j\omega} \tag{5.160}$$

5.41. Consider a real signal $x(t)$ and let

$$X(\omega) = \mathcal{F}[x(t)] = A(\omega) + jB(\omega)$$

and

$$x(t) = x_e(t) + x_o(t)$$

where $x_e(t)$ and $x_o(t)$ are the even and odd components of $x(t)$, respectively. Show that

$$x_e(t) \leftrightarrow A(\omega) \tag{5.161a}$$
$$x_o(t) \leftrightarrow jB(\omega) \tag{5.161b}$$

From Eqs. (1.5) and (1.6) we have

$$x_e(t) = \frac{1}{2}[x(t) + x(-t)]$$

$$x_o(t) = \frac{1}{2}[x(t) - x(-t)]$$

Now if $x(t)$ is real, then by Eq. (5.158) we have

$$x(t) \leftrightarrow X(\omega) = A(\omega) + jB(\omega)$$
$$x(-t) \leftrightarrow X(-\omega) = X^*(\omega) = A(\omega) - jB(\omega)$$

Thus, we conclude that

$$x_e(t) \leftrightarrow \frac{1}{2}X(\omega) + \frac{1}{2}X^*(\omega) = A(\omega)$$

$$x_o(t) \leftrightarrow \frac{1}{2}X(\omega) - \frac{1}{2}X^*(\omega) = jB(\omega)$$

Equations (5.161a) and (5.161b) show that the Fourier transform of a real even signal is a real function of ω, and that of a real odd signal is an imaginary function of ω, respectively.

5.42. Using Eqs. (5.161a) and (5.155), find the Fourier transform of $e^{-a|t|}$ $(a > 0)$.

From Eq. (5.155) we have

$$e^{-at}u(t) \leftrightarrow \frac{1}{a+j\omega} = \frac{a}{a^2+\omega^2} - j\frac{\omega}{a^2+\omega^2}$$

By Eq. (1.5) the even component of $e^{-at}u(t)$ is given by

$$\frac{1}{2}e^{-at}u(t) + \frac{1}{2}e^{at}u(-t) = \frac{1}{2}e^{-a|t|}$$

Thus, by Eq. (5.161a) we have

$$\frac{1}{2}e^{-a|t|} \leftrightarrow \mathrm{Re}\left(\frac{1}{a+j\omega}\right) = \frac{a}{a^2+\omega^2}$$

or

$$e^{-a|t|} \leftrightarrow \frac{2a}{a^2+\omega^2}$$

which is the same result obtained in Prob. 5.21 [Eq. (5.138)].

5.43. Find the Fourier transform of a Gaussian pulse signal

$$x(t) = e^{-at^2} \qquad a > 0$$

By definition (5.31)

$$X(\omega) = \int_{-\infty}^{\infty} e^{-at^2} e^{-j\omega t}\, dt \tag{5.162}$$

Taking the derivative of both sides of Eq. (5.162) with respect to ω, we have

$$\frac{dX(\omega)}{d\omega} = -j\int_{-\infty}^{\infty} te^{-at^2} e^{-j\omega t}\, dt$$

Now, using the integration by parts formula

$$\int_{\alpha}^{\beta} u\, dv = uv\Big|_{\alpha}^{\beta} - \int_{\alpha}^{\beta} v\, du$$

and letting

$$u = e^{-j\omega t} \qquad \text{and} \qquad dv = te^{-at^2}\, dt$$

we have

$$du = -j\omega e^{-j\omega t}\, dt \qquad \text{and} \qquad v = -\frac{1}{2a}e^{-at^2}$$

and

$$\int_{-\infty}^{\infty} te^{-at^2} e^{-j\omega t}\, dt = -\frac{1}{2a}e^{-at^2} e^{-j\omega t}\Big|_{-\infty}^{\infty} - j\frac{\omega}{2a}\int_{-\infty}^{\infty} e^{-at^2} e^{-j\omega t}\, dt$$

$$= -j\frac{\omega}{2a}\int_{-\infty}^{\infty} e^{-at^2} e^{-j\omega t}\, dt$$

since $a > 0$. Thus, we get

$$\frac{dX(\omega)}{d\omega} = -\frac{\omega}{2a}X(\omega)$$

Solving the above separable differential equation for $X(\omega)$, we obtain

$$X(\omega) = Ae^{-\omega^2/4a} \tag{5.163}$$

where A is an arbitrary constant. To evaluate A, we proceed as follows. Setting $\omega = 0$ in Eq. (5.162) and by a change of variable, we have

$$X(0) = A = \int_{-\infty}^{\infty} e^{-at^2}\, dt = 2\int_0^{\infty} e^{-at^2}\, dt = \frac{2}{\sqrt{a}} \int_0^{\infty} e^{-\lambda^2}\, d\lambda = \sqrt{\frac{\pi}{a}}$$

Substituting this value of A into Eq. (5.163), we get

$$X(\omega) = \sqrt{\frac{\pi}{a}}\, e^{-\omega^2/4a} \tag{5.164}$$

Hence, we have

$$e^{-at^2}, a > 0 \leftrightarrow \sqrt{\frac{\pi}{a}}\, e^{-\omega^2/4a} \tag{5.165}$$

Note that the Fourier transform of a Gaussian pulse signal is also a Gaussian pulse in the frequency domain. Fig. 5-26 shows the relationship in Eq. (5.165).

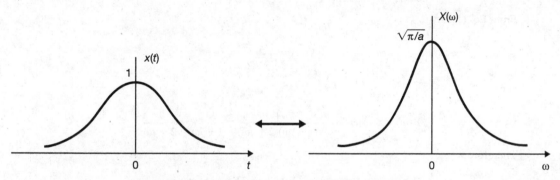

Fig. 5-26 Gaussian pulse and its Fourier transform.

Frequency Response

5.44. Using the Fourier transform, redo Prob. 2.25.

The system is described by

$$y'(t) + 2y(t) = x(t) + x'(t)$$

Taking the Fourier transforms of the above equation, we get

$$j\omega Y(\omega) + 2Y(\omega) = X(\omega) + j\omega X(\omega)$$

or

$$(j\omega + 2)\, Y(\omega) = (1 + j\omega)\, X(\omega)$$

Hence, by Eq. (5.67) the frequency response $H(\omega)$ is

$$H(\omega) = \frac{Y(\omega)}{X(\omega)} = \frac{1 + j\omega}{2 + j\omega} = \frac{2 + j\omega - 1}{2 + j\omega} = 1 - \frac{1}{2 + j\omega}$$

Taking the inverse Fourier transform of $H(\omega)$, the impulse response $h(t)$ is

$$h(t) = \delta(t) - e^{-2t}u(t)$$

Note that the procedure is identical to that of the Laplace transform method with s replaced by $j\omega$ (Prob. 3.29).

5.45. Consider a continuous-time LTI system described by

$$\frac{dy(t)}{dt} + 2y(t) = x(t) \qquad\qquad (5.166)$$

Using the Fourier transform, find the output $y(t)$ to each of the following input signals:

(a) $x(t) = e^{-t}u(t)$

(b) $x(t) = u(t)$

(a) Taking the Fourier transforms of Eq. (5.166), we have

$$j\omega Y(\omega) + 2Y(\omega) = X(\omega)$$

Hence,

$$H(\omega) = \frac{Y(\omega)}{X(\omega)} = \frac{1}{2 + j\omega}$$

From Eq. (5.155)

$$X(\omega) = \frac{1}{1 + j\omega}$$

and

$$Y(\omega) = X(\omega)H(\omega) = \frac{1}{(1 + j\omega)(2 + j\omega)} = \frac{1}{1 + j\omega} - \frac{1}{2 + j\omega}$$

Therefore,

$$y(t) = (e^{-t} - e^{-2t})\,u(t)$$

(b) From Eq. (5.154)

$$X(\omega) = \pi\delta(\omega) + \frac{1}{j\omega}$$

Thus, by Eq. (5.66) and using the partial-fraction expansion technique, we have

$$Y(\omega) = X(\omega)H(\omega) = \left[\pi\delta(\omega) + \frac{1}{j\omega}\right]\frac{1}{2 + j\omega}$$

$$= \pi\delta(\omega)\frac{1}{2 + j\omega} + \frac{1}{j\omega(2 + j\omega)}$$

$$= \frac{\pi}{2}\delta(\omega) + \frac{1}{2}\frac{1}{j\omega} - \frac{1}{2}\frac{1}{2 + j\omega}$$

$$= \frac{1}{2}\left[\pi\delta(\omega) + \frac{1}{j\omega}\right] - \frac{1}{2}\frac{1}{2 + j\omega}$$

where we used the fact that $f(\omega)\delta(\omega) = f(0)\delta(\omega)$ [Eq. (1.25)]. Thus,

$$y(t) = \frac{1}{2}u(t) - \frac{1}{2}e^{-2t}u(t) = \frac{1}{2}(1 - e^{-2t})u(t)$$

We observe that the Laplace transform method is easier in this case because of the Fourier transform of $u(t)$.

5.46. Consider the LTI system in Prob. 5.45. If the input $x(t)$ is the periodic square waveform shown in Fig. 5-27, find the amplitude of the first and third harmonics in the output $y(t)$.

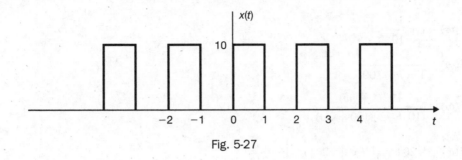

Fig. 5-27

Note that $x(t)$ is the same $x(t)$ shown in Fig. 5-8 [Prob. 5.5]. Thus, setting $A = 10$, $T_0 = 2$, and $\omega_0 = 2\pi/T_0 = \pi$ in Eq. (5.106), we have

$$x(t) = 5 + \frac{10}{j\pi} \sum_{m=-\infty}^{\infty} \frac{1}{2m+1} e^{j(2m+1)\pi t}$$

Next, from Prob. 5.45

$$H(\omega) = \frac{1}{2 + j\omega} \rightarrow H(k\omega_0) = H(k\pi) = \frac{1}{2 + jk\pi}$$

Thus, by Eq. (5.74) we obtain

$$y(t) = 5H(0) + \frac{10}{j\pi} \sum_{m=-\infty}^{\infty} \frac{1}{2m+1} H[(2m+1)\pi] e^{j(2m+1)\pi t}$$

$$= \frac{5}{2} + \frac{10}{j\pi} \sum_{m=-\infty}^{\infty} \frac{1}{(2m+1)[2 + j(2m+1)\pi]} e^{j(2m+1)\pi t} \tag{5.167}$$

Let

$$y(t) = \sum_{k=-\infty}^{\infty} d_k e^{jk\omega_0 t}$$

The harmonic form of $y(t)$ is given by [Eq. (5.15)]

$$y(t) = D_0 + \sum_{k=1}^{\infty} D_k \cos(k\omega_0 t - \phi_k)$$

where D_k is the amplitude of the kth harmonic component of $y(t)$. By Eqs. (5.11) and (5.16), D_k and d_k are related by

$$D_k = 2|d_k| \tag{5.168}$$

Thus, from Eq. (5.167), with $m = 0$, we obtain

$$D_1 = 2|d_1| = 2\left|\frac{10}{j\pi(2 + j\pi)}\right| = 1.71$$

With $m = 1$, we obtain

$$D_3 = 2|d_3| = 2\left|\frac{10}{j\pi(3)(2 + j3\pi)}\right| = 0.22$$

5.47. The most widely used graphical representation of the frequency response $H(\omega)$ is the *Bode plot* in which the quantities $20 \log_{10}|H(\omega)|$ and $\theta_H(\omega)$ are plotted versus ω, with ω plotted on a logarithmic scale. The quantity $20 \log_{10}|H(\omega)|$ is referred to as the magnitude expressed in *decibels* (dB), denoted by $|H(\omega)|_{dB}$. Sketch the Bode plots for the following frequency responses:

(a) $H(\omega) = 1 + \dfrac{j\omega}{10}$

(b) $H(\omega) = \dfrac{1}{1 + j\omega/100}$

(c) $H(\omega) = \dfrac{10^4(1+j\omega)}{(10+j\omega)(100+j\omega)}$

(a) $|H(\omega)|_{dB} = 20 \log_{10}|H(\omega)| = 20 \log_{10}\left|1 + j\dfrac{\omega}{10}\right|$

For $\omega \ll 10$,

$$|H(\omega)|_{dB} = 20 \log_{10}\left|1 + j\dfrac{\omega}{10}\right| \rightarrow 20 \log_{10} 1 = 0 \qquad \text{as } \omega \rightarrow 0$$

For $\omega \gg 10$,

$$|H(\omega)|_{dB} = 20 \log_{10}\left|1 + j\dfrac{\omega}{10}\right| \rightarrow 20 \log_{10}\left(\dfrac{\omega}{10}\right) \qquad \text{as } \omega \rightarrow 0$$

On a log frequency scale, $20 \log_{10}(\omega/10)$ is a straight line with a slope of 20 dB/decade (a decade is a 10-to-1 change in frequency). This straight line intersects the 0-dB axis at $\omega = 10$ [Fig. 5-28(a)]. (This value of ω is called the *corner frequency*.) At the corner frequency $\omega = 10$

$$H(10)|_{dB} = 20 \log_{10}|1 + j1| = 20 \log_{10}\sqrt{2} \approx 3 \text{ dB}$$

The plot of $|H(\omega)|_{dB}$ is sketched in Fig. 5-28(a). Next,

$$\theta_H(\omega) = \tan^{-1}\dfrac{\omega}{10}$$

Then

$$\theta_H(\omega) = \tan^{-1}\dfrac{\omega}{10} \rightarrow 0 \qquad \text{as } \omega \rightarrow 0$$

$$\theta_H(\omega) = \tan^{-1}\dfrac{\omega}{10} \rightarrow \dfrac{\pi}{2} \qquad \text{as } \omega \rightarrow \infty$$

At $\omega = 10$, $\theta_H(10) = \tan^{-1} 1 = \pi/4$ radian (rad). The plot of $\theta_H(\omega)$ is sketched in Fig. 5-28(b). Note that the dotted lines represent the straight-line approximation of the Bode plots.

(b) $|H(\omega)|_{dB} = 20 \log_{10}\left|\dfrac{1}{1 + j\omega/100}\right| = -20 \log_{10}\left|1 + j\dfrac{\omega}{100}\right|$

For $\omega \ll 100$,

$$|H(\omega)|_{dB} = -20 \log_{10}\left|1 + j\dfrac{\omega}{100}\right| \rightarrow -20 \log_{10} 1 = 0 \qquad \text{as } \omega \rightarrow 0$$

For $\omega \gg 100$,

$$|H(\omega)|_{dB} = -20 \log_{10}\left|1 - j\dfrac{\omega}{100}\right| \rightarrow -20 \log_{10}\left(\dfrac{\omega}{100}\right) \qquad \text{as } \omega \rightarrow \infty$$

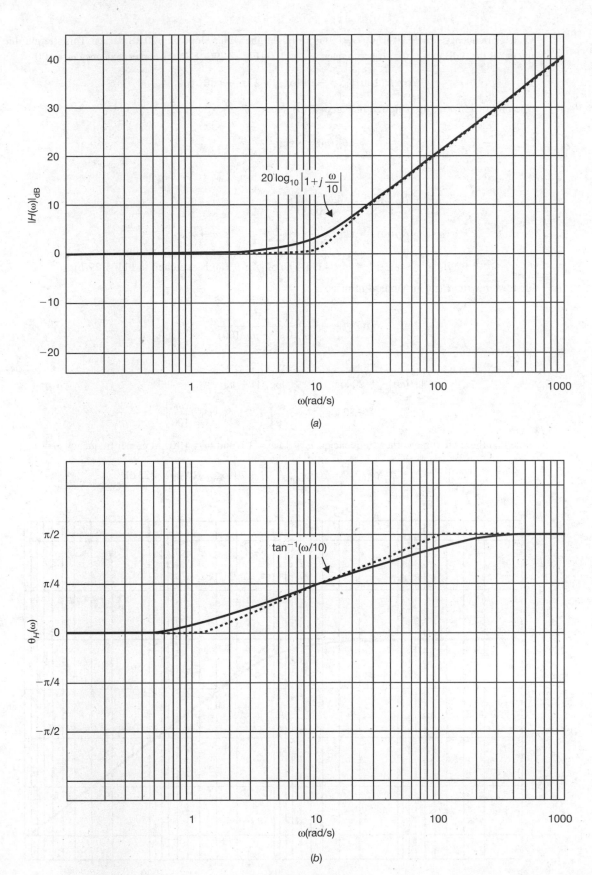

Fig. 5-28 Bode plots.

On a log frequency scale $-20 \log_{10}(\omega/100)$ is a straight line with a slope of -20 dB/decade. This straight line intersects the 0-dB axis at the corner frequency $\omega = 100$ [Fig. 5-29(a)]. At the corner frequency $\omega = 100$

$$H(100)\big|_{dB} = -20 \log_{10} \sqrt{2} \approx -3 \text{ dB}$$

The plot of $|H(\omega)|_{dB}$ is sketched in Fig. 5-29(a). Next,

$$\theta_H(\omega) = -\tan^{-1} \frac{\omega}{100}$$

Then

$$\theta_H(\omega) = -\tan^{-1} \frac{\omega}{100} \to 0 \qquad \text{as } \omega \to 0$$

$$\theta_H(\omega) = -\tan^{-1} \frac{\omega}{100} \to -\frac{\pi}{2} \qquad \text{as } \omega \to \infty$$

At $\omega = 100$, $\theta_H(100) = -\tan^{-1} 1 = -\pi/4$ rad. The plot of $\theta_H(\omega)$ is sketched in Fig. 5-29(b).

(c)   First, we rewrite $H(\omega)$ in standard form as

$$H(\omega) = \frac{10(1 + j\omega)}{(1 + j\omega/10)(1 + j\omega/100)}$$

Then

$$|H(\omega)|_{dB} = 20 \log_{10} 10 + 20 \log_{10} |1 + j\omega|$$
$$- 20 \log_{10} \left|1 + j\frac{\omega}{10}\right| - 20 \log_{10} \left|1 + j\frac{\omega}{100}\right|$$

Note that there are three corner frequencies, $\omega = 1$, $\omega = 10$, and $\omega = 100$. At corner frequency $\omega = 1$

$$H(1)\big|_{dB} = 20 + 20 \log_{10} \sqrt{2} - 20 \log_{10} \sqrt{1.01} - 20 \log_{10} \sqrt{1.0001} \approx 23 \text{ dB}$$

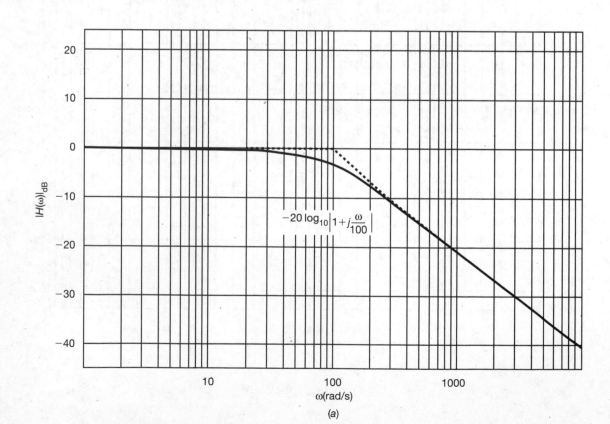

(a)

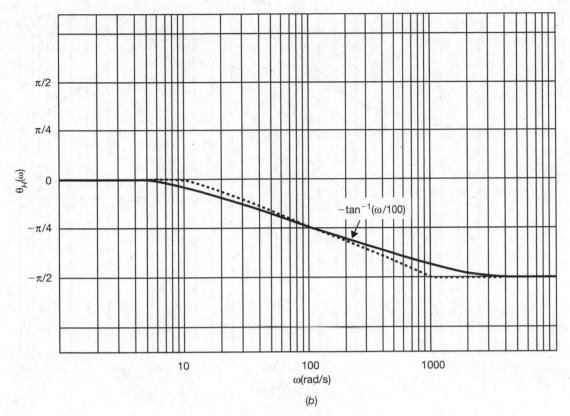

Fig. 5-29 Bode plots.

At corner frequency $\omega = 10$

$$H(10)\big|_{dB} = 20 + 20 \log_{10} \sqrt{101} - 20 \log_{10} \sqrt{2} - 20 \log_{10} \sqrt{1.01} \approx 37 \text{ dB}$$

At corner frequency $\omega = 100$

$$H(100)\big|_{dB} = 20 + 20 \log_{10} \sqrt{10{,}001} - 20 \log_{10} \sqrt{101} - 20 \log_{10} \sqrt{2} \approx 37 \text{ dB}$$

The Bode amplitude plot is sketched in Fig. 5-30(a). Each term contributing to the overall amplitude is also indicated. Next,

$$\theta_H(\omega) = \tan^{-1} \omega - \tan^{-1} \frac{\omega}{10} - \tan^{-1} \frac{\omega}{100}$$

Then

$$\theta_H(\omega) = \rightarrow 0 - 0 - 0 = 0 \qquad \text{as } \omega \rightarrow 0$$
$$\theta_H(\omega) = \rightarrow \frac{\pi}{2} - \frac{\pi}{2} - \frac{\pi}{2} = -\frac{\pi}{2} \qquad \text{as } \omega \rightarrow \infty$$

and

$$\theta_H(1) = \tan^{-1}(1) - \tan^{-1}(0.1) - \tan^{-1}(0.01) = 0.676 \text{ rad}$$
$$\theta_H(10) = \tan^{-1}(10) - \tan^{-1}(1) - \tan^{-1}(0.1) = 0.586 \text{ rad}$$
$$\theta_H(100) = \tan^{-1}(100) - \tan^{-1}(10) - \tan^{-1}(1) = -0.696 \text{ rad}$$

The plot of $\theta_H(\omega)$ is sketched in Fig. 5-30(b).

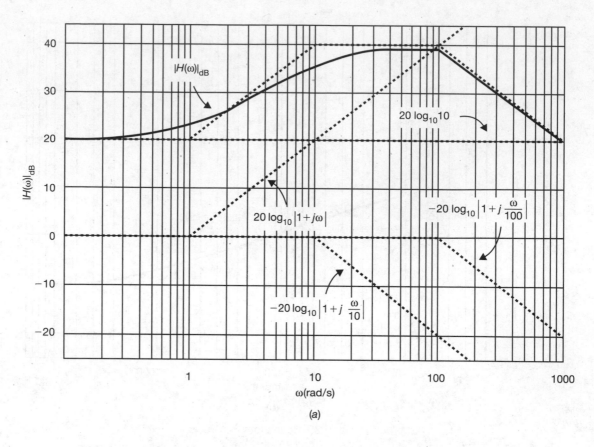

(a)

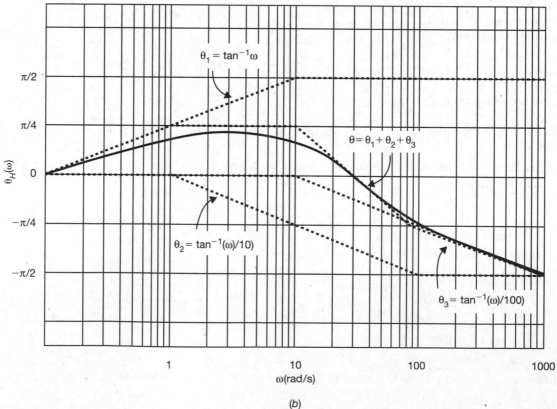

(b)

Fig. 5-30 Bode plots.

5.48. An ideal $(-\pi/2)$ radian (or $-90°$) phase shifter (Fig. 5-31) is defined by the frequency response

$$H(\omega) = \begin{cases} e^{-j(\pi/2)} & \omega > 0 \\ e^{j(\pi/2)} & \omega < 0 \end{cases} \tag{5.169}$$

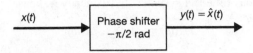

Fig. 5-31 $-\pi/2$ rad phase shifter.

(a) Find the impulse response $h(t)$ of this phase shifter.

(b) Find the output $y(t)$ of this phase shifter due to an arbitrary input $x(t)$.

(c) Find the output $y(t)$ when $x(t) = \cos \omega_0 t$.

(a) Since $e^{-j\pi/2} = -j$ and $e^{j\pi/2} = j$, $H(\omega)$ can be rewritten as

$$H(\omega) = -j \, \text{sgn}(\omega) \tag{5.170}$$

where

$$\text{sgn}(\omega) = \begin{cases} 1 & \omega > 0 \\ -1 & \omega < 0 \end{cases} \tag{5.171}$$

Now from Eq. (5.153)

$$\text{sgn}(t) \leftrightarrow \frac{2}{j\omega}$$

and by the duality property (5.54) we have

$$\frac{2}{jt} \leftrightarrow 2\pi \, \text{sgn}(-\omega) = -2\pi \, \text{sgn}(\omega)$$

or

$$\frac{1}{\pi t} \leftrightarrow -j \, \text{sgn}(\omega) \tag{5.172}$$

since $\text{sgn}(\omega)$ is an odd function of ω. Thus, the impulse response $h(t)$ is given by

$$h(t) = \mathscr{F}^{-1}[H(\omega)] = \mathscr{F}^{-1}[-j \, \text{sgn}(\omega)] = \frac{1}{\pi t} \tag{5.173}$$

(b) By Eq. (2.6)

$$y(t) = x(t) * \frac{1}{\pi t} = \frac{1}{\pi} \int_{-\infty}^{\infty} \frac{x(\tau)}{t - \tau} \, d\tau \tag{5.174}$$

The signal $y(t)$ defined by Eq. (5.174) is called the *Hilbert transform* of $x(t)$ and is usually denoted by $\hat{x}(t)$.

(c) From Eq. (5.144)

$$\cos \omega_0 t \leftrightarrow \pi[\delta(\omega - \omega_0) + \delta(\omega + \omega_0)]$$

Then

$$\begin{aligned} Y(\omega) = X(\omega)H(\omega) &= \pi[\delta(\omega - \omega_0) + \delta(\omega + \omega_0)][-j \, \text{sgn}(\omega)] \\ &= -j\pi \, \text{sgn}(\omega_0)\delta(\omega - \omega_0) - j\pi \, \text{sgn}(-\omega_0)\delta(\omega + \omega_0) \\ &= -j\pi\delta(\omega - \omega_0) + j\pi\delta(\omega + \omega_0) \end{aligned}$$

since $\operatorname{sgn}(\omega_0) = 1$ and $\operatorname{sgn}(-\omega_0) = -1$. Thus, from Eq. (5.145) we get

$$y(t) = \sin \omega_0 t$$

Note that $\cos(\omega_0 t - \pi/2) = \sin \omega_0 t$.

5.49. Consider a causal continuous-time LTI system with frequency response

$$H(\omega) = A(\omega) + jB(\omega)$$

Show that the impulse response $h(t)$ of the system can be obtained in terms of $A(\omega)$ or $B(\omega)$ alone.

Since the system is causal, by definition

$$h(t) = 0 \qquad t < 0$$

Accordingly,

$$h(-t) = 0 \qquad t > 0$$

Let

$$h(t) = h_e(t) + h_o(t)$$

where $h_e(t)$ and $h_o(t)$ are the even and odd components of $h(t)$, respectively. Then from Eqs. (1.5) and (1.6) we can write

$$h(t) = 2h_e(t) = 2h_o(t) \tag{5.175}$$

From Eqs. (5.61b) and (5.61c) we have

$$h_e(t) \leftrightarrow A(\omega) \qquad \text{and} \qquad h_o(t) \leftrightarrow jB(\omega)$$

Thus, by Eq. (5.175)

$$h(t) = 2h_e(t) = 2\mathcal{F}^{-1}[A(\omega)] \qquad t > 0 \tag{5.176a}$$
$$h(t) = 2h_o(t) = 2\mathcal{F}^{-1}[jB(\omega)] \qquad t > 0 \tag{5.176b}$$

Equation (5.176a) and (5.176b) indicate that $h(t)$ can be obtained in terms of $A(\omega)$ or $B(\omega)$ alone.

5.50. Consider a causal continuous-time LTI system with frequency response

$$H(\omega) = A(\omega) + jB(\omega)$$

If the impulse response $h(t)$ of the system contains no impulses at the origin, then show that $A(\omega)$ and $B(\omega)$ satisfy the following equation:

$$A(\omega) = \frac{1}{\pi} \int_{-\infty}^{\infty} \frac{B(\lambda)}{\omega - \lambda} \, d\lambda \tag{5.177a}$$

$$B(\omega) = -\frac{1}{\pi} \int_{-\infty}^{\infty} \frac{A(\lambda)}{\omega - \lambda} \, d\lambda \tag{5.177b}$$

As in Prob. 5.49, let

$$h(t) = h_e(t) + h_o(t)$$

Since $h(t)$ is causal, that is, $h(t) = 0$ for $t < 0$, we have

$$h_e(t) = -h_o(t) \qquad t < 0$$

Also from Eq. (5.175) we have

$$h_e(t) = h_o(t) \qquad t > 0$$

Thus, using Eq. (5.152), we can write

$$h_e(t) = h_o(t)\ \text{sgn}(t) \tag{5.178a}$$
$$h_o(t) = h_e(t)\ \text{sgn}(t) \tag{5.178b}$$

Now, from Eqs. (5.61b), (5.61c), and (5.153) we have

$$h_e(t) \leftrightarrow A(\omega) \qquad h_o(t) \leftrightarrow jB(\omega) \qquad \text{sgn}(t) \leftrightarrow \frac{2}{j\omega}$$

Thus, by the frequency convolution theorem (5.59) we obtain

$$A(\omega) = \frac{1}{2\pi} jB(\omega) * \frac{2}{j\omega} = \frac{1}{\pi} B(\omega) * \frac{1}{\omega} = \frac{1}{\pi} \int_{-\infty}^{\infty} \frac{B(\lambda)}{\omega - \lambda}\, d\lambda$$

and

$$jB(\omega) = \frac{1}{2\pi} A(\omega) * \frac{2}{j\omega} = -j\frac{1}{\pi} A(\omega) * \frac{1}{\omega}$$

or

$$B(\omega) = -\frac{1}{\pi} A(\omega) * \frac{1}{\omega} = -\frac{1}{\pi} \int_{-\infty}^{\infty} \frac{A(\lambda)}{\omega - \lambda}\, d\lambda$$

Note that $A(\omega)$ is the Hilbert transform of $B(\omega)$ [Eq. (5.174)] and that $B(\omega)$ is the negative of the Hilbert transform of $A(\omega)$.

5.51. The real part of the frequency response $H(\omega)$ of a causal LTI system is known to be $\pi\delta(\omega)$. Find the frequency response $H(\omega)$ and the impulse function $h(t)$ of the system.

Let

$$H(\omega) = A(\omega) + jB(\omega)$$

Using Eq. (5.177b), with $A(\omega) = \pi\delta(\omega)$, we obtain

$$B(\omega) = -\frac{1}{\pi} \int_{-\infty}^{\infty} \frac{\pi\delta(\lambda)}{\omega - \lambda}\, d\lambda = -\int_{-\infty}^{\infty} \delta(\lambda)\frac{1}{\omega - \lambda}\, d\lambda = -\frac{1}{\omega}$$

Hence,

$$H(\omega) = \pi\delta(\omega) - j\frac{1}{\omega} = \pi\delta(\omega) + \frac{1}{j\omega}$$

and by Eq. (5.154)

$$h(t) = u(t)$$

Filtering

5.52. Consider an ideal low-pass filter with frequency response

$$H(\omega) = \begin{cases} 1 & |\omega| < \omega_c \\ 0 & |\omega| > \omega_c \end{cases}$$

The input to this filter is

$$x(t) = \frac{\sin at}{\pi t}$$

(a) Find the output $y(t)$ for $a < \omega_c$.

(b) Find the output $y(t)$ for $a > \omega_c$.

(c) In which case does the output suffer distortion?

(a) From Eq. (5.137) (Prob. 5.20) we have

$$x(t) = \frac{\sin at}{\pi t} \leftrightarrow X(\omega) = p_a(\omega) = \begin{cases} 1 & |\omega| < a \\ 0 & |\omega| > a \end{cases}$$

Then when $a < \omega_c$, we have

$$Y(\omega) = X(\omega)H(\omega) = X(\omega)$$

Thus,

$$y(t) = x(t) = \frac{\sin at}{\pi t}$$

(b) When $a > \omega_c$, we have

$$Y(\omega) = X(\omega)H(\omega) = H(\omega)$$

Thus,

$$y(t) = h(t) = \frac{\sin \omega_c t}{\pi t}$$

(c) In case (a), that is, when $\omega_c > a$, $y(t) = x(t)$ and the filter does not produce any distortion. In case (b), that is, when $\omega_c < a$, $y(t) = h(t)$ and the filter produces distortion.

5.53. Consider an ideal low-pass filter with frequency response

$$H(\omega) = \begin{cases} 1 & |\omega| < 4\pi \\ 0 & |\omega| > 4\pi \end{cases}$$

The input to this filter is the periodic square wave shown in Fig. 5-27. Find the output $y(t)$.

Setting $A = 10$, $T_0 = 2$, and $\omega_0 = 2\pi/T_0 = \pi$ in Eq. (5.107) (Prob. 5.5), we get

$$x(t) = 5 + \frac{20}{\pi}\left(\sin \pi t + \frac{1}{3}\sin 3\pi t + \frac{1}{5}\sin 5\pi t + \cdots\right)$$

Since the cutoff frequency ω_c of the filter is 4π rad, the filter passes all harmonic components of $x(t)$ whose angular frequencies are less than 4π rad and rejects all harmonic components of $x(t)$ whose angular frequencies are greater than 4π rad. Therefore,

$$y(t) = 5 + \frac{20}{\pi}\sin \pi t + \frac{20}{3\pi}\sin 3\pi t$$

5.54. Consider an ideal low-pass filter with frequency response

$$H(\omega) = \begin{cases} 1 & |\omega| < \omega_c \\ 0 & |\omega| > \omega_c \end{cases}$$

The input to this filter is

$$x(t) = e^{-2t}u(t)$$

Find the value of ω_c such that this filter passes exactly one-half of the normalized energy of the input signal $x(t)$.

From Eq. (5.155)

$$X(\omega) = \frac{1}{2 + j\omega}$$

Then

$$Y(\omega) = X(\omega)H(\omega) = \begin{cases} \dfrac{1}{2 + j\omega} & |\omega| < \omega_c \\ 0 & |\omega| > \omega_c \end{cases}$$

The normalized energy of $x(t)$ is

$$E_x = \int_{-\infty}^{\infty} |x(t)|^2 \, dt = \int_{0}^{\infty} e^{-4t} \, dt = \frac{1}{4}$$

Using Parseval's identity (5.64), the normalized energy of $y(t)$ is

$$E_y = \int_{-\infty}^{\infty} |y(t)|^2 \, dt = \frac{1}{2\pi} \int_{-\infty}^{\infty} |Y(\omega)|^2 \, d\omega = \frac{1}{2\pi} \int_{-\omega_c}^{\omega_c} \frac{d\omega}{4 + \omega^2}$$

$$= \frac{1}{\pi} \int_{0}^{\omega_c} \frac{d\omega}{4 + \omega^2} = \frac{1}{2\pi} \tan^{-1} \frac{\omega_c}{2} = \frac{1}{2} E_x = \frac{1}{8}$$

from which we obtain

$$\frac{\omega_c}{2} = \tan \frac{\pi}{4} = 1 \qquad \text{and} \qquad \omega_c = 2 \text{ rad/s}$$

5.55. The *equivalent bandwidth* of a filter with frequency response $H(\omega)$ is defined by

$$W_{eq} = \frac{1}{|H(\omega)|^2_{max}} \int_{0}^{\infty} |H(\omega)|^2 \, d\omega \tag{5.179}$$

where $|H(\omega)|_{max}$ denotes the maximum value of the magnitude spectrum. Consider the low-pass *RC* filter shown in Fig. 5-6(*a*).

(*a*) Find its 3-dB bandwidth $W_{3\,dB}$.

(*b*) Find its equivalent bandwidth W_{eq}.

(*a*) From Eq. (5.91) the frequency response $H(\omega)$ of the *RC* filter is given by

$$H(\omega) = \frac{1}{1 + j\omega RC} = \frac{1}{1 + j(\omega/\omega_0)}$$

where $\omega_0 = 1/RC$. Now

$$|H(\omega)| = \frac{1}{[1 + (\omega/\omega_0)^2]^{1/2}}$$

The amplitude spectrum $|H(\omega)|$ is plotted in Fig. 5-6(*b*). When $\omega = \omega_0 = 1/RC$, $|H(\omega_0)| = 1/\sqrt{2}$. Thus, the 3-dB bandwidth of the *RC* filter is given by

$$W_{3dB} = \omega_0 = \frac{1}{RC}$$

(b)   From Fig. 5-6(b) we see that $|H(0)| = 1$ is the maximum of the magnitude spectrum. Rewriting $H(\omega)$ as

$$H(\omega) = \frac{1}{1 + j\omega RC} = \frac{1}{RC}\frac{1}{1/RC + j\omega}$$

and using Eq. (5.179), the equivalent bandwidth of the RC filter is given by (Fig. 5-32)

$$W_{eq} = \frac{1}{(RC)^2}\int_0^\infty \frac{d\omega}{(1/RC)^2 + \omega^2} = \frac{1}{(RC)^2}\frac{\pi}{2/RC} = \frac{\pi}{2RC}$$

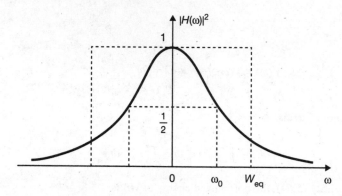

Fig. 5-32 Filter bandwidth.

5.56. The risetime t_r of the low-pass RC filter in Fig. 5-6(a) is defined as the time required for a unit step response to go from 10 to 90 percent of its final value. Show that

$$t_r = \frac{0.35}{f_{3\,dB}}$$

where $f_{3\,dB} = W_{3\,dB}/2\pi = 1/2\pi RC$ is the 3-dB bandwidth (in hertz) of the filter.

From the frequency response $H(\omega)$ of the RC filter, the impulse response is

$$h(t) = \frac{1}{RC}e^{-t/RC}\,u(t)$$

Then, from Eq. (2.12) the unit step response $s(t)$ is found to be

$$s(t) = \int_0^t h(\tau)\,d\tau = \int_0^t \frac{1}{RC}e^{-\tau/RC}\,d\tau = (1 - e^{-t/RC})\,u(t)$$

which is sketched in Fig. 5-33. By definition of the risetime

$$t_r = t_2 - t_1$$

where

$$s(t_1) = 1 - e^{-t_1/RC} = 0.1 \rightarrow e^{-t_1/RC} = 0.9$$
$$s(t_2) = 1 - e^{-t_2/RC} = 0.9 \rightarrow e^{-t_2/RC} = 0.1$$

Dividing the first equation by the second equation on the right-hand side, we obtain

$$e^{(t_2 - t_1)/RC} = 9$$

and

$$t_r = t_2 - t_1 = RC\ln(9) = 2.197RC = \frac{2.197}{2\pi f_{3\,dB}} = \frac{0.35}{f_{3\,dB}}$$

which indicates the inverse relationship between bandwidth and risetime.

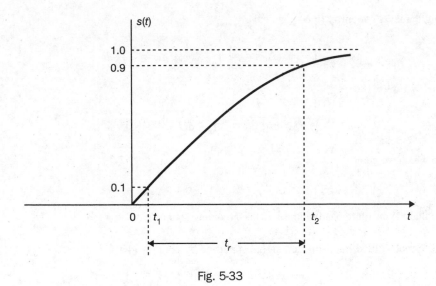

Fig. 5-33

5.57. Another definition of bandwidth for a signal $x(t)$ is the 90 percent *energy containment bandwidth* W_{90}, defined by

$$\frac{1}{2\pi}\int_{-W_{90}}^{W_{90}}|X(\omega)|^2 d\omega = \frac{1}{\pi}\int_0^{W_{90}}|X(\omega)|^2 d\omega = 0.9E_x \tag{5.180}$$

where E_x is the normalized energy content of signal $x(t)$. Find the W_{90} for the following signals:

(a) $x(t) = e^{-at}u(t), a > 0$

(b) $x(t) = \dfrac{\sin at}{\pi t}$

(a)   From Eq. (5.155)

$$x(t) = e^{-at}u(t) \leftrightarrow X(\omega) = \frac{1}{a + j\omega}$$

From Eq. (1.14)

$$E_x = \int_{-\infty}^{\infty}|x(t)|^2 dt = \int_0^{\infty} e^{-2at}dt = \frac{1}{2a}$$

Now, by Eq. (5.180)

$$\frac{1}{\pi}\int_0^{W_{90}}|X(\omega)|^2 d\omega = \frac{1}{\pi}\int_0^{W_{90}}\frac{d\omega}{a^2 + \omega^2} = \frac{1}{a\pi}\tan^{-1}\left(\frac{W_{90}}{a}\right) = 0.9\frac{1}{2a}$$

from which we get

$$\tan^{-1}\left(\frac{W_{90}}{a}\right) = 0.45\pi$$

Thus,

$$W_{90} = a\tan(0.45\pi) = 6.31a \qquad \text{rad/s}$$

(b)   From Eq. (5.137)

$$x(t) = \frac{\sin at}{\pi t} \leftrightarrow X(\omega) = p_a(\omega) = \begin{cases} 1 & |\omega| < a \\ 0 & |\omega| > a \end{cases}$$

Using Parseval's identity (5.64), we have

$$E_x = \frac{1}{2\pi}\int_{-\infty}^{\infty}|X(\omega)|^2\,dt = \frac{1}{\pi}\int_{0}^{\infty}|X(\omega)|^2\,d\omega = \frac{1}{\pi}\int_{0}^{a}d\omega = \frac{a}{\pi}$$

Then, by Eq. (5.180)

$$\frac{1}{\pi}\int_{0}^{W_{90}}|X(\omega)|^2\,d\omega = \frac{1}{\pi}\int_{0}^{W_{90}}d\omega = \frac{W_{90}}{\pi} = 0.9\frac{a}{\pi}$$

from which we get

$$W_{90} = 0.9a \qquad \text{rad/s}$$

Note that the absolute bandwidth of $x(t)$ is a (radians/second).

5.58. Let $x(t)$ be a real-valued band-limited signal specified by [Fig. 5-34(b)]

$$X(\omega) = 0 \qquad |\omega| > \omega_M$$

Let $x_s(t)$ be defined by

$$x_s(t) = x(t)\delta_{T_s}(t) = x(t)\sum_{k=-\infty}^{\infty}\delta(t - kT_s) \tag{5.181}$$

(a) Sketch $x_s(t)$ for $T_s < \pi/\omega_M$ and for $T_s > \pi/\omega_M$.
(b) Find and sketch the Fourier spectrum $X_s(\omega)$ of $x_s(t)$ for $T_s < \pi/\omega_M$ and for $T_s > \pi/\omega_M$.

(a) Using Eq. (1.26), we have

$$x_s(t) = x(t)\delta_{T_s}(t) = x(t)\sum_{k=-\infty}^{\infty}\delta(t - kT_s)$$

$$= \sum_{k=-\infty}^{\infty}x(t)\delta(t - kT_s) = \sum_{k=-\infty}^{\infty}x(kT_s)\,\delta(t - kT_s) \tag{5.182}$$

The sampled signal $x_s(t)$ is sketched in Fig. 5-34(c) for $T_s < \pi/\omega_M$, and in Fig. 5-34(i) for $T_s > \pi/\omega_M$.

The signal $x_s(t)$ is called the *ideal sampled signal*, T_s is referred to as the *sampling interval* (or *period*), and $f_s = 1/T_s$ is referred to as the *sampling rate* (or *frequency*).

(b) From Eq. (5.147) (Prob. 5.25) we have

$$\delta_{T_s}(t) \leftrightarrow \omega_s\sum_{k=-\infty}^{\infty}\delta(\omega - k\omega_s) \qquad \omega_s = \frac{2\pi}{T_s}$$

Let

$$x_s(t) \leftrightarrow X_s(\omega)$$

Then, according to the frequency convolution theorem (5.59), we have

$$X_s(\omega) = \mathcal{F}[x(t)\delta_{T_s}(t)] = \frac{1}{2\pi}\left[X(\omega) * \omega_s\sum_{k=-\infty}^{\infty}\delta(\omega - k\omega_s)\right]$$

$$= \frac{1}{T_s}\sum_{k=-\infty}^{\infty}X(\omega) * \delta(\omega - k\omega_s)$$

Using Eq. (1.26), we obtain

$$X_s(\omega) = \frac{1}{T_s} \sum_{k=-\infty}^{\infty} X(\omega - k\omega_s) \tag{5.183}$$

which shows that $X_s(\omega)$ consists of periodically repeated replicas of $X(\omega)$ centered about $k\omega_s$ for all k. The Fourier spectrum $X_s(\omega)$ is shown in Fig. 5-34(f) for $T_s < \pi/\omega_M$ (or $\omega_s > 2\omega_M$), and in Fig. 5-34(j) for $T_s > \pi/\omega_M$ (or $\omega_s < 2\omega_M$), where $\omega_s = 2\pi/T_s$. It is seen that no overlap of the replicas $X(\omega - k\omega_s)$ occurs in $X_s(\omega)$ for $\omega_s \geq 2\omega_M$ and that overlap of the spectral replicas is produced for $\omega_s < 2\omega_M$. This effect is known as *aliasing*.

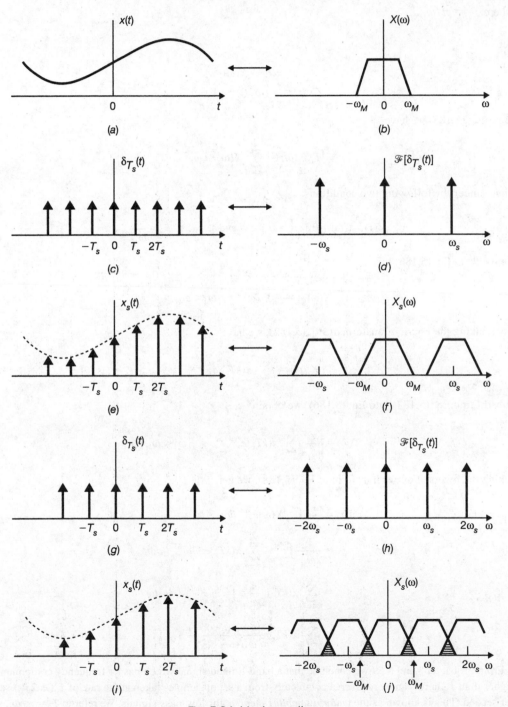

Fig. 5-34 Ideal sampling.

5.59. Let $x(t)$ be a real-valued band-limited signal specified by

$$X(\omega) = 0 \qquad |\omega| > \omega_M$$

Show that $x(t)$ can be expressed as

$$x(t) = \sum_{k=-\infty}^{\infty} x(kT_s) \frac{\sin \omega_M (t - kT_s)}{\omega_M (t - kT_s)} \tag{5.184}$$

where $T_s = \pi/\omega_M$.

Let

$$x(t) \leftrightarrow X(\omega)$$

$$x_s(t) = x(t)\delta_{T_s}(t) \leftrightarrow X_s(\omega)$$

From Eq. (5.183) we have

$$T_s X_s(\omega) = \sum_{k=-\infty}^{\infty} X(\omega - k\omega_s) \tag{5.185}$$

Then, under the following two conditions,

$$(1) \quad X(\omega) = 0, |\omega| > \omega_M \qquad \text{and} \qquad (2) \quad T_s = \frac{\pi}{\omega_M}$$

we see from Eq. (5.185) that

$$X(\omega) = \frac{\pi}{\omega_M} X_s(\omega) \qquad |\omega| < \omega_M \tag{5.186}$$

Next, taking the Fourier transform of Eq. (5.182), we have

$$X_s(\omega) = \sum_{k=-\infty}^{\infty} x(kT_s) e^{-jkT_s\omega} \tag{5.187}$$

Substituting Eq. (5.187) into Eq. (5.186), we obtain

$$X(\omega) = \frac{\pi}{\omega_M} \sum_{k=-\infty}^{\infty} x(kT_s) e^{-jkT_s\omega} \qquad |\omega| < \omega_M \tag{5.188}$$

Taking the inverse Fourier transform of Eq. (5.188), we get

$$x(t) = \frac{1}{2\pi} \int_{-\infty}^{\infty} X(\omega) e^{j\omega t} \, d\omega$$

$$= \frac{1}{2\omega_M} \int_{-\omega_M}^{\omega_M} \sum_{k=-\infty}^{\infty} x(kT_s) e^{j\omega(t - kT_s)} \, d\omega$$

$$= \sum_{k=-\infty}^{\infty} x(kT_s) \frac{1}{2\omega_M} \int_{-\omega_M}^{\omega_M} e^{j\omega(t - kT_s)} \, d\omega$$

$$= \sum_{k=-\infty}^{\infty} x(kT_s) \frac{\sin \omega_M (t - kT_s)}{\omega_M (t - kT_s)}$$

From Probs. 5.58 and 5.59 we conclude that a band-limited signal which has no frequency components higher than f_M hertz can be recovered completely from a set of samples taken at the rate of $f_s (\geq 2f_M)$ samples per second. This is known as the *uniform sampling theorem* for low-pass signals. We refer to $T_s = \pi/\omega_M = 1/2f_M$ ($\omega_M = 2\pi f_M$) as the *Nyquist sampling interval* and $f_s = 1/T_s = 2f_M$ as the *Nyquist sampling rate*.

5.60. Consider the system shown in Fig. 5-35(a). The frequency response $H(\omega)$ of the ideal low-pass filter is given by [Fig. 5-35(b)]

$$H(\omega) = T_s p_{\omega_c}(\omega) = \begin{cases} T_s & |\omega| < \omega_c \\ 0 & |\omega| > \omega_c \end{cases}$$

Show that if $\omega_c = \omega_s/2$, then for any choice of T_s,

$$y(mT_s) = x(mT_s) \qquad m = 0, \pm 1, \pm 2, \dots$$

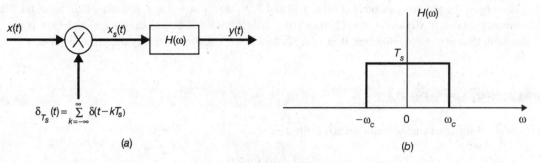

(a)   (b)

Fig. 5-35

From Eq. (5.137) the impulse response $h(t)$ of the ideal low-pass filter is given by

$$h(t) = T_s \frac{\sin \omega_c t}{\pi t} = \frac{T_s \omega_c}{\pi} \frac{\sin \omega_c t}{\omega_c t} \tag{5.189}$$

From Eq. (5.182) we have

$$x_s(t) = x(t)\delta_{T_s}(t) = \sum_{k=-\infty}^{\infty} x(kT_s)\delta(t - kT_s)$$

By Eq. (2.6) and using Eqs. (2.7) and (1.26), the output $y(t)$ is given by

$$y(t) = x_s(t) * h(t) = \left[\sum_{k=-\infty}^{\infty} x(kT_s)\delta(t - kT_s) \right] * h(t)$$

$$= \sum_{k=-\infty}^{\infty} x(kT_s)[h(t) * \delta(t - kT_s)]$$

$$= \sum_{k=-\infty}^{\infty} x(kT_s)h(t - kT_s)$$

Using Eq. (5.189), we get

$$y(t) = \sum_{k=-\infty}^{\infty} x(kT_s) \frac{T_s \omega_c}{\pi} \frac{\sin \omega_c(t - kT_s)}{\omega_c(t - kT_s)}$$

If $\omega_c = \omega_s/2$, then $T_s\omega_c/\pi = 1$ and we have

$$y(t) = \sum_{k=-\infty}^{\infty} x(kT_s) \frac{\sin[\omega_s(t - kT_s)/2]}{\omega_s(t - kT_s)/2}$$

Setting $t = mT_s$ (m = integer) and using the fact that $\omega_s T_s = 2\pi$, we get

$$y(mT_s) = \sum_{k=-\infty}^{\infty} x(kT_s) \frac{\sin \pi(m - k)}{\pi(m - k)}$$

Since

$$\frac{\sin \pi(m-k)}{\pi(m-k)} = \begin{cases} 0 & m \neq k \\ 0 & m = k \end{cases}$$

we have

$$y(mT_s) = x(mT_s) \qquad m = 0, \pm 1, \pm 2, \ldots$$

which shows that without any restriction on $x(t)$, $y(mT_s) = x(mT_s)$ for any integer value of m.

Note from the sampling theorem (Probs. 5.58 and 5.59) that if $\omega_s = 2\pi/T_s$ is greater than twice the highest frequency present in $x(t)$ and $\omega_c = \omega_s/2$, then $y(t) = x(t)$. If this condition on the bandwidth of $x(t)$ is not satisfied, then $y(t) \neq x(t)$. However, if $\omega_c = \omega_s/2$, then $y(mT_s) = x(mT_s)$ for any integer value of m.

SUPPLEMENTARY PROBLEMS

5.61. Consider a rectified sine wave signal $x(t)$ defined by

$$x(t) = |A \sin \pi t|$$

    (*a*)   Sketch $x(t)$ and find its fundamental period.

    (*b*)   Find the complex exponential Fourier series of $x(t)$.

    (*c*)   Find the trigonometric Fourier series of $x(t)$.

5.62. Find the trigonometric Fourier series of a periodic signal $x(t)$ defined by

$$x(t) = t^2, \quad -\pi < t < \pi \qquad \text{and} \qquad x(t + 2\pi) = x(t)$$

5.63. Using the result from Prob. 5.10, find the trigonometric Fourier series of the signal $x(t)$ shown in Fig. 5-36.

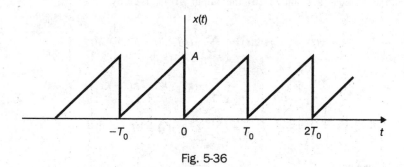

Fig. 5-36

5.64. Derive the harmonic form Fourier series representation (5.15) from the trigonometric Fourier series representation (5.8).

5.65. Show that the mean-square value of a real periodic signal $x(t)$ is the sum of the mean-square values of its harmonics.

5.66. Show that if

$$x(t) \leftrightarrow X(\omega)$$

then

$$x^{(n)}(t) = \frac{d^n x(t)}{dt^n} \leftrightarrow (j\omega)^n X(\omega)$$

5.67. Using the differentiation technique, find the Fourier transform of the triangular pulse signal shown in Fig. 5-37.

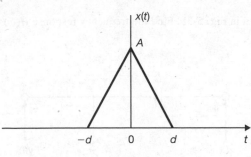

Fig. 5-37

5.68. Find the inverse Fourier transform of

$$X(\omega) = \frac{1}{(a + j\omega)^N}$$

5.69. Find the inverse Fourier transform of

$$X(\omega) = \frac{1}{2 - \omega^2 + j3\omega}$$

5.70. Verify the frequency differentiation property (5.56); that is,

$$(-jt)x(t) \leftrightarrow \frac{dX(\omega)}{d\omega}$$

5.71. Find the Fourier transform of each of the following signals:

    (a) $x(t) = \cos \omega_0 t u(t)$

    (b) $x(t) = \sin \omega_0 t u(t)$

    (c) $x(t) = e^{-at} \cos \omega_0 t u(t), a > 0$

    (d) $x(t) = e^{-at} \sin \omega_0 t u(t), a > 0$

5.72. Let $x(t)$ be a signal with Fourier transform $X(\omega)$ given by

$$X(\omega) = \begin{cases} 1 & |\omega| < 1 \\ 0 & |\omega| > 1 \end{cases}$$

Consider the signal

$$y(t) = \frac{d^2 x(t)}{dt^2}$$

Find the value of

$$\int_{-\infty}^{\infty} |y(t)|^2 \, dt$$

5.73. Let $x(t)$ be a real signal with the Fourier transform $X(\omega)$. The *analytical signal* $x_+(t)$ associated with $x(t)$ is a complex signal defined by

$$x_+(t) = x(t) + j\hat{x}(t)$$

where $\hat{x}(t)$ is the Hilbert transform of $x(t)$.

    (a)   Find the Fourier transform $X_+(\omega)$ of $x_+(t)$.

    (b)   Find the analytical signal $x_+(t)$ associated with $\cos \omega_0 t$ and its Fourier transform $X_+(\omega)$.

5.74. Consider a continuous-time LTI system with frequency response $H(\omega)$. Find the Fourier transform $S(\omega)$ of the unit step response $s(t)$ of the system.

5.75. Consider the *RC* filter shown in Fig. 5-38. Find the frequency response $H(\omega)$ of this filter and discuss the type of filter.

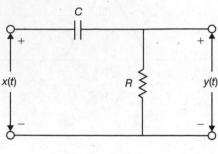

Fig. 5-38

5.76. Determine the 99 percent energy containment bandwidth for the signal

$$x(t) = \frac{1}{t^2 + a^2}$$

5.77. The *sampling theorem in the frequency domain* states that if a real signal $x(t)$ is a duration-limited signal, that is,

$$x(t) = 0 \qquad |t| > t_M$$

then its Fourier transform $X(\omega)$ can be uniquely determined from its values $X(n\pi/t_M)$ at a series of equidistant points spaced π/t_M apart. In fact, $X(\omega)$ is given by

$$X(\omega) = \sum_{n=-\infty}^{\infty} X\left(\frac{n\pi}{t_M}\right) \frac{\sin(\omega t_M - n\pi)}{\omega t_M - n\pi}$$

Verify the above sampling theorem in the frequency domain.

ANSWERS TO SUPPLEMENTARY PROBLEMS

5.61. (*a*) $X(t)$ is sketched in Fig. 5-39 and   1 .

    (*b*) $x(t) = -\dfrac{2A}{\pi} \displaystyle\sum_{k=-\infty}^{\infty} \frac{1}{4k^2 - 1} e^{jk2\pi t}$

    (*c*) $x(t) = \dfrac{2A}{\pi} - \dfrac{4A}{\pi} \displaystyle\sum_{k=1}^{\infty} \frac{1}{4k^2 - 1} \cos k2\pi t$

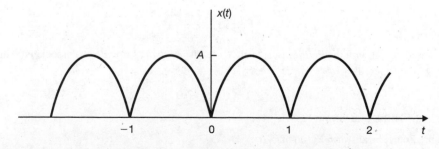

Fig. 5-39

5.62. $x(t) = \dfrac{\pi^2}{3} + 4 \displaystyle\sum_{k=1}^{\infty} \dfrac{(-1)^k}{k^2} \cos kt$

5.63. $x(t) = \dfrac{A}{2} - \dfrac{A}{\pi} \displaystyle\sum_{k=1}^{\infty} \dfrac{1}{k} \sin k\omega_0 t \qquad \omega_0 = \dfrac{2\pi}{T_0}$

5.64. *Hint:* Rewrite $a_k \cos k\omega_0 t + b_k \sin k\omega_0 t$ as

$$\sqrt{a_k^2 + b_k^2} \left[\dfrac{a_k}{(a_k^2 + b_k^2)^{1/2}} \cos k\omega_0 t + \dfrac{b_k}{(a_k^2 + b_k^2)^{1/2}} \sin k\omega_0 t \right]$$

and use the trigonometric formula $\cos(A - B) = \cos A \cos B + \sin A \sin B$.

5.65. *Hint:* Use Parseval's identity (5.21) for the Fourier series and Eq. (5.168).

5.66. *Hint:* Repeat the time-differentiation property (5.55).

5.67. $Ad \left[\dfrac{\sin(\omega d/2)}{\omega d/2} \right]^2$

5.68. *Hint:* Differentiate Eq. (5.155) N times with respect to (a).

$$\dfrac{t^{N-1}}{(N-1)!} e^{-at} u(t)$$

5.69. *Hint:* Note that

$$2 - \omega^2 + j3\omega = 2 + (j\omega)^2 + j3\omega = (1 + j\omega)(2 + j\omega)$$

and apply the technique of partial-fraction expansion.

$x(t) = (e^{-t} - e^{-2t}) u(t)$

5.70. *Hint:* Use definition (5.31) and proceed in a manner similar to Prob. 5.28.

5.71. *Hint:* Use multiplication property (5.59).

(a) $X(\omega) = \dfrac{\pi}{2} \delta(\omega - \omega_0) + \dfrac{\pi}{2} \delta(\omega + \omega_0) + \dfrac{j\omega}{(j\omega)^2 + \omega_0^2}$

(b) $X(\omega) = \dfrac{\pi}{2j} \delta(\omega - \omega_0) - \dfrac{\pi}{2j} \delta(\omega + \omega_0) + \dfrac{\omega_0}{(j\omega)^2 + \omega_0^2}$

(c) $X(\omega) = \dfrac{a + j\omega}{(a + j\omega)^2 + \omega_0^2}$

(d) $X(\omega) = \dfrac{\omega_0}{(a + j\omega)^2 + \omega_0^2}$

5.72. *Hint:* Use Parseval's identity (5.64) for the Fourier transform.

$1/3\pi$

5.73. (a) $X_+(\omega) = 2X(\omega)u(\omega) = \begin{cases} 2X(\omega) & \omega > 0 \\ 0 & \omega < 0 \end{cases}$

(b) $x_+(t) = e^{j\omega_0 t}, \ X_+(\omega) = 2\pi \, \delta(\omega - \omega_0)$

5.74. *Hint:*   Use Eq. (2.12) and the integration property (5.57).

$$S(\omega) = \pi H(0)\delta(\omega) + (1/j\omega)\,H(\omega)$$

5.75. $H(\omega) = \dfrac{j\omega}{(1/RC) + j\omega}$,   high-pass filter

5.76. $W_{99} = 2.3/a$ radians / second or $f_{99} = 0.366/a$ hertz

5.77. *Hint:*   Expand $x(t)$ in a complex Fourier series and proceed in a manner similar to that for Prob. 5.59.

CHAPTER 6

Fourier Analysis of Discrete-Time Signals and Systems

6.1 Introduction

In this chapter we present the Fourier analysis in the context of discrete-time signals (sequences) and systems. The Fourier analysis plays the same fundamental role in discrete time as in continuous time. As we will see, there are many similarities between the techniques of discrete-time Fourier analysis and their continuous-time counterparts, but there are also some important differences.

6.2 Discrete Fourier Series

A. Periodic Sequences:

In Chap. 1 we defined a discrete-time signal (or sequence) $x[n]$ to be periodic if there is a positive integer N for which

$$x[n + N] = x[n] \qquad \text{all } n \tag{6.1}$$

The fundamental period N_0 of $x[n]$ is the smallest positive integer N for which Eq. (6.1) is satisfied.

As we saw in Sec. 1.4, the complex exponential sequence

$$x[n] = e^{j(2\pi/N_0)n} = e^{j\Omega_0 n} \tag{6.2}$$

where $\Omega_0 = 2\pi/N_0$, is a periodic sequence with fundamental period N_0. As we discussed in Sec. 1.4C, one very important distinction between the discrete-time and the continuous-time complex exponential is that the signals $e^{j\omega_0 t}$ are distinct for distinct values of ω_0, but the sequences $e^{j\Omega_0 n}$, which differ in frequency by a multiple of 2π, are identical. That is,

$$e^{j(\Omega_0 + 2\pi k)n} = e^{j\Omega_0 n} \, e^{j2\pi kn} = e^{j\Omega_0 n} \tag{6.3}$$

Let

$$\Psi_k[n] = e^{jk\Omega_0 n} \qquad \Omega_0 = \frac{2\pi}{N_0} \qquad k = 0, \pm 1, \pm 2, \dots \tag{6.4}$$

Then by Eq. (6.3) we have

$$\Psi_0[n] = \Psi_{N_0}[n] \qquad \Psi_1[n] = \Psi_{N_0 + 1}[n] \qquad \dots \qquad \Psi_k[n] = \Psi_{N_0 + k}[n] \qquad \dots \tag{6.5}$$

and more generally,

$$\Psi_k[n] = \Psi_{k+mN_0}[n] \qquad m = \text{integer} \tag{6.6}$$

Thus, the sequences $\Psi_k[n]$ are distinct only over a range of N_0 successive values of k.

B.   Discrete Fourier Series Representation:

The discrete Fourier series representation of a periodic sequence $x[n]$ with fundamental period N_0 is given by

$$x[n] = \sum_{k=0}^{N_0-1} c_k\, e^{jk\Omega_0 n} \qquad \Omega_0 = \frac{2\pi}{N_0} \tag{6.7}$$

where c_k are the Fourier coefficients and are given by (Prob. 6.2)

$$c_k = \frac{1}{N_0} \sum_{n=0}^{N_0-1} x[n] e^{-jk\Omega_0 n} \tag{6.8}$$

Because of Eq. (6.5) [or Eq. (6.6)], Eqs. (6.7) and (6.8) can be rewritten as

$$x[n] = \sum_{k=\langle N_0 \rangle} c_k\, e^{jk\Omega_0 n} \qquad \Omega_0 = \frac{2\pi}{N_0} \tag{6.9}$$

$$c_k = \frac{1}{N_0} \sum_{n=\langle N_0 \rangle} x[n] e^{-jk\Omega_0 n} \tag{6.10}$$

where $\sum_{k=\langle N_0 \rangle}$ denotes that the summation is on k as k varies over a range of N_0 successive integers. Setting $k = 0$ in Eq. (6.10), we have

$$c_0 = \frac{1}{N_0} \sum_{n=\langle N_0 \rangle} x[n] \tag{6.11}$$

which indicates that c_0 equals the average value of $x[n]$ over a period.

The Fourier coefficients c_k are often referred to as the *spectral coefficients* of $x[n]$.

C.   Convergence of Discrete Fourier Series:

Since the discrete Fourier series is a finite series, in contrast to the continuous-time case, there are no convergence issues with discrete Fourier series.

D.   Properties of Discrete Fourier Series:

1.   Periodicity of Fourier Coefficients:
From Eqs. (6.5) and (6.7) [or (6.9)], we see that

$$c_{k+N_0} = c_k \tag{6.12}$$

which indicates that the Fourier series coefficients c_k are periodic with fundamental period N_0.

2.   Duality:
From Eq. (6.12) we see that the Fourier coefficients c_k form a periodic sequence with fundamental period N_0. Thus, writing c_k as $c[k]$, Eq. (6.10) can be rewritten as

$$c[k] = \sum_{n=\langle N_0 \rangle} \frac{1}{N_0} x[n] e^{-jk\Omega_0 n} \tag{6.13}$$

Let $n = -m$ in Eq. (6.13). Then

$$c[k] = \sum_{m=\langle N_0 \rangle} \frac{1}{N_0} x[-m] e^{jk\Omega_0 m}$$

Letting $k = n$ and $m = k$ in the above expression, we get

$$c[n] = \sum_{k=\langle N_0 \rangle} \frac{1}{N_0} x[-k] e^{jk\Omega_0 n} \tag{6.14}$$

Comparing Eq. (6.14) with Eq. (6.9), we see that $(1/N_0)x[-k]$ are the Fourier coefficients of $c[n]$. If we adopt the notation

$$x[n] \xleftarrow{\text{DFS}} c_k = c[k] \tag{6.15}$$

to denote the discrete Fourier series pair, then by Eq. (6.14) we have

$$c[n] \xleftarrow{\text{DFS}} \frac{1}{N_0} x[-k] \tag{6.16}$$

Equation (6.16) is known as the *duality* property of the discrete Fourier series.

3. Other Properties:

When $x[n]$ is real, then from Eq. (6.8) or [Eq. (6.10)] and Eq. (6.12) it follows that

$$c_{-k} = c_{N_0-k} = c_k^* \tag{6.17}$$

where $*$ denotes the complex conjugate.

Even and Odd Sequences:

When $x[n]$ is real, let

$$x[n] = x_e[n] + x_o[n]$$

where $x_e[n]$ and $x_o[n]$, are the even and odd components of $x[n]$, respectively. Let

$$x[n] \xleftarrow{\text{DFS}} c_k$$

Then

$$x_e[n] \xleftarrow{\text{DFS}} \text{Re}[c_k] \tag{6.18a}$$

$$x_o[n] \xleftarrow{\text{DFS}} j \, \text{Im}[c_k] \tag{6.18b}$$

Thus, we see that if $x[n]$ is real and even, then its Fourier coefficients are real, while if $x[n]$ is real and odd, its Fourier coefficients are imaginary.

E. Parseval's Theorem:

If $x[n]$ is represented by the discrete Fourier series in Eq. (6.9), then it can be shown that (Prob. 6.10)

$$\frac{1}{N_0} \sum_{n=\langle N_0 \rangle} |x[n]|^2 = \sum_{k=\langle N_0 \rangle} |c_k|^2 \tag{6.19}$$

Equation (6.19) is called *Parseval's identity* (or *Parseval's theorem*) for the discrete Fourier series.

6.3 The Fourier Transform

A. From Discrete Fourier Series to Fourier Transform:

Let $x[n]$ be a nonperiodic sequence of finite duration. That is, for some positive integer N_1,

$$x[n] = 0 \qquad |n| > N_1$$

Such a sequence is shown in Fig. 6-1(*a*). Let $x_{N_0}[n]$ be a periodic sequence formed by repeating $x[n]$ with fundamental period N_0 as shown in Fig. 6-1(*b*). If we let $N_0 \to \infty$, we have

$$\lim_{N_0 \to \infty} x_{N_0}[n] = x[n] \tag{6.20}$$

The discrete Fourier series of $x_{N_0}[n]$ is given by

$$x_{N_0}[n] = \sum_{k=\langle N_0 \rangle} c_k e^{jk\Omega_0 n} \qquad \Omega_0 = \frac{2\pi}{N_0} \tag{6.21}$$

where

$$c_k = \frac{1}{N_0} \sum_{n=\langle N_0 \rangle} x_{N_0}[n] \, e^{-jk\Omega_0 n} \tag{6.22a}$$

Since $x_{N_0}[n] = x[n]$ for $|n| \le N_1$ and also since $x[n] = 0$ outside this interval, Eq. (6.22*a*) can be rewritten as

$$c_k = \frac{1}{N_0} \sum_{n=-N_1}^{N_1} x[n] \, e^{-jk\Omega_0 n} = \frac{1}{N_0} \sum_{n=-\infty}^{\infty} x[n] \, e^{-jk\Omega_0 n} \tag{6.22b}$$

Let us define $X(\Omega)$ as

$$X(\Omega) = \sum_{n=-\infty}^{\infty} x[n] \, e^{-j\Omega n} \tag{6.23}$$

Then, from Eq. (6.22*b*) the Fourier coefficients c_k can be expressed as

$$c_k = \frac{1}{N_0} X(k\Omega_0) \tag{6.24}$$

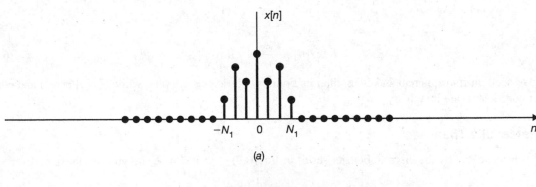

(*a*)

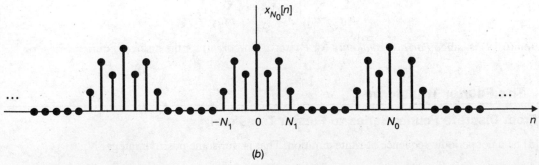

(*b*)

Fig. 6-1 (*a*) Nonperiodic finite sequence $x[n]$; (*b*) periodic sequence formed by periodic extension of $x[n]$.

Substituting Eq. (6.24) into Eq. (6.21), we have

$$x_{N_0}[n] = \sum_{k=\langle N_0 \rangle} \frac{1}{N_0} X(k\Omega_0) e^{jk\Omega_0 n}$$

or

$$x_{N_0}[n] = \frac{1}{2\pi} \sum_{k=\langle N_0 \rangle} X(k\Omega_0) e^{jk\Omega_0 n} \Omega_0 \qquad (6.25)$$

From Eq. (6.23), $X(\Omega)$ is periodic with period 2π and so is $e^{j\Omega n}$. Thus, the product $X(\Omega) e^{j\Omega n}$ will also be periodic with period 2π. As shown in Fig. 6-2, each term in the summation in Eq. (6.25) represents the area of a rectangle of height $X(k\Omega_0)e^{jk\Omega 0 n}$ and width Ω_0. As $N_0 \rightarrow \infty$, $\Omega_0 = 2\pi/N_0$ becomes infinitesimal ($\Omega_0 \rightarrow 0$) and Eq. (6.25) passes to an integral. Furthermore, since the summation in Eq. (6.25) is over N_0 consecutive intervals of width $\Omega_0 = 2\pi/N_0$, the total interval of integration will always have a width 2π. Thus, as $N_0 \rightarrow \infty$ and in view of Eq. (6.20), Eq. (6.25) becomes

$$x[n] = \frac{1}{2\pi} \int_{2\pi} X(\Omega) e^{j\Omega n} \, d\Omega \qquad (6.26)$$

Since $X(\Omega)e^{j\Omega n}$ is periodic with period 2π, the interval of integration in Eq. (6.26) can be taken as any interval of length 2π.

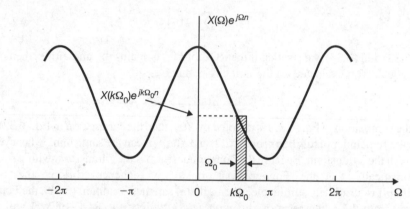

Fig. 6-2 Graphical interpretation of Eq. (6.25).

B. Fourier Transform Pair:

The function $X(\Omega)$ defined by Eq. (6.23) is called the *Fourier transform* of $x[n]$, and Eq. (6.26) defines the *inverse Fourier transform* of $X(\Omega)$. Symbolically they are denoted by

$$X(\Omega) = \mathscr{F}\{x[n]\} = \sum_{n=-\infty}^{\infty} x[n] e^{-j\Omega n} \qquad (6.27)$$

$$x[n] = \mathscr{F}^{-1}\{X(\Omega)\} = \frac{1}{2\pi} \int_{2\pi} X(\Omega) e^{j\Omega n} \, d\Omega \qquad (6.28)$$

and we say that $x[n]$ and $X(\Omega)$ form a Fourier transform pair denoted by

$$x[n] \leftrightarrow X(\Omega) \qquad (6.29)$$

Equations (6.27) and (6.28) are the discrete-time counterparts of Eqs. (5.31) and (5.32).

C. Fourier Spectra:

The Fourier transform $X(\Omega)$ of $x[n]$ is, in general, complex and can be expressed as

$$X(\Omega) = |X(\Omega)| e^{j\phi(\Omega)} \qquad (6.30)$$

As in continuous time, the Fourier transform $X(\Omega)$ of a nonperiodic sequence $x[n]$ is the frequency-domain specification of $x[n]$ and is referred to as the *spectrum* (or *Fourier spectrum*) of $x[n]$. The quantity $|X(\Omega)|$ is called the *magnitude spectrum* of $x[n]$, and $\phi(\Omega)$ is called the *phase spectrum* of $x[n]$. Furthermore, if $x[n]$ is real, the amplitude spectrum $|X(\Omega)|$ is an even function and the phase spectrum $\phi(\Omega)$ is an odd function of Ω.

D.   Convergence of $X(\Omega)$:

Just as in the case of continuous time, the sufficient condition for the convergence of $X(\Omega)$ is that $x[n]$ is absolutely summable, that is,

$$\sum_{n=-\infty}^{\infty} |x[n]| < \infty \tag{6.31}$$

E.   Connection between the Fourier Transform and the *z*-Transform:

Equation (6.27) defines the Fourier transform of $x[n]$ as

$$X(\Omega) = \sum_{n=-\infty}^{\infty} x[n] e^{-j\Omega n} \tag{6.32}$$

The z-transform of $x[n]$, as defined in Eq. (4.3), is given by

$$X(z) = \sum_{n=-\infty}^{\infty} x[n] z^{-n} \tag{6.33}$$

Comparing Eqs. (6.32) and (6.33), we see that if the ROC of $X(z)$ contains the unit circle, then the Fourier transform $X(\Omega)$ of $x[n]$ equals $X(z)$ evaluated on the unit circle, that is,

$$X(\Omega) = X(z)\big|_{z = e^{j\Omega}} \tag{6.34}$$

   Note that since the summation in Eq. (6.33) is denoted by $X(z)$, then the summation in Eq. (6.32) may be denoted as $X(e^{j\Omega})$. Thus, in the remainder of this book, both $X(\Omega)$ and $X(e^{j\Omega})$ mean the same thing whenever we connect the Fourier transform with the z-transform. Because the Fourier transform is the z-transform with $z = e^{j\Omega}$, it should not be assumed automatically that the Fourier transform of a sequence $x[n]$ is the z-transform with z replaced by $e^{j\Omega}$. If $x[n]$ is absolutely summable, that is, if $x[n]$ satisfies condition (6.31), the Fourier transform of $x[n]$ can be obtained from the z-transform of $x[n]$ with $z = e^{j\Omega}$ since the ROC of $X(z)$ will contain the unit circle; that is, $|e^{j\Omega}| = 1$. This is not generally true of sequences which are not absolutely summable. The following examples illustrate the above statements.

EXAMPLE 6.1   Consider the unit impulse sequence $\delta[n]$.

From Eq. (4.14) the z-transform of $\delta[n]$ is

$$\mathcal{Z}\{\delta[n]\} = 1 \qquad \text{all } z \tag{6.35}$$

By definitions (6.27) and (1.45), the Fourier transform of $\delta[n]$ is

$$\mathcal{F}\{\delta[n]\} = \sum_{n=-\infty}^{\infty} \delta[n] e^{-j\Omega n} = 1 \tag{6.36}$$

Thus, the z-transform and the Fourier transform of $\delta[n]$ are the same. Note that $\delta[n]$ is absolutely summable and that the ROC of the z-transform of $\delta[n]$ contains the unit circle.

EXAMPLE 6.2   Consider the causal exponential sequence

$$x[n] = a^n u[n] \qquad a \text{ real}$$

From Eq. (4.9) the z-transform of $x[n]$ is given by

$$X(z) = \frac{1}{1 - az^{-1}} \qquad |z| > |a|$$

Thus, $X(e^{j\Omega})$ exists for $|a| < 1$ because the ROC of $X(z)$ then contains the unit circle. That is,

$$X(e^{j\Omega}) = \frac{1}{1 - ae^{-j\Omega}} \qquad |a| < 1 \tag{6.37}$$

Next, by definition (6.27) and Eq. (1.91) the Fourier transform of $x[n]$ is

$$X(\Omega) = \sum_{n=-\infty}^{\infty} a^n u[n] \, e^{-j\Omega n} = \sum_{n=0}^{\infty} a^n e^{-j\Omega n} = \sum_{n=0}^{\infty} (ae^{-j\Omega})^n$$

$$= \frac{1}{1 - ae^{-j\Omega}} \qquad |ae^{-j\Omega}| = |a| < 1 \tag{6.38}$$

Thus, comparing Eqs. (6.37) and (6.38), we have

$$X(\Omega) = X(z)\big|_{z = e^{j\Omega}}$$

Note that $x[n]$ is absolutely summable.

EXAMPLE 6.3   Consider the unit step sequence $u[n]$.
From Eq. (4.16) the z-transform of $u[n]$ is

$$\mathfrak{Z}\{u[n]\} = \frac{1}{1 - z^{-1}} \qquad |z| > 1 \tag{6.39}$$

The Fourier transform of $u[n]$ cannot be obtained from its z-transform because the ROC of the z-transform of $u[n]$ does not include the unit circle. Note that the unit step sequence $u[n]$ is not absolutely summable. The Fourier transform of $u[n]$ is given by (Prob. 6.28)

$$\mathscr{F}\{u[n]\} = \pi \, \delta(\Omega) + \frac{1}{1 - e^{-j\Omega}} \qquad |\Omega| \le \pi \tag{6.40}$$

6.4 Properties of the Fourier Transform

Basic properties of the Fourier transform are presented in the following. There are many similarities to and several differences from the continuous-time case. Many of these properties are also similar to those of the z-transform when the ROC of $X(z)$ includes the unit circle.

A. Periodicity:

$$X(\Omega + 2\pi) = X(\Omega) \tag{6.41}$$

As a consequence of Eq. (6.41), in the discrete-time case we have to consider values of Ω (radians) only over the range $0 \le \Omega < 2\pi$ or $-\pi \le \Omega < \pi$, while in the continuous-time case we have to consider values of ω (radians / second) over the entire range $-\infty < \omega < \infty$.

B. Linearity:

$$a_1 x_1[n] + a_2 x_2[n] \leftrightarrow a_1 X_1(\Omega) + a_2 X_2(\Omega) \tag{6.42}$$

C. Time Shifting:

$$x[n - n_0] \leftrightarrow e^{-j\Omega n_0} X(\Omega) \tag{6.43}$$

D. Frequency Shifting:

$$e^{j\Omega_0 n} x[n] \leftrightarrow X(\Omega - \Omega_0) \tag{6.44}$$

E. Conjugation:

$$x^*[n] \leftrightarrow X^*(-\Omega) \tag{6.45}$$

where $*$ denotes the complex conjugate.

F. Time Reversal:

$$x[-n] \leftrightarrow X(-\Omega) \tag{6.46}$$

G. Time Scaling:

In Sec. 5.4D the scaling property of a continuous-time Fourier transform is expressed as [Eq. (5.52)]

$$x(at) \leftrightarrow \frac{1}{|a|} X\left(\frac{\omega}{a}\right) \tag{6.47}$$

However, in the discrete-time case, $x[an]$ is not a sequence if a is not an integer. On the other hand, if a is an integer, say $a = 2$, then $x[2n]$ consists of only the even samples of $x[n]$. Thus, time scaling in discrete time takes on a form somewhat different from Eq. (6.47).

Let m be a positive integer and define the sequence

$$x_{(m)}[n] = \begin{cases} x[n/m] = x[k] & \text{if } n = km, k = \text{integer} \\ 0 & \text{if } n \neq km \end{cases} \tag{6.48}$$

Then we have

$$x_{(m)}[n] \leftrightarrow X(m\Omega) \tag{6.49}$$

Equation (6.49) is the discrete-time counterpart of Eq. (6.47). It states again the inverse relationship between time and frequency. That is, as the signal spreads in time ($m > 1$), its Fourier transform is compressed (Prob. 6.22). Note that $X(m\Omega)$ is periodic with period $2\pi/m$ since $X(\Omega)$ is periodic with period 2π.

H. Duality:

In Sec. 5.4F the duality property of a continuous-time Fourier transform is expressed as [Eq. (5.54)]

$$X(t) \leftrightarrow 2\pi x(-\omega) \tag{6.50}$$

There is no discrete-time counterpart of this property. However, there is a duality between the discrete-time Fourier transform and the continuous-time Fourier series. Let

$$x[n] \leftrightarrow X(\Omega)$$

From Eqs. (6.27) and (6.41)

$$X(\Omega) = \sum_{n=-\infty}^{\infty} x[n] e^{-j\Omega n} \tag{6.51}$$

$$X(\Omega + 2\pi) = X(\Omega) \tag{6.52}$$

Since Ω is a continuous variable, letting $\Omega = t$ and $n = -k$ in Eq. (6.51), we have

$$X(t) = \sum_{k=-\infty}^{\infty} x[-k] e^{jkt} \tag{6.53}$$

Since $X(t)$ is periodic with period $T_0 = 2\pi$ and the fundamental frequency $\omega_0 = 2\pi/T_0 = 1$, Eq. (6.53) indicates that the Fourier series coefficients of $X(t)$ will be $x[-k]$. This duality relationship is denoted by

$$X(t) \xleftarrow{\text{FS}} c_k = x[-k] \tag{6.54}$$

where FS denotes the Fourier series and c_k are its Fourier coefficients.

I. Differentiation in Frequency:

$$nx[n] \leftrightarrow j\frac{dX(\Omega)}{d\Omega} \tag{6.55}$$

J. Differencing:

$$x[n] - x[n-1] \leftrightarrow (1 - e^{-j\Omega})X(\Omega) \tag{6.56}$$

The sequence $x[n] - x[n-1]$ is called the *first difference* sequence. Equation (6.56) is easily obtained from the linearity property (6.42) and the time-shifting property (6.43).

K. Accumulation:

$$\sum_{k=-\infty}^{n} x[k] \leftrightarrow \pi X(0)\,\delta(\Omega) + \frac{1}{1 - e^{-j\Omega}}X(\Omega) \qquad |\Omega| \le \pi \tag{6.57}$$

Note that accumulation is the discrete-time counterpart of integration. The impulse term on the right-hand side of Eq. (6.57) reflects the dc or average value that can result from the accumulation.

L. Convolution:

$$x_1[n] * x_2[n] \leftrightarrow X_1(\Omega)\,X_2(\Omega) \tag{6.58}$$

As in the case of the z-transform, this convolution property plays an important role in the study of discrete-time LTI systems.

M. Multiplication:

$$x_1[n]\,x_2[n] \leftrightarrow \frac{1}{2\pi}X_1(\Omega) \otimes X_2(\Omega) \tag{6.59}$$

where $\otimes$ denotes the periodic convolution defined by [Eq. (2.70)]

$$X_1(\Omega) \otimes X_2(\Omega) = \int_{2\pi} X_1(\theta)\,X_2(\Omega - \theta)\,d\theta \tag{6.60}$$

The multiplication property (6.59) is the dual property of Eq. (6.58).

N. Additional Properties:

If $x[n]$ is real, let

$$x[n] = x_e[n] + x_o[n]$$

where $x_e[n]$ and $x_o[n]$ are the even and odd components of $x[n]$, respectively. Let

$$x[n] \leftrightarrow X(\Omega) = A(\Omega) + jB(\Omega) = |X(\Omega)|e^{j\theta(\Omega)} \tag{6.61}$$

Then

$$X(-\Omega) = X^*(\Omega) \tag{6.62}$$

$$x_e[n] \leftrightarrow \text{Re}\{X(\Omega)\} = A(\Omega) \tag{6.63a}$$

$$x_o[n] \leftrightarrow j\text{Im}\{X(\Omega)\} = jB(\Omega) \tag{6.63b}$$

Equation (6.62) is the necessary and sufficient condition for $x[n]$ to be real. From Eqs. (6.62) and (6.61) we have

$$A(-\Omega) = A(\Omega) \qquad B(-\Omega) = -B(\Omega) \tag{6.64a}$$

$$|X(-\Omega)| = |X(\Omega)| \qquad \theta(-\Omega) = -\theta(\Omega) \tag{6.64b}$$

From Eqs. (6.63a), (6.63b), and (6.64a) we see that if $x[n]$ is real and even, then $X(\Omega)$ is real and even, while if $x[n]$ is real and odd, $X(\Omega)$ is imaginary and odd.

O. Parseval's Relations:

$$\sum_{n=-\infty}^{\infty} x_1[n]\, x_2[n] = \frac{1}{2\pi} \int_{2\pi} X_1(\Omega)\, X_2(-\Omega)\, d\Omega \tag{6.65}$$

$$\sum_{n=-\infty}^{\infty} |x[n]|^2 = \frac{1}{2\pi} \int_{2\pi} |X(\Omega)|^2\, d\Omega \tag{6.66}$$

Equation (6.66) is known as *Parseval's identity* (or *Parseval's theorem*) for the discrete-time Fourier transform.

Table 6-1 contains a summary of the properties of the Fourier transform presented in this section. Some common sequences and their Fourier transforms are given in Table 6-2.

TABLE 6-1 Properties of the Fourier Transform

PROPERTY	SEQUENCE	FOURIER TRANSFORM
	$x[n]$	$X(\Omega)$
	$x_1[n]$	$X_1(\Omega)$
	$x_2[n]$	$X_2(\Omega)$
Periodicity	$x[n]$	$X(\Omega + 2\pi) = X(\Omega)$
Linearity	$a_1 x_1[n] + a_2 x_2[n]$	$a_1 X_1(\Omega) + a_2 X_2(\Omega)$
Time shifting	$x[n - n_0]$	$e^{-j\Omega n_0} X(\Omega)$
Frequency shifting	$e^{j\Omega_0 n} x[n]$	$X(\Omega - \Omega_0)$
Conjugation	$x^*[n]$	$X^*(-\Omega)$
Time reversal	$x[-n]$	$X(-\Omega)$
Time scaling	$x_{(m)}[n] = \begin{cases} x[n/m] & \text{if } n = km \\ 0 & \text{if } n \neq km \end{cases}$	$X(m\Omega)$
Frequency differentiation	$nx[n]$	$j\dfrac{dX(\Omega)}{d\Omega}$
First difference	$x[n] - x[n-1]$	$(1 - e^{-j\Omega})X(\Omega)$
Accumulation	$\displaystyle\sum_{k=-\infty}^{n} x[k]$	$\pi X(0)\delta(\Omega) + \dfrac{1}{1 - e^{-j\Omega}} X(\Omega)$ $|\Omega| \leq \pi$
Convolution	$x_1[n] * x_2[n]$	$X_1(\Omega) X_2(\Omega)$
Multiplication	$x_1[n] x_2[n]$	$\dfrac{1}{2\pi} X_1(\Omega) \otimes X_2(\Omega)$
Real sequence	$x[n] = x_e[n] + x_o[n]$	$X(\Omega) = A(\Omega) + jB(\Omega)$ $X(-\Omega) = X^*(\Omega)$
Even component	$x_e[n]$	$\text{Re}\{X(\Omega)\} = A(\Omega)$
Odd component	$x_o[n]$	$j\,\text{Im}\{X(\Omega)\} = jB(\Omega)$
Parseval's theorem		

$$\sum_{n=-\infty}^{\infty} x_1[n] x_2[n] = \frac{1}{2\pi} \int_{2\pi} X_1(\Omega) X_2(-\Omega)\, d\Omega$$

$$\sum_{n=-\infty}^{\infty} |x[n]|^2 = \frac{1}{2\pi} \int_{2\pi} |X(\Omega)|^2\, d\Omega$$

TABLE 6-2 **Common Fourier Transform Pairs**

$x[n]$	$X(\Omega)$
$\delta[n]$	1
$\delta(n - n_0)$	$e^{-j\Omega n_0}$
$x[n] = 1$	$2\pi\delta(\Omega), \|\Omega\| \leq \pi$
$e^{j\Omega_0 n}$	$2\pi\delta(\Omega - \Omega_0), \|\Omega\|, \|\Omega_0\| \leq \pi$
$\cos\Omega_0 n$	$\pi[\delta(\Omega - \Omega_0) + \delta(\Omega + \Omega_0)], \|\Omega\|, \|\Omega_0\| \leq \pi$
$\sin\Omega_0 n$	$-j\pi[\delta(\Omega - \Omega_0) - \delta(\Omega + \Omega_0)], \|\Omega\|, \|\Omega_0\| \leq \pi$
$u[n]$	$\pi\delta(\Omega) + \dfrac{1}{1 - e^{-j\Omega}}, \|\Omega\| \leq \pi$
$-u[-n-1]$	$-\pi\delta(\Omega) + \dfrac{1}{1 - e^{-j\Omega}}, \|\Omega\| \leq \pi$
$a^n u[n], \|a\| < 1$	$\dfrac{1}{1 - ae^{-j\Omega}}$
$-a^n u[-n-1], \|a\| > 1$	$\dfrac{1}{1 - ae^{-j\Omega}}$
$(n+1)a^n u[n], \|a\| < 1$	$\dfrac{1}{\left(1 - ae^{-j\Omega}\right)^2}$
$a^{\|n\|}, \|a\| < 1$	$\dfrac{1 - a^2}{1 - 2a\cos\Omega + a^2}$
$x[n] = \begin{cases} 1 & \|n\| \leq N_1 \\ 0 & \|n\| > N_1 \end{cases}$	$\dfrac{\sin\left[\Omega\left(N_1 + \dfrac{1}{2}\right)\right]}{\sin(\Omega/2)}$
$\dfrac{\sin Wn}{\pi n}, 0 < W < \pi$	$X(\Omega) = \begin{cases} 1 & 0 \leq \|\Omega\| \leq W \\ 0 & W < \|\Omega\| \leq \pi \end{cases}$
$\displaystyle\sum_{k=-\infty}^{\infty} \delta[n - kN_0]$	$\Omega_0 \displaystyle\sum_{k=-\infty}^{\infty} \delta(\Omega - k\Omega_0), \Omega_0 = \dfrac{2\pi}{N_0}$

6.5 The Frequency Response of Discrete-Time LTI Systems

A. Frequency Response:

In Sec. 2.6 we showed that the output $y[n]$ of a discrete-time LTI system equals the convolution of the input $x[n]$ with the impulse response $h[n]$; that is,

$$y[n] = x[n] * h[n] \tag{6.67}$$

Applying the convolution property (6.58), we obtain

$$Y(\Omega) = X(\Omega)H(\Omega) \tag{6.68}$$

where $Y(\Omega)$, $X(\Omega)$, and $H(\Omega)$ are the Fourier transforms of $y[n]$, $x[n]$, and $h[n]$, respectively. From Eq. (6.68) we have

$$H(\Omega) = \frac{Y(\Omega)}{X(\Omega)} \tag{6.69}$$

Relationship represented by Eqs. (6.67) and (6.68) are depicted in Fig. 6-3. Let

$$H(\Omega) = |H(\Omega)| e^{j\theta_H(\Omega)} \tag{6.70}$$

As in the continuous-time case, the function $H(\Omega)$ is called the *frequency response* of the system, $|H(\Omega)|$ the *magnitude response* of the system, and $\theta_H(\Omega)$ the *phase response* of the system.

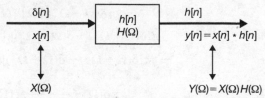

Fig. 6-3 Relationships between inputs and outputs in an LTI discrete-time system.

Consider the complex exponential sequence

$$x[n] = e^{j\Omega_0 n} \tag{6.71}$$

Then, setting $z = e^{j\Omega_0}$ in Eq. (4.1), we obtain

$$y[n] = H(e^{j\Omega_0}) e^{j\Omega_0 n} = H(\Omega_0) e^{j\Omega_0 n} \tag{6.72}$$

which indicates that the complex exponential sequence $e^{j\Omega_0 n}$ is an eigenfunction of the LTI system with corresponding eigenvalue $H(\Omega_0)$, as previously observed in Chap. 2 (Sec. 2.8). Furthermore, by the linearity property (6.42), if the input $x[n]$ is periodic with the discrete Fourier series

$$x[n] = \sum_{k=\langle N_0 \rangle} c_k e^{jk\Omega_0 n} \qquad \Omega_0 = \frac{2\pi}{N_0} \tag{6.73}$$

then the corresponding output $y[n]$ is also periodic with the discrete Fourier series

$$y[n] = \sum_{k=\langle N_0 \rangle} c_k H(k\Omega_0) e^{jk\Omega_0 n} \tag{6.74}$$

If $x[n]$ is not periodic, then from Eqs. (6.68) and (6.28) the corresponding output $y[n]$ can be expressed as

$$y[n] = \frac{1}{2\pi} \int_{\langle 2\pi \rangle} H(\Omega) X(\Omega) e^{j\Omega n} \, d\Omega \tag{6.75}$$

B. LTI Systems Characterized by Difference Equations:

As discussed in Sec. 2.9, many discrete-time LTI systems of practical interest are described by linear constant-coefficient difference equations of the form

$$\sum_{k=0}^{N} a_k y[n-k] = \sum_{k=0}^{M} b_k x[n-k] \tag{6.76}$$

with $M \leq N$. Taking the Fourier transform of both sides of Eq. (6.76) and using the linearity property (6.42) and the time-shifting property (6.43), we have

$$\sum_{k=0}^{N} a_k e^{-jk\Omega} Y(\Omega) = \sum_{k=0}^{M} b_k e^{-jk\Omega} X(\Omega)$$

or, equivalently,

$$H(\Omega) = \frac{Y(\Omega)}{X(\Omega)} = \frac{\displaystyle\sum_{k=0}^{M} b_k e^{-jk\Omega}}{\displaystyle\sum_{k=0}^{N} a_k e^{-jk\Omega}} \tag{6.77}$$

The result (6.77) is the same as the z-transform counterpart $H(z) = Y(z)/X(z)$ with $z = e^{j\Omega}$ [Eq. (4.44)]; that is,

$$H(\Omega) = H(z)\big|_{z=e^{j\Omega}} = H(e^{j\Omega})$$

C. Periodic Nature of the Frequency Response:

From Eq. (6.41) we have

$$H(\Omega) = H(\Omega + 2\pi) \tag{6.78}$$

Thus, unlike the frequency response of continuous-time systems, that of all discrete-time LTI systems is periodic with period 2π. Therefore, we need observe the frequency response of a system only over the frequency range $0 \le \Omega < 2\pi$ or $-\pi \le \Omega < \pi$.

6.6 System Response to Sampled Continuous-Time Sinusoids

A. System Responses:

We denote by $y_c[n]$, $y_s[n]$, and $y[n]$ the system responses to $\cos \Omega n$, $\sin \Omega n$, and $e^{j\Omega n}$, respectively (Fig. 6-4). Since $e^{j\Omega n} = \cos \Omega n + j \sin \Omega n$, it follows from Eq. (6.72) and the linearity property of the system that

$$y[n] = y_c[n] + jy_s[n] = H(\Omega)\, e^{j\Omega n} \tag{6.79a}$$

$$y_c[n] = \text{Re}\{y[n]\} = \text{Re}\{H(\Omega)\, e^{j\Omega n}\} \tag{6.79b}$$

$$y_s[n] = \text{Im}\{y[n]\} = \text{Im}\{H(\Omega)\, e^{j\Omega n}\} \tag{6.79c}$$

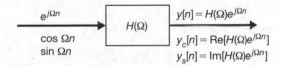

Fig. 6-4 System responses to $e^{j\Omega n}$, $\cos \Omega n$, and $\sin \Omega n$.

When a sinusoid $\cos \Omega n$ is obtained by sampling a continuous-time sinusoid $\cos \omega t$ with sampling interval T_s, that is,

$$\cos \Omega n = \cos \omega t\big|_{t=nT_s} = \cos \omega T_s n \tag{6.80}$$

all the results developed in this section apply if we substitute ωT_s for Ω:

$$\Omega = \omega T_s \tag{6.81}$$

For a continuous-time sinusoid $\cos \omega t$ there is a unique waveform for every value of ω in the range 0 to ∞. Increasing ω results in a sinusoid of ever-increasing frequency. On the other hand, the discrete-time sinusoid $\cos \Omega n$ has a unique waveform only for values of Ω in the range 0 to 2π because

$$\cos[(\Omega + 2\pi m)n] = \cos(\Omega n + 2\pi mn) = \cos \Omega n \qquad m = \text{integer} \tag{6.82}$$

This range is further restricted by the fact that

$$\cos(\pi \pm \Omega)n = \cos \pi n \cos \Omega n \mp \sin \pi n \sin \Omega n$$

$$= (-1)^n \cos \Omega n \tag{6.83}$$

Therefore,

$$\cos(\pi + \Omega)n = \cos(\pi - \Omega)n \tag{6.84}$$

Equation (6.84) shows that a sinusoid of frequency $(\pi + \Omega)$ has the same waveform as one with frequency $(\pi - \Omega)$. Therefore, a sinusoid with any value of Ω outside the range 0 to π is identical to a sinusoid with Ω in the range 0 to π. Thus, we conclude that every discrete-time sinusoid with a frequency in the range $0 \leq \Omega < \pi$ has a distinct waveform, and we need observe only the frequency response of a system over the frequency range $0 \leq \Omega < \pi$.

B. Sampling Rate:

Let $\omega_M \ (= 2\pi f_M)$ be the highest frequency of the continuous-time sinusoid. Then from Eq. (6.81) the condition for a sampled discrete-time sinusoid to have a unique waveform is

$$\omega_M T_s < \pi \rightarrow T_s < \frac{\pi}{\omega_M} \qquad \text{or} \qquad f_s > 2 f_M \tag{6.85}$$

where $f_s = 1/T_s$ is the sampling rate (or frequency). Equation (6.85) indicates that to process a continuous-time sinusoid by a discrete-time system, the sampling rate must not be less than twice the frequency (in hertz) of the sinusoid. This result is a special case of the sampling theorem we discussed in Prob. 5.59.

6.7 Simulation

Consider a continuous-time LTI system with input $x(t)$ and output $y(t)$. We wish to find a discrete-time LTI system with input $x[n]$ and output $y[n]$ such that

$$\text{if } x[n] = x(nT_s) \text{ then } y[n] = y(nT_s) \tag{6.86}$$

where T_s is the sampling interval.

Let $H_c(s)$ and $H_d(z)$ be the system functions of the continuous-time and discrete-time systems, respectively (Fig. 6-5). Let

$$x(t) = e^{j\omega t} \qquad x[n] = x(nT_s) = e^{jn\omega T_s} \tag{6.87}$$

Then from Eqs. (3.1) and (4.1) we have

$$y(t) = H_c(j\omega) e^{j\omega t} \qquad y[n] = H_d(e^{j\omega T_s}) e^{jn\omega T_s} \tag{6.88}$$

Thus, the requirement $y[n] = y(nT_s)$ leads to the condition

$$H_c(j\omega) e^{jn\omega T_s} = H_d(e^{j\omega T_s}) e^{jn\omega T_s}$$

from which it follows that

$$H_c(j\omega) = H_d(e^{j\omega T_s}) \tag{6.89}$$

In terms of the Fourier transform, Eq. (6.89) can be expressed as

$$H_c(\omega) = H_d(\Omega) \qquad \Omega = \omega T_s \tag{6.90}$$

Note that the frequency response $H_d(\Omega)$ of the discrete-time system is a periodic function of ω (with period $2\pi/T_s$), but that the frequency response $H_c(\omega)$ of the continuous-time system is not. Therefore, Eq. (6.90) or

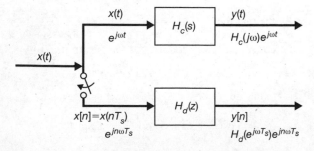

Fig. 6-5 Digital simulation of analog systems.

Eq. (6.89) cannot, in general, be true for every ω. If the input $x(t)$ is band-limited [Eq. (5.94)], then it is possible, in principle, to satisfy Eq. (6.89) for every ω in the frequency range $(-\pi/T_s, \pi/T_s)$ (Fig. 6-6). However, from Eqs. (5.85) and (6.77), we see that $H_c(\omega)$ is a rational function of ω, whereas $H_d(\Omega)$ is a rational function of $e^{j\Omega}$ ($\Omega = \omega T_s$). Therefore, Eq. (6.89) is impossible to satisfy. However, there are methods for determining a discrete-time system so as to satisfy Eq. (6.89) with reasonable accuracy for every ω in the band of the input (Probs. 6.43 to 6.47).

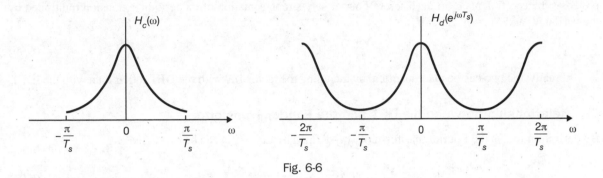

Fig. 6-6

6.8 The Discrete Fourier Transform

In this section we introduce the technique known as the *discrete Fourier transform* (DFT) for finite-length sequences. It should be noted that the DFT should not be confused with the Fourier transform.

A. Definition:

Let $x[n]$ be a finite-length sequence of length N, that is,

$$x[n] = 0 \qquad \text{outside the range } 0 \le n \le N - 1 \tag{6.91}$$

The DFT of $x[n]$, denoted as $X[k]$, is defined by

$$X[k] = \sum_{n=0}^{N-1} x[n] W_N^{kn} \qquad k = 0, 1, \ldots, N-1 \tag{6.92}$$

where W_N is the Nth root of unity given by

$$W_N = e^{-j(2\pi/N)} \tag{6.93}$$

The inverse DFT (IDFT) is given by

$$x[n] = \frac{1}{N} \sum_{k=0}^{N-1} X[k] W_N^{-kn} \qquad n = 0, 1, \ldots, N-1 \tag{6.94}$$

The DFT pair is denoted by

$$x[n] \leftrightarrow X[k] \tag{6.95}$$

Important features of the DFT are the following:

1. There is a one-to-one correspondence between $x[n]$ and $X[k]$.
2. There is an extremely fast algorithm, called the fast Fourier transform (FFT) for its calculation.
3. The DFT is closely related to the discrete Fourier series and the Fourier transform.
4. The DFT is the appropriate Fourier representation for digital computer realization because it is discrete and of finite length in both the time and frequency domains.

Note that the choice of N in Eq. (6.92) is not fixed. If $x[n]$ has length $N_1 < N$, we want to assume that $x[n]$ has length N by simply adding ($N - N_1$) samples with a value of 0. This addition of dummy samples is known as *zero padding*. Then the resultant $x[n]$ is often referred to as an *N-point sequence*, and $X[k]$ defined in Eq. (6.92) is referred to as an *N-point DFT*. By a judicious choice of N, such as choosing it to be a power of 2, computational efficiencies can be gained.

B. Relationship between the DFT and the Discrete Fourier Series:

Comparing Eqs. (6.94) and (6.92) with Eqs. (6.7) and (6.8), we see that $X[k]$ of finite sequence $x[n]$ can be interpreted as the coefficients c_k in the discrete Fourier series representation of its periodic extension multiplied by the period N_0 and $N_0 = N$. That is,

$$X[k] = Nc_k \qquad (6.96)$$

Actually, the two can be made identical by including the factor $1/N$ with the DFT rather than with the IDFT.

C. Relationship between the DFT and the Fourier Transform:

By definition (6.27) the Fourier transform of $x[n]$ defined by Eq. (6.91) can be expressed as

$$X(\Omega) = \sum_{n=0}^{N-1} x[n]\, e^{-j\Omega n} \qquad (6.97)$$

Comparing Eq. (6.97) with Eq. (6.92), we see that

$$X[k] = X(\Omega)\big|_{\Omega = k2\pi/N} = X\!\left(\frac{k2\pi}{N}\right) \qquad (6.98)$$

Thus, $X[k]$ corresponds to the sampled $X(\Omega)$ at the uniformly spaced frequencies $\Omega = k2\pi/N$ for integer k.

D. Properties of the DFT:

Because of the relationship (6.98) between the DFT and the Fourier transform, we would expect their properties to be quite similar, except that the DFT $X[k]$ is a function of a discrete variable while the Fourier transform $X(\Omega)$ is a function of a continuous variable. Note that the DFT variables n and k must be restricted to the range $0 \le n, k < N$, the DFT shifts $x[n - n_0]$ or $X[k - k_0]$ imply $x[n - n_0]_{\mathrm{mod}\,N}$ or $X[k - k_0]_{\mathrm{mod}\,N}$, where the modulo notation $[m]_{\mathrm{mod}\,N}$ means that

$$[m]_{\mathrm{mod}\,N} = m + iN \qquad (6.99)$$

for some integer i such that

$$0 \le [m]_{\mathrm{mod}\,N} < N \qquad (6.100)$$

For example, if $x[n] = \delta[n - 3]$, then

$$x[n - 4]_{\mathrm{mod}\,6} = \delta[n - 7]_{\mathrm{mod}\,6} = \delta[n - 7 + 6] = \delta[n - 1]$$

The DFT shift is also known as a *circular shift*. Basic properties of the DFT are the following:

1. Linearity:

$$a_1 x_1[n] + a_2 x_2[n] \leftrightarrow a_1 X_1[k] + a_2 X_2[k] \qquad (6.101)$$

2. Time Shifting:

$$x[n - n_0]_{\mathrm{mod}\,N} \leftrightarrow W_N^{kn_0} X[k] \qquad W_N = e^{-j(2\pi/N)} \qquad (6.102)$$

3. Frequency Shifting:

$$W_N^{-kn_0} x[n] \leftrightarrow X[k - k_0]_{\mathrm{mod}\,N} \qquad (6.103)$$

4. **Conjugation:**

$$x^*[n] \leftrightarrow X^*[-k]_{\text{mod } N} \tag{6.104}$$

where $*$ denotes the complex conjugate.

5. **Time Reversal:**

$$x[-n]_{\text{mod } N} \leftrightarrow X[-k]_{\text{mod } N} \tag{6.105}$$

6. **Duality:**

$$X[n] \leftrightarrow Nx[-k]_{\text{mod } N} \tag{6.106}$$

7. **Circular Convolution:**

$$x_1[n] \otimes x_2[n] \leftrightarrow X_1[k] \, X_2[k] \tag{6.107}$$

where

$$x_1[n] \otimes x_2[n] = \sum_{i=0}^{N-1} x_1[i] x_2[n-i]_{\text{mod } N} \tag{6.108}$$

The convolution sum in Eq. (6.108) is known as the *circular convolution* of $x_1[n]$ and $x_2[n]$.

8. **Multiplication:**

$$x_1[n] \, x_2[n] \leftrightarrow \frac{1}{N} X_1[k] \otimes X_2[k] \tag{6.109}$$

where

$$X_1[k] \otimes X_2[k] = \sum_{i=1}^{N-1} X_1[i] \, X_2[k-i]_{\text{mod } N}$$

9. **Additional Properties:**

When $x[n]$ is real, let

$$x[n] = x_e[n] + x_o[n]$$

where $x_e[n]$ and $x_o[n]$ are the even and odd components of $x[n]$, respectively. Let

$$x[n] \leftrightarrow X[k] = A[k] + jB[k] = |X[k]| e^{j\theta[k]}$$

Then

$$X[-k]_{\text{mod } N} = X^*[k] \tag{6.110}$$

$$x_e[n] \leftrightarrow \text{Re}\{X[k]\} = A[k] \tag{6.111a}$$

$$x_o[n] \leftrightarrow j \, \text{Im}\{X[k]\} = jB[k] \tag{6.111b}$$

From Eq. (6.110) we have

$$A[-k]_{\text{mod } N} = A[k] \qquad B[-k]_{\text{mod } N} = -B[k] \tag{6.112a}$$

$$|X[-k]|_{\text{mod } N} = |X[k]| \qquad \theta[-k]_{\text{mod } N} = -\theta[k] \tag{6.112b}$$

10. **Parseval's Relation:**

$$\sum_{n=0}^{N-1} |x[n]|^2 = \frac{1}{N} \sum_{n=0}^{N-1} |X[k]|^2 \tag{6.113}$$

Equation (6.113) is known as *Parseval's identity* (or *Parseval's theorem*) for the DFT.

SOLVED PROBLEMS

Discrete Fourier Series

6.1. We call a set of sequences $\{\Psi_k[n]\}$ orthogonal on an interval $[N_1, N_2]$ if any two signals $\Psi_m[n]$ and $\Psi_k[n]$ in the set satisfy the condition

$$\sum_{n=N_1}^{N_2} \Psi_m[n]\,\Psi_k^*[n] = \begin{cases} 0 & m \neq k \\ \alpha & m = k \end{cases} \tag{6.114}$$

where $*$ denotes the complex conjugate and $\alpha \neq 0$. Show that the set of complex exponential sequences

$$\Psi_k[n] = e^{jk(2\pi/N)n} \qquad k = 0, 1, ..., N-1 \tag{6.115}$$

is orthogonal on any interval of length N.

From Eq. (1.90) we note that

$$\sum_{n=0}^{N-1} \alpha^n = \begin{cases} N & \alpha = 1 \\ \dfrac{1-\alpha^N}{1-\alpha} & \alpha \neq 1 \end{cases} \tag{6.116}$$

Applying Eq. (6.116), with $\alpha = e^{jk(2\pi/N)}$, we obtain

$$\sum_{n=0}^{N-1} e^{jk(2\pi/N)n} = \begin{cases} N & k = 0, \pm N, \pm 2N, ... \\ \dfrac{1-e^{jk(2\pi/N)N}}{1-e^{jk(2\pi/N)}} = 0 & \text{otherwise} \end{cases} \tag{6.117}$$

since $e^{jk(2\pi/N)N} = e^{jk2\pi} = 1$. Since each of the complex exponentials in the summation in Eq. (6.117) is periodic with period N, Eq. (6.117) remains valid with a summation carried over any interval of length N. That is,

$$\sum_{n=\langle N \rangle} e^{jk(2\pi/N)n} = \begin{cases} N & k = 0, \pm N, \pm 2N, ... \\ 0 & \text{otherwise} \end{cases} \tag{6.118}$$

Now, using Eq. (6.118), we have

$$\sum_{n=\langle N \rangle} \Psi_m[n]\,\Psi_k^*[n] = \sum_{n=\langle N \rangle} e^{jm(2\pi/N)n}\, e^{-jk(2\pi/N)n}$$

$$= \sum_{n=\langle N \rangle} e^{j(m-k)(2\pi/N)n} = \begin{cases} N & m = k \\ 0 & m \neq k \end{cases} \tag{6.119}$$

where $m, k < N$. Equation (6.119) shows that the set $\{e^{jk(2\pi/N)n}: k = 0, 1, ..., N-1\}$ is orthogonal over any interval of length N. Equation (6.114) is the discrete-time counterpart of Eq. (5.95) introduced in Prob. 5.1.

6.2. Using the orthogonality condition Eq. (6.119), derive Eq. (6.8) for the Fourier coefficients.

Replacing the summation variable k by m in Eq. (6.7), we have

$$x[n] = \sum_{m=0}^{N-1} c_m\, e^{jm(2\pi/N_0)n} \tag{6.120}$$

Using Eq. (6.115) with $N = N_0$, Eq. (6.120) can be rewritten as

$$x[n] = \sum_{m=0}^{N_0-1} c_m \Psi_m[n] \tag{6.121}$$

Multiplying both sides of Eq. (6.121) by $\Psi_k^*[n]$ and summing over $n = 0$ to $(N_0 - 1)$, we obtain

$$\sum_{n=0}^{N_0-1} x[n]\Psi_k^*[n] = \sum_{n=0}^{N_0-1} \left(\sum_{m=0}^{N_0-1} c_m \Psi_m[n]\right)\Psi_k^*[n]$$

Interchanging the order of the summation and using Eq. (6.119), we get

$$\sum_{n=0}^{N_0-1} x[n]\Psi_k^*[n] = \sum_{m=0}^{N_0-1} c_m \left(\sum_{n=0}^{N_0-1} \Psi_m[n]\Psi_k^*[n]\right) = N_0 c_k \qquad (6.122)$$

Thus,

$$c_k = \frac{1}{N_0} \sum_{n=0}^{N_0-1} x[n]\Psi_k^*[n] = \frac{1}{N_0} \sum_{n=0}^{N_0-1} x[n] e^{-jk(2\pi/N_0)n}$$

6.3. Determine the Fourier coefficients for the periodic sequence $x[n]$ shown in Fig. 6-7.

From Fig. 6-7 we see that $x[n]$ is the periodic extension of $\{0, 1, 2, 3\}$ with fundamental period $N_0 = 4$. Thus,

$$\Omega_0 = \frac{2\pi}{4} \qquad \text{and} \qquad e^{-j\Omega_0} = e^{-j2\pi/4} = e^{-j\pi/2} = -j$$

By Eq. (6.8) the discrete-time Fourier coefficients c_k are

$$c_0 = \frac{1}{4}\sum_{n=0}^{3} x[n] = \frac{1}{4}(0 + 1 + 2 + 3) = \frac{3}{2}$$

$$c_1 = \frac{1}{4}\sum_{n=0}^{3} x[n](-j)^n = \frac{1}{4}(0 - j1 - 2 + j3) = -\frac{1}{2} + j\frac{1}{2}$$

$$c_2 = \frac{1}{4}\sum_{n=0}^{3} x[n](-j)^{2n} = \frac{1}{4}(0 - 1 + 2 - 3) = -\frac{1}{2}$$

$$c_3 = \frac{1}{4}\sum_{n=0}^{3} x[n](-j)^{3n} = \frac{1}{4}(0 + j1 - 2 - j3) = -\frac{1}{2} - j\frac{1}{2}$$

Note that $c_3 = c_{4-1} = c_1^*$ [Eq. (6.17)].

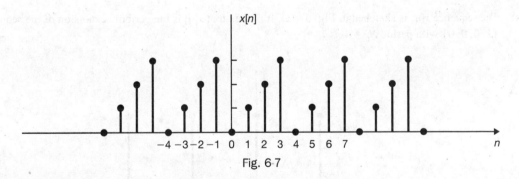

Fig. 6-7

6.4. Consider the periodic sequence $x[n]$ shown in Fig. 6-8(a). Determine the Fourier coefficients c_k and sketch the magnitude spectrum $|c_k|$.

From Fig. 6-8(a) we see that the fundamental period of $x[n]$ is $N_0 = 10$ and $\Omega_0 = 2\pi/N_0 = \pi/5$. By Eq. (6.8) and using Eq. (1.90), we get

$$c_k = \frac{1}{10}\sum_{n=0}^{4} e^{-jk(\pi/5)n} = \frac{1}{10}\frac{1 - e^{-jk\pi}}{1 - e^{-jk(\pi/5)}}$$

$$= \frac{1}{10}\frac{e^{-jk\pi/2}(e^{jk\pi/2} - e^{-jk\pi/2})}{e^{-jk\pi/10}(e^{jk\pi/10} - e^{-jk\pi/10})}$$

$$= \frac{1}{10}e^{-jk(2\pi/5)}\frac{\sin(k\pi/2)}{\sin(k\pi/10)} \qquad k = 0, 1, 2, \ldots, 9$$

The magnitude spectrum $|c_k|$ is plotted in Fig. 6-8(b)

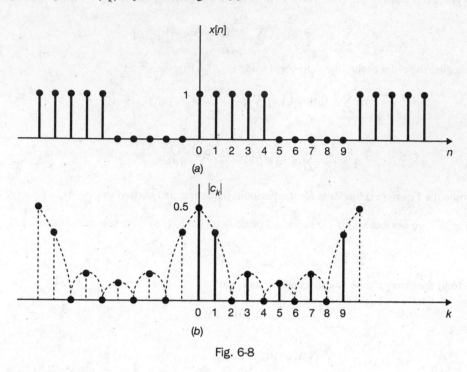

Fig. 6-8

6.5. Consider a sequence

$$x[n] = \sum_{k=-\infty}^{\infty} \delta[n - 4k]$$

(a)   Sketch $x[n]$.

(b)   Find the Fourier coefficients c_k of $x[n]$.

(a)   The sequence $x[n]$ is sketched in Fig. 6-9(a). It is seen that $x[n]$ is the periodic extension of the sequence $\{1, 0, 0, 0\}$ with period $N_0 = 4$.

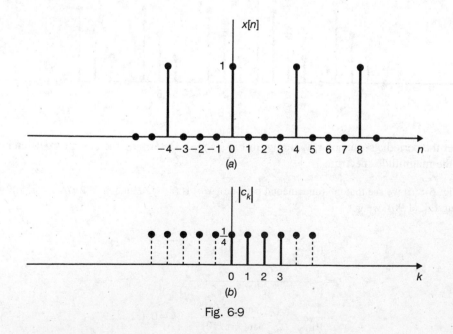

Fig. 6-9

(b) From Eqs. (6.7) and (6.8) and Fig. 6-9(a) we have

$$x[n] = \sum_{k=0}^{3} c_k \, e^{jk(2\pi/4)n} = \sum_{k=0}^{3} c_k \, e^{jk(\pi/2)n}$$

and

$$c_k = \frac{1}{4} \sum_{n=0}^{3} x[n] e^{-jk(2\pi/4)n} = \frac{1}{4} x[0] = \frac{1}{4} \quad \text{all } k$$

since $x[1] = x[2] = x[3] = 0$. The Fourier coefficients of $x[n]$ are sketched in Fig. 6-9(b).

6.6. Determine the discrete Fourier series representation for each of the following sequences:

(a) $x[n] = \cos\dfrac{\pi}{4} n$

(b) $x[n] = \cos\dfrac{\pi}{3} n + \sin\dfrac{\pi}{4} n$

(c) $x[n] = \cos^2\left(\dfrac{\pi}{8} n\right)$

(a) The fundamental period of $x[n]$ is $N_0 = 8$, and $\Omega_0 = 2\pi/N_0 = \pi/4$. Rather than using Eq. (6.8) to evaluate the Fourier coefficients c_k, we use Euler's formula and get

$$\cos\frac{\pi}{4} n = \frac{1}{2}(e^{j(\pi/4)n} + e^{-j(\pi/4)n}) = \frac{1}{2} e^{j\Omega_0 n} + \frac{1}{2} e^{-j\Omega_0 n}$$

Thus, the Fourier coefficients for $x[n]$ are $c_1 = \frac{1}{2}$, $c_{-1} = c_{-1+8} = c_7 = \frac{1}{2}$, and all other $c_k = 0$. Hence, the discrete Fourier series of $x[n]$ is

$$x[n] = \cos\frac{\pi}{4} n = \frac{1}{2} e^{j\Omega_0 n} + \frac{1}{2} e^{j7\Omega_0 n} \qquad \Omega_0 = \frac{\pi}{4}$$

(b) From Prob. 1.16(i) the fundamental period of $x[n]$ is $N_0 = 24$, and $\Omega_0 = 2\pi/N_0 = \pi/12$. Again by Euler's formula we have

$$x[n] = \frac{1}{2}(e^{j(\pi/3)n} + e^{-j(\pi/3)n}) + \frac{1}{2j}(e^{j(\pi/4)n} - e^{-j(\pi/4)n})$$

$$= \frac{1}{2} e^{-j4\Omega_0 n} + j\frac{1}{2} e^{-j3\Omega_0 n} - j\frac{1}{2} e^{j3\Omega_0 n} + \frac{1}{2} e^{j4\Omega_0 n}$$

Thus, $c_3 = -j(\frac{1}{2})$, $c_4 = \frac{1}{2}$, $c_{-4} = c_{-4+24} = c_{20} = \frac{1}{2}$, $c_{-3} = c_{-3+24} = c_{21} = j(\frac{1}{2})$, and all other $c_k = 0$. Hence, the discrete Fourier series of $x[n]$ is

$$x[n] = -j\frac{1}{2} e^{j3\Omega_0 n} + \frac{1}{2} e^{j4\Omega_0 n} + \frac{1}{2} e^{j20\Omega_0 n} + j\frac{1}{2} e^{j21\Omega_0 n} \qquad \Omega_0 = \frac{\pi}{12}$$

(c) From Prob. 1.16(j) the fundamental period of $x[n]$ is $N_0 = 8$, and $\Omega_0 = 2\pi/N_0 = \pi/4$. Again by Euler's formula we have

$$x[n] = \left(\frac{1}{2} e^{j(\pi/8)n} + \frac{1}{2} e^{-j(\pi/8)n}\right)^2 = \frac{1}{4} e^{j(\pi/4)n} + \frac{1}{2} + \frac{1}{4} e^{-j(\pi/4)n}$$

$$= \frac{1}{4} e^{j\Omega_0 n} + \frac{1}{2} + \frac{1}{4} e^{-j\Omega_0 n}$$

Thus, $c_0 = \frac{1}{2}$, $c_1 = \frac{1}{4}$, $c_{-1} = c_{-1+8} = c_7 = \frac{1}{4}$, and all other $c_k = 0$. Hence, the discrete Fourier series of $x[n]$ is

$$x[n] = \frac{1}{2} + \frac{1}{4} e^{j\Omega_0 n} + \frac{1}{4} e^{j7\Omega_0 n} \qquad \Omega_0 = \frac{\pi}{4}$$

6.7. Let $x[n]$ be a real periodic sequence with fundamental period N_0 and Fourier coefficients $c_k = a_k + jb_k$, where a_k and b_k are both real.

(a) Show that $a_{-k} = a_k$ and $b_{-k} = -b_k$.

(b) Show that $c_{N_0/2}$ is real if N_0 is even.

(c) Show that $x[n]$ can also be expressed as a discrete trigonometric Fourier series of the form

$$x[n] = c_0 + 2 \sum_{k=1}^{(N_0-1)/2} (a_k \cos k\Omega_0 n - b_k \sin k\Omega_0 n) \qquad \Omega_0 = \frac{2\pi}{N_0} \qquad (6.123)$$

if N_0 is odd or

$$x[n] = c_0 + (-1)^n c_{N_0/2} + 2 \sum_{k=1}^{(N_0-2)/2} (a_k \cos k\Omega_0 n - b_k \sin k\Omega_0 n) \qquad (6.124)$$

if N_0 is even.

(a) If $x[n]$ is real, then from Eq. (6.8) we have

$$c_{-k} = \frac{1}{N_0} \sum_{n=0}^{N_0-1} x[n]\, e^{jk\Omega_0 n} = \left(\frac{1}{N_0} \sum_{n=0}^{N_0-1} x[n]\, e^{-jk\Omega_0 n} \right)^* = c_k^*$$

Thus,

$$c_{-k} = a_{-k} + jb_{-k} = (a_k + jb_k)^* = a_k - jb_k$$

and we have

$$a_{-k} = a_k \qquad \text{and} \qquad b_{-k} = -b_k$$

(b) If N_0 is even, then from Eq. (6.8)

$$c_{N_0/2} = \frac{1}{N_0} \sum_{n=0}^{N_0-1} x[n]\, e^{-j(N_0/2)(2\pi/N_0)n} = \frac{1}{N_0} \sum_{n=0}^{N_0-1} x[n]\, e^{-j\pi n}$$

$$= \frac{1}{N_0} \sum_{n=0}^{N_0-1} (-1)^n x[n] = \text{real} \qquad (6.125)$$

(c) Rewrite Eq. (6.7) as

$$x[n] = \sum_{k=0}^{N_0-1} c_k\, e^{jk\Omega_0 n} = c_0 + \sum_{k=1}^{N_0-1} c_k\, e^{jk\Omega_0 n}$$

If N_0 is odd, then $(N_0 - 1)$ is even and we can write $x[n]$ as

$$x[n] = c_0 + \sum_{k=1}^{(N_0-1)/2} \left(c_k\, e^{jk\Omega_0 n} + c_{N_0-k}\, e^{j(N_0-k)\Omega_0 n} \right)$$

Now, from Eq. (6.17)

$$c_{N_0-k} = c_k^*$$

and

$$e^{j(N_0-k)\Omega_0 n} = e^{jN_0\Omega_0 n} e^{-jk\Omega_0 n} = e^{j2\pi n} e^{-jk\Omega_0 n} = e^{-jk\Omega_0 n}$$

Thus,

$$x[n] = c_0 + \sum_{k=1}^{(N_0-1)/2} (c_k e^{jk\Omega_0 n} + c_k^* e^{-jk\Omega_0 n})$$

$$= c_0 + \sum_{k=1}^{(N_0-1)/2} 2\operatorname{Re}(c_k e^{jk\Omega_0 n})$$

$$= c_0 + 2 \sum_{k=1}^{(N_0-1)/2} \operatorname{Re}(a_k + jb_k)(\cos k\Omega_0 n + j\sin k\Omega_0 n)$$

$$= c_0 + 2 \sum_{k=1}^{(N_0-1)/2} (a_k \cos k\Omega_0 n - b_k \sin k\Omega_0 n)$$

If N_0 is even, we can write $x[n]$ as

$$x[n] = c_0 + \sum_{k=1}^{N_0-1} c_k e^{jk\Omega_0 n}$$

$$= c_0 + \sum_{k=1}^{(N_0-2)/2} \left(c_k e^{jk\Omega_0 n} + c_{N_0-k} e^{j(N_0-k)\Omega_0 n} \right) + c_{N_0/2} e^{j(N_0/2)\Omega_0 n}$$

Again from Eq. (6.17)

$$c_{N_0-k} = c_k^* \qquad \text{and} \qquad e^{j(N_0-k)\Omega_0 n} = e^{-jk\Omega_0 n}$$

and

$$e^{j(N_0/2)\Omega_0 n} = e^{j(N_0/2)(2\pi/N_0)n} = e^{j\pi n} = (-1)^n$$

Then

$$x[n] = c_0 + (-1)^n c_{N_0/2} + \sum_{k=1}^{(N_0-2)/2} 2\operatorname{Re}(c_k e^{jk\Omega_0 n})$$

$$= c_0 + (-1)^n c_{N_0/2} + 2 \sum_{k=1}^{(N_0-2)/2} (a_k \cos k\Omega_0 n - b_k \sin k\Omega_0 n)$$

6.8. Let $x_1[n]$ and $x_2[n]$ be periodic sequences with fundamental period N_0 and their discrete Fourier series given by

$$x_1[n] = \sum_{k=0}^{N_0-1} d_k e^{jk\Omega_0 n} \qquad x_2[n] = \sum_{k=0}^{N_0-1} e_k e^{jk\Omega_0 n} \qquad \Omega_0 = \frac{2\pi}{N_0}$$

Show that the sequence $x[n] = x_1[n]x_2[n]$ is periodic with the same fundamental period N_0 and can be expressed as

$$x[n] = \sum_{k=0}^{N_0-1} c_k e^{jk\Omega_0 n} \qquad \Omega_0 = \frac{2\pi}{N_0}$$

where c_k is given by

$$c_k = \sum_{m=0}^{N_0-1} d_m e_{k-m} \tag{6.126}$$

Now note that

$$x[n + N_0] = x_1[n + N_0]x_2[n + N_0] = x_1[n]x_2[n] = x[n]$$

Thus, $x[n]$ is periodic with fundamental period N_0. Let

$$x[n] = \sum_{k=0}^{N_0-1} c_k e^{jk\Omega_0 n} \qquad \Omega_0 = \frac{2\pi}{N_0}$$

Then
$$c_k = \frac{1}{N_0} \sum_{n=0}^{N_0-1} x[n] e^{-jk\Omega_0 n} = \frac{1}{N_0} \sum_{n=0}^{N_0-1} x_1[n] x_2[n] e^{-jk\Omega_0 n}$$

$$= \frac{1}{N_0} \sum_{n=0}^{N_0-1} \left(\sum_{m=0}^{N_0-1} d_m e^{jm\Omega_0 n} \right) x_2[n] e^{-jk\Omega_0 n}$$

$$= \sum_{m=0}^{N_0-1} d_m \left(\frac{1}{N_0} \sum_{n=0}^{N_0-1} x_2[n] e^{-j(k-m)\Omega_0 n} \right) = \sum_{m=0}^{N_0-1} d_m e_{k-m}$$

since
$$e_k = \frac{1}{N_0} \sum_{n=0}^{N_0-1} x_2[n] e^{-jk\Omega_0 n}$$

and the term in parentheses is equal to e_{k-m}.

6.9. Let $x_1[n]$ and $x_2[n]$ be the two periodic signals in Prob. 6.8. Show that

$$\frac{1}{N_0} \sum_{n=0}^{N_0-1} x_1[n] x_2[n] = \sum_{k=0}^{N_0-1} d_k \, e_{-k} \tag{6.127}$$

Equation (6.127) is known as *Parseval's relation* for periodic sequences.

From Eq. (6.126) we have

$$c_k = \frac{1}{N_0} \sum_{n=0}^{N_0-1} x_1[n] x_2[n] e^{-jk\Omega_0 n} = \sum_{m=0}^{N_0-1} d_m e_{k-m}$$

Setting $k = 0$ in the above expression, we get

$$\frac{1}{N_0} \sum_{n=0}^{N_0-1} x_1[n] x_2[n] = \sum_{m=0}^{N_0-1} d_m e_{-m} = \sum_{k=0}^{N_0-1} d_k e_{-k}$$

6.10. (a) Verify Parseval's identity [Eq. (6.19)] for the discrete Fourier series; that is,

$$\frac{1}{N_0} \sum_{n=0}^{N_0-1} |x[n]|^2 = \sum_{k=0}^{N_0-1} |c_k|^2$$

(b) Using $x[n]$ in Prob. 6.3, verify Parseval's identity [Eq. (6.19)].

(a) Let

$$x[n] = \sum_{k=0}^{N_0-1} c_k e^{jk\Omega_0 n}$$

and
$$x^*[n] = \sum_{k=0}^{N_0-1} d_k e^{jk\Omega_0 n}$$

Then
$$d_k = \frac{1}{N_0} \sum_{n=0}^{N_0-1} x^*[n] e^{-jk\Omega_0 n} = \left(\frac{1}{N_0} \sum_{n=0}^{N_0-1} x[n] e^{jk\Omega_0 n} \right)^* = c_{-k}^* \tag{6.128}$$

Equation (6.128) indicates that if the Fourier coefficients of $x[n]$ are c_k, then the Fourier coefficients of $x^*[n]$ are c_{-k}^*. Setting $x_1[n] = x[n]$ and $x_2[n] = x^*[n]$ in Eq. (6.127), we have $d_k = c_k$ and $e_k = c_{-k}^*$ (or $e^{-k} = c_k^*$) and we obtain

$$\frac{1}{N_0} \sum_{n=0}^{N_0-1} x[n] x^*[n] = \sum_{k=0}^{N_0-1} c_k c_k^* \tag{6.129}$$

and
$$\frac{1}{N_0} \sum_{n=0}^{N_0-1} |x[n]|^2 = \sum_{k=0}^{N_0-1} |c_k|^2$$

(b) From Fig. 6-7 and the results from Prob. 6.3, we have

$$\frac{1}{N_0}\sum_{n=0}^{N_0-1}|x[n]|^2 = \frac{1}{4}(0+1^2+2^2+3^2)=\frac{14}{4}=\frac{7}{2}$$

$$\sum_{n=0}^{N_0-1}|c_k|^2 = \left(\frac{3}{2}\right)^2+\left[\left(\frac{1}{2}\right)^2+\left(\frac{1}{2}\right)^2\right]+\left(-\frac{1}{2}\right)^2+\left[\left(\frac{1}{2}\right)^2+\left(\frac{1}{2}\right)^2\right]=\frac{14}{4}=\frac{7}{2}$$

and Parseval's identity is verified.

Fourier Transform

6.11. Find the Fourier transform of

$$x[n] = -a^n u[-n-1] \qquad a \text{ real}$$

From Eq. (4.12) the z-transform of $x[n]$ is given by

$$X(z)=\frac{1}{1-az^{-1}} \qquad |z|<|a|$$

Thus, $X(e^{j\Omega})$ exists for $|a|>1$ because the ROC of $X(z)$ then contains the unit circle. Thus,

$$X(\Omega)=X(e^{j\Omega})=\frac{1}{1-ae^{-j\Omega}} \qquad |a|>1 \tag{6.130}$$

6.12. Find the Fourier transform of the rectangular pulse sequence (Fig. 6-10)

$$x[n] = u[n] - u[n-N]$$

Using Eq. (1.90), the z-transform of $x[n]$ is given by

$$X(z)=\sum_{n=0}^{N-1}z^n=\frac{1-z^N}{1-z} \qquad |z|>0 \tag{6.131}$$

Thus, $X(e^{j\Omega})$ exists because the ROC of $X(z)$ includes the unit circle. Hence,

$$X(\Omega)=X(e^{j\Omega})=\frac{1-e^{-j\Omega N}}{1-e^{-j\Omega}}=\frac{e^{-j\Omega N/2}(e^{j\Omega N/2}-e^{-j\Omega N/2})}{e^{-j\Omega/2}(e^{j\Omega/2}-e^{-j\Omega/2})}$$

$$=e^{-j\Omega(N-1)/2}\frac{\sin(\Omega N/2)}{\sin(\Omega/2)} \tag{6.132}$$

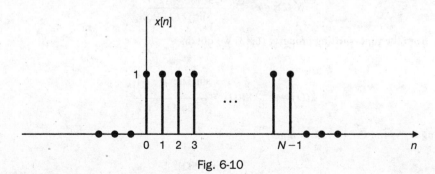

Fig. 6-10

6.13. Verify the time-shifting property (6.43); that is,

$$x[n-n_0] \leftrightarrow e^{-j\Omega n_0}X(\Omega)$$

By definition (6.27)

$$\mathcal{F}\{x[n-n_0]\}=\sum_{n=-\infty}^{\infty}x[n-n_0]e^{-j\Omega n}$$

By the change of variable $m = n - n_0$, we obtain

$$\mathcal{F}\{x[n-n_0]\} = \sum_{m=-\infty}^{\infty} x[m] e^{-j\Omega(m+n_0)}$$

$$= e^{-j\Omega n_0} \sum_{m=-\infty}^{\infty} x[m] e^{-j\Omega m} = e^{-j\Omega n_0} X(\Omega)$$

Hence,

$$x[n-n_0] \leftrightarrow e^{-j\Omega n_0} X(\Omega)$$

6.14. (*a*)   Find the Fourier transform $X(\Omega)$ of the rectangular pulse sequence shown in Fig. 6-11(*a*).

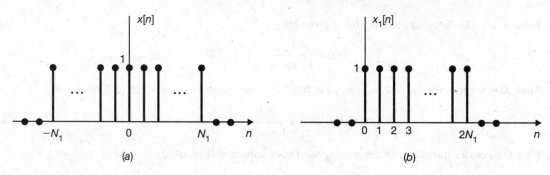

Fig. 6-11

(*b*)   Plot $X(\Omega)$ for $N_1 = 4$ and $N_1 = 8$.

(*a*)   From Fig. 6-11 we see that

$$x[n] = x_1[n + N_1]$$

where $x_1[n]$ is shown in Fig. 6-11(*b*). Setting $N = 2N_1 + 1$ in Eq. (6.132), we have

$$X_1(\Omega) = e^{-j\Omega N_1} \frac{\sin\left[\Omega\left(N_1 + \frac{1}{2}\right)\right]}{\sin(\Omega/2)}$$

Now, from the time-shifting property (6.43) we obtain

$$X(\Omega) = e^{j\Omega N_1} X_1(\Omega) = \frac{\sin\left[\Omega\left(N_1 + \frac{1}{2}\right)\right]}{\sin(\Omega/2)} \tag{6.133}$$

(*b*)   Setting $N_1 = 4$ in Eq. (6.133), we get

$$X(\Omega) = \frac{\sin(4.5\Omega)}{\sin(0.5\Omega)}$$

which is plotted in Fig. 6-12(*a*). Similarly, for $N_1 = 8$ we get

$$X(\Omega) = \frac{\sin(8.5\Omega)}{\sin(0.5\Omega)}$$

which is plotted in Fig. 6-12(*b*).

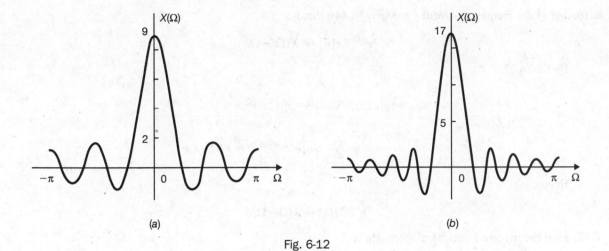

Fig. 6-12

6.15. (a) Find the inverse Fourier transform $x[n]$ of the rectangular pulse spectrum $X(\Omega)$ defined by [Fig. 6-13(a)]

$$X(\Omega) = \begin{cases} 1 & |\Omega| \leq W \\ 0 & W < |\Omega| \leq \pi \end{cases}$$

(b) Plot $x[n]$ for $W = \pi/4$.

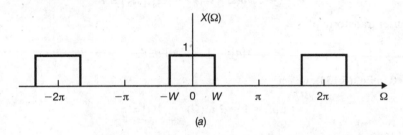

(a)

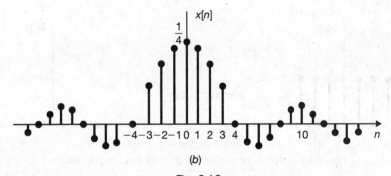

(b)

Fig. 6-13

(a) From Eq. (6.28)

$$x[n] = \frac{1}{2\pi} \int_{-\pi}^{\pi} X(\Omega) e^{j\Omega n} d\Omega = \frac{1}{2\pi} \int_{-W}^{W} e^{j\Omega n} d\Omega = \frac{\sin Wn}{\pi n}$$

Thus, we obtain

$$\frac{\sin Wn}{\pi n} \leftrightarrow X(\Omega) = \begin{cases} 1 & |\Omega| \leq W \\ 0 & W < |\Omega| \leq \pi \end{cases} \tag{6.134}$$

(b) The sequence $x[n]$ is plotted in Fig. 6-13(b) for $W = \pi/4$.

6.16. Verify the frequency-shifting property (6.44); that is,

$$e^{j\Omega_0 n}x[n] \leftrightarrow X(\Omega - \Omega_0)$$

By Eq. (6.27)

$$\mathcal{F}\{e^{j\Omega_0 n}x[n]\} = \sum_{n=-\infty}^{\infty} e^{j\Omega_0 n}x[n]e^{-j\Omega n}$$

$$= \sum_{n=-\infty}^{\infty} x[n]e^{-j(\Omega-\Omega_0)n} = X(\Omega - \Omega_0)$$

Hence,

$$e^{j\Omega_0 n}x[n] \leftrightarrow X(\Omega - \Omega_0)$$

6.17. Find the inverse Fourier transform $x[n]$ of

$$X(\Omega) = 2\pi\delta(\Omega - \Omega_0) \qquad |\Omega|, |\Omega_0| \le \pi$$

From Eqs. (6.28) and (1.22) we have

$$x[n] = \frac{1}{2\pi}\int_{-\pi}^{\pi} 2\pi\,\delta(\Omega - \Omega_0)e^{j\Omega n}\,d\Omega = e^{j\Omega_0 n}$$

Thus, we have

$$e^{j\Omega_0 n} \leftrightarrow 2\pi\,\delta(\Omega - \Omega_0) \qquad |\Omega|,|\Omega_0|\le \pi \tag{6.135}$$

6.18. Find the Fourier transform of

$$x[n] = 1 \qquad \text{all } n$$

Setting $\Omega_0 = 0$ in Eq. (6.135), we get

$$x[n] = 1 \leftrightarrow 2\pi\delta(\Omega) \qquad |\Omega| \le \pi \tag{6.136}$$

Equation (6.136) is depicted in Fig. 6-14.

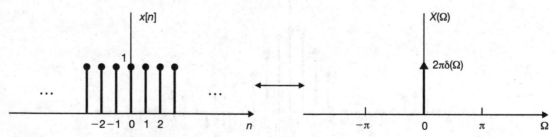

Fig. 6-14 A constant sequence and its Fourier transform.

6.19. Find the Fourier transform of the sinusoidal sequence

$$x[n] = \cos\Omega_0 n \qquad |\Omega_0| \le \pi$$

From Euler's formula we have

$$\cos\Omega_0 n = \frac{1}{2}(e^{j\Omega_0 n} + e^{-j\Omega_0 n})$$

Thus, using Eq. (6.135) and the linearity property (6.42), we get

$$X(\Omega) = \pi[\delta(\Omega - \Omega_0) + \delta(\Omega + \Omega_0)] \qquad |\Omega|, |\Omega_0| \le \pi$$

which is illustrated in Fig. 6-15. Thus,

$$\cos \Omega_0 n \leftrightarrow \pi[\delta(\Omega - \Omega_0) + \delta(\Omega + \Omega_0)] \qquad |\Omega|, |\Omega_0| \leq \pi \tag{6.137}$$

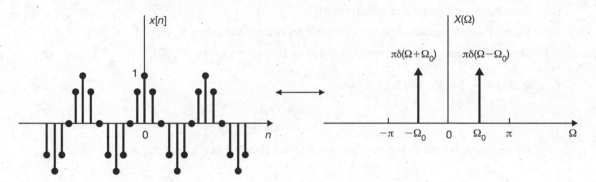

Fig. 6-15 A cosine sequence and its Fourier transform.

6.20. Verify the conjugation property (6.45); that is,

$$x^*[n] \leftrightarrow X^*(-\Omega)$$

From Eq. (6.27)

$$\mathcal{F}\{x^*[n]\} = \sum_{n=-\infty}^{\infty} x^*[n]\, e^{-j\Omega n} = \left(\sum_{n=-\infty}^{\infty} x[n]\, e^{j\Omega n} \right)^*$$

$$= \left(\sum_{n=-\infty}^{\infty} x[n]\, e^{-j(-\Omega)n} \right)^* = X^*(-\Omega)$$

Hence,

$$x^*[n] \leftrightarrow X^*(-\Omega)$$

6.21. Verify the time-scaling property (6.49); that is,

$$x_{(m)}[n] \leftrightarrow X(m\Omega)$$

From Eq. (6.48)

$$x_{(m)}[n] = \begin{cases} x[n/m] = x[k] & \text{if } n = km,\, k = \text{integer} \\ 0 & \text{if } n \neq km \end{cases}$$

Then, by Eq. (6.27)

$$\mathcal{F}\{x_{(m)}[n]\} = \sum_{n=-\infty}^{\infty} x_{(m)}[n]\, e^{-j\Omega n}$$

Changing the variable $n = km$ on the right-hand side of the above expression, we obtain

$$\mathcal{F}\{x_{(m)}[n]\} = \sum_{k=-\infty}^{\infty} x_{(m)}[km]\, e^{-j\Omega km} = \sum_{k=-\infty}^{\infty} x[k]\, e^{-j(m\Omega)k} = X(m\Omega)$$

Hence,

$$x_{(m)}[n] \leftrightarrow X(m\Omega)$$

6.22. Consider the sequence $x[n]$ defined by

$$x[n] = \begin{cases} 1 & |n| \leq 2 \\ 0 & \text{otherwise} \end{cases}$$

(a)   Sketch $x[n]$ and its Fourier transform $X(\Omega)$.

(b)   Sketch the time-scaled sequence $x_{(2)}[n]$ and its Fourier transform $X_{(2)}(\Omega)$.

(c)   Sketch the time-scaled sequence $x_{(3)}[n]$ and its Fourier transform $X_{(3)}(\Omega)$.

(a)   Setting $N_1 = 2$ in Eq. (6.133), we have

$$X(\Omega) = \frac{\sin(2.5\Omega)}{\sin(0.5\Omega)} \tag{6.138}$$

The sequence $x[n]$ and its Fourier transform $X(\Omega)$ are sketched in Fig. 6-16(a).

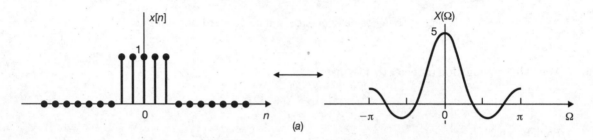

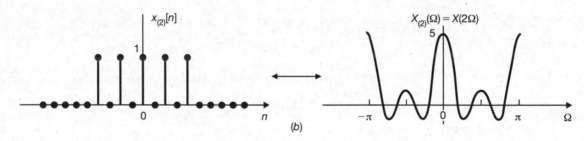

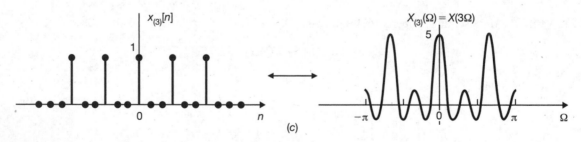

Fig. 6-16

(b)   From Eqs. (6.49) and (6.138) we have

$$X_{(2)}(\Omega) = X(2\Omega) = \frac{\sin(5\Omega)}{\sin(\Omega)}$$

The time-scaled sequence $x_{(2)}[n]$ and its Fourier transform $X_{(2)}(\Omega)$ are sketched in Fig. 6-16(b).

(c)   In a similar manner we get

$$X_{(3)}(\Omega) = X(3\Omega) = \frac{\sin(7.5\Omega)}{\sin(1.5\Omega)}$$

The time-scaled sequence $x_{(3)}[n]$ and its Fourier transform $X_{(3)}(\Omega)$ are sketched in Fig. 6-16(c).

6.23. Verify the differentiation in frequency property (6.55); that is,

$$nx[n] \leftrightarrow j \frac{dX(\Omega)}{d\Omega}$$

From definition (6.27)

$$X(\Omega) = \sum_{n=-\infty}^{\infty} x[n] e^{-j\Omega n}$$

Differentiating both sides of the above expression with respect to Ω and interchanging the order of differentiation and summation, we obtain

$$\frac{dX(\Omega)}{d\Omega} = \frac{d}{d\Omega} \left(\sum_{n=-\infty}^{\infty} x[n] e^{-j\Omega n} \right) = \sum_{n=-\infty}^{\infty} x[n] \frac{d}{d\Omega} (e^{-j\Omega n})$$

$$= -j \sum_{n=-\infty}^{\infty} nx[n] e^{-j\Omega n}$$

Multiplying both sides by j, we see that

$$\mathcal{F}\{nx[n]\} = \sum_{n=-\infty}^{\infty} nx[n] e^{-j\Omega n} = j \frac{dX(\Omega)}{d\Omega}$$

Hence,

$$nx[n] \leftrightarrow j \frac{dX(\Omega)}{d\Omega}$$

6.24. Verify the convolution theorem (6.58); that is,

$$x_1[n] * x_2[n] \leftrightarrow X_1(\Omega) X_2(\Omega)$$

By definitions (2.35) and (6.27), we have

$$\mathcal{F}\{x_1[n] * x_2[n]\} = \sum_{n=-\infty}^{\infty} \left(\sum_{k=-\infty}^{\infty} x_1[k] x_2[n-k] \right) e^{-j\Omega n}$$

Changing the order of summation, we get

$$\mathcal{F}\{x_1[n] * x_2[n]\} = \sum_{k=-\infty}^{\infty} x_1[k] \left(\sum_{n=-\infty}^{\infty} x_2[n-k] e^{-j\Omega n} \right)$$

By the time-shifting property Eq. (6.43)

$$\sum_{n=-\infty}^{\infty} x_2[n-k] e^{-j\Omega n} = e^{-j\Omega k} X_2(\Omega)$$

Thus, we have

$$\mathcal{F}\{x_1[n] * x_2[n]\} = \sum_{k=-\infty}^{\infty} x_1[k] e^{-j\Omega k} X_2(\Omega)$$

$$= \left(\sum_{k=-\infty}^{\infty} x_1[k] e^{-j\Omega k} \right) X_2(\Omega) = X_1(\Omega) X_2(\Omega)$$

Hence,

$$x_1[n] * x_2[n] \leftrightarrow X_1(\Omega) X_2(\Omega)$$

6.25. Using the convolution theorem (6.58), find the inverse Fourier transform $x[n]$ of

$$X(\Omega) = \frac{1}{\left(1 - ae^{-j\Omega}\right)^2} \qquad |a| < 1$$

From Eq. (6.37) we have

$$a^n u[n] \leftrightarrow \frac{1}{1 - ae^{-j\Omega}} \qquad |a| < 1$$

Now

$$X(\Omega) = \frac{1}{\left(1 - ae^{-j\Omega}\right)^2} = \left(\frac{1}{1 - ae^{-j\Omega}}\right)\left(\frac{1}{1 - ae^{-j\Omega}}\right)$$

Thus, by the convolution theorem Eq. (6.58) we get

$$x[n] = a^n u[n] * a^n u[n] = \sum_{k=-\infty}^{\infty} a^k u[k] a^{n-k} u[n-k]$$

$$= a^n \sum_{k=0}^{\infty} 1 = (n+1) a^n u[n]$$

Hence,

$$(n+1) a^n u[n] \leftrightarrow \frac{1}{\left(1 - ae^{-j\Omega}\right)^2} \qquad |a| < 1 \tag{6.139}$$

6.26. Verify the multiplication property (6.59); that is,

$$x_1[n] x_2[n] \leftrightarrow \frac{1}{2\pi} X_1(\Omega) \otimes X_2(\Omega)$$

Let $x[n] = x_1[n] x_2[n]$. Then by definition (6.27)

$$X(\Omega) = \sum_{n=-\infty}^{\infty} x_1[n] x_2[n] e^{-j\Omega n}$$

By Eq. (6.28)

$$x_1[n] = \frac{1}{2\pi} \int_{2\pi} X_1(\theta) e^{j\theta n} \, d\theta$$

Then

$$X(\Omega) = \sum_{n=-\infty}^{\infty} \left[\frac{1}{2\pi} \int_{2\pi} X_1(\theta) e^{j\theta n} \, d\theta \right] x_2[n] e^{-j\Omega n}$$

Interchanging the order of summation and integration, we get

$$X(\Omega) = \frac{1}{2\pi} \int_{2\pi} X_1(\theta) \left(\sum_{n=-\infty}^{\infty} x_2[n] e^{-j(\Omega - \theta)n} \right) d\theta$$

$$= \frac{1}{2\pi} \int_{2\pi} X_1(\theta) X_2(\Omega - \theta) \, d\theta = \frac{1}{2\pi} X_1(\Omega) \otimes X_2(\Omega)$$

Hence,

$$x_1[n] x_2[n] \leftrightarrow \frac{1}{2\pi} X_1(\Omega) \otimes X_2(\Omega)$$

6.27. Verify the properties (6.62), (6.63a), and (6.63b); that is, if $x[n]$ is real and

$$x[n] = x_e[n] + x_o[n] \leftrightarrow X(\Omega) = A(\Omega) + jB(\Omega) \tag{6.140}$$

where $x_e[n]$ and $x_o[n]$ are the even and odd components of $x[n]$, respectively, then

$$X(-\Omega) = X^*(\Omega)$$

$$x_e[n] \leftrightarrow \operatorname{Re}\{X(\Omega)\} = A(\Omega)$$

$$x_o[n] \leftrightarrow j \operatorname{Im}\{X(\Omega)\} = jB(\Omega)$$

If $x[n]$ is real, then $x^*[n], = x[n]$, and by Eq. (6.45) we have

$$x^*[n] \leftrightarrow X^*(-\Omega)$$

from which we get

$$X(\Omega) = X^*(-\Omega) \quad \text{or} \quad X(-\Omega) = X^*(\Omega)$$

Next, using Eq. (6.46) and Eqs. (1.2) and (1.3), we have

$$x[-n] = x_e[n] - x_o[n] \leftrightarrow X(-\Omega) = X^*(\Omega) = A(\Omega) - jB(\Omega) \qquad (6.141)$$

Adding (subtracting) Eq. (6.141) to (from) Eq. (6.140), we obtain

$$x_e[n] \leftrightarrow A(\Omega) = \text{Re}\{X(\Omega)\}$$
$$x_o[n] \leftrightarrow jB(\Omega) = j\,\text{Im}\{X(\Omega)\}$$

6.28. Show that

$$u[n] \leftrightarrow \pi\delta(\Omega) + \frac{1}{1 - e^{-j\Omega}} \qquad |\Omega| \le \pi \qquad (6.142)$$

Let

$$u[n] \leftrightarrow X(\Omega)$$

Now, note that

$$\delta[n] = u[n] - u[n-1]$$

Taking the Fourier transform of both sides of the above expression and by Eqs. (6.36) and (6.43), we have

$$1 = (1 - e^{-j\Omega})\,X(\Omega)$$

Noting that $(1 - e^{-j\Omega}) = 0$ for $\Omega = 0$, $X(\Omega)$ must be of the form

$$X(\Omega) = A\,\delta(\Omega) + \frac{1}{1 - e^{-j\Omega}} \qquad |\Omega| \le \pi$$

where A is a constant. To determine A we proceed as follows. From Eq. (1.5) the even component of $u[n]$ is given by

$$u_e[n] = \frac{1}{2} + \frac{1}{2}\delta[n]$$

Then the odd component of $u[n]$ is given by

$$u_o[n] = u[n] - u_e[n] = u[n] - \frac{1}{2} - \frac{1}{2}\delta[n]$$

and

$$\mathcal{F}\{u_o[n]\} = A\,\delta(\Omega) + \frac{1}{1 - e^{-j\Omega}} - \pi\,\delta(\Omega) - \frac{1}{2}$$

From Eq. (6.63b) the Fourier transform of an odd real sequence must be purely imaginary. Thus, we must have $A = \pi$, and

$$u[n] \leftrightarrow \pi\,\delta(\Omega) + \frac{1}{1 - e^{-j\Omega}} \qquad |\Omega| \le \pi$$

6.29. Verify the accumulation property (6.57); that is,

$$\sum_{k=-\infty}^{n} x[k] \leftrightarrow \pi X(0)\,\delta(\Omega) + \frac{1}{1 - e^{-j\Omega}}\,X(\Omega) \qquad |\Omega| \le \pi$$

From Eq. (2.132)

$$\sum_{k=-\infty}^{n} x[k] = x[n] * u[n]$$

Thus, by the convolution theorem (6.58) and Eq. (6.142) we get

$$\sum_{k=-\infty}^{n} x[k] \leftrightarrow X(\Omega)\left[\pi\,\delta(\Omega) + \frac{1}{1-e^{-j\Omega}}\right] \qquad |\Omega| \leq \pi$$

$$= \pi X(0)\,\delta(\Omega) + \frac{1}{1-e^{-j\Omega}}\,X(\Omega)$$

since $X(\Omega)\delta(\Omega) = X(0)\delta(\Omega)$ by Eq. (1.25).

6.30. Using the accumulation property (6.57) and Eq. (1.50), find the Fourier transform of $u[n]$.

From Eq. (1.50)

$$u[n] = \sum_{k=-\infty}^{n} \delta[k]$$

Now, from Eq. (6.36) we have

$$\delta[n] \leftrightarrow 1$$

Setting $x[k] = \delta[k]$ in Eq. (6.57), we have

$$x[n] = \delta[n] \leftrightarrow X(\Omega) = 1 \qquad \text{and} \qquad X(0) = 1$$

and

$$u[n] = \sum_{k=-\infty}^{n} \delta[k] \leftrightarrow \pi\,\delta(\Omega) + \frac{1}{1-e^{-j\Omega}} \qquad |\Omega| \leq \pi$$

Frequency Response

6.31. A causal discrete-time LTI system is described by

$$y[n] - \frac{3}{4}y[n-1] + \frac{1}{8}y[n-2] = x[n] \qquad\qquad (6.143)$$

where $x[n]$ and $y[n]$ are the input and output of the system, respectively (Prob. 4.32).

(a) Determine the frequency response $H(\Omega)$ of the system.

(b) Find the impulse response $h[n]$ of the system.

(a) Taking the Fourier transform of Eq. (6.143), we obtain

$$Y(\Omega) - \frac{3}{4}e^{-j\Omega}\,Y(\Omega) + \frac{1}{8}e^{-j2\Omega}Y(\Omega) = X(\Omega)$$

or

$$\left(1 - \frac{3}{4}e^{-j\Omega} + \frac{1}{8}e^{-j2\Omega}\right)Y(\Omega) = X(\Omega)$$

Thus,

$$H(\Omega) = \frac{Y(\Omega)}{X(\Omega)} = \frac{1}{1 - \dfrac{3}{4}e^{-j\Omega} + \dfrac{1}{8}e^{-j2\Omega}} = \frac{1}{\left(1 - \dfrac{1}{2}e^{-j\Omega}\right)\left(1 - \dfrac{1}{4}e^{-j\Omega}\right)}$$

(b) Using partial-fraction expansions, we have

$$H(\Omega) = \frac{1}{\left(1 - \dfrac{1}{2}e^{-j\Omega}\right)\left(1 - \dfrac{1}{4}e^{-j\Omega}\right)} = \frac{2}{1 - \dfrac{1}{2}e^{-j\Omega}} - \frac{1}{1 - \dfrac{1}{4}e^{-j\Omega}}$$

Taking the inverse Fourier transform of $H(\Omega)$, we obtain

$$h[n] = \left[2\left(\frac{1}{2}\right)^n - \left(\frac{1}{4}\right)^n \right] u[n]$$

which is the same result obtained in Prob. 4.32(*b*).

6.32. Consider a discrete-time LTI system described by

$$y[n] - \frac{1}{2} y[n-1] = x[n] + \frac{1}{2} x[n-1] \tag{6.144}$$

(*a*) Determine the frequency response $H(\Omega)$ of the system.

(*b*) Find the impulse response $h[n]$ of the system.

(*c*) Determine its response $y[n]$ to the input

$$x[n] = \cos \frac{\pi}{2} n$$

(*a*) Taking the Fourier transform of Eq. (6.144), we obtain

$$Y(\Omega) - \frac{1}{2} e^{-j\Omega} Y(\Omega) = X(\Omega) + \frac{1}{2} e^{-j\Omega} X(\Omega)$$

Thus,

$$H(\Omega) = \frac{Y(\Omega)}{X(\Omega)} = \frac{1 + \dfrac{1}{2} e^{-j\Omega}}{1 - \dfrac{1}{2} e^{-j\Omega}}$$

(*b*)
$$H(\Omega) = \frac{1}{1 - \dfrac{1}{2} e^{-j\Omega}} + \frac{1}{2} \frac{e^{-j\Omega}}{1 - \dfrac{1}{2} e^{-j\Omega}}$$

Taking the inverse Fourier transform of $H(\Omega)$, we obtain

$$h[n] = \left(\frac{1}{2}\right)^n u[n] + \frac{1}{2} \left(\frac{1}{2}\right)^{n-1} u[n-1] = \begin{cases} 1 & n = 0 \\ \left(\dfrac{1}{2}\right)^{n-1} & n \geq 1 \end{cases}$$

(*c*) From Eq. (6.137)

$$X(\Omega) = \pi \left[\delta\left(\Omega - \frac{\pi}{2}\right) + \delta\left(\Omega + \frac{\pi}{2}\right) \right] \qquad |\Omega| \leq \pi$$

Then

$$Y(\Omega) = X(\Omega) H(\Omega) = \pi \left[\delta\left(\Omega - \frac{\pi}{2}\right) + \delta\left(\Omega + \frac{\pi}{2}\right) \right] \frac{1 + \dfrac{1}{2} e^{-j\Omega}}{1 - \dfrac{1}{2} e^{-j\Omega}}$$

$$= \pi \left(\frac{1 + \dfrac{1}{2} e^{-j\pi/2}}{1 - \dfrac{1}{2} e^{-j\pi/2}} \right) \delta\left(\Omega - \frac{\pi}{2}\right) + \pi \left(\frac{1 + \dfrac{1}{2} e^{+j\pi/2}}{1 - \dfrac{1}{2} e^{+j\pi/2}} \right) \delta\left(\Omega + \frac{\pi}{2}\right)$$

$$= \pi \left(\frac{1 - j\dfrac{1}{2}}{1 + j\dfrac{1}{2}} \right) \delta\left(\Omega - \frac{\pi}{2}\right) + \pi \left(\frac{1 + j\dfrac{1}{2}}{1 - j\dfrac{1}{2}} \right) \delta\left(\Omega + \frac{\pi}{2}\right)$$

$$= \pi \, \delta\left(\Omega - \frac{\pi}{2}\right) e^{-j2\tan^{-1}(1/2)} + \pi \, \delta\left(\Omega + \frac{\pi}{2}\right) e^{j2\tan^{-1}(1/2)}$$

Taking the inverse Fourier transform of $Y(\Omega)$ and using Eq. (6.135), we get

$$y[n] = \frac{1}{2} e^{j(\pi/2)n} e^{-j2\tan^{-1}(1/2)} + \frac{1}{2} e^{-j(\pi/2)} e^{j2\tan^{-1}(1/2)}$$

$$= \cos\left(\frac{\pi}{2}n - 2\tan^{-1}\frac{1}{2}\right)$$

6.33. Consider a discrete-time LTI system with impulse response

$$h[n] = \frac{\sin(\pi n / 4)}{\pi n}$$

Find the output $y[n]$ if the input $x[n]$ is a periodic sequence with fundamental period $N_0 = 5$ as shown in Fig. 6-17.

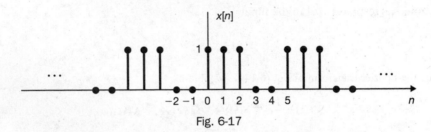

Fig. 6-17

From Eq. (6.134) we have

$$H(\Omega) = \begin{cases} 1 & |\Omega| \leq \pi/4 \\ 0 & \pi/4 < |\Omega| \leq \pi \end{cases}$$

Since $\Omega_0 = 2\pi/N_0 = 2\pi/5$ and the filter passes only frequencies in the range $|\Omega| \leq \pi/4$, only the dc term is passed through. From Fig. 6-17 and Eq. (6.11)

$$c_0 = \frac{1}{5}\sum_{n=0}^{4} x[n] = \frac{3}{5}$$

Thus, the output $y[n]$ is given by

$$y[n] = \frac{3}{5} \qquad \text{all } n$$

6.34. Consider the discrete-time LTI system shown in Fig. 6-18.

 (a) Find the frequency response $H(\Omega)$ of the system.

 (b) Find the impulse response $h[n]$ of the system.

 (c) Sketch the magnitude response $|H(\Omega)|$ and the phase response $\theta(\Omega)$.

 (d) Find the 3-dB bandwidth of the system.

 (a) From Fig. 6-18 we have

$$y[n] = x[n] + x[n-1] \tag{6.145}$$

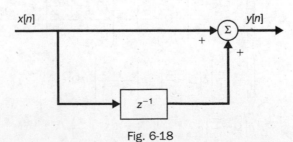

Fig. 6-18

Taking the Fourier transform of Eq. (6.145) and by Eq. (6.77), we have

$$H(\Omega) = \frac{Y(\Omega)}{X(\Omega)} = 1 + e^{-j\Omega} = e^{-j\Omega/2}\left(e^{j\Omega/2} + e^{-j\Omega/2}\right)$$

$$= 2e^{-j\Omega/2}\cos\left(\frac{\Omega}{2}\right) \qquad |\Omega| \le \pi \tag{6.146}$$

(*b*) By the definition of $h[n]$ [Eq. (2.30)] and Eq. (6.145) we obtain

$$h[n] = \delta[n] + \delta[n-1]$$

or

$$h[n] = \begin{cases} 1 & 0 \le n \le 1 \\ 0 & \text{otherwise} \end{cases}$$

(*c*) From Eq. (6.146)

$$|H(\Omega)| = 2\cos\left(\frac{\Omega}{2}\right) \qquad |\Omega| \le \pi$$

and

$$\theta(\Omega) = -\frac{\Omega}{2} \qquad |\Omega| \le \pi$$

which are sketched in Fig. 6-19.

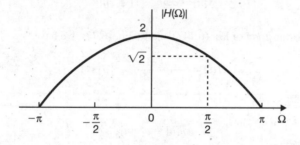

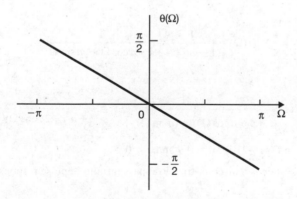

Fig. 6-19

(*d*) Let $\Omega_{3\,\mathrm{db}}$ be the 3-dB bandwidth of the system. Then by definition (Sec. 5.7)

$$\left|H(\Omega_{3\,\mathrm{dB}})\right| = \frac{1}{\sqrt{2}}\left|H(\Omega)\right|_{\max}$$

we obtain

$$\cos\left(\frac{\Omega_{3\,\mathrm{dB}}}{2}\right) = \frac{1}{\sqrt{2}} \qquad \text{and} \qquad \Omega_{3\,\mathrm{dB}} = \frac{\pi}{2}$$

We see that the system is a discrete-time wideband low-pass finite impulse response (FIR) filter (Sec. 2.9C).

6.35. Consider the discrete-time LTI system shown in Fig. 6-20, where a is a constant and $0 < a < 1$.

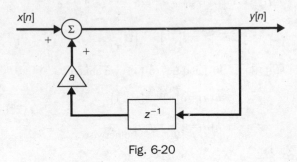

Fig. 6-20

(*a*) Find the frequency response $H(\Omega)$ of the system.

(*b*) Find the impulse response $h[n]$ of the system.

(*c*) Sketch the magnitude response $|H(\Omega)|$ of the system for $a = 0.9$ and $a = 0.5$.

(*a*) From Fig. 6-20 we have

$$y[n] - ay[n-1] = x[n] \tag{6.147}$$

Taking the Fourier transform of Eq. (6.147) and by Eq. (6.77), we have

$$H(\Omega) = \frac{1}{1 - ae^{-j\Omega}} \qquad |a| < 1 \tag{6.148}$$

(*b*) Using Eq. (6.37), we obtain

$$h[n] = a^n u[n]$$

(*c*) From Eq. (6.148)

$$H(\Omega) = \frac{1}{1 - ae^{-j\Omega}} = \frac{1}{1 - a\cos\Omega + ja\sin\Omega}$$

and

$$|H(\Omega)| = \frac{1}{\left[(1 - a\cos\Omega)^2 + (a\sin\Omega)^2\right]^{1/2}} = \frac{1}{(1 + a^2 - 2a\cos\Omega)^{1/2}} \tag{6.149}$$

which is sketched in Fig. 6-21 for $a = 0.9$ and $a = 0.5$.

We see that the system is a discrete-time low-pass infinite impulse response (IIR) filter (Sec. 2.9C).

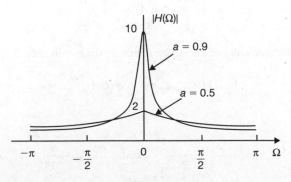

Fig. 6-21

6.36. Let $h_{\text{LPF}}[n]$ be the impulse response of a discrete-time low-pass filter with frequency response $H_{\text{LPF}}(\Omega)$. Show that a discrete-time filter whose impulse response $h[n]$ is given by

$$h[n] = (-1)^n h_{\text{LPF}}[n] \tag{6.150}$$

is a high-pass filter with the frequency response

$$H(\Omega) = H_{\text{LPF}}(\Omega - \pi) \tag{6.151}$$

Since $-1 = e^{j\pi}$, we can write

$$h[n] = (-1)^n h_{\text{LPF}}[n] = e^{j\pi n} h_{\text{LPF}}[n] \tag{6.152}$$

Taking the Fourier transform of Eq. (6.152) and using the frequency-shifting property (6.44), we obtain

$$H(\Omega) = H_{\text{LPF}}(\Omega - \pi)$$

which represents the frequency response of a high-pass filter. This is illustrated in Fig. 6-22.

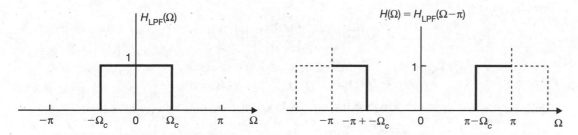

Fig. 6-22 Transformation of a low-pass filter to a high-pass filter.

6.37. Show that if a discrete-time low-pass filter is described by the difference equation

$$y[n] = -\sum_{k=1}^{N} a_k \, y[n-k] + \sum_{k=0}^{M} b_k \, x[n-k] \tag{6.153}$$

then the discrete-time filter described by

$$y[n] = -\sum_{k=1}^{N} (-1)^k a_k \, y[n-k] + \sum_{k=0}^{M} (-1)^k b_k \, x[n-k] \tag{6.154}$$

is a high-pass filter.

Taking the Fourier transform of Eq. (6.153), we obtain the frequency response $H_{\text{LPF}}(\Omega)$ of the low-pass filter as

$$H_{\text{LPF}}(\Omega) = \frac{Y(\Omega)}{X(\Omega)} = \frac{\displaystyle\sum_{k=0}^{M} b_k e^{-jk\Omega}}{1 + \displaystyle\sum_{k=1}^{N} a_k e^{-jk\Omega}} \tag{6.155}$$

If we replace Ω by $(\Omega - \pi)$ in Eq. (6.155), then we have

$$H_{\text{HPF}}(\Omega) = H_{\text{LPF}}(\Omega - \pi) = \frac{\displaystyle\sum_{k=0}^{M} b_k e^{-jk(\Omega-\pi)}}{1 + \displaystyle\sum_{k=1}^{N} a_k e^{-jk(\Omega-\pi)}} = \frac{\displaystyle\sum_{k=0}^{M} b_k (-1)^k e^{-jk\Omega}}{1 + \displaystyle\sum_{k=1}^{N} (-1)^k a_k e^{-jk\Omega}} \tag{6.156}$$

which corresponds to the difference equation

$$y[n] = -\sum_{k=1}^{N} (-1)^k a_k\, y[n-k] + \sum_{k=0}^{M} (-1)^k b_k\, x[n-k]$$

6.38. Convert the discrete-time low-pass filter shown in Fig. 6-18 (Prob. 6.34) to a high-pass filter.

From Prob. 6.34 the discrete-time low-pass filter shown in Fig. 6-18 is described by [Eq. (6.145)]

$$y[n] = x[n] + x[n-1]$$

Using Eq. (6.154), the converted high-pass filter is described by

$$y[n] = x[n] - x[n-1] \tag{6.157}$$

which leads to the circuit diagram in Fig. 6-23. Taking the Fourier transform of Eq. (6.157) and by Eq. (6.77), we have

$$H(\Omega) = 1 - e^{-j\Omega} = e^{-j\Omega/2}(e^{j\Omega/2} - e^{-j\Omega/2})$$
$$= j2e^{-j\Omega/2}\sin\frac{\Omega}{2} = 2e^{j(\pi-\Omega)/2}\sin\frac{\Omega}{2} \qquad |\Omega| \le \pi \tag{6.158}$$

From Eq. (6.158)

$$|H(\Omega)| = 2\left|\sin\left(\frac{\Omega}{2}\right)\right| \qquad |\Omega| \le \pi$$

and

$$\theta(\Omega) = \begin{cases} (\pi - \Omega)/2 & 0 < \Omega < \pi \\ (-\pi - \Omega)/2 & -\pi \le \Omega < 0 \end{cases}$$

which are sketched in Fig. 6-24. We see that the system is a discrete-time high-pass FIR filter.

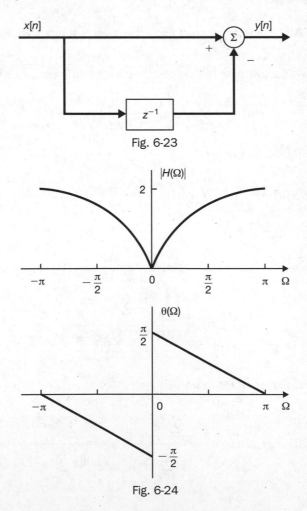

Fig. 6-23

Fig. 6-24

6.39. The system function $H(z)$ of a causal discrete-time LTI system is given by

$$H(z) = \frac{b + z^{-1}}{1 - az^{-1}} \tag{6.159}$$

where a is real and $|a| < 1$. Find the value of b so that the frequency response $H(\Omega)$ of the system satisfies the condition

$$|H(\Omega)| = 1 \qquad \text{all } \Omega \tag{6.160}$$

Such a system is called an *all-pass* filter.

By Eq. (6.34) the frequency response of the system is

$$H(\Omega) = H(z)\big|_{z=e^{j\Omega}} = \frac{b + e^{-j\Omega}}{1 - ae^{-j\Omega}} \tag{6.161}$$

Then, by Eq. (6.160)

$$|H(\Omega)| = \left| \frac{b + e^{-j\Omega}}{1 - ae^{-j\Omega}} \right| = 1$$

which leads to

$$|b + e^{-j\Omega}| = |1 - ae^{-j\Omega}|$$

or
$$|b + \cos\Omega - j\sin\Omega| = |1 - a\cos\Omega + ja\sin\Omega|$$

or
$$1 + b^2 + 2b\cos\Omega = 1 + a^2 - 2a\cos\Omega \tag{6.162}$$

and we see that if $b = -a$, Eq. (6.162) holds for all Ω and Eq. (6.160) is satisfied.

6.40. Let $h[n]$ be the impulse response of an FIR filter so that

$$h[n] = 0 \qquad n < 0, n \geq N$$

Assume that $h[n]$ is real and let the frequency response $H(\Omega)$ be expressed as

$$H(\Omega) = |H(\Omega)| e^{j\theta(\Omega)}$$

(a) Find the phase response $\theta(\Omega)$ when $h[n]$ satisfies the condition [Fig. 6-25(a)]

$$h[n] = h[N - 1 - n] \tag{6.163}$$

(b) Find the phase response $\theta(\Omega)$ when $h[n]$ satisfies the condition [Fig. 6-25(b)]

$$h[n] = -h[N - 1 - n] \tag{6.164}$$

(a) Taking the Fourier transform of Eq. (6.163) and using Eqs. (6.43), (6.46), and (6.62), we obtain

$$H(\Omega) = H^*(\Omega)\, e^{-j(N-1)\Omega}$$

or
$$|H(\Omega)|\, e^{j\theta(\Omega)} = |H(\Omega)|\, e^{-j\theta(\Omega)}\, e^{-j(N-1)\Omega}$$

Thus,

$$\theta(\Omega) = -\theta(\Omega) - (N-1)\,\Omega$$

and
$$\theta(\Omega) = -\frac{1}{2}(N-1)\,\Omega \tag{6.165}$$

which indicates that the phase response is linear.

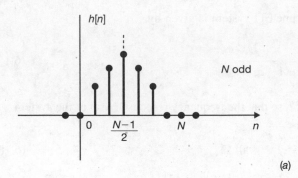

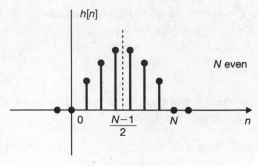

(a)

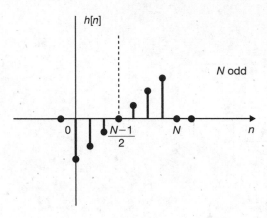

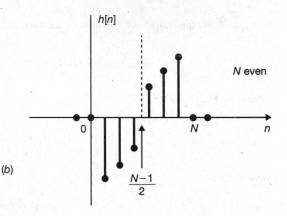

(b)

Fig. 6-25

(b)   Similarly, taking the Fourier transform of Eq. (6.164), we get

$$H(\Omega) = -H^*(\Omega)\, e^{-j(N-1)\Omega}$$

or

$$|H(\Omega)|\, e^{j\theta(\Omega)} = |H(\Omega)|\, e^{j\pi}\, e^{-j\theta(\Omega)}\, e^{-j(N-1)\Omega}$$

Thus,

$$\theta(\Omega) = \pi - \theta(\Omega) - (N-1)\,\Omega$$

and

$$\theta(\Omega) = \frac{\pi}{2} - \frac{1}{2}(N-1)\,\Omega \tag{6.166}$$

which indicates that the phase response is also linear.

6.41. Consider a three-point moving-average discrete-time filter described by the difference equation

$$y[n] = \frac{1}{3}\{x[n] + x[n-1] + x[n-2]\} \tag{6.167}$$

(a)   Find and sketch the impulse response $h[n]$ of the filter.

(b)   Find the frequency response $H(\Omega)$ of the filter.

(c)   Sketch the magnitude response $|H(\Omega)|$ and the phase response $\theta(\Omega)$ of the filter.

(a)   By the definition of $h[n]$ [Eq. (2.30)] we have

$$h[n] = \frac{1}{3}\{\delta[n] + \delta[n-1] + \delta[n-2]\} \tag{6.168}$$

or

$$h[n] = \begin{cases} \dfrac{1}{3} & 0 \le n \le 2 \\ 0 & \text{otherwise} \end{cases}$$

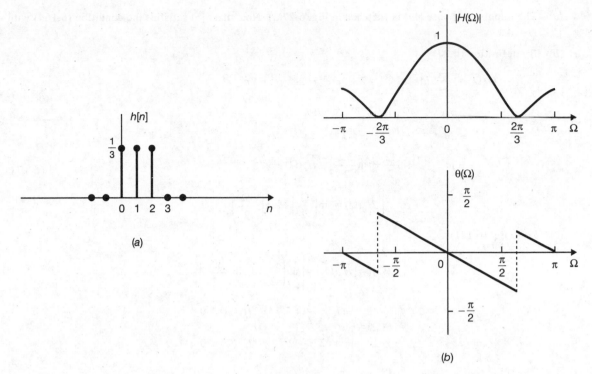

Fig. 6-26

which is sketched in Fig. 6-26(*a*). Note that $h[n]$ satisfies the condition (6.163) with $N = 3$.

(*b*) Taking the Fourier transform of Eq. (6.168), we have

$$H(\Omega) = \frac{1}{3}\{1 + e^{-j\Omega} + e^{-2j\Omega}\}$$

By Eq. (1.90), with $\alpha = e^{-j\Omega}$, we get

$$H(\Omega) = \frac{1}{3}\frac{1 - e^{-j3\Omega}}{1 - e^{-j\Omega}} = \frac{1}{3}\frac{e^{-j3\Omega}(e^{j3\Omega/2} - e^{-j3\Omega/2})}{e^{-j\Omega/2}(e^{j\Omega/2} - e^{-j\Omega/2})}$$

$$= \frac{1}{3}e^{-j\Omega}\frac{\sin(3\Omega/2)}{\sin(\Omega/2)} = H_r(\Omega)e^{-j\Omega} \tag{6.169}$$

where

$$H_r(\Omega) = \frac{1}{3}\frac{\sin(3\Omega/2)}{\sin(\Omega/2)} \tag{6.170}$$

(*c*) From Eq. (6.169)

$$|H(\Omega)| = |H_r(\Omega)| = \frac{1}{3}\left|\frac{\sin(3\Omega/2)}{\sin(\Omega/2)}\right|$$

and

$$\theta(\Omega) = \begin{cases} -\Omega & \text{when } H_r(\Omega) > 0 \\ -\Omega + \pi & \text{when } H_r(\Omega) < 0 \end{cases}$$

which are sketched in Fig. 6-26(*b*). We see that the system is a low-pass FIR filter with linear phase.

6.42. Consider a causal discrete-time FIR filter described by the impulse response

$$h[n] = \{2, 2, -2, -2\}$$

(*a*) Sketch the impulse response $h[n]$ of the filter.

(*b*) Find the frequency response $H(\Omega)$ of the filter.

(*c*) Sketch the magnitude response $|H(\Omega)|$ and the phase response $\theta(\Omega)$ of the filter.

(a) The impulse response $h[n]$ is sketched in Fig. 6-27(a). Note that $h[n]$ satisfies the condition (6.164) with $N=4$.

(b) By definition (6.27)

$$H(\Omega) = \sum_{n=-\infty}^{\infty} h[n]\,e^{-j\Omega n} = 2 + 2e^{-j\Omega} - 2e^{-j2\Omega} - 2e^{-j3\Omega}$$

$$= 2(1 - e^{-j3\Omega}) + 2(e^{-j\Omega} - e^{-j2\Omega})$$

$$= 2e^{-j3\Omega/2}(e^{j3\Omega/2} - e^{-j3\Omega/2}) + 2e^{-j3\Omega/2}(e^{j\Omega/2} - e^{-j\Omega/2})$$

$$= je^{-j3\Omega/2}\left(\sin\frac{\Omega}{2} + \sin\frac{3\Omega}{2}\right) = H_r(\Omega)\,e^{j[(\pi/2)-(3\Omega/2)]} \tag{6.171}$$

where
$$H_r(\Omega) = \sin\left(\frac{\Omega}{2}\right) + \sin\left(\frac{3\Omega}{2}\right)$$

(c) From Eq. (6.171)

$$\left| H(\Omega) \right| = \left| H_r(\Omega) \right| = \left| \sin\left(\frac{\Omega}{2}\right) + \sin\left(\frac{3\Omega}{2}\right) \right|$$

$$\theta(\Omega) = \begin{cases} \pi/2 - \dfrac{3}{2}\Omega & H_r(\Omega) > 0 \\[2mm] -\pi/2 - \dfrac{3}{2}\Omega & H_r(\Omega) < 0 \end{cases}$$

which are sketched in Fig. 6-27(b). We see that the system is a bandpass FIR filter with linear phase.

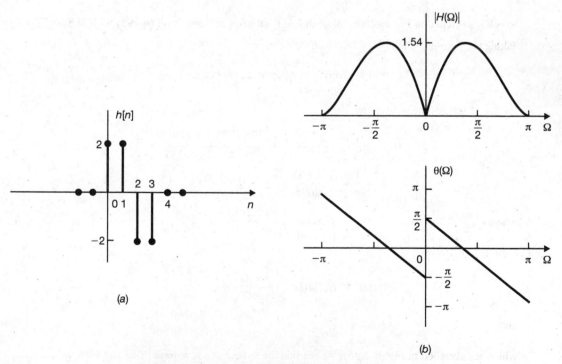

(a)

(b)

Fig. 6-27

Simulation

6.43. Consider the *RC* low-pass filter shown in Fig. 6-28(a) with $RC = 1$.

(a) Construct a discrete-time filter such that

$$h_d[n] = h_c(t)\big|_{t=nT_s} = h_c(nT_s) \tag{6.172}$$

where $h_c(t)$ is the impulse response of the *RC* filter, $h_d[n]$ is the impulse response of the discrete-time filter, and T_s is a positive number to be chosen as part of the design procedures.

(b) Plot the magnitude response $|H_c(\omega)|$ of the *RC* filter and the magnitude response $|H_d(\omega T_s)|$ of the discrete-time filter for $T_s = 1$ and $T_s = 0.1$.

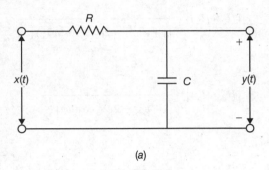

(a)

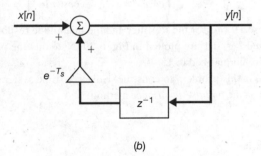

(b)

Fig. 6-28 Simulation of an *RC* filter by the impulse invariance method.

(a) The system function $H_c(s)$ of the *RC* filter is given by (Prob. 3.23)

$$H_c(s) = \frac{1}{s+1} \qquad (6.173)$$

and the impulse response $h_c(t)$ is

$$h_c(t) = e^{-t}u(t) \qquad (6.174)$$

By Eq. (6.172) the corresponding $h_d[n]$ is given by

$$h_d[n] = e^{-nT_s}u[n] = (e^{-T_s})^n\, u[n] \qquad (6.175)$$

Then, taking the z-transform of Eq. (6.175), the system function $H_d(z)$ of the discrete-time filter is given by

$$H_d(z) = \frac{1}{1 - e^{-T_s}z^{-1}}$$

from which we obtain the difference equation describing the discrete-time filter as

$$y[n] - e^{-T_s}y[n-1] = x[n] \qquad (6.176)$$

from which the discrete-time filter that simulates the *RC* filter is shown in Fig. 6-28(b).

(b) By Eq. (5.40)

$$H_c(\omega) = H_c(s)\big|_{s=j\omega} = \frac{1}{j\omega + 1}$$

Then

$$|H_c(\omega)| = \frac{1}{(1+\omega^2)^{1/2}}$$

By Eqs. (6.34) and (6.81)

$$H_d(\omega T_s) = H_d(z)\big|_{z=e^{j\omega T_s}} = \frac{1}{1 - e^{-T_s}e^{-j\omega T_s}}$$

From Eq. (6.149)

$$|H_d(\omega T_s)| = \frac{1}{\left[1 + e^{-2T_s} - 2e^{-T_s}\cos(\omega T_s)\right]^{1/2}}$$

From $T_s = 1$,

$$|H_d(\omega T_s)| = \frac{1}{\left[1 + e^{-2} - 2e^{-1}\cos(\omega)\right]^{1/2}}$$

For $T_s = 0.1$,

$$|H_d(\omega T_s)| = \frac{1}{\left[1 + e^{-0.2} - 2e^{-0.1}\cos(0.1\omega)\right]^{1/2}}$$

The magnitude response $|H_c(\omega)|$ of the *RC* filter and the magnitude response $|H_d(\omega T_s)|$ of the discrete-time filter for $T_s = 1$ and $T_s = 0.1$ are plotted in Fig. 6-29. Note that the plots are scaled such that the magnitudes at $\omega = 0$ are normalized to 1.

The method utilized in this problem to construct a discrete-time system to simulate the continuous-time system is known as the *impulse-invariance* method.

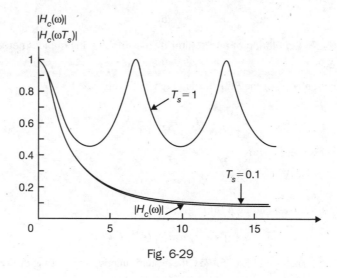

Fig. 6-29

6.44. By applying the impulse-invariance method, determine the frequency response $H_d(\Omega)$ of the discrete-time system to simulate the continuous-time LTI system with the system function

$$H_c(s) = \frac{1}{(s+1)(s+2)}$$

Using the partial-fraction expansion, we have

$$H_c(s) = \frac{1}{s+1} - \frac{1}{s+2}$$

Thus, by Table 3-1 the impulse response of the continuous-time system is

$$h_c(t) = (e^{-t} - e^{-2t})u(t) \tag{6.177}$$

Let $h_d[n]$ be the impulse response of the discrete-time system. Then, by Eq. (6.177)

$$h_d[n] = h_c(nT_s) = (e^{-nT_s} - e^{-2nT_s})u[n]$$

and the system function of the discrete-time system is given by

$$H_d(z) = \frac{1}{1 - e^{-nT_s}z^{-1}} - \frac{1}{1 - e^{-2nT_s}z^{-1}} \tag{6.178}$$

Thus, the frequency response $H_d(\Omega)$ of the discrete-time system is

$$H_d(\Omega) = H_d(z)\big|_{z=e^{j\Omega}} = \frac{1}{1 - e^{-nT_s}e^{-j\Omega}} - \frac{1}{1 - e^{-2nT_s}e^{-j\Omega}} \tag{6.179}$$

Note that if the system function of a continuous-time LTI system is given by

$$H_c(s) = \sum_{k=1}^{N} \frac{A_k}{s + \alpha_k} \tag{6.180}$$

then the impulse-invariance method yields the corresponding discrete-time system with the system function $H_d(z)$ given by

$$H_d(z) = \sum_{k=1}^{N} \frac{A_k}{1 - e^{-\alpha_k nT_s}z^{-1}} \tag{6.181}$$

6.45. A differentiator is a continuous-time LTI system with the system function [Eq. (3.20)]

$$H_c(s) = s \tag{6.182}$$

A discrete-time LTI system is constructed by replacing s in $H_c(s)$ by the following transformation known as the *bilinear transformation*:

$$s = \frac{2}{T_s}\frac{1 - z^{-1}}{1 + z^{-1}} \tag{6.183}$$

to simulate the differentiator. Again T_s in Eq. (6.183) is a positive number to be chosen as part of the design procedure.

(a) Draw a diagram for the discrete-time system.

(b) Find the frequency response $H_d(\Omega)$ of the discrete-time system and plot its magnitude and phase responses.

(a) Let $H_d(z)$ be the system function of the discrete-time system. Then, from Eqs. (6.182) and (6.183) we have

$$H_d(z) = \frac{2}{T_s}\frac{1 - z^{-1}}{1 + z^{-1}} \tag{6.184}$$

Writing $H_d(z)$ as

$$H_d(z) = \frac{2}{T_s}\left(\frac{1}{1 + z^{-1}}\right)(1 - z^{-1})$$

then, from Probs. (6.35) and (6.38) the discrete-time system can be constructed as a cascade connection of two systems as shown in Fig. 6-30(a). From Fig. 6-30(a) it is seen that we can replace two unit-delay elements by one unit-delay element as shown in Fig. 6-30(b).

(b) By Eq. (6.184) the frequency response $H_d(\Omega)$ of the discrete-time system is given by

$$H_d(\Omega) = \frac{2}{T_s}\frac{1 - e^{-j\Omega}}{1 + e^{-j\Omega}} = \frac{2}{T_s}\frac{e^{-j\Omega/2}(e^{j\Omega/2} - e^{-j\Omega/2})}{e^{-j\Omega/2}(e^{j\Omega/2} + e^{-j\Omega/2})}$$

$$= j\frac{2}{T_s}\frac{\sin\Omega/2}{\cos\Omega/2} = j\frac{2}{T_s}\tan\frac{\Omega}{2} = \frac{2}{T_s}\tan\frac{\Omega}{2}e^{j\pi/2} \tag{6.185}$$

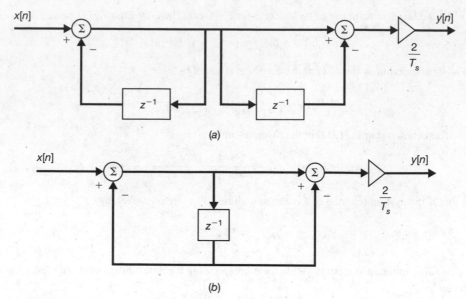

(a)

(b)

Fig. 6-30 Simulation of a differentiator.

Note that when $\Omega \ll 1$, we have

$$H_d(\Omega) = j\frac{2}{T_s}\tan\frac{\Omega}{2} \approx j\frac{\Omega}{T_s} = j\omega \qquad (6.186)$$

If $\Omega = \omega T_s$ (Fig. 6-31).

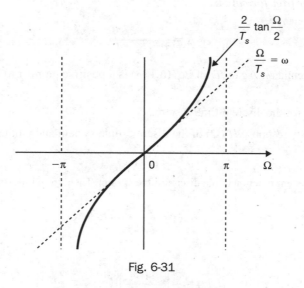

Fig. 6-31

6.46. Consider designing a discrete-time LTI system with system function $H_d(z)$ obtained by applying the bilinear transformation to a continuous-time LTI system with rational system function $H_c(s)$. That is,

$$H_d(z) = H_c(s)\big|_{s = (2/T_s)(1 - z^{-1})/(1 + z^{-1})} \qquad (6.187)$$

Show that a stable, causal continuous-time system will always lead to a stable, causal discrete-time system.

Consider the bilinear transformation of Eq. (6.183)

$$s = \frac{2}{T_s}\frac{1 - z^{-1}}{1 + z^{-1}} \qquad (6.188)$$

Solving Eq. (6.188) for z, we obtain

$$z = \frac{1 + (T_s/2)s}{1 - (T_s/2)s} \qquad (6.189)$$

Setting $s = j\omega$ in Eq. (6.189), we get

$$|z| = \left| \frac{1 + j\omega(T_s/2)}{1 - j\omega(T_s/2)} \right| = 1 \qquad (6.190)$$

Thus, we see that the $j\omega$-axis of the s-plane is transformed into the unit circle of the z-plane. Let

$$z = re^{j\Omega} \qquad \text{and} \qquad s = \sigma + j\omega$$

Then from Eq. (6.188)

$$s = \frac{2}{T_s} \frac{z-1}{z+1} = \frac{2}{T_s} \frac{re^{j\Omega} - 1}{re^{j\Omega} + 1}$$

$$= \frac{2}{T_s} \left(\frac{r^2 - 1}{1 + r^2 + 2r\cos\Omega} + j\frac{2r\sin\Omega}{1 + r^2 + 2r\cos\Omega} \right)$$

Hence,

$$\sigma = \frac{2}{T_s} \frac{r^2 - 1}{1 + r^2 + 2r\cos\Omega} \qquad (6.191a)$$

$$\omega = \frac{2}{T_s} \frac{2r\sin\Omega}{1 + r^2 + 2r\cos\Omega} \qquad (6.191b)$$

From Eq. (6.191a) we see that if $r < 1$, then $\sigma < 0$, and if $r > 1$, then $\sigma > 0$. Consequently, the left-hand plane (LHP) in s maps into the inside of the unit circle in the z-plane, and the right-hand plane (RHP) in s maps into the outside of the unit circle (Fig. 6-32). Thus, we conclude that a stable, causal continuous-time system will lead to a stable, causal discrete-time system with a bilinear transformation (see Sec. 3.6B and Sec. 4.6B). When $r = 1$, then $\sigma = 0$ and

$$\omega = \frac{2}{T_s} \frac{\sin\Omega}{1 + \cos\Omega} = \frac{2}{T_s} \tan\frac{\Omega}{2} \qquad (6.192)$$

or

$$\Omega = 2\tan^{-1}\frac{\omega T_s}{2} \qquad (6.193)$$

From Eq. (6.193) we see that the entire range $-\infty < \omega < \infty$ is mapped only into the range $-\pi \le \Omega \le \pi$.

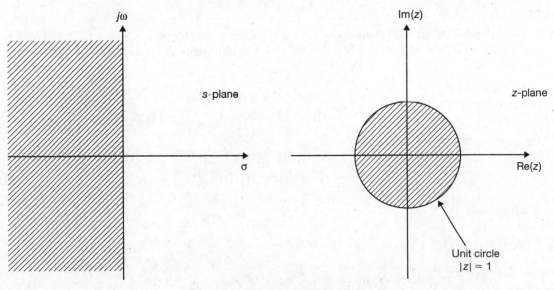

Fig. 6-32 Bilinear transformation.

6.47. Consider the low-pass *RC* filter in Fig. 6-28(*a*). Design a low-pass discrete-time filter by the bilinear transformation method such that its 3-dB bandwidth is $\pi/4$.

Using Eq. (6.192), $\Omega_{3\,\text{dB}} = \pi/4$ corresponds to

$$\omega_{3\,\text{dB}} = \frac{2}{T_s}\tan\frac{\Omega_{3\,\text{dB}}}{2} = \frac{2}{T_s}\tan\frac{\pi}{8} = \frac{0.828}{T_s} \tag{6.194}$$

From Prob. 5.55(*a*), $\omega_{3\,\text{dB}} = 1/RC$. Thus, the system function $H_c(s)$ of the *RC* filter is given by

$$H_c(s) = \frac{0.828/T_s}{s + 0.828/T_s} \tag{6.195}$$

Let $H_d(z)$ be the system function of the desired discrete-time filter. Applying the bilinear transformation (6.183) to Eq. (6.195), we get

$$H_d(z) = \frac{0.828/T_s}{\dfrac{2}{T_s}\dfrac{1-z^{-1}}{1+z^{-1}} + \dfrac{0.828}{T_s}} = \frac{0.293(1+z^{-1})}{1-0.414z^{-1}} \tag{6.196}$$

from which the system in Fig. 6-33 results. The frequency response of the discrete-time filter is

$$H_d(\Omega) = \frac{0.293(1+e^{-j\Omega})}{1-0.414e^{-j\Omega}} \tag{6.197}$$

At $\Omega = 0$, $H_d(0) = 1$, and at $\Omega = \pi/4$, $|H_d(\pi/4)| = 0.707 = 1/\sqrt{2}$, which is the desired response.

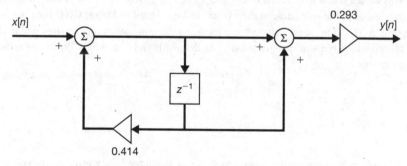

Fig. 6-33 Simulation of an *RC* filter by the bilinear transformation method.

6.48. Let $h[n]$ denote the impulse response of a desired IIR filter with frequency response $H(\Omega)$ and let $h_o[n]$ denote the impulse response of an FIR filter of length N with frequency response $H_o(\Omega)$. Show that when

$$h_o[n] = \begin{cases} h[n] & 0 \le n \le N-1 \\ 0 & \text{otherwise} \end{cases} \tag{6.198}$$

the mean-square error ε^2 defined by

$$\varepsilon^2 = \frac{1}{2\pi}\int_{-\pi}^{\pi}\left| H(\Omega) - H_o(\Omega)\right|^2 d\Omega \tag{6.199}$$

is minimized.

By definition (6.27)

$$H(\Omega) = \sum_{n=-\infty}^{\infty} h[n]e^{-j\Omega n} \quad \text{and} \quad H_o(\Omega) = \sum_{n=-\infty}^{\infty} h_o[n]e^{-j\Omega n}$$

Let
$$E(\Omega) = H(\Omega) - H_o(\Omega) = \sum_{n=-\infty}^{\infty} (h[n] - h_o[n]) e^{-j\Omega n}$$

$$= \sum_{n=-\infty}^{\infty} e[n] e^{-j\Omega n} \tag{6.200}$$

where $e[n] = h[n] - h_o[n]$. By Parseval's theorem (6.66) we have

$$\varepsilon^2 = \frac{1}{2\pi} \int_{-\pi}^{\pi} |E(\Omega)|^2 \, d\Omega = \sum_{n=-\infty}^{\infty} |e[n]|^2 = \sum_{n=-\infty}^{\infty} |h[n] - h_o[n]|^2$$

$$= \sum_{n=0}^{N-1} |h[n] - h_0[n]|^2 + \sum_{n=-\infty}^{-1} |h[n]|^2 + \sum_{n=N}^{\infty} |h[n]|^2 \tag{6.201}$$

The last two terms in Eq. (6.201) are two positive constants. Thus, ε^2 is minimized when

$$h[n] - h_o[n] = 0 \qquad 0 \le n \le N - 1$$

that is,

$$h[n] = h_o[n] \qquad 0 \le n \le N - 1$$

Note that Eq. (6.198) can be expressed as

$$h_o[n] = h[n]w[n] \tag{6.202}$$

where $w[n]$ is known as a rectangular *window function* given by

$$w[n] = \begin{cases} 1 & 0 \le n \le N - 1 \\ 0 & \text{otherwise} \end{cases} \tag{6.203}$$

Discrete Fourier Transform

6.49. Find the N-point DFT of the following sequences $x[n]$:

(a) $x[n] = \delta[n]$

(b) $x[n] = u[n] - u[n - N]$

(a) From definitions (6.92) and (1.45), we have

$$X[k] = \sum_{n=0}^{N-1} \delta[n]w_N^{kn} = 1 \qquad k = 0, 1, \ldots, N - 1$$

Fig. 6-34 shows $x[n]$ and its N-point DFT $X[k]$.

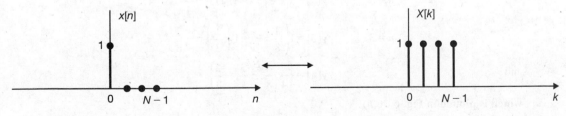

Fig. 6-34

(b) Again from definitions (6.92) and (1.44) and using Eq. (1.90), we obtain

$$X[k] = \sum_{n=0}^{N-1} W_N^{kn} = \frac{1 - W_N^{kN}}{1 - W_N^k} = 0 \qquad k \ne 0$$

since $W_N^{kN} = e^{-j(2\pi/N)kN} = e^{-jk2\pi} = 1$.

$$X[0] = \sum_{n=0}^{N-1} W_N^0 = \sum_{n=0}^{N-1} 1 = N$$

Fig. 6-35 shows $x[n]$ and its N-point DFT $X[k]$.

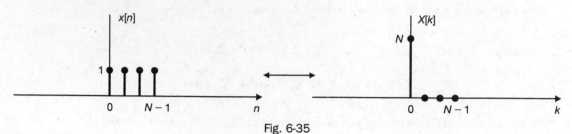

Fig. 6-35

6.50. Consider two sequences $x[n]$ and $h[n]$ of length 4 given by

$$x[n] = \cos\left(\frac{\pi}{2}n\right) \qquad n = 0, 1, 2, 3$$

$$h[n] = \left(\frac{1}{2}\right)^n \qquad n = 0, 1, 2, 3$$

(*a*) Calculate $y[n] = x[n] \otimes h[n]$ by doing the circular convolution directly.

(*b*) Calculate $y[n]$ by DFT.

(*a*) The sequences $x[n]$ and $h[n]$ can be expressed as

$$x[n] = \{1, 0, -1, 0\} \qquad \text{and} \qquad h[n] = \{1, \tfrac{1}{2}, \tfrac{1}{4}, \tfrac{1}{8}\}$$

By Eq. (6.108)

$$y[n] = x[n] \otimes h[n] = \sum_{i=0}^{3} x[i] h[n-i]_{\text{mod}4}$$

The sequences $x[i]$ and $h[n-i]_{\text{mod}4}$ for $n = 0, 1, 2, 3$ are plotted in Fig. 6-36(*a*). Thus, by Eq. (6.108) we get

$$n = 0 \qquad y[0] = 1(1) + (-1)\left(\frac{1}{4}\right) = \frac{3}{4}$$

$$n = 1 \qquad y[1] = 1\left(\frac{1}{2}\right) + (-1)\left(\frac{1}{8}\right) = \frac{3}{8}$$

$$n = 2 \qquad y[2] = 1\left(\frac{1}{4}\right) + (-1)(1) = -\frac{3}{4}$$

$$n = 3 \qquad y[3] = 1\left(\frac{1}{8}\right) + (-1)\left(\frac{1}{2}\right) = -\frac{3}{8}$$

and

$$y[n] = \left\{\frac{3}{4}, \frac{3}{8}, -\frac{3}{4}, -\frac{3}{8}\right\}$$

which is plotted in Fig. 6-36(*b*).

(*b*) By Eq. (6.92)

$$X[k] = \sum_{n=0}^{3} x[n] W_4^{kn} = 1 - W_4^{2k} \qquad k = 0, 1, 2, 3$$

$$H[k] = \sum_{n=0}^{3} h[n] W_4^{kn} = 1 + \frac{1}{2} W_4^k + \frac{1}{4} W_4^{2k} + \frac{1}{8} W_4^{3k} \qquad k = 0, 1, 2, 3$$

Then by Eq. (6.107) the DFT of $y[n]$ is

$$Y[k] = X[k]H[k] = \left(1 - W_4^{2k}\right)\left(1 + \frac{1}{2}W_4^k + \frac{1}{4}W_4^{2k} + \frac{1}{8}W_4^{3k}\right)$$

$$= 1 + \frac{1}{2}W_4^k - \frac{3}{4}W_4^{2k} - \frac{3}{8}W_4^{3k} - \frac{1}{4}W_4^{4k} - \frac{1}{8}W_4^{5k}$$

Since $W_4^{4k} = (W_4^4)^k = 1^k$ and $W_4^{5k} = W_4^{(4+1)k} = W_4^k$, we obtain

$$Y[k] = \frac{3}{4} + \frac{3}{8}W_4^k - \frac{3}{4}W_4^{2k} - \frac{3}{8}W_4^{3k} \qquad k = 0, 1, 2, 3$$

Thus, by the definition of DFT [Eq. (6.92)] we get

$$y[n] = \left\{\frac{3}{4}, \frac{3}{8}, -\frac{3}{4}, -\frac{3}{8}\right\}$$

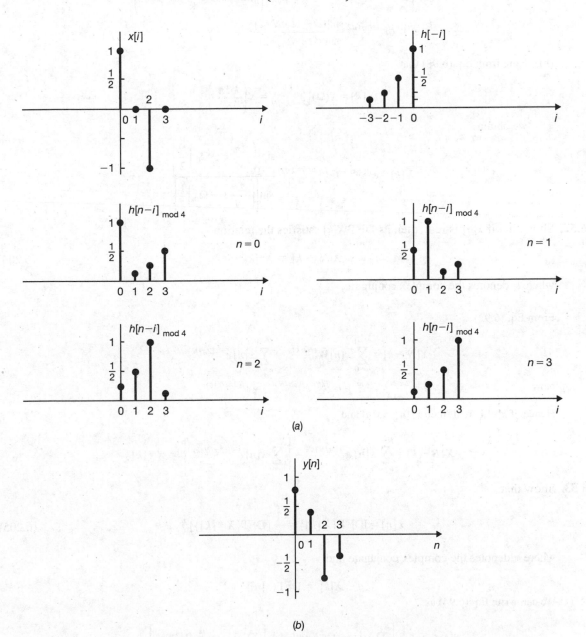

(a)

(b)

Fig. 6-36

6.51. Consider the finite-length complex exponential sequence

$$x[n] = \begin{cases} e^{j\Omega_0 n} & 0 \le n \le N-1 \\ 0 & \text{otherwise} \end{cases}$$

(a) Find the Fourier transform $X(\Omega)$ of $x[n]$.

(b) Find the N-point DFT $X[k]$ of $x[n]$.

(a) From Eq. (6.27) and using Eq. (1.90), we have

$$X(\Omega) = \sum_{n=-\infty}^{\infty} x[n] e^{-j\Omega n} = \sum_{n=0}^{N-1} e^{j\Omega_0 n} e^{-j\Omega n} = \sum_{n=0}^{N-1} e^{-j(\Omega-\Omega_0)n}$$

$$= \frac{1-e^{-j(\Omega-\Omega_0)N}}{1-e^{-j(\Omega-\Omega_0)}} = \frac{e^{-j(\Omega-\Omega_0)N/2} \left(e^{j(\Omega-\Omega_0)N/2} - e^{-j(\Omega-\Omega_0)N/2} \right)}{e^{-j(\Omega-\Omega_0)/2} \left(e^{j(\Omega-\Omega_0)/2} - e^{-j(\Omega-\Omega_0)/2} \right)}$$

$$= e^{j(\Omega-\Omega_0)(N-1)/2} \frac{\sin[(\Omega-\Omega_0)N/2]}{\sin[(\Omega-\Omega_0)/2]}$$

(b) Note from Eq. (6.98) that

$$X[k] = X(\Omega)\big|_{\Omega=k2\pi/N} = X\left(\frac{k2\pi}{N}\right)$$

we obtain

$$X[k] = e^{j[(2\pi/N)k-\Omega_0][(N-1)/2]} \frac{\sin\left[\left(\dfrac{2\pi}{N}k-\Omega_0\right)\dfrac{N}{2}\right]}{\sin\left[\left(\dfrac{2\pi}{N}k-\Omega_0\right)\dfrac{1}{2}\right]}$$

6.52. Show that if $x[n]$ is real, then its DFT $X[k]$ satisfies the relation

$$X[N-k] = X^*[k] \qquad (6.204)$$

where $*$ denotes the complex conjugate.

From Eq. (6.92)

$$X[N-k] = \sum_{n=0}^{N-1} x[n] W_N^{(N-k)n} = \sum_{n=0}^{N-1} x[n] e^{-j(2\pi/N)(N-k)n}$$

Now $e^{-j(2\pi/N)(N-k)n} = e^{-j2\pi n} e^{j(2\pi/N)kn} = e^{j(2\pi/N)kn}$

Hence, if $x[n]$ is real, then $x^*[n] = x[n]$ and

$$X[N-k] = \sum_{n=0}^{N-1} x[n] e^{j(2\pi/N)kn} = \left[\sum_{n=0}^{N-1} x[n] e^{-j(2\pi/N)kn} \right]^* = X^*[k]$$

6.53. Show that

$$x[n] = \text{IDFT}\{X[k]\} = \frac{1}{N}[\text{DFT}\{X^*[k]\}]^* \qquad (6.205)$$

where $*$ denotes the complex conjugate and

$$X[k] = \text{DFT}\{x[n]\}$$

We can write Eq. (6.94) as

$$x[n] = \frac{1}{N}\left[\sum_{n=0}^{N-1} X[k] e^{j(2\pi/N)kn} \right] = \frac{1}{N}\left[\sum_{n=0}^{N-1} X^*[k] e^{-j(2\pi/N)nk} \right]^*$$

Noting that the term in brackets in the last term is the DFT of $X^*[k]$, we get

$$x[n] = \text{IDFT}\{X[k]\} = \frac{1}{N}[\text{DFT}\{X^*[k]\}]^*$$

which shows that the same algorithm used to evaluate the DFT can be used to evaluate the IDFT.

6.54. The DFT definition in Eq. (6.92) can be expressed in a matrix operation form as

$$\mathbf{X} = \mathbf{W}_N \mathbf{x} \tag{6.206}$$

where

$$\mathbf{x} = \begin{bmatrix} x[0] \\ x[1] \\ \vdots \\ x[N-1] \end{bmatrix} \qquad \mathbf{X} = \begin{bmatrix} X[0] \\ X[1] \\ \vdots \\ X[N-1] \end{bmatrix}$$

$$\mathbf{W}_N = \begin{bmatrix} 1 & 1 & 1 & \cdots & 1 \\ 1 & W_N & W_N^2 & \cdots & W_N^{N-1} \\ 1 & W_N^2 & W_N^4 & \cdots & W_N^{2(N-1)} \\ \vdots & \vdots & \vdots & \ddots & \vdots \\ 1 & W_N^{N-1} & W_N^{2(N-1)} & \cdots & W_N^{(N-1)(N-1)} \end{bmatrix} \tag{6.207}$$

The $N \times N$ matrix $\mathbf{W}_N$ is known as the DFT matrix. Note that $\mathbf{W}_N$ is symmetric; that is, $\mathbf{W}_N^T = \mathbf{W}_N$, where $\mathbf{W}_N^T$ is the transpose of $\mathbf{W}_N$.

(*a*) Show that

$$\mathbf{W}_N^{-1} = \frac{1}{N}\mathbf{W}_N^* \tag{6.208}$$

where $\mathbf{W}_N^{-1}$ is the inverse of $\mathbf{W}_N$ and $\mathbf{W}_N^*$ is the complex conjugate of $\mathbf{W}_N$.

(*b*) Find $\mathbf{W}_4$ and $\mathbf{W}_4^{-1}$ explicitly.

(*a*) If we assume that the inverse of $\mathbf{W}_N$ exists, then multiplying both sides of Eq. (6.206) by $\mathbf{W}_N^{-1}$, we obtain

$$\mathbf{x} = \mathbf{W}_N^{-1}\mathbf{X} \tag{6.209}$$

which is just an expression for the IDFT. The IDFT as given by Eq. (6.94) can be expressed in matrix form as

$$\mathbf{x} = \frac{1}{N}\mathbf{W}_N^*\mathbf{X} \tag{6.210}$$

Comparing Eq. (6.210) with Eq. (6.209), we conclude that

$$\mathbf{W}_N^{-1} = \frac{1}{N}\mathbf{W}_N^*$$

(*b*) Let $W_{n+1,\,k+1}$ denote the entry in the $(n+1)$st row and $(k+1)$st column of the $\mathbf{W}_4$ matrix. Then, from Eq. (6.207)

$$W_{n+1,\,k+1} = W_4^{nk} = e^{-j(2\pi/4)nk} = e^{-j(\pi/2)nk} = (-j)^{nk} \tag{6.211}$$

and we have

$$\mathbf{W}_4 = \begin{bmatrix} 1 & 1 & 1 & 1 \\ 1 & -j & -1 & j \\ 1 & -1 & 1 & -1 \\ 1 & j & -1 & -j \end{bmatrix} \qquad \mathbf{W}_4^{-1} = \frac{1}{4}\begin{bmatrix} 1 & 1 & 1 & 1 \\ 1 & j & -1 & -j \\ 1 & -1 & 1 & -1 \\ 1 & -j & -1 & j \end{bmatrix} \tag{6.212}$$

6.55. (a) Find the DFT $X[k]$ of $x[n] = \{0, 1, 2, 3\}$.

(b) Find the IDFT $x[n]$ from $X[k]$ obtained in part (a).

(a) Using Eqs. (6.206) and (6.212), the DFT $X[k]$ of $x[n]$ is given by

$$\begin{bmatrix} X[0] \\ X[1] \\ X[2] \\ X[3] \end{bmatrix} = \begin{bmatrix} 1 & 1 & 1 & 1 \\ 1 & -j & -1 & j \\ 1 & -1 & 1 & -1 \\ 1 & j & -1 & -j \end{bmatrix} \begin{bmatrix} 0 \\ 1 \\ 2 \\ 3 \end{bmatrix} = \begin{bmatrix} 6 \\ -2+j2 \\ -2 \\ -2-j2 \end{bmatrix}$$

(b) Using Eqs. (6.209) and (6.212), the IDFT $x[n]$ of $X[k]$ is given by

$$\begin{bmatrix} x[0] \\ x[1] \\ x[2] \\ x[3] \end{bmatrix} = \frac{1}{4} \begin{bmatrix} 1 & 1 & 1 & 1 \\ 1 & j & -1 & -j \\ 1 & -1 & 1 & -1 \\ 1 & -j & -1 & j \end{bmatrix} \begin{bmatrix} 6 \\ -2+j2 \\ -2 \\ -2-j2 \end{bmatrix} = \frac{1}{4} \begin{bmatrix} 0 \\ 4 \\ 8 \\ 12 \end{bmatrix} = \begin{bmatrix} 0 \\ 1 \\ 2 \\ 3 \end{bmatrix}$$

6.56. Let $x[n]$ be a sequence of finite length N such that

$$x[n] = 0 \qquad n < 0, n \geq N \tag{6.213}$$

Let the N-point DFT $X[k]$ of $x[n]$ be given by [Eq. (6.92)]

$$X[k] = \sum_{n=0}^{N-1} x[n] W_N^{kn} \qquad W_N = e^{-j(2\pi/N)} \qquad k = 0, 1, \dots, N-1 \tag{6.214}$$

Suppose N is even and let

$$f[n] = x[2n] \tag{6.215a}$$
$$g[n] = x[2n+1] \tag{6.215b}$$

The sequences $f[n]$ and $g[n]$ represent the even-numbered and odd-numbered samples of $x[n]$, respectively.

(a) Show that

$$f[n] = g[n] = 0 \qquad \text{outside } 0 \leq n \leq \frac{N}{2} - 1 \tag{6.216}$$

(b) Show that the N-point DFT $X[k]$ of $x[n]$ can be expressed as

$$X[k] = F[k] + W_N^k G[k] \qquad k = 0, 1, \dots, \frac{N}{2} - 1 \tag{6.217a}$$

$$X\left[k + \frac{N}{2}\right] = F[k] - W_N^k G[k] \qquad k = 0, 1, \dots, \frac{N}{2} - 1 \tag{6.217b}$$

where

$$F[k] = \sum_{n=0}^{(N/2)-1} f[n] W_{N/2}^{kn} \qquad k = 0, 1, \dots, \frac{N}{2} - 1 \tag{6.218a}$$

$$G[k] = \sum_{n=0}^{(N/2)-1} g[n] W_{N/2}^{kn} \qquad k = 0, 1, \dots, \frac{N}{2} - 1 \tag{6.218b}$$

(c) Draw a flow graph to illustrate the evaluation of $X[k]$ from Eqs. (6.217a) and (6.217b) with $N = 8$.

(d) Assume that $x[n]$ is complex and W_N^{nk} have been precomputed. Determine the numbers of complex multiplications required to evaluate $X[k]$ from Eq. (6.214) and from Eqs. (6.217a) and (6.217b) and compare the results for $N = 2^{10} = 1024$.

(a)　From Eq. (6.213)

$$f[n] = x[2n] = 0, n < 0 \quad \text{and} \quad f\left[\frac{N}{2}\right] = x[N] = 0$$

Thus　　　　　　　　　　$f[n] = 0 \qquad n < 0, n \ge \dfrac{N}{2}$

Similarly

$$g[n] = x[2n+1] = 0, n < 0 \quad \text{and} \quad g\left[\frac{N}{2}\right] = x[N+1] = 0$$

Thus,　　　　　　　　　　$g[n] = 0 \qquad n < 0, n \ge \dfrac{N}{2}$

(b)　We rewrite Eq. (6.214) as

$$
\begin{aligned}
X[k] &= \sum_{n\,\text{even}} x[n] W_N^{kn} + \sum_{n\,\text{odd}} x[n] W_N^{kn} \\
&= \sum_{m=0}^{(N/2)-1} x[2m] W_N^{2mk} + \sum_{m=0}^{(N/2)-1} x[2m+1] W_N^{(2m+1)k}
\end{aligned}
\tag{6.219}
$$

But　　　　$W_N^2 = (e^{-j(2\pi/N)})^2 = e^{-j(4\pi/N)} = e^{-j(2\pi/N/2)} = W_{N/2} \tag{6.220}$

With this substitution Eq. (6.219) can be expressed as

$$
\begin{aligned}
X[k] &= \sum_{m=0}^{(N/2)-1} f[m] W_{N/2}^{mk} + W_N^k \sum_{m=0}^{(N/2)-1} g[m] W_{N/2}^{mk} \\
&= F[k] + W_N^k G[k] \qquad k = 0, 1, \ldots, N-1
\end{aligned}
\tag{6.221}
$$

where　　　$F[k] = \sum_{n=0}^{(N/2)-1} f[n] W_{N/2}^{kn} \qquad k = 0, 1, \ldots, \dfrac{N}{2} - 1$

$$G[k] = \sum_{n=0}^{(N/2)-1} g[n] W_{N/2}^{kn} \qquad k = 0, 1, \ldots, \frac{N}{2} - 1$$

Note that $F[k]$ and $G[k]$ are the $(N/2)$-point DFTs of $f[n]$ and $g[n]$, respectively. Now

$$W_N^{k+N/2} = W_N^k W_N^{N/2} = -W_N^k \tag{6.222}$$

since　　　$W_N^{N/2} = (e^{-j(2\pi/N)})^{(N/2)} = e^{-j\pi} = -1 \tag{6.223}$

Hence, Eq. (6.221) can be expressed as

$$X[k] = F[k] + W_N^k G[k] \qquad k = 0, 1, \ldots, \frac{N}{2} - 1$$

$$X\left[k + \frac{N}{2}\right] = F[k] - W_N^k G[k] \qquad k = 0, 1, \ldots, \frac{N}{2} - 1$$

(c)　The flow graph illustrating the steps involved in determining $X[k]$ by Eqs. (6.217a) and (6.217b) is shown in Fig. 6-37.

(d)　To evaluate a value of $X[k]$ from Eq. (6.214) requires N complex multiplications. Thus, the total number of complex multiplications based on Eq. (6.214) is N^2. The number of complex multiplications in evaluating $F[k]$ or $G[k]$ is $(N/2)^2$. In addition, there are N multiplications involved in the evaluation of $W_N^k G[k]$. Thus, the total number of complex multiplications based on Eqs. (6.217a) and (6.217b) is $2(N/2)^2 + N = N^2/2 + N$. For $N = 2^{10} = 1024$ the total number of complex multiplications based on Eq. (6.214) is $2^{20} \approx 10^6$ and is $10^6/2 + 1024 \approx 10^6/2$ based on Eqs. (6.217a) and (6.217b). So we see that the number of multiplications is reduced approximately by a factor of 2 based on Eqs. (6.217a) and (6.217b).

　　The method of evaluating $X[k]$ based on Eqs. (6.217a) and (6.217b) is known as the *decimation-in-time fast Fourier transform* (FFT) algorithm. Note that since $N/2$ is even, using the same procedure, $F[k]$ and $G[k]$ can be found by first determining the $(N/4)$-point DFTs of appropriately chosen sequences and combining them.

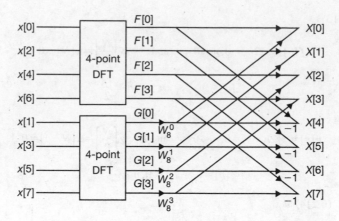

Fig. 6-37 Flow graph for an 8-point decimation-in-time FFT algorithm.

6.57. Consider a sequence

$$x[n] = \{1, 1, -1, -1, -1, 1, 1, -1\}$$

Determine the DFT $X[k]$ of $x[n]$ using the decimation-in-time FFT algorithm.

From Figs. 6-38(a) and (b), the phase factors W_4^k and W_8^k are easily found as follows:

$$W_4^0 = 1 \qquad W_4^1 = -j \qquad W_4^2 = -1 \qquad W_4^3 = j$$

and

$$W_8^0 = 1 \qquad W_8^1 = \frac{1}{\sqrt{2}} - j\frac{1}{\sqrt{2}} \qquad W_8^2 = -j \qquad W_8^3 = -\frac{1}{\sqrt{2}} - j\frac{1}{\sqrt{2}}$$

$$W_8^4 = -1 \qquad W_8^5 = -\frac{1}{\sqrt{2}} + j\frac{1}{\sqrt{2}} \qquad W_8^6 = j \qquad W_8^7 = \frac{1}{\sqrt{2}} + j\frac{1}{\sqrt{2}}$$

Next, from Eqs. (6.215a) and (6.215b)

$$f[n] = x[2n] = \{x[0], x[2], x[4], x[6]\} = \{1, -1, -1, 1\}$$
$$g[n] = x[2n + 1] = \{x[1], x[3], x[5], x[7]\} = \{1, -1, 1, -1\}$$

Then, using Eqs. (6.206) and (6.212), we have

$$\begin{bmatrix} F[0] \\ F[1] \\ F[2] \\ F[3] \end{bmatrix} = \begin{bmatrix} 1 & 1 & 1 & 1 \\ 1 & -j & -1 & j \\ 1 & -1 & 1 & -1 \\ 1 & j & -1 & -j \end{bmatrix} \begin{bmatrix} 1 \\ -1 \\ -1 \\ 1 \end{bmatrix} = \begin{bmatrix} 0 \\ 2+j2 \\ 0 \\ 2-j2 \end{bmatrix}$$

$$\begin{bmatrix} G[0] \\ G[1] \\ G[2] \\ G[3] \end{bmatrix} = \begin{bmatrix} 1 & 1 & 1 & 1 \\ 1 & -j & -1 & j \\ 1 & -1 & 1 & -1 \\ 1 & j & -1 & -j \end{bmatrix} \begin{bmatrix} 1 \\ -1 \\ 1 \\ -1 \end{bmatrix} = \begin{bmatrix} 0 \\ 0 \\ 4 \\ 0 \end{bmatrix}$$

and by Eqs. (6.217a) and (6.217b) we obtain

$$X[0] = F[0] + W_8^0 G[0] = 0 \qquad X[4] = F[0] - W_8^0 G[0] = 0$$
$$X[1] = F[1] + W_8^1 G[1] = 2 + j2 \qquad X[5] = F[1] - W_8^1 G[1] = 2 + j2$$
$$X[2] = F[2] + W_8^2 G[2] = -j4 \qquad X[6] = F[2] - W_8^2 G[2] = j4$$
$$X[3] = F[3] + W_8^3 G[3] = 2 - j2 \qquad X[7] = F[3] - W_8^3 G[3] = 2 - j2$$

Noting that since $x[n]$ is real and using Eq. (6.204), $X[7]$, $X[6]$, and $X[5]$ can be easily obtained by taking the conjugates of $X[1]$, $X[2]$, and $X[3]$, respectively.

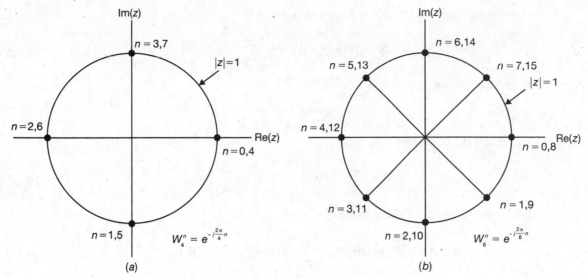

Fig. 6-38 Phase factors W_4^n and W_8^n.

6.58. Let $x[n]$ be a sequence of finite length N such that

$$x[n] = 0 \qquad n < 0, n \geq N$$

Let the N-point DFT $X[k]$ of $x[n]$ be given by [Eq. (6.92)]

$$X[k] = \sum_{n=0}^{N-1} x[n] W_N^{kn} \qquad W_N = e^{-j(2\pi/N)} \qquad k = 0, 1, \ldots, N-1 \qquad (6.224)$$

Suppose N is even and let

$$p[n] = x[n] + x\left[n + \frac{N}{2}\right] \qquad 0 \leq n < \frac{N}{2} \qquad (6.225a)$$

$$q[n] = \left(x[n] - x\left[n + \frac{N}{2}\right]\right) W_N^n \qquad 0 \leq n < \frac{N}{2} \qquad (6.225b)$$

(a) Show that the N-point DFT $X[k]$ of $x[n]$ can be expressed as

$$X[2k] = P[k] \qquad k = 0, 1, \ldots, \frac{N}{2} - 1 \qquad (6.226a)$$

$$X[2k + 1] = Q[k] \qquad k = 0, 1, \ldots, \frac{N}{2} - 1 \qquad (6.226b)$$

where $$P[k] = \sum_{n=0}^{(N/2)-1} p[n] W_{N/2}^{kn} \qquad k = 0, 1, \ldots, \frac{N}{2} - 1 \qquad (6.227a)$$

$$Q[k] = \sum_{n=0}^{(N/2)-1} q[n] W_{N/2}^{kn} \qquad k = 0, 1, \ldots, \frac{N}{2} - 1 \qquad (6.227b)$$

(b) Draw a flow graph to illustrate the evaluation of $X[k]$ from Eqs. (6.226a) and (6.226b) with $N = 8$.

(a) We rewrite Eq. (6.224) as

$$X[k] = \sum_{n=0}^{(N/2)-1} x[n] W_N^{kn} + \sum_{n=N/2}^{N-1} x[n] W_N^{kn} \qquad (6.228)$$

Changing the variable $n = m + N/2$ in the second term of Eq. (6.228), we have

$$X[k] = \sum_{n=0}^{(N/2)-1} x[n] W_N^{kn} + W_N^{(N/2)k} \sum_{m=0}^{(N/2)-1} x\left[m + \frac{N}{2}\right] W_N^{km} \tag{6.229}$$

Noting that [Eq. (6.223)]

$$W_N^{(N/2)k} = (-1)^k$$

Eq. (6.229) can be expressed as

$$X[k] = \sum_{n=0}^{(N/2)-1} \left\{ x[n] + (-1)^k x\left[n + \frac{N}{2}\right] \right\} W_N^{kn} \tag{6.230}$$

For k even, setting $k = 2r$ in Eq. (6.230), we have

$$X[2r] = \sum_{m=0}^{(N/2)-1} p[n] W_N^{2rn} = \sum_{n=0}^{(N/2)-1} p[n] W_{N/2}^{rn} \qquad r = 0, 1, \ldots, \frac{N}{2} - 1 \tag{6.231}$$

where the relation in Eq. (6.220) has been used. Similarly, for k odd, setting $k = 2r + 1$ in Eq. (6.230), we get

$$X[2r+1] = \sum_{m=0}^{(N/2)-1} q[n] W_N^{2rn} = \sum_{n=0}^{(N/2)-1} q[n] W_{N/2}^{rn} \qquad r = 0, 1, \ldots, \frac{N}{2} - 1 \tag{6.232}$$

Equations (6.231) and (6.232) represent the $(N/2)$-point DFT of $p[n]$ and $q[n]$, respectively. Thus, Eqs. (6.231) and (6.232) can be rewritten as

$$X[2k] = P[k] \qquad k = 0, 1, \ldots, \frac{N}{2} - 1$$

$$X[2k+1] = Q[k] \qquad k = 0, 1, \ldots, \frac{N}{2} - 1$$

where

$$P[k] = \sum_{n=0}^{(N/2)-1} p[n] W_{N/2}^{kn} \qquad k = 0, 1, \ldots, \frac{N}{2} - 1$$

$$Q[k] = \sum_{n=0}^{(N/2)-1} q[n] W_{N/2}^{kn} \qquad k = 0, 1, \ldots, \frac{N}{2} - 1$$

(b)   The flow graph illustrating the steps involved in determining $X[k]$ by Eqs. (6.227a) and (6.227b) is shown in Fig. 6-39.

     The method of evaluating $X[k]$ based on Eqs. (6.227a) and (6.227b) is known as the *decimation-in-frequency fast Fourier transform* (FFT) algorithm.

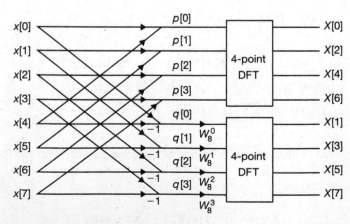

Fig. 6-39 Flow graph for an 8-point decimation-in-frequency FFT algorithm.

6.59. Using the decimation-in-frequency FFT technique, redo Prob. 6.57.

From Prob. 6.57

$$x[n] = \{1, 1, -1, -1, -1, 1, 1, -1\}$$

By Eqs. (6.225a) and (6.225b) and using the values of W_8^n obtained in Prob. 6.57, we have

$$p[n] = x[n] + x\left[n + \frac{N}{2}\right]$$
$$= \{(1-1), (1+1), (-1+1), (-1-1)\} = \{0, 2, 0, 2\}$$
$$q[n] = \left(x[n] - x\left[n + \frac{N}{2}\right]\right) W_8^n$$
$$= \{(1+1) W_8^0, (1-1) W_8^1, (-1-1) w_8^2, (-1+1) W_8^3\}$$
$$= \{2, 0, j2, 0\}$$

Then using Eqs. (6.206) and (6.212), we have

$$\begin{bmatrix} P[0] \\ P[1] \\ P[2] \\ P[3] \end{bmatrix} = \begin{bmatrix} 1 & 1 & 1 & 1 \\ 1 & -j & -1 & j \\ 1 & -1 & 1 & -1 \\ 1 & j & -1 & -j \end{bmatrix} \begin{bmatrix} 0 \\ 2 \\ 0 \\ -2 \end{bmatrix} = \begin{bmatrix} 0 \\ -j4 \\ 0 \\ j4 \end{bmatrix}$$

$$\begin{bmatrix} Q[0] \\ Q[1] \\ Q[2] \\ Q[3] \end{bmatrix} = \begin{bmatrix} 1 & 1 & 1 & 1 \\ 1 & -j & -1 & j \\ 1 & -1 & 1 & -1 \\ 1 & j & -1 & -j \end{bmatrix} \begin{bmatrix} 2 \\ 0 \\ j2 \\ 0 \end{bmatrix} = \begin{bmatrix} 2 + j2 \\ 2 - j2 \\ 2 + j2 \\ 2 - j2 \end{bmatrix}$$

and by Eqs. (6.226a) and (6.226b) we get

$X[0] = P[0] = 0$	$X[4] = P[2] = 0$
$X[1] = Q[0] = 2 + j2$	$X[5] = Q[2] = 2 + j2$
$X[2] = P[1] = -j4$	$X[6] = P[3] = j4$
$X[3] = Q[1] = 2 - j2$	$X[7] = Q[3] = 2 - j2$

which are the same results obtained in Prob. 6.57.

6.60. Consider a causal continuous-time band-limited signal $x(t)$ with the Fourier transform $X(\omega)$. Let

$$x[n] = T_s x(nT_s) \tag{6.233}$$

where T_s is the sampling interval in the time domain. Let

$$X[k] = X(k \Delta \omega) \tag{6.234}$$

where $\Delta\omega$ is the sampling interval in the frequency domain known as the *frequency resolution*. Let T_1 be the record length of $x(t)$, and let ω_M be the highest frequency of $x(t)$. Show that $x[n]$ and $X[k]$ form an N-point DFT pair if

$$\frac{T_1}{T_s} = \frac{2\omega_M}{\Delta\omega} = N \quad \text{and} \quad N \geq \frac{\omega_M T_1}{\pi} \tag{6.235}$$

Since $x(t) = 0$ for $t < 0$, the Fourier transform $X(\omega)$ of $x(t)$ is given by [Eq. (5.31)]

$$X(\omega) = \int_{-\infty}^{\infty} x(t) e^{-j\omega t} \, dt = \int_{0}^{\infty} x(t) e^{-j\omega t} \, dt \tag{6.236}$$

Let T_1 be the total recording time of $x(t)$ required to evaluate $X(\omega)$. Then the above integral can be approximated by a finite series as

$$X(\omega) = \Delta t \sum_{n=0}^{N-1} x(t_n) e^{-j\omega t_n}$$

where $t_n = n\,\Delta t$ and $T_1 = N\,\Delta t$. Setting $\omega = \omega_k$ in the above expression, we have

$$X(\omega_k) = \Delta t \sum_{n=0}^{N-1} x(t_n) e^{-j\omega_k t_n} \tag{6.237}$$

Next, since the highest frequency of $x(t)$ is ω_M, the inverse Fourier transform of $X(\omega)$ is given by [Eq. (5.32)]

$$x(t) = \frac{1}{2\pi} \int_{-\infty}^{\infty} X(\omega) e^{j\omega t}\, d\omega = \frac{1}{2\pi} \int_{-\omega_M}^{\omega_M} X(\omega) e^{j\omega t}\, d\omega \tag{6.238}$$

Dividing the frequency range $-\omega_M \le \omega \le \omega_M$ into N (even) intervals of length $\Delta\omega$, the above integral can be approximated by

$$x(t) = \frac{\Delta\omega}{2\pi} \sum_{k=-N/2}^{(N/2)-1} X(\omega_k) e^{j\omega_k t}$$

where $2\omega_M = N\,\Delta\omega$. Setting $t = t_n$ in the above expression, we have

$$x(t_n) = \frac{\Delta\omega}{2\pi} \sum_{k=-N/2}^{(N/2)-1} X(\omega_k) e^{j\omega_k t_n} \tag{6.239}$$

Since the highest frequency in $x(t)$ is ω_M, then from the sampling theorem (Prob. 5.59) we should sample $x(t)$ so that

$$\frac{2\pi}{T_s} \ge 2\omega_M$$

where T_s is the sampling interval. Since $T_s = \Delta t$, selecting the largest value of Δt (the Nyquist interval), we have

$$\Delta t = \frac{\pi}{\omega_M}$$

and

$$\omega_M = \frac{\pi}{\Delta t} = \frac{\pi N}{T_1} \tag{6.240}$$

Thus, N is a suitable even integer for which

$$\frac{T_1}{T_s} = \frac{2\omega_M}{\Delta\omega} = N \quad \text{and} \quad N \ge \frac{\omega_M T_1}{\pi} \tag{6.241}$$

From Eq. (6.240) the frequency resolution $\Delta\omega$ is given by

$$\Delta\omega = \frac{2\omega_M}{N} = \frac{2\pi N}{NT_1} = \frac{2\pi}{T_1} \tag{6.242}$$

Let $t_n = n\,\Delta t$ and $\omega_k = k\,\Delta\omega$. Then

$$t_n \omega_k = (n\,\Delta t)(k\,\Delta\omega) = nk \frac{T_1}{N} \frac{2\pi}{T_1} = \frac{2\pi}{N} nk \tag{6.243}$$

Substituting Eq. (6.243) into Eqs. (6.237) and (6.239), we get

$$X(k\,\Delta\omega) = \sum_{n=0}^{N-1} \Delta t\, x(n\,\Delta t) e^{-j(2\pi/N)nk} \tag{6.244}$$

and

$$x(n\,\Delta t) = \frac{\Delta\omega}{2\pi} \sum_{k=-N/2}^{(N/2)-1} X(k\,\Delta\omega) e^{(2\pi/N)nk} \tag{6.245}$$

Rewrite Eq. (6.245) as

$$x(n\,\Delta t) = \frac{\Delta\omega}{2\pi}\left[\sum_{k=0}^{(N/2)-1} X(k\,\Delta\omega)\,e^{j(2\pi/N)nk} + \sum_{k=-N/2}^{-1} X(k\,\Delta\omega)\,e^{j(2\pi/N)nk}\right]$$

Then from Eq. (6.244) we note that $X(k\Delta\omega)$ is periodic in k with period N. Thus, changing the variable $k = m - N$ in the second sum in the above expression, we get

$$x(n\,\Delta t) = \frac{\Delta\omega}{2\pi}\left[\sum_{k=0}^{(N/2)-1} X(k\,\Delta\omega)\,e^{j(2\pi/N)nk} + \sum_{m=N/2}^{N-1} X(m\,\Delta\omega)\,e^{j(2\pi/N)nm}\right]$$

$$= \frac{\Delta\omega}{2\pi}\sum_{k=0}^{N-1} X(k\,\Delta\omega)\,e^{j(2\pi/N)nk} \tag{6.246}$$

Multiplying both sides of Eq. (6.246) by Δt and noting that $\Delta\omega\,\Delta t = 2\pi/N$, we have

$$x(n\,\Delta t)\,\Delta t = \frac{1}{N}\sum_{n=0}^{N-1} X(k\,\Delta\omega)\,e^{j(2\pi/N)nk} \tag{6.247}$$

Now if we define

$$x[n] = \Delta t\,x(n\,\Delta t) = T_s x(nT_s) \tag{6.248}$$

$$X[k] = X(k\,\Delta\omega) \tag{6.249}$$

then Eqs. (6.244) and (6.247) reduce to the DFT pair; that is,

$$X[k] = \sum_{n=0}^{N-1} x[n]\,W_N^{kn} \qquad k = 0,1,\dots,N-1$$

$$x[n] = \frac{1}{N}\sum_{n=0}^{N-1} X[k]\,W_N^{-kn} \qquad n = 0,1,\dots,N-1$$

6.61. (a) Using the DFT, estimate the Fourier spectrum $X(\omega)$ of the continuous-time signal

$$x(t) = e^{-t}u(t)$$

Assume that the total recording time of $x(t)$ is $T_1 = 10$ s and the highest frequency of $x(t)$ is $\omega_M = 100$ rad/s.

(b) Let $X[k]$ be the DFT of the sampled sequence of $x(t)$. Compare the values of $X[0]$, $X[1]$, and $X[10]$ with the values of $X(0)$, $X(\Delta\omega)$, and $X(10\Delta\omega)$.

(a) From Eq. (6.241)

$$N \geq \frac{\omega_M T_1}{\pi} = \frac{100(10)}{\pi} = 318.3$$

Thus, choosing $N = 320$, we obtain

$$\Delta\omega = \frac{200}{320} = \frac{5}{8} = 0.625 \text{ rad}$$

$$\Delta t = \frac{10}{320} = \frac{1}{32} = 0.031 \text{ s}$$

and

$$W_N = W_{320} = e^{-j(2\pi/230)}$$

Then from Eqs. (6.244), (6.249), and (1.92), we have

$$X[k] = \sum_{n=0}^{N-1} \Delta t\,x(n\,\Delta t)\,e^{-j(2\pi/N)nk}$$

$$= \frac{1}{32}\sum_{n=0}^{319} e^{-n(0.031)}e^{-j(2\pi/320)nk} = \frac{1}{32}\frac{1-e^{320(0.031)}}{1-e^{-0.031}e^{-j(2\pi/320)k}}$$

$$= \frac{0.031}{[1-0.969\cos(k\pi/160)]+j0.969\sin(k\pi/160)} \tag{6.250}$$

which is the estimate of $X(k\,\Delta\,\omega)$.

(*b*)   Setting $k = 0$, $k = 1$, and $k = 10$ in Eq. (6.250), we have

$$X[0] = \frac{0.031}{1 - 0.969} = 1$$

$$X[1] = \frac{0.031}{0.0312 + j0.019} = 0.855e^{-j0.547}$$

$$X[10] = \frac{0.031}{0.0496 - j0.189} = 0.159e^{-j1.314}$$

From Table 5-2

$$x(t) = e^{-t}u(t) \leftrightarrow X(\omega) = \frac{1}{j\omega + 1}$$

and $$X(0) = 1$$

$$X(\Delta\omega) = X(0.625) = \frac{1}{1 + j0.625} = 0.848e^{-j0.559}$$

$$X(10\,\Delta\,\omega) = X(6.25) = \frac{1}{1 + j6.25} = 0.158e^{-j1.412}$$

Even though $x(t)$ is not band-limited, we see that $X[k]$ offers a quite good approximation to $X(\omega)$ for the frequency range we specified.

SUPPLEMENTARY PROBLEMS

6.62. Find the discrete Fourier series for each of the following periodic sequences:

(*a*) $x[n] = \cos(0, 1\pi n)$

(*b*) $x[n] = \sin(0, 1\pi n)$

(*c*) $x[n] = 2\cos(1.6\pi n) + \sin(2.4\pi n)$

6.63. Find the discrete Fourier series for the sequence $x[n]$ shown in Fig. 6-40.

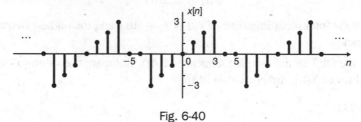

Fig. 6-40

6.64. Find the trigonometric form of the discrete Fourier series for the periodic sequence $x[n]$ shown in Fig. 6-7 in Prob. 6.3.

6.65. Find the Fourier transform of each of the following sequences:

(*a*) $x[n] = a^{|n|}$, $|a| < 1$

(*b*) $x[n] = \sin(\Omega_0 n)$, $|\Omega_0| < \pi$

(*c*) $x[n] = u[-n - 1]$

6.66. Find the Fourier transform of the sequence $x[n]$ shown in Fig. 6-41.

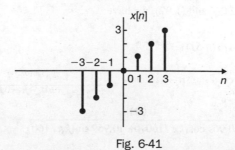

Fig. 6-41

6.67. Find the inverse Fourier transform of each of the following Fourier transforms:

(a) $X(\Omega) = \cos(2\Omega)$

(b) $X(\Omega) = j\Omega$

6.68. Consider the sequence $y[n]$ given by

$$y[n] = \begin{cases} x[n] & n \text{ even} \\ 0 & n \text{ odd} \end{cases}$$

Express $y(\Omega)$ in terms of $X(\Omega)$.

6.69. Let

$$x[n] = \begin{cases} 1 & |n| \le 2 \\ 0 & |n| > 2 \end{cases}$$

(a) Find $y[n] = x[n] * x[n]$.

(b) Find the Fourier transform $Y(\Omega)$ of $y[n]$.

6.70. Verify Parseval's theorem [Eq. (6.66)] for the discrete-time Fourier transform, that is,

$$\sum_{n=-\infty}^{\infty} |x[n]|^2 = \frac{1}{2\pi} \int_{2\pi} |X(\Omega)|^2 \, d\Omega$$

6.71. A causal discrete-time LTI system is described by

$$y[n] - \frac{3}{4} y[n-1] + \frac{1}{8} y[n-2] = x[n]$$

where $x[n]$ and $y[n]$ are the input and output of the system, respectively.

(a) Determine the frequency response $H(\Omega)$ of the system.

(b) Find the impulse response $h[n]$ of the system.

(c) Find $y[n]$ if $x[n] = (\frac{1}{2})^n u[n]$.

6.72. Consider a causal discrete-time LTI system with frequency response

$$H(\Omega) = \text{Re}\{H(\Omega)\} + j \,\text{Im}\{H(\Omega)\} = A(\Omega) + jB(\Omega)$$

(a) Show that the impulse response $h[n]$ of the system can be obtained in terms of $A(\Omega)$ or $B(\Omega)$ alone.

(b) Find $H(\Omega)$ and $h[n]$ if

$$\text{Re}\{H(\Omega)\} = A(\Omega) = 1 + \cos \Omega$$

6.73. Find the impulse response $h[n]$ of the ideal discrete-time HPF with cutoff frequency Ω_c $(0 < \Omega_c < \pi)$ shown in Fig. 6-42.

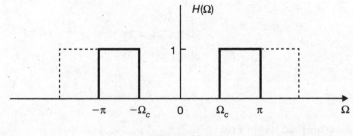

Fig. 6-42

6.74. Show that if $H_{\text{LPF}}(z)$ is the system function of a discrete-time low-pass filter, then the discrete-time system whose system function $H(z)$ is given by $H(z) = H_{\text{LPF}}(-z)$ is a high-pass filter.

6.75. Consider a continuous-time LTI system with the system function

$$H_c(s) = \frac{1}{(s+1)^2}$$

Determine the frequency response $H_d(\Omega)$ of the discrete-time system designed from this system based on the impulse invariance method.

6.76. Consider a continuous-time LTI system with the system function

$$H_c(s) = \frac{1}{s+1}$$

Determine the frequency response $H_d(\Omega)$ of the discrete-time system designed from this system based on the step response invariance; that is,

$$s_d[n] = s_c(nT_s)$$

where $s_c(t)$ and $s_d[n]$ are the step response of the continuous-time and the discrete-time systems, respectively.

6.77. Let $H_p(z)$ be the system function of a discrete-time prototype low-pass filter. Consider a new discrete-time low-pass filter whose system function $H(z)$ is obtained by replacing z in $H_p(z)$ with $(z - \alpha)/(1 - \alpha z)$, where α is real.

(a) Show that

$$H_p(z)\big|_{z=1+j0} = H(z)\big|_{z=1+j0}$$
$$H_p(z)\big|_{z=-1+j0} = H(z)\big|_{z=-1+j0}$$

(b) Let Ω_{p1} and Ω_1 be the specified frequencies ($< \pi$) of the prototype low-pass filter and the new low-pass filter, respectively. Then show that

$$\alpha = \frac{\sin\left[\left(\Omega_{p1} - \Omega_1\right)/2\right]}{\sin\left[\left(\Omega_{p1} + \Omega_1\right)/2\right]}$$

6.78. Consider a discrete-time prototype low-pass filter with system function

$$H_p(z) = 0.5(1 + z^{-1})$$

(a) Find the 3-dB bandwidth of the prototype filter.

(b) Design a discrete-time low-pass filter from this prototype filter so that the 3-dB bandwidth of the new filter is $2\pi/3$.

6.79. Determine the DFT of the sequence

$$x[n] = a^n \qquad 0 \leq n \leq N - 1$$

6.80. Evaluate the circular convolution

$$y[n] = x[n] \otimes h[n]$$

where
$$x[n] = u[n] - u[n - 4]$$
$$h[n] = u[n] - u[n - 3]$$

(a) Assuming $N = 4$.

(b) Assuming $N = 8$.

6.81. Consider the sequences $x[n]$ and $h[n]$ in Prob. 6.80.

(a) Find the 4-point DFT of $x[n]$, $h[n]$, and $y[n]$.

(b) Find $y[n]$ by taking the IDFT of $Y[k]$.

6.82. Consider a continuous-time signal $x(t)$ that has been prefiltered by a low-pass filter with a cutoff frequency of 10 kHz. The spectrum of $x(t)$ is estimated by use of the N-point DFT. The desired frequency resolution is 0.1 Hz. Determine the required value of N (assuming a power of 2) and the necessary data length T_1.

ANSWERS TO SUPPLEMENTARY PROBLEMS

6.62. (a) $x[n] = \dfrac{1}{2} e^{j\Omega_0 n} + \dfrac{1}{2} e^{j19\Omega_0 n}, \Omega_0 = 0.1\pi$

(b) $x[n] = \dfrac{1}{2j} e^{j\Omega_0 n} - \dfrac{1}{2j} e^{j19\Omega_0 n}, \Omega_0 = 0.1\pi$

(c) $x[n] = (1 - j0.5) e^{j\Omega_0 n} + (1 + j0.5) e^{j4\Omega_0 n}, \Omega_0 = 0.4\pi$

6.63. $x[n] = \displaystyle\sum_{k=0}^{8} c_k \, e^{j\Omega_0 kn}, \Omega_0 = \dfrac{2\pi}{9}$

$c_k = -j\dfrac{2}{9}\left[\sin\left(\dfrac{2\pi}{9}\right)k + 2\sin\left(\dfrac{4\pi}{9}\right)k + 3\sin\left(\dfrac{6\pi}{9}\right)k \right]$

6.64. $x[n] = \dfrac{3}{2} - \cos\dfrac{\pi}{2}n - \sin\dfrac{\pi}{2}n - \dfrac{1}{2}\cos\pi n$

6.65. (a) $X(\Omega) = \dfrac{1 - a^2}{1 - 2a\cos\Omega + a^2}$

(b) $X(\Omega) = -j\pi[\delta(\Omega - \Omega_0) - \delta(\Omega - \Omega_0)], |\Omega|, |\Omega_0| \le \pi$

(c) $X(\Omega) = \pi\,\delta(\Omega) - \dfrac{1}{1 - e^{-j\Omega}}, |\Omega| \le \pi$

6.66. $X(\Omega) = j2(\sin\Omega + 2\sin 2\Omega + 3\sin 3\Omega)$

6.67. (a) $x[n] = \dfrac{1}{2}\delta[n - 2] + \dfrac{1}{2}\delta[n + 2]$

(b) $x[n] = \begin{cases} (-1)^n / n & n \ne 0 \\ 0 & n = 0 \end{cases}$

6.68. $Y(\Omega) = \frac{1}{2}X(\Omega) + \frac{1}{2}X(\Omega - \pi)$

6.69. (a) $y[n] = \begin{cases} 5(1 - |n|/5) & |n| \le 5 \\ 0 & |n| > 5 \end{cases}$

(b) $Y(\Omega) = \left(\dfrac{\sin(2.5\Omega)}{\sin(0.5\Omega)}\right)^2$

6.70. *Hint:* Proceed in a manner similar to that for solving Prob. 5.38.

6.71. (a) $H[\Omega] = \dfrac{1}{1 - \dfrac{3}{4}e^{-j\Omega} + \dfrac{1}{8}e^{-2j\Omega}}$

(b) $h[n] = \left[2\left(\dfrac{1}{2}\right)^n - \left(\dfrac{1}{4}\right)^n \right] u[n]$

(c) $y[n] = \left[\left(\dfrac{1}{4}\right)^n + n\left(\dfrac{1}{2}\right)^{n-1} \right] u[n]$

6.72. (a) *Hint:* Process in a manner similar to that for Prob. 5.49.

(b) $H(\Omega) = 1 + e^{-j\Omega}$, $h[n] = \delta[n] + \delta[n-1]$

6.73. $h[n] = \delta[n] - \dfrac{\sin \Omega_c n}{\pi n}$

6.74. *Hint:* Use Eq. (6.156) in Prob. 6.37.

6.75. $H(\Omega) = T_s\, e^{-T_s}\, \dfrac{e^{-j\Omega}}{(1 - e^{-T_s}\, e^{-j\Omega})^2}$, where T_s is the sampling interval of $h_c(t)$.

6.76. *Hint:* $h_d[n] = s_d[n] - s_d[n-1]$

$$H_d(\Omega) = \dfrac{(1 - e^{-T_s})\, e^{-j\Omega}}{1 - e^{-T_s}\, e^{-j\Omega}}$$

6.77. *Hint:* Set $e^{j\Omega_{p1}} = \dfrac{e^{j\Omega_1} - \alpha}{1 - \alpha\, e^{j\Omega_1}}$ and solve for α.

6.78. *Hint:* Use the result from Prob. 6.77.

(a) $\Omega_{3\,db} = \dfrac{\pi}{2}$

(b) $H(z) = 0.634 \dfrac{1 + z^{-1}}{1 + 0.268 z^{-1}}$

6.79. $X[k] = \dfrac{1 - a^N}{1 - a e^{-j(2\pi/N)k}}$ $k = 0.1, \ldots, N-1$

6.80. (a) $y[n] = \{3, 3, 3, 3\}$

(b) $y[n] = \{1, 2, 3, 3, 2, 1, 0, 0\}$

6.81. (a) $[X[0], X[1], X[2], X[3]] = [4, 0, 0, 0]$
$[H[0], H[1], H[2], H[3]] = [3, -j, 1, j]$
$[Y[0], Y[1], Y[2], Y[3]] = [12, 0, 0, 0]$

(b) $y[n] = \{3, 3, 3, 3\}$

6.82. $N = 2^{18}$ and $T_1 = 13.1072$ s

CHAPTER 7

State Space Analysis

7.1 Introduction

So far we have studied linear time-invariant systems based on their input-output relationships, which are known as the external descriptions of the systems. In this chapter we discuss the method of *state space* representations of systems, which are known as the internal descriptions of the systems. The representation of systems in this form has many advantages:

1. It provides an insight into the behavior of the system.
2. It allows us to handle systems with multiple inputs and outputs in a unified way.
3. It can be extended to nonlinear and time-varying systems.

Since the state space representation is given in terms of matrix equations, the reader should have some familiarity with matrix or linear algebra. A brief review is given in App. A.

7.2 The Concept of State

A. Definition:

The *state* of a system at time t_0 (or n_0) is defined as the minimal information that is sufficient to determine the state and the output of the system for all times $t \geq t_0$ (or $n \geq n_0$) when the input to the system is also known for all times $t \geq t_0$ (or $n \geq n_0$). The variables that contain this information are called the *state variables*. Note that this definition of the state of the system applies only to causal systems.

Consider a single-input single-output LTI electric network whose structure is known. Then the complete knowledge of the input $x(t)$ over the time interval $-\infty$ to t is sufficient to determine the output $y(t)$ over the same time interval. However, if the input $x(t)$ is known over only the time interval t_0 to t, then the current through the inductors and the voltage across the capacitors at some time t_0 must be known in order to determine the output $y(t)$ over the time interval t_0 to t. These currents and voltages constitute the "state" of the network at time t_0. In this sense, the state of the network is related to the memory of the network.

B. Selection of State Variables:

Since the state variables of a system can be interpreted as the "memory elements" of the system, for discrete-time systems which are formed by unit-delay elements, amplifiers, and adders, we choose the outputs of the unit-delay elements as the state variables of the system (Prob. 7.1). For continuous-time systems which are formed by integrators, amplifiers, and adders, we choose the outputs of the integrators as the state variables of the system (Prob. 7.3). For a continuous-time system containing physical energy-storing elements, the outputs of these memory elements can be chosen to be the state variables of the system (Probs. 7.4 and 7.5). If the system is described by the difference or differential equation, the state variables can be chosen as shown in the following sections.

Note that the choice of state variables of a system is not unique. There are infinitely many choices for any given system.

7.3　State Space Representation of Discrete-Time LTI Systems

A.　Systems Described by Difference Equations:

Suppose that a single-input single-output discrete-time LTI system is described by an Nth-order difference equation

$$y[n] + a_1 y[n-1] + \cdots + a_N y[n-N] = x[n] \tag{7.1}$$

We know from previous discussion that if $x[n]$ is given for $n \geq 0$, Eq. (7.1) requires N initial conditions $y[-1]$, $y[-2], \ldots, y[-N]$ to uniquely determine the complete solution for $n > 0$. That is, N values are required to specify the state of the system at any time.

Let us define N state variables $q_1[n], q_2[n], \ldots, q_N[n]$ as

$$
\begin{aligned}
q_1[n] &= y[n-N] \\
q_2[n] &= y[n-(N-1)] = y[n-N+1] \\
&\vdots \\
q_N[n] &= y[n-1]
\end{aligned}
\tag{7.2}
$$

Then from Eqs. (7.2) and (7.1) we have

$$
\begin{aligned}
q_1[n+1] &= q_2[n] \\
q_2[n+1] &= q_3[n] \\
&\vdots
\end{aligned}
$$

$$q_N[n+1] = -a_N q_1[n] - a_{N-1} q_2[n] - \cdots - a_1 q_N[n] + x[n] \tag{7.3a}$$

and
$$y[n] = -a_N q_1[n] - a_{N-1} q_2[n] - \cdots - a_1 q_N[n] + x[n] \tag{7.3b}$$

In matrix form Eqs. (7.3a) and (7.3b) can be expressed as

$$
\begin{bmatrix} q_1[n+1] \\ q_2[n+1] \\ \vdots \\ q_N[n+1] \end{bmatrix}
=
\begin{bmatrix}
0 & 1 & 0 & \cdots & 0 \\
0 & 0 & 1 & \cdots & 0 \\
\vdots & \vdots & \vdots & \ddots & \vdots \\
-a_N & -a_{N-1} & -a_{N-2} & \cdots & -a_1
\end{bmatrix}
\begin{bmatrix} q_1[n] \\ q_2[n] \\ \vdots \\ q_N[n] \end{bmatrix}
+
\begin{bmatrix} 0 \\ 0 \\ \vdots \\ 1 \end{bmatrix}
x[n]
\tag{7.4a}
$$

$$
y[n] = \begin{bmatrix} -a_N & -a_{N-1} & \cdots & -a_1 \end{bmatrix}
\begin{bmatrix} q_1[n] \\ q_2[n] \\ \vdots \\ q_N[n] \end{bmatrix}
+ [1] x[n]
\tag{7.4b}
$$

Now we define an $N \times 1$ matrix (or N-dimensional vector) $\mathbf{q}[n]$, which we call the *state vector*:

$$
\mathbf{q}[n] = \begin{bmatrix} q_1[n] \\ q_2[n] \\ \vdots \\ q_N[n] \end{bmatrix}
\tag{7.5}
$$

Then Eqs. (7.4a) and (7.4b) can be rewritten compactly as

$$\mathbf{q}[n + 1] = \mathbf{A}\mathbf{q}[n] + \mathbf{b}x[n] \tag{7.6a}$$

$$y[n] = \mathbf{c}\mathbf{q}[n] + dx[n] \tag{7.6b}$$

where

$$\mathbf{A} = \begin{bmatrix} 0 & 1 & 0 & \cdots & 0 \\ 0 & 0 & 1 & \cdots & 0 \\ \vdots & \vdots & \vdots & \ddots & \vdots \\ -a_N & -a_{N-1} & -a_{N-2} & \cdots & -a_1 \end{bmatrix} \qquad \mathbf{b} = \begin{bmatrix} 0 \\ 0 \\ \vdots \\ 1 \end{bmatrix}$$

$$\mathbf{c} = \begin{bmatrix} -a_N & -a_{N-1} & \cdots & -a_1 \end{bmatrix} \qquad d = 1$$

Equations (7.6a) and (7.6b) are called an *N-dimensional state space representation* (or *state equations*) of the system, and the $N \times N$ matrix $\mathbf{A}$ is termed the *system matrix*. The solution of Eqs. (7.6a) and (7.6b) for a given initial state is discussed in Sec. 7.5.

B. Similarity Transformation:

As mentioned before, the choice of state variables is not unique and there are infinitely many choices of the state variables for any given system. Let $\mathbf{T}$ be any $N \times N$ *nonsingular* matrix (App. A) and define a new state vector

$$\mathbf{v}[n] = \mathbf{T}\mathbf{q}[n] \tag{7.7}$$

where $\mathbf{q}[n]$ is the old state vector which satisfies Eqs. (7.6a) and (7.6b). Since $\mathbf{T}$ is nonsingular; that is, $\mathbf{T}^{-1}$ exists, and we have

$$\mathbf{q}[n] = \mathbf{T}^{-1}\mathbf{v}[n] \tag{7.8}$$

Now

$$\mathbf{v}[n + 1] = \mathbf{T}\mathbf{q}[n + 1] = \mathbf{T}(\mathbf{A}\mathbf{q}[n] + \mathbf{b}x[n])$$

$$= \mathbf{T}\mathbf{A}\mathbf{q}[n] + \mathbf{T}\mathbf{b}x[n] = \mathbf{T}\mathbf{A}\mathbf{T}^{-1}\,\mathbf{v}[n] + \mathbf{T}\mathbf{b}x[n] \tag{7.9a}$$

$$y[n] = \mathbf{c}\mathbf{q}[n] + dx[n] = \mathbf{c}\mathbf{T}^{-1}\,\mathbf{v}[n] + dx[n] \tag{7.9b}$$

Thus, if we let

$$\hat{\mathbf{A}} = \mathbf{T}\mathbf{A}\mathbf{T}^{-1} \tag{7.10a}$$

$$\hat{\mathbf{b}} = \mathbf{T}\mathbf{b} \qquad \hat{\mathbf{c}} = \mathbf{c}\mathbf{T}^{-1} \qquad \hat{d} = d \tag{7.10b}$$

then Eqs. (7.9a) and (7.9b) become

$$\mathbf{v}[n + 1] = \hat{\mathbf{A}}\mathbf{v}[n] + \hat{\mathbf{b}}x[n] \tag{7.11a}$$

$$y[n] = \hat{\mathbf{c}}\mathbf{v}[n] + \hat{d}x[n] \tag{7.11b}$$

Equations (7.11a) and (7.11b) yield the same output $y[n]$ for a given input $x[n]$ with different state equations. In matrix algebra, Eq. (7.10a) is known as the *similarity transformation* and matrices $\mathbf{A}$ and $\hat{\mathbf{A}}$ are called *similar matrices* (App. A).

C. Multiple-Input Multiple-Output Systems:

If a discrete-time LTI system has m inputs and p outputs and N state variables, then a state space representation of the system can be expressed as

$$\mathbf{q}[n + 1] = \mathbf{A}\mathbf{q}[n] + \mathbf{B}\mathbf{x}[n] \tag{7.12a}$$

$$\mathbf{y}[n] = \mathbf{C}\mathbf{q}[n] + \mathbf{D}\mathbf{x}[n] \tag{7.12b}$$

where

$$\mathbf{q}[n] = \begin{bmatrix} q_1[n] \\ q_2[n] \\ \vdots \\ q_N[n] \end{bmatrix} \qquad \mathbf{x}[n] = \begin{bmatrix} x_1[n] \\ x_2[n] \\ \vdots \\ x_m[n] \end{bmatrix} \qquad \mathbf{y}[n] = \begin{bmatrix} y_1[n] \\ y_2[n] \\ \vdots \\ y_p[n] \end{bmatrix}$$

and

$$\mathbf{A} = \begin{bmatrix} a_{11} & a_{12} & \cdots & a_{1N} \\ a_{21} & a_{22} & \cdots & a_{2N} \\ \vdots & \vdots & \ddots & \vdots \\ a_{N1} & a_{N2} & \cdots & a_{NN} \end{bmatrix}_{N \times N} \qquad \mathbf{B} = \begin{bmatrix} b_{11} & b_{12} & \cdots & b_{1m} \\ b_{21} & b_{22} & \cdots & b_{2m} \\ \vdots & \vdots & \ddots & \vdots \\ b_{N1} & b_{N2} & \cdots & b_{Nm} \end{bmatrix}_{N \times m}$$

$$\mathbf{C} = \begin{bmatrix} c_{11} & c_{12} & \cdots & c_{1N} \\ c_{21} & c_{22} & \cdots & c_{2N} \\ \vdots & \vdots & \ddots & \vdots \\ c_{p1} & a_{p2} & \cdots & c_{pN} \end{bmatrix}_{p \times N} \qquad \mathbf{D} = \begin{bmatrix} d_{11} & d_{12} & \cdots & d_{1m} \\ d_{21} & d_{22} & \cdots & d_{2m} \\ \vdots & \vdots & \ddots & \vdots \\ d_{p1} & d_{p2} & \cdots & d_{pm} \end{bmatrix}_{p \times m}$$

7.4 State Space Representation of Continuous-Time LTI Systems

A. Systems Described by Differential Equations:

Suppose that a single-input single-output continuous-time LTI system is described by an N th-order differential equation

$$\frac{d^N y(t)}{dt^N} + a_1 \frac{d^{N-1} y(t)}{dt^{N-1}} + \cdots + a_N y(t) = x(t) \tag{7.13}$$

One possible set of initial conditions is $y(0), y^{(1)}(0), \ldots, y^{(N-1)}(0)$, where $y^{(k)}(t) = d^k y(t)/dt^k$. Thus, let us define N state variables $q_1(t), q_2(t), \ldots, q_N(t)$ as

$$\begin{aligned} q_1(t) &= y(t) \\ q_2(t) &= y^{(1)}(t) \\ &\vdots \\ q_N(t) &= y^{(N-1)}(t) \end{aligned} \tag{7.14}$$

Then from Eqs. (7.14) and (7.13) we have

$$\begin{aligned} \dot{q}_1(t) &= q_2(t) \\ \dot{q}_2(t) &= q_3(t) \\ &\vdots \\ \dot{q}_N(t) &= -a_N q_1(t) - a_{N-1} q_2(t) - \cdots - a_1 q_N(t) + x(t) \end{aligned} \tag{7.15a}$$

and

$$y(t) = q_1(t) \tag{7.15b}$$

where $\dot{q}_k(t) = dq_k(t)/dt$.

In matrix form Eqs. (7.15a) and (7.15b) can be expressed as

$$\begin{bmatrix} \dot{q}_1(t) \\ \dot{q}_2(t) \\ \vdots \\ \dot{q}_N(t) \end{bmatrix} = \begin{bmatrix} 0 & 1 & 0 & \cdots & 0 \\ 0 & 0 & 1 & \cdots & 0 \\ \vdots & \vdots & \vdots & \ddots & \vdots \\ -a_N & -a_{N-1} & -a_{N-2} & \cdots & -a_1 \end{bmatrix} \begin{bmatrix} q_1(t) \\ q_2(t) \\ \vdots \\ q_N(t) \end{bmatrix} + \begin{bmatrix} 0 \\ 0 \\ \vdots \\ 1 \end{bmatrix} x(t) \tag{7.16a}$$

$$y(t) = \begin{bmatrix} 1 & 0 & \cdots & 0 \end{bmatrix} \begin{bmatrix} q_1(t) \\ q_2(t) \\ \vdots \\ q_N(t) \end{bmatrix} \tag{7.16b}$$

Now we define an $N \times 1$ matrix (or N-dimensional vector) $\mathbf{q}(t)$ which we call the state vector:

$$\mathbf{q}(t) = \begin{bmatrix} q_1(t) \\ q_2(t) \\ \vdots \\ q_N(t) \end{bmatrix} \tag{7.17}$$

The derivative of a matrix is obtained by taking the derivative of each element of the matrix. Thus,

$$\frac{d\mathbf{q}(t)}{dt} = \dot{\mathbf{q}}(t) = \begin{bmatrix} \dot{q}_1(t) \\ \dot{q}_2(t) \\ \vdots \\ \dot{q}_N(t) \end{bmatrix} \tag{7.18}$$

Then Eqs. (7.16a) and (7.16b) can be rewritten compactly as

$$\dot{\mathbf{q}}(t) = \mathbf{A}\mathbf{q}(t) + \mathbf{b}x(t) \tag{7.19a}$$

$$y(t) = \mathbf{c}\mathbf{q}(t) \tag{7.19b}$$

where

$$\mathbf{A} = \begin{bmatrix} 0 & 1 & 0 & \cdots & 0 \\ 0 & 0 & 1 & \cdots & 0 \\ \vdots & \vdots & \vdots & \ddots & \vdots \\ -a_N & -a_{N-1} & -a_{N-2} & \cdots & -a_1 \end{bmatrix} \qquad \mathbf{b} = \begin{bmatrix} 0 \\ 0 \\ \vdots \\ 1 \end{bmatrix} \qquad \mathbf{c} = \begin{bmatrix} 1 & 0 & \cdots & 0 \end{bmatrix}$$

As in the discrete-time case, Eqs. (7.19a) and (7.19b) are called an N-dimensional state space representation (or state equations) of the system, and the $N \times N$ matrix $\mathbf{A}$ is termed the system matrix. In general, state equations of a single-input single-output continuous time LTI system are given by

$$\dot{\mathbf{q}}(t) = \mathbf{A}\mathbf{q}(t) + \mathbf{b}x(t) \tag{7.20a}$$

$$y(t) = \mathbf{c}\mathbf{q}(t) + dx(t) \tag{7.20b}$$

As in the discrete-time case, there are infinitely many choices of state variables for any given system. The solution of Eqs. (7.20a) and (7.20b) for a given initial state are discussed in Sec. 7.6.

B. Multiple-Input Multiple-Output Systems:

If a continuous-time LTI system has m inputs, p outputs, and N state variables, then a state space representation of the system can be expressed as

$$\dot{\mathbf{q}}(t) = \mathbf{A}\mathbf{q}(t) + \mathbf{B}\mathbf{x}(t) \tag{7.21a}$$

$$\mathbf{y}(t) = \mathbf{C}\mathbf{q}(t) + \mathbf{D}\mathbf{x}(t) \tag{7.21b}$$

where

$$\mathbf{q}(t) = \begin{bmatrix} q_1(t) \\ q_2(t) \\ \vdots \\ q_N(t) \end{bmatrix} \qquad \mathbf{x}(t) = \begin{bmatrix} x_1(t) \\ x_2(t) \\ \vdots \\ x_m(t) \end{bmatrix} \qquad \mathbf{y}(t) = \begin{bmatrix} y_1(t) \\ y_2(t) \\ \vdots \\ y_p(t) \end{bmatrix}$$

and

$$\mathbf{A} = \begin{bmatrix} a_{11} & a_{12} & \cdots & a_{1N} \\ a_{21} & a_{22} & \cdots & a_{2N} \\ \vdots & \vdots & \ddots & \vdots \\ a_{N1} & a_{N2} & \cdots & a_{NN} \end{bmatrix}_{N \times N} \qquad \mathbf{B} = \begin{bmatrix} b_{11} & b_{12} & \cdots & b_{1m} \\ b_{21} & b_{22} & \cdots & b_{2m} \\ \vdots & \vdots & \ddots & \vdots \\ b_{N1} & b_{N2} & \cdots & b_{Nm} \end{bmatrix}_{N \times m}$$

$$\mathbf{C} = \begin{bmatrix} c_{11} & c_{12} & \cdots & c_{1N} \\ c_{21} & c_{22} & \cdots & c_{2N} \\ \vdots & \vdots & \ddots & \vdots \\ c_{p1} & a_{p2} & \cdots & c_{pN} \end{bmatrix}_{p \times N} \qquad \mathbf{D} = \begin{bmatrix} d_{11} & d_{12} & \cdots & d_{1m} \\ d_{21} & d_{22} & \cdots & d_{2m} \\ \vdots & \vdots & \ddots & \vdots \\ d_{p1} & d_{p2} & \cdots & d_{pm} \end{bmatrix}_{p \times m}$$

7.5 Solutions of State Equations for Discrete-Time LTI Systems

A. Solution in the Time Domain:

Consider an N-dimensional state representation

$$\mathbf{q}[n + 1] = \mathbf{A}\mathbf{q}[n] + \mathbf{b}x[n] \qquad (7.22a)$$

$$y[n] = \mathbf{c}\mathbf{q}[n] + dx[n] \qquad (7.22b)$$

where $\mathbf{A}$, $\mathbf{b}$, $\mathbf{c}$, and d are $N \times N$, $N \times 1$, $1 \times N$, and 1×1 matrices, respectively. One method of finding $\mathbf{q}[n]$, given the initial state $\mathbf{q}[0]$, is to solve Eq. (7.22a) iteratively. Thus,

$$\mathbf{q}[1] = \mathbf{A}\mathbf{q}[0] + \mathbf{b}x[0]$$

$$\mathbf{q}[2] = \mathbf{A}\mathbf{q}[1] + \mathbf{b}x[1] = \mathbf{A}\{\mathbf{A}\mathbf{q}[0] + \mathbf{b}x[0]\} + \mathbf{b}x[1]$$

$$= \mathbf{A}^2\mathbf{q}[0] + \mathbf{A}\mathbf{b}x[0] + \mathbf{b}x[1]$$

By continuing this process, we obtain

$$\mathbf{q}[n] = \mathbf{A}^n\mathbf{q}[0] + \mathbf{A}^{n-1}\mathbf{b}x[0] + \cdots + \mathbf{b}x[n-1]$$

$$= \mathbf{A}^n\mathbf{q}[0] + \sum_{k=0}^{n-1} \mathbf{A}^{n-1-k}\mathbf{b}x[k] \qquad n > 0 \qquad (7.23)$$

If the initial state is $\mathbf{q}[n_0]$ and $x[n]$ is defined for $n \geq n_0$, then, proceeding in a similar manner, we obtain

$$\mathbf{q}[n] = \mathbf{A}^{n-n_0}\mathbf{q}[n_0] + \sum_{k=0}^{n-1} \mathbf{A}^{n-1-k}\mathbf{b}x[n_0 + k] \qquad n > n_0 \qquad (7.24)$$

The matrix $\mathbf{A}^n$ is the n-fold product

$$\mathbf{A}^n = \underbrace{\mathbf{A}\mathbf{A}\cdots\mathbf{A}}_{n}$$

and is known as the *state-transition* matrix of the discrete-time system. Substituting Eq. (7.23) into Eq. (7.22b), we obtain

$$y[n] = \mathbf{c}\mathbf{A}^n\mathbf{q}[0] + \sum_{k=0}^{n-1} \mathbf{c}\mathbf{A}^{n-1-k}\mathbf{b}x[k] + dx[n] \qquad n > 0 \tag{7.25}$$

The first term $\mathbf{c}\mathbf{A}^n\mathbf{q}[0]$ is the zero-input response, and the second and third terms together form the zero-state response.

B. Determination of $\mathbf{A}^n$:

Method 1: Let $\mathbf{A}$ be an $N \times N$ matrix. The *characteristic equation* of $\mathbf{A}$ is defined to be (App. A)

$$c(\lambda) = |\lambda\mathbf{I} - \mathbf{A}| = 0 \tag{7.26}$$

where $|\lambda\mathbf{I} - \mathbf{A}|$ means the determinant of $\lambda\mathbf{I} - \mathbf{A}$ and $\mathbf{I}$ is the *identity matrix* (or *unit matrix*) of Nth order. The roots of $c(\lambda) = 0$, λ_k $(k = 1, 2, \ldots, N)$, are known as the *eigenvalues* of $\mathbf{A}$. By the *Cayley-Hamilton theorem* $\mathbf{A}^n$ can be expressed as [App. A, Eq. (A.57)]

$$\mathbf{A}^n = b_0\mathbf{I} + b_1\mathbf{A} + \cdots + b_{N-1}\mathbf{A}^{N-1} \tag{7.27}$$

When the eigenvalues λ_k are all distinct, the coefficients b_0, b_1, $\ldots$, b_{N-1} can be found from the conditions

$$b_0 + b_1\lambda_k + \cdots + b_{N-1}\lambda_k^{N-1} = \lambda_k^n \qquad k = 1, 2, \ldots, N \tag{7.28}$$

For the case of repeated eigenvalues, see Prob. 7.25.

Method 2: The second method of finding $\mathbf{A}^n$ is based on the *diagonalization* of a matrix $\mathbf{A}$. If eigenvalues λ_k of $\mathbf{A}$ are all distinct, then $\mathbf{A}^n$ can be expressed as [App. A, Eq. (A.53)]

$$\mathbf{A}^n = \mathbf{P}\begin{bmatrix} \lambda_1^n & 0 & \cdots & 0 \\ 0 & \lambda_2^n & \cdots & 0 \\ \vdots & \vdots & \ddots & \vdots \\ 0 & 0 & \cdots & \lambda_N^n \end{bmatrix}\mathbf{P}^{-1} \tag{7.29}$$

where matrix $\mathbf{P}$ is known as the *diagonalization matrix* and is given by [App. A, Eq. (A.36)]

$$\mathbf{P} = [\mathbf{x}_1 \quad \mathbf{x}_2 \quad \cdots \quad \mathbf{x}_N] \tag{7.30}$$

and $\mathbf{x}_k(k = 1, 2, \ldots, N)$ are the *eigenvectors* of $\mathbf{A}$ defined by

$$\mathbf{A}\mathbf{x}_k = \lambda_k\mathbf{x}_k \qquad k = 1, 2, \ldots, N \tag{7.31}$$

Method 3: The third method of finding $\mathbf{A}^n$ is based on the *spectral decomposition* of a matrix $\mathbf{A}$. When all eigenvalues of $\mathbf{A}$ are distinct, then $\mathbf{A}$ can be expressed as

$$\mathbf{A} = \lambda_1\mathbf{E}_1 + \lambda_2\mathbf{E}_2 + \cdots + \lambda_N\mathbf{E}_N = \sum_{k=1}^{N} \lambda_k\mathbf{E}_k \tag{7.32}$$

where λ_k $(k = 1, 2, \ldots, N)$ are the distinct eigenvalues of $\mathbf{A}$ and $\mathbf{E}_k$ $(k = 1, 2, \ldots, N)$ are called *constituent matrices*, which can be evaluated as [App. A, Eq. (A.67)]

$$\mathbf{E}_k = \frac{\displaystyle\prod_{\substack{m=1 \\ m \neq k}}^{N} (\mathbf{A} - \lambda_m \mathbf{I})}{\displaystyle\prod_{\substack{m=1 \\ m \neq k}}^{N} (\lambda_k - \lambda_m)} \tag{7.33}$$

Then we have

$$\mathbf{A}^n = \lambda_1^n \mathbf{E}_1 + \lambda_2^n \mathbf{E}_2 + \cdots + \lambda_N^n \mathbf{E}_N \tag{7.34}$$

Method 4:   The fourth method of finding $\mathbf{A}^n$ is based on the z-transform.

$$\mathbf{A}^n = \mathfrak{Z}_I^{-1}\left\{(z\mathbf{I} - \mathbf{A})^{-1} z\right\} \tag{7.35}$$

which is derived in the following section [Eq. (7.41)].

C.   The z-Transform Solution:

Taking the unilateral z-transform of Eqs. (7.22a) and (7.22b) and using Eq. (4.51), we get

$$z\mathbf{Q}(z) - z\mathbf{q}(0) = \mathbf{A}\mathbf{Q}(z) + \mathbf{b}X(z) \tag{7.36a}$$
$$Y(z) = \mathbf{c}\mathbf{Q}(z) + dX(z) \tag{7.36b}$$

where $X(z) = \mathfrak{Z}_I\{x[n]\}$, $Y(z) = \mathfrak{Z}_I\{y[n]\}$, and

$$\mathbf{Q}(z) = \mathfrak{Z}_I\{\mathbf{q}[n]\} = \begin{bmatrix} Q_1(z) \\ Q_2(z) \\ \vdots \\ Q_N(z) \end{bmatrix} \qquad \text{where } Q_k(z) = \mathfrak{Z}_I\{q_k[n]\}$$

Rearranging Eq. (7.36a), we have

$$(z\mathbf{I} - \mathbf{A})\mathbf{Q}(z) = z\mathbf{q}(0) + \mathbf{b}X(z) \tag{7.37}$$

Premultiplying both sides of Eq. (7.37) by $(z\mathbf{I} - \mathbf{A})^{-1}$ yields

$$\mathbf{Q}(z) = (z\mathbf{I} - \mathbf{A})^{-1} z\mathbf{q}(0) + (z\mathbf{I} - \mathbf{A})^{-1} \mathbf{b}X(z) \tag{7.38}$$

Hence, taking the inverse unilateral z-transform of Eq. (7.38), we get

$$\mathbf{q}[n] = \mathfrak{Z}_I^{-1}\left\{(z\mathbf{I} - \mathbf{A})^{-1} z\right\}\mathbf{q}(0) + \mathfrak{Z}_I^{-1}\left\{(z\mathbf{I} - \mathbf{A})^{-1}\mathbf{b}X(z)\right\} \tag{7.39}$$

Substituting Eq. (7.39) into Eq. (7.22b), we get

$$y[n] = \mathbf{c}\mathfrak{Z}_I^{-1}\left\{(z\mathbf{I} - \mathbf{A})^{-1} z\right\}\mathbf{q}(0) + \mathbf{c}\mathfrak{Z}_I^{-1}\left\{(z\mathbf{I} - \mathbf{A})^{-1}\mathbf{b}X(z)\right\} + dx[n] \tag{7.40}$$

A comparison of Eq. (7.39) with Eq. (7.23) shows that

$$\mathbf{A}^n = \mathfrak{Z}_I^{-1}\left\{(z\mathbf{I} - \mathbf{A})^{-1} z\right\} \tag{7.41}$$

D. System Function *H*(*z*):

In Sec. 4.6 the system function $H(z)$ of a discrete-time LTI system is defined by $H(z) = Y(z)/X(z)$ with zero initial conditions. Thus, setting $\mathbf{q}[0] = \mathbf{0}$ in Eq. (7.38), we have

$$\mathbf{Q}(z) = (z\mathbf{I} - \mathbf{A})^{-1}\mathbf{b}X(z) \tag{7.42}$$

The substitution of Eq. (7.42) into Eq. (7.36b) yields

$$Y(z) = [\mathbf{c}(z\mathbf{I} - \mathbf{A})^{-1}\mathbf{b} + d]\,X(z) \tag{7.43}$$

Thus,

$$H(z) = [\mathbf{c}(z\mathbf{I} - \mathbf{A})^{-1}\mathbf{b} + d] \tag{7.44}$$

E. Stability:

From Eqs. (7.25) and (7.29) or (7.34) we see that if the magnitudes of all eigenvalues λ_k of the system matrix $\mathbf{A}$ are less than unity, that is,

$$|\lambda_k| < 1 \qquad \text{all } k \tag{7.45}$$

then the system is said to be *asymptotically stable*; that is, if, undriven, its state tends to zero from any finite initial state $\mathbf{q}_0$. It can be shown that if all eigenvalues of $\mathbf{A}$ are distinct and satisfy the condition (7.45), then the system is also BIBO stable.

7.6 Solutions of State Equations for Continuous-Time LTI Systems

A. Laplace Transform Method:

Consider an N-dimensional state space representation

$$\dot{\mathbf{q}}(t) = \mathbf{A}\mathbf{q}(t) + \mathbf{b}x(t) \tag{7.46a}$$
$$y(t) = \mathbf{c}\mathbf{q}(t) + dx(t) \tag{7.46b}$$

where $\mathbf{A}$, $\mathbf{b}$, $\mathbf{c}$, and d are $N \times N$, $N \times 1$, $1 \times N$, and 1×1 matrices, respectively. In the following we solve Eqs. (7.46a) and (7.46b) with some initial state $\mathbf{q}(0)$ by using the unilateral Laplace transform. Taking the unilateral Laplace transform of Eqs. (7.46a) and (7.46b) and using Eq. (3.44), we get

$$s\mathbf{Q}(s) - \mathbf{q}(0) = \mathbf{A}\mathbf{Q}(s) + \mathbf{b}X(s) \tag{7.47a}$$
$$Y(s) = \mathbf{c}\mathbf{Q}(s) + dX(s) \tag{7.47b}$$

where $X(s) = \mathscr{L}_I\{x(t)\}$, $Y(s) = \mathscr{L}_I\{y(t)\}$, and

$$\mathbf{Q}(s) = \mathscr{L}_I\{\mathbf{q}(t)\} = \begin{bmatrix} Q_1(s) \\ Q_2(s) \\ \vdots \\ Q_N(s) \end{bmatrix} \qquad \text{where } Q_k(s) = \mathscr{L}_I\{q_k(t)\}$$

Rearranging Eq. (7.47a), we have

$$(s\mathbf{I} - \mathbf{A})\mathbf{Q}(s) = \mathbf{q}(0) + \mathbf{b}X(s) \tag{7.48}$$

Premultiplying both sides of Eq. (7.48) by $(s\mathbf{I} - \mathbf{A})^{-1}$ yields

$$\mathbf{Q}(s) = (s\mathbf{I} - \mathbf{A})^{-1}\mathbf{q}(0) + (s\mathbf{I} - \mathbf{A})^{-1}\mathbf{b}X(s) \tag{7.49}$$

Substituting Eq. (7.49) into Eq. (7.47b), we get

$$Y(s) = \mathbf{c}(s\mathbf{I} - \mathbf{A})^{-1}\mathbf{q}(0) + [\mathbf{c}(s\mathbf{I} - \mathbf{A})^{-1}\mathbf{b} + d]\,X(s) \tag{7.50}$$

Taking the inverse Laplace transform of Eq. (7.50), we obtain the output $y(t)$. Note that $\mathbf{c}(s\mathbf{I} - \mathbf{A})^{-1}\mathbf{q}(0)$ corresponds to the zero-input response and that the second term corresponds to the zero-state response.

B. System Function $H(s)$:

As in the discrete-time case, the system function $H(s)$ of a continuous-time LTI system is defined by $H(s) = Y(s)/X(s)$ with zero initial conditions. Thus, setting $\mathbf{q}(0) = \mathbf{0}$ in Eq. (7.50), we have

$$Y(s) = [\mathbf{c}(s\mathbf{I} - \mathbf{A})^{-1}\mathbf{b} + d]\,X(s) \tag{7.51}$$

Thus,

$$H(s) = \mathbf{c}(s\mathbf{I} - \mathbf{A})^{-1}\mathbf{b} + d \tag{7.52}$$

C. Solution in the Time Domain:

Following

$$e^{at} = 1 + at + \frac{a^2}{2!}t^2 + \cdots + \frac{a^k}{k!}t^k + \cdots$$

we define

$$e^{\mathbf{A}t} = \mathbf{I} + \mathbf{A}t + \frac{\mathbf{A}^2}{2!}t^2 + \cdots + \frac{\mathbf{A}^k}{k!}t^k + \cdots \tag{7.53}$$

where $k! = k(k-1)\cdots 2 \cdot 1$. If $t = 0$, then Eq. (7.53) reduces to

$$e^{\mathbf{0}} = \mathbf{I} \tag{7.54}$$

where $\mathbf{0}$ is an $N \times N$ zero matrix whose entries are all zeros. As in $e^{a(t-\tau)} = e^{at}e^{-a\tau} = e^{-a\tau}e^{at}$, we can show that

$$e^{\mathbf{A}(t-\tau)} = e^{\mathbf{A}t}e^{-\mathbf{A}\tau} = e^{-\mathbf{A}\tau}e^{\mathbf{A}t} \tag{7.55}$$

Setting $\tau = t$ in Eq. (7.55), we have

$$e^{\mathbf{A}t}e^{-\mathbf{A}t} = e^{-\mathbf{A}t}e^{\mathbf{A}t} = e^{\mathbf{0}} = \mathbf{I} \tag{7.56}$$

Thus,

$$e^{-\mathbf{A}t} = (e^{\mathbf{A}t})^{-1} \tag{7.57}$$

which indicates that $e^{-\mathbf{A}t}$ is the inverse of $e^{\mathbf{A}t}$.

The differentiation of Eq. (7.53) with respect to t yields

$$\frac{d}{dt}e^{\mathbf{A}t} = \mathbf{0} + \mathbf{A} + \frac{\mathbf{A}^2}{2!}2t + \cdots + \frac{\mathbf{A}^k}{k!}kt^{k-1} + \cdots$$

$$= \mathbf{A}\left[\mathbf{I} + \mathbf{A}t + \frac{\mathbf{A}^2}{2!}t^2 + \cdots\right]$$

$$= \left[\mathbf{I} + \mathbf{A}t + \frac{\mathbf{A}^2}{2!}t^2 + \cdots\right]\mathbf{A}$$

which implies

$$\frac{d}{dt}e^{\mathbf{A}t} = \mathbf{A}e^{\mathbf{A}t} = e^{\mathbf{A}t}\mathbf{A} \tag{7.58}$$

Now using the relationship [App. A, Eq. (A.70)]

$$\frac{d}{dt}(\mathbf{AB}) = \frac{d\mathbf{A}}{dt}\mathbf{B} + \mathbf{A}\frac{d\mathbf{B}}{dt}$$

and Eq. (7.58), we have

$$\frac{d}{dt}\left[e^{-\mathbf{A}t}\mathbf{q}(t)\right] = \left[\frac{d}{dt}e^{-\mathbf{A}t}\right]\mathbf{q}(t) + e^{-\mathbf{A}t}\dot{\mathbf{q}}(t)$$

$$= -e^{-\mathbf{A}t}\mathbf{A}\mathbf{q}(t) + e^{-\mathbf{A}t}\dot{\mathbf{q}}(t) \tag{7.59}$$

Now premultiplying both sides of Eq. (7.46a) by $e^{-\mathbf{A}t}$, we obtain

$$e^{-\mathbf{A}t}\dot{\mathbf{q}}(t) = e^{-\mathbf{A}t}\mathbf{A}\mathbf{q}(t) + e^{-\mathbf{A}t}\mathbf{b}x(t)$$

or

$$e^{-\mathbf{A}t}\dot{\mathbf{q}}(t) - e^{-\mathbf{A}t}\mathbf{A}\mathbf{q}(t) = e^{-\mathbf{A}t}\mathbf{b}x(t) \tag{7.60}$$

From Eq. (7.59) Eq. (7.60) can be rewritten as

$$\frac{d}{dt}\left[e^{-\mathbf{A}t}\mathbf{q}(t)\right] = e^{-\mathbf{A}t}\mathbf{b}x(t) \tag{7.61}$$

Integrating both sides of Eq. (7.61) from 0 to t, we get

$$\left.e^{-\mathbf{A}t}\mathbf{q}(t)\right|_0^t = \int_0^t e^{-\mathbf{A}\tau}\mathbf{b}x(\tau)\,d\tau$$

or

$$e^{-\mathbf{A}t}\mathbf{q}(t) - \mathbf{q}(0) = \int_0^t e^{-\mathbf{A}\tau}\mathbf{b}x(\tau)\,d\tau$$

Hence

$$e^{-\mathbf{A}t}\mathbf{q}(t) = \mathbf{q}(0) + \int_0^t e^{-\mathbf{A}\tau}\mathbf{b}x(\tau)\,d\tau \tag{7.62}$$

Premultiplying both sides of Eq. (7.62) by $e^{\mathbf{A}t}$ and using Eqs. (7.55) and (7.56), we obtain

$$\mathbf{q}(t) = e^{\mathbf{A}t}\mathbf{q}(0) + \int_0^t e^{\mathbf{A}(t-\tau)}\mathbf{b}x(\tau)\,d\tau \tag{7.63}$$

If the initial state is $\mathbf{q}(t_0)$ and we have $x(t)$ for $t \geq t_0$, then

$$\mathbf{q}(t) = e^{\mathbf{A}(t-t_0)}\mathbf{q}(t_0) + \int_{t_0}^t e^{\mathbf{A}(t-\tau)}\mathbf{b}x(\tau)\,d\tau \tag{7.64}$$

which is obtained easily by integrating both sides of Eq. (7.61) from t_0 to t. The matrix function $e^{\mathbf{A}t}$ is known as the state-transition matrix of the continuous-time system. Substituting Eq. (7.63) into Eq. (7.46b), we obtain

$$y(t) = \mathbf{c}e^{\mathbf{A}t}\mathbf{q}(0) + \int_0^t \mathbf{c}e^{\mathbf{A}(t-\tau)}\mathbf{b}x(\tau)\,d\tau + dx(t) \tag{7.65}$$

D. Evaluation of $e^{\mathbf{A}t}$:

Method 1: As in the evaluation of $\mathbf{A}^n$, by the *Cayley-Hamilton* theorem we have

$$e^{\mathbf{A}t} = b_0\mathbf{I} + b_1\mathbf{A} + \cdots + b_{N-1}\mathbf{A}^{N-1} \tag{7.66}$$

When the eigenvalues λ_k of $\mathbf{A}$ are all distinct, the coefficients $b_0, b_1, \ldots, b_{N-1}$ can be found from the conditions

$$b_0 + b_1\lambda_k + \cdots + b_{N-1}\,\lambda_k^{N-1} = e^{\lambda_k t} \qquad k = 1, 2, \ldots, N \tag{7.67}$$

For the case of repeated eigenvalues see Prob. 7.45.

Method 2: Again, as in the evaluation of $\mathbf{A}^n$, we can also evaluate $e^{\mathbf{A}t}$ based on the diagonalization of $\mathbf{A}$. If all eigenvalues λ_k of $\mathbf{A}$ are distinct, we have

$$e^{\mathbf{A}t} = \mathbf{P} \begin{bmatrix} e^{\lambda_1 t} & 0 & \cdots & 0 \\ 0 & e^{\lambda_2 t} & \cdots & 0 \\ \vdots & \vdots & \ddots & \vdots \\ 0 & 0 & \cdots & e^{\lambda_N t} \end{bmatrix} \mathbf{P}^{-1} \tag{7.68}$$

where $\mathbf{P}$ is given by Eq. (7.30).

Method 3: We could also evaluate $e^{\mathbf{A}t}$ using the spectral decomposition of $\mathbf{A}$, that is, find constituent matrices $\mathbf{E}_k$ $(k = 1, 2, \ldots, N)$ for which

$$\mathbf{A} = \lambda_1\mathbf{E}_1 + \lambda_2\mathbf{E}_2 + \cdots + \lambda_N\mathbf{E}_N \tag{7.69}$$

where λ_k $(k = 1, 2, \ldots, N)$ are the distinct eigenvalues of $\mathbf{A}$. Then, when eigenvalues λ_k of $\mathbf{A}$ are all distinct, we have

$$e^{\mathbf{A}t} = e^{\lambda_1 t}\mathbf{E}_1 + e^{\lambda_2 t}\mathbf{E}_2 + \cdots + e^{\lambda_N t}\mathbf{E}_N \tag{7.70}$$

Method 4: Using the Laplace transform, we can calculate $e^{\mathbf{A}t}$. Comparing Eqs. (7.63) and (7.49), we see that

$$e^{\mathbf{A}t} = \mathcal{L}_I^{-1}\left\{(s\mathbf{I} - \mathbf{A})^{-1}\right\} \tag{7.71}$$

E. Stability:

From Eqs. (7.63) and (7.68) or (7.70), we see that if all eigenvalues λ_k of the system matrix $\mathbf{A}$ have negative real parts, that is,

$$\mathrm{Re}\{\lambda_k\} < 0 \qquad \text{all } k \tag{7.72}$$

then the system is said to be *asymptotically stable*. As in the discrete-time case, if all eigenvalues of $\mathbf{A}$ are distinct and satisfy the condition (7.72), then the system is also BIBO stable.

SOLVED PROBLEMS

State Space Representation

7.1. Consider the discrete-time LTI system shown in Fig. 7-1. Find the state space representation of the system by choosing the outputs of unit-delay elements 1 and 2 as state variables $q_1[n]$ and $q_2[n]$, respectively.

From Fig. 7-1 we have

$$q_1[n + 1] = q_2[n]$$
$$q_2[n + 1] = 2q_1[n] + 3q_2[n] + x[n]$$
$$y[n] = 2q_1[n] + 3q_2[n] + x[n]$$

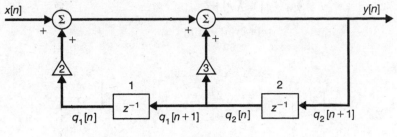

Fig. 7-1

In matrix form

$$\begin{bmatrix} q_1[n+1] \\ q_2[n+1] \end{bmatrix} = \begin{bmatrix} 0 & 1 \\ 2 & 3 \end{bmatrix} \begin{bmatrix} q_1[n] \\ q_2[n] \end{bmatrix} + \begin{bmatrix} 0 \\ 1 \end{bmatrix} x[n]$$

$$y[n] = \begin{bmatrix} 2 & 3 \end{bmatrix} \begin{bmatrix} q_1[n] \\ q_2[n] \end{bmatrix} + x[n] \tag{7.73a}$$

or

$$\mathbf{q}[n+1] = \mathbf{A}\mathbf{q}[n] + \mathbf{b}x[n]$$

$$y[n] = \mathbf{c}\mathbf{q}[n] + dx[n] \tag{7.73b}$$

where

$$\mathbf{q}[n] = \begin{bmatrix} q_1[n] \\ q_2[n] \end{bmatrix} \qquad \mathbf{A} = \begin{bmatrix} 0 & 1 \\ 2 & 3 \end{bmatrix} \qquad \mathbf{b} = \begin{bmatrix} 0 \\ 1 \end{bmatrix} \qquad \mathbf{c} = \begin{bmatrix} 2 & 3 \end{bmatrix} \qquad d = 1$$

7.2. Redo Prob. 7.1 by choosing the outputs of unit-delay elements 2 and 1 as state variables $v_1[n]$ and $v_2[n]$, respectively, and verify the relationships in Eqs. (7.10a) and (7.10b).

We redraw Fig. 7-1 with the new state variables as shown in Fig. 7-2. From Fig. 7-2 we have

$$v_1[n+1] = 3v_1[n] + 2v_2[n] + x[n]$$
$$v_2[n+1] = v_1[n]$$
$$y[n] = 3v_1[n] + 2v_2[n] + x[n]$$

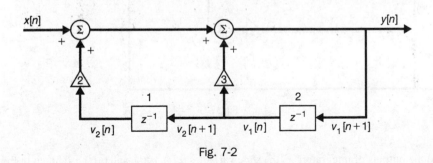

Fig. 7-2

In matrix form

$$\begin{bmatrix} v_1[n+1] \\ v_2[n+1] \end{bmatrix} = \begin{bmatrix} 3 & 2 \\ 1 & 0 \end{bmatrix} \begin{bmatrix} v_1[n] \\ v_2[n] \end{bmatrix} + \begin{bmatrix} 1 \\ 0 \end{bmatrix} x[n]$$

$$y[n] = \begin{bmatrix} 3 & 2 \end{bmatrix} \begin{bmatrix} v_1[n] \\ v_2[n] \end{bmatrix} + x[n] \tag{7.74a}$$

or $$\mathbf{v}[n+1] = \mathbf{\hat{A}}\mathbf{v}[n] + \mathbf{\hat{b}}x[n]$$

$$y[n] = \mathbf{\hat{c}}\mathbf{v}[n] + \hat{d}x[n] \tag{7.74b}$$

where

$$\mathbf{v}[n] = \begin{bmatrix} v_1[n] \\ v_2[n] \end{bmatrix} \qquad \mathbf{\hat{A}} = \begin{bmatrix} 3 & 2 \\ 1 & 0 \end{bmatrix} \qquad \mathbf{\hat{b}} = \begin{bmatrix} 1 \\ 0 \end{bmatrix} \qquad \mathbf{\hat{c}} = \begin{bmatrix} 3 & 2 \end{bmatrix} \qquad \hat{d} = 1$$

Note that $v_1[n] = q_2[n]$ and $v_2[n] = q_1[n]$. Thus, we have

$$\mathbf{v}[n] = \begin{bmatrix} 0 & 1 \\ 1 & 0 \end{bmatrix} \mathbf{q}[n] = \mathbf{T}\mathbf{q}[n]$$

Now using the results from Prob. 7.1, we have

$$\mathbf{TAT}^{-1} = \begin{bmatrix} 0 & 1 \\ 1 & 0 \end{bmatrix}\begin{bmatrix} 0 & 1 \\ 2 & 3 \end{bmatrix}\begin{bmatrix} 0 & 1 \\ 1 & 0 \end{bmatrix}^{-1} = \begin{bmatrix} 0 & 1 \\ 1 & 0 \end{bmatrix}\begin{bmatrix} 0 & 1 \\ 2 & 3 \end{bmatrix}\begin{bmatrix} 0 & 1 \\ 1 & 0 \end{bmatrix} = \begin{bmatrix} 3 & 2 \\ 1 & 0 \end{bmatrix} = \mathbf{\hat{A}}$$

$$\mathbf{Tb} = \begin{bmatrix} 0 & 1 \\ 1 & 0 \end{bmatrix}\begin{bmatrix} 0 \\ 1 \end{bmatrix} = \begin{bmatrix} 1 \\ 0 \end{bmatrix} = \mathbf{\hat{b}}$$

$$\mathbf{cT}^{-1} = \begin{bmatrix} 2 & 3 \end{bmatrix}\begin{bmatrix} 0 & 1 \\ 1 & 0 \end{bmatrix} = \begin{bmatrix} 3 & 2 \end{bmatrix} = \mathbf{\hat{c}} \qquad d = 1 = \hat{d}$$

which are the relationships in Eqs. (7.10a) and (7.10b).

7.3.   Consider the continuous-time LTI system shown in Fig. 7-3. Find a state space representation of the system

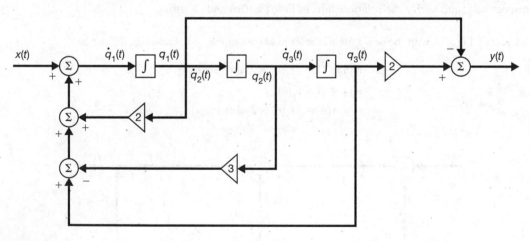

Fig. 7-3

We choose the outputs of integrators as the state variables $q_1(t)$, $q_2(t)$, and $q_3(t)$ as shown in Fig. 7-3. Then from Fig. 7-3 we obtain

$$\dot{q}_1(t) = 2q_1(t) - 3q_2(t) + q_3(t) + x(t)$$

$$\dot{q}_2(t) = q_1(t)$$

$$\dot{q}_3(t) = q_2(t)$$

$$y(t) = -q_1(t) + 2q_3(t)$$

In matrix form

$$\dot{\mathbf{q}}(t) = \begin{bmatrix} 2 & -3 & 1 \\ 1 & 0 & 0 \\ 0 & 1 & 0 \end{bmatrix} \mathbf{q}(t) + \begin{bmatrix} 1 \\ 0 \\ 0 \end{bmatrix} x(t)$$

$$y(t) = \begin{bmatrix} -1 & 0 & 2 \end{bmatrix} \mathbf{q}(t)$$

(7.75)

7.4. Consider the mechanical system shown in Fig. 7-4. It consists of a block with mass m connected to a wall by a spring. Let k_1 be the spring constant and k_2 be the viscous friction coefficient. Let the output $y(t)$ be the displacement of the block and the input $x(t)$ be the applied force. Find a state space representation of the system.

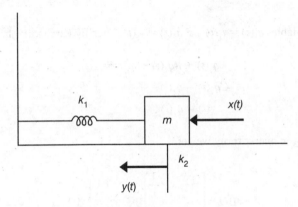

Fig. 7-4 Mechanical system.

By Newton's law we have

$$m\ddot{y}(t) = -k_1 y(t) - k_2 \dot{y}(t) + x(t)$$

or

$$m\ddot{y}(t) + k_2 \dot{y}(t) + k_1 y(t) = x(t)$$

The potential energy and kinetic energy of a mass are stored in its position and velocity. Thus, we select the state variables $q_1(t)$ and $q_2(t)$ as

$$q_1(t) = y(t)$$
$$q_2(t) = \dot{y}(t)$$

Then we have

$$\dot{q}_1(t) = q_2(t)$$
$$\dot{q}_2(t) = -\frac{k_1}{m} q_1(t) - \frac{k_2}{m} q_2(t) + \frac{1}{m} x(t)$$
$$y(t) = q_1(t)$$

In matrix form

$$\dot{\mathbf{q}}(t) = \begin{bmatrix} 0 & 1 \\ -\dfrac{k_1}{m} & -\dfrac{k_2}{m} \end{bmatrix} \mathbf{q}(t) + \begin{bmatrix} 0 \\ \dfrac{1}{m} \end{bmatrix} x(t)$$

$$y(t) = \begin{bmatrix} 1 & 0 \end{bmatrix} \mathbf{q}(t)$$

(7.76)

7.5. Consider the *RLC* circuit shown in Fig. 7-5. Let the output $y(t)$ be the loop current. Find a state space representation of the circuit.

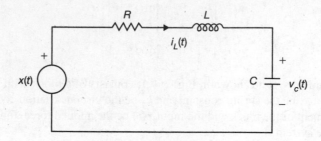

Fig. 7-5 *RLC* circuit.

We choose the state variables $q_1(t) = i_L(t)$ and $q_2(t) = v_c(t)$. Then by Kirchhoff's law we get

$$L\dot{q}_1(t) + Rq_1(t) + q_2(t) = x(t)$$

$$C\dot{q}_2(t) = q_1(t)$$

$$y(t) = q_1(t)$$

Rearranging and writing in matrix form, we get

$$\dot{\mathbf{q}}(t) = \begin{bmatrix} -\dfrac{R}{L} & -\dfrac{1}{L} \\ \dfrac{1}{C} & 0 \end{bmatrix} \mathbf{q}(t) + \begin{bmatrix} \dfrac{1}{L} \\ 0 \end{bmatrix} x(t) \qquad (7.77)$$

$$y(t) = \begin{bmatrix} 1 & 0 \end{bmatrix} \mathbf{q}(t)$$

7.6. Find a state space representation of the circuit shown in Fig. 7-6, assuming that the outputs are the currents flowing in R_1 and R_2.

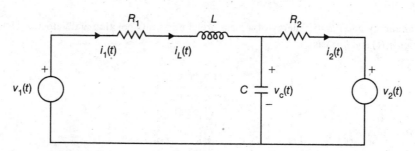

Fig. 7-6

We choose the state variables $q_1(t) = i_L(t)$ and $q_2(t) = v_c(t)$. There are two voltage sources and let $x_1(t) = v_1(t)$ and $x_2(t) = v_2(t)$. Let $y_1(t) = i_1(t)$ and $y_2(t) = i_2(t)$. Applying Kirchhoff's law to each loop, we obtain

$$L\dot{q}_1(t) + R_1 q_1(t) + q_2(t) = x_1(t)$$

$$q_2(t) - [q_1(t) - C\dot{q}_2(t)] R_2 = x_2(t)$$

$$y_1(t) = q_1(t)$$

$$y_2(t) = \frac{1}{R_2}[q_2(t) - x_2(t)]$$

Rearranging and writing in matrix form, we get

$$\dot{\mathbf{q}}(t) = \begin{bmatrix} -\dfrac{R_1}{L} & -\dfrac{1}{L} \\ \dfrac{1}{C} & -\dfrac{1}{R_2 C} \end{bmatrix} \mathbf{q}(t) + \begin{bmatrix} \dfrac{1}{L} & 0 \\ 0 & \dfrac{1}{R_2 C} \end{bmatrix} \mathbf{x}(t)$$

$$\mathbf{y}(t) = \begin{bmatrix} 1 & 0 \\ 0 & \dfrac{1}{R_2} \end{bmatrix} \mathbf{q}(t) + \begin{bmatrix} 0 & 0 \\ 0 & -\dfrac{1}{R_2} \end{bmatrix} \mathbf{x}(t)$$

(7.78)

where

$$\mathbf{q}(t) = \begin{bmatrix} q_1(t) \\ q_2(t) \end{bmatrix} \qquad \mathbf{x}(t) = \begin{bmatrix} x_1(t) \\ x_2(t) \end{bmatrix} \qquad \mathbf{y}(t) = \begin{bmatrix} y_1(t) \\ y_2(t) \end{bmatrix}$$

State Equations of Discrete-Time LTI Systems Described by Difference Equations

7.7. Find state equations of a discrete-time system described by

$$y[n] - \frac{3}{4} y[n-1] + \frac{1}{8} y[n-2] = x[n]$$

(7.79)

Choose the state variables $q_1[n]$ and $q_2[n]$ as

$$q_1[n] = y[n-2]$$
$$q_2[n] = y[n-1]$$

(7.80)

Then from Eqs. (7.79) and (7.80) we have

$$q_1[n+1] = q_2[n]$$

$$q_2[n+1] = -\frac{1}{8} q_1[n] + \frac{3}{4} q_2[n] + x[n]$$

$$y[n] = -\frac{1}{8} q_1[n] + \frac{3}{4} q_2[n] + x[n]$$

In matrix form

$$\mathbf{q}[n+1] = \begin{bmatrix} 0 & 1 \\ -\dfrac{1}{8} & \dfrac{3}{4} \end{bmatrix} \mathbf{q}[n] + \begin{bmatrix} 0 \\ 1 \end{bmatrix} x[n]$$

$$y[n] = \begin{bmatrix} -\dfrac{1}{8} & \dfrac{3}{4} \end{bmatrix} \mathbf{q}[n] + x[n]$$

(7.81)

7.8. Find state equations of a discrete-time system described by

$$y[n] - \frac{3}{4} y[n-1] + \frac{1}{8} y[n-2] = x[n] + \frac{1}{2} x[n-1]$$

(7.82)

Because of the existence of the term $\frac{1}{2} x[n-1]$ on the right-hand side of Eq. (7.82), the selection of $y[n-2]$ and $y[n-1]$ as state variables will not yield the desired state equations of the system. Thus, in order to find suitable state variables, we construct a simulation diagram of Eq. (7.82) using unit-delay elements, amplifiers, and adders. Taking the z-transforms of both sides of Eq. (7.82) and rearranging, we obtain

$$Y(z) = \frac{3}{4} z^{-1} Y(z) - \frac{1}{8} z^{-2} Y(z) + X(z) + \frac{1}{2} z^{-1} X(z)$$

from which (noting that z^{-k} corresponds to k unit time delays) the simulation diagram in Fig. 7-7 can be drawn. Choosing the outputs of unit-delay elements as state variables as shown in Fig. 7-7, we get

$$y[n] = q_1[n] + x[n]$$

$$q_1[n+1] = q_2[n] + \frac{3}{4}y[n] + \frac{1}{2}x[n]$$

$$= \frac{3}{4}q_1[n] + q_2[n] + \frac{5}{4}x[n]$$

$$q_2[n+1] = -\frac{1}{8}y[n] = -\frac{1}{8}q_1[n] - \frac{1}{8}x[n]$$

In matrix form

$$\mathbf{q}[n+1] = \begin{bmatrix} \dfrac{3}{4} & 1 \\ -\dfrac{1}{8} & 0 \end{bmatrix} \mathbf{q}[n] + \begin{bmatrix} \dfrac{5}{4} \\ -\dfrac{1}{8} \end{bmatrix} x[n]$$

$$y[n] = \begin{bmatrix} 1 & 0 \end{bmatrix} \mathbf{q}[n] + x[n] \tag{7.83}$$

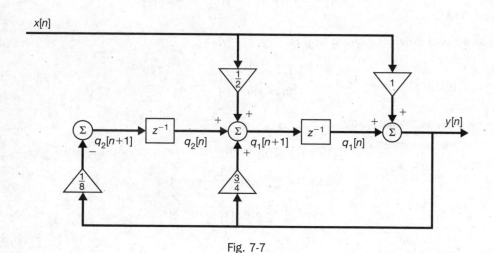

Fig. 7-7

7.9. Find state equations of a discrete-time LTI system with system function

$$H(z) = \frac{b_0 + b_1 z^{-1} + b_2 z^{-2}}{1 + a_1 z^{-1} + a_2 z^{-2}} \tag{7.84}$$

From the definition of the system function [Eq. (4.41)]

$$H(z) = \frac{Y(z)}{X(z)} = \frac{b_0 + b_1 z^{-1} + b_2 z^{-2}}{1 + a_1 z^{-1} + a_2 z^{-2}}$$

we have

$$(1 + a_1 z^{-1} + a_2 z^{-2})Y(z) = (b_0 + b_1 z^{-1} + b_2 z^{-2})X(z)$$

Rearranging the above equation, we get

$$Y(z) = -a_1 z^{-1} Y(z) - a_2 z^{-2} Y(z) + b_0 X(z) + b_1 z^{-1} X(z) + b_2 z^{-2} X(z)$$

from which the simulation diagram in Fig. 7-8 can be drawn. Choosing the outputs of unit-delay elements as state variables as shown in Fig. 7-8, we get

$$y[n] = q_1[n] + b_0 x[n]$$
$$q_1[n+1] = -a_1 y[n] + q_2[n] + b_1 x[n]$$
$$= -a_1 q_1[n] + q_2[n] + (b_1 - a_1 b_0) x[n]$$
$$q_2[n+1] = -a_2 y[n] + b_2 x[n]$$
$$= -a_2 q_1[n] + (b_2 - a_2 b_0) x[n]$$

In matrix form

$$\mathbf{q}[n+1] = \begin{bmatrix} -a_1 & 1 \\ -a_2 & 0 \end{bmatrix} \mathbf{q}[n] + \begin{bmatrix} b_1 - a_1 b_0 \\ b_2 - a_2 b_0 \end{bmatrix} x[n] \tag{7.85}$$

$$y[n] = \begin{bmatrix} 1 & 0 \end{bmatrix} \mathbf{q}[n] + b_0 x[n]$$

Note that in the simulation diagram in Fig. 7-8 the number of unit-delay elements is 2 (the order of the system) and is the minimum number required. Thus, Fig. 7-8 is known as the *canonical simulation of the first form* and Eq. (7.85) is known as the *canonical state representation of the first form*.

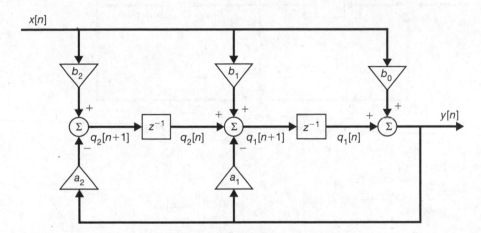

Fig. 7-8 Canonical simulation of the first form.

7.10. Redo Prob. 7.9 by expressing $H(z)$ as

$$H(z) = H_1(z) H_2(z)$$

where

$$H_1(z) = \frac{1}{1 + a_1 z^{-1} + a_2 z^{-2}} \qquad H_2(z) = b_0 + b_1 z^{-1} + b_2 z^{-2}$$

Let

$$H_1(z) = \frac{W(z)}{X(z)} = \frac{1}{1 + a_1 z^{-1} + a_2 z^{-2}} \tag{7.86}$$

$$H_2(z) = \frac{Y(z)}{W(z)} = b_0 + b_1 z^{-1} + b_2 z^{-2} \tag{7.87}$$

Then we have

$$W(z) + a_1 z^{-1} W(z) + a_2 z^{-2} W(z) = X(z) \tag{7.88}$$

$$Y(z) = b_0 W(z) + b_1 z^{-1} W(z) + b_2 z^{-2} W(z) \tag{7.89}$$

Rearranging Eq. (7.88), we get

$$W(z) = -a_1 z^{-1} W(z) - a_2 z^{-2} W(z) + X(z) \qquad (7.90)$$

From Eqs. (7.89) and (7.90) the simulation diagram in Fig. 7-9 can be drawn. Choosing the outputs of unit-delay elements as state variables as shown in Fig. 7-9, we have

$$v_1[n+1] = v_2[n]$$
$$v_2[n+1] = -a_2 v_1[n] - a_1 v_2[n] + x[n]$$
$$y[n] = b_2 v_1[n] + b_1 v_2[n] + b_0 v_2[n+1]$$
$$= (b_2 - b_0 a_2) v_1[n] + (b_1 - b_0 a_1) v_2[n] + b_0 x[n]$$

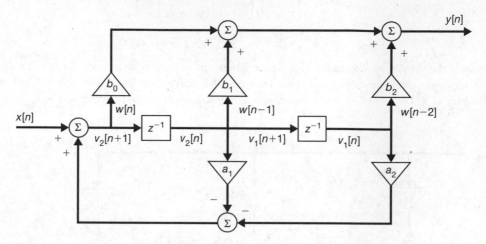

Fig. 7-9 Canonical simulation of the second form.

In matrix form

$$\mathbf{v}[n+1] = \begin{bmatrix} 0 & 1 \\ -a_2 & -a_1 \end{bmatrix} \mathbf{v}[n] + \begin{bmatrix} 0 \\ 1 \end{bmatrix} x[n] \qquad (7.91)$$
$$y[n] = \begin{bmatrix} b_2 - b_0 a_2 & b_1 - b_0 a_1 \end{bmatrix} \mathbf{v}[n] + b_0 x[n]$$

The simulation in Fig. 7-9 is known as the *canonical simulation of the second form,* and Eq. (7.91) is known as the *canonical state representation of the second form.*

7.11. Consider a discrete-time LTI system with system function

$$H(z) = \frac{z}{2z^2 - 3z + 1} \qquad (7.92)$$

Find a state representation of the system.

Rewriting $H(z)$ as

$$H(z) = \frac{z}{2z^2 \left(1 - \dfrac{3}{2} z^{-1} + \dfrac{1}{2} z^{-2}\right)} = \frac{\dfrac{1}{2} z^{-1}}{1 - \dfrac{3}{2} z^{-1} + \dfrac{1}{2} z^{-2}} \qquad (7.93)$$

Comparing Eq. (7.93) with Eq. (7.84) in Prob. 7.9, we see that

$$a_1 = -\frac{3}{2} \qquad a_2 = \frac{1}{2} \qquad b_0 = 0 \qquad b_1 = \frac{1}{2} \qquad b_2 = 0$$

Substituting these values into Eq. (7.85) in Prob. 7.9, we get

$$\mathbf{q}[n+1] = \begin{bmatrix} \dfrac{3}{2} & 1 \\ -\dfrac{1}{2} & 0 \end{bmatrix} \mathbf{q}[n] + \begin{bmatrix} \dfrac{1}{2} \\ 0 \end{bmatrix} x[n] \tag{7.94}$$

$$y[n] = \begin{bmatrix} 1 & 0 \end{bmatrix} \mathbf{q}[n]$$

7.12. Consider a discrete-time LTI system with system function

$$H(z) = \frac{z}{2z^2 - 3z + 1} = \frac{z}{2(z-1)\left(z - \dfrac{1}{2}\right)} \tag{7.95}$$

Find a state representation of the system such that its system matrix $\mathbf{A}$ is diagonal.

First we expand $H(z)$ in partial fractions as

$$H(z) = \frac{z}{2(z-1)\left(z - \dfrac{1}{2}\right)} = \frac{z}{z-1} - \frac{z}{z - \dfrac{1}{2}}$$

$$= \frac{1}{1 - z^{-1}} - \frac{1}{1 - \dfrac{1}{2}z^{-1}} = H_1(z) + H_2(z)$$

where

$$H_1(z) = \frac{1}{1 - z^{-1}} \qquad H_2(z) = \frac{-1}{1 - \dfrac{1}{2}z^{-1}}$$

Let

$$H_k(z) = \frac{\alpha_k}{1 - p_k z^{-1}} = \frac{Y_k(z)}{X(z)} \tag{7.96}$$

Then

$$(1 - p_k z^{-1})Y_k(z) = \alpha_k X(z)$$

or

$$Y_k(z) = p_k z^{-1} Y_k(z) + \alpha_k X(z)$$

from which the simulation diagram in Fig. 7-10 can be drawn. Thus, $H(z) = H_1(z) + H_2(z)$ can be simulated by the diagram in Fig. 7-11 obtained by parallel connection of two systems. Choosing the outputs of unit-delay elements as state variables as shown in Fig. 7-11, we have

$$q_1[n+1] = q_1[n] + x[n]$$

$$q_2[n+1] = \frac{1}{2}q_2[n] - x[n]$$

$$y[n] = q_1[n+1] + q_2[n+1] = q_1[n] + \frac{1}{2}q_2[n]$$

In matrix form

$$\mathbf{q}[n+1] = \begin{bmatrix} 1 & 0 \\ 0 & \dfrac{1}{2} \end{bmatrix} \mathbf{q}[n] + \begin{bmatrix} 1 \\ -1 \end{bmatrix} x[n]$$

$$y[n] = \begin{bmatrix} 1 & \dfrac{1}{2} \end{bmatrix} \mathbf{q}[n] \tag{7.97}$$

Note that the system matrix $\mathbf{A}$ is a diagonal matrix whose diagonal elements consist of the poles of $H(z)$.

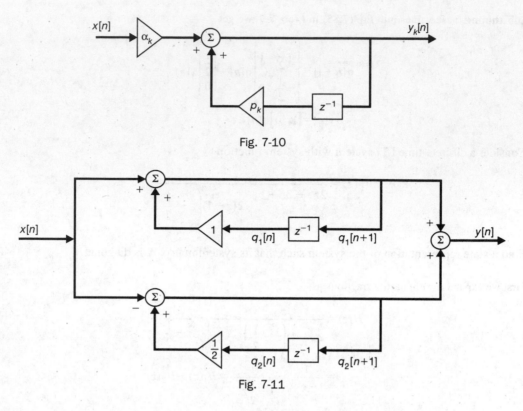

Fig. 7-10

Fig. 7-11

7.13. Sketch a block diagram of a discrete-time system with the state representation

$$\mathbf{q}[n+1] = \begin{bmatrix} 0 & 1 \\ \dfrac{1}{2} & \dfrac{2}{3} \end{bmatrix} \mathbf{q}[n] + \begin{bmatrix} 0 \\ 1 \end{bmatrix} x[n]$$

$$y[n] = \begin{bmatrix} 3 & -2 \end{bmatrix} \mathbf{q}[n] \tag{7.98}$$

We rewrite Eq. (7.98) as

$$q_1[n+1] = q_2[n]$$
$$q_2[n+1] = \frac{1}{2}q_1[n] + \frac{2}{3}q_2[n] + x[n]$$
$$y[n] = 3q_1[n] - 2q_2[n] \tag{7.99}$$

from which we can draw the block diagram in Fig. 7-12.

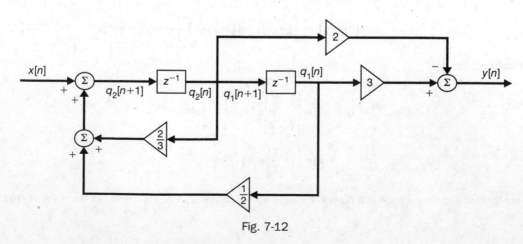

Fig. 7-12

State Equations of Continuous-Time LTI Systems Described by Differential Equations

7.14. Find state equations of a continuous-time LTI system described by

$$\ddot{y}(t) + 3\dot{y}(t) + 2y(t) = x(t) \tag{7.100}$$

Choose the state variables as

$$\begin{aligned} q_1(t) &= y(t) \\ q_2(t) &= \dot{y}(t) \end{aligned} \tag{7.101}$$

Then from Eqs. (7.100) and (7.101) we have

$$\begin{aligned} \dot{q}_1(t) &= q_2(t) \\ \dot{q}_2(t) &= -2q_1(t) - 3q_2(t) + x(t) \\ y(t) &= q_1(t) \end{aligned}$$

In matrix form

$$\begin{aligned} \dot{\mathbf{q}}(t) &= \begin{bmatrix} 0 & 1 \\ -2 & -3 \end{bmatrix} \mathbf{q}(t) + \begin{bmatrix} 0 \\ 1 \end{bmatrix} x(t) \\ y(t) &= \begin{bmatrix} 1 & 0 \end{bmatrix} \mathbf{q}(t) \end{aligned} \tag{7.102}$$

7.15. Find state equations of a continuous-time LTI system described by

$$\ddot{y}(t) + 3\dot{y}(t) + 2y(t) = 4\dot{x}(t) + x(t) \tag{7.103}$$

Because of the existence of the term $4\dot{x}(t)$ on the right-hand side of Eq. (7.103), the selection of $y(t)$ and $\dot{y}(t)$ as state variables will not yield the desired state equations of the system. Thus, in order to find suitable state variables, we construct a simulation diagram of Eq. (7.103) using integrators, amplifiers, and adders. Taking the Laplace transforms of both sides of Eq. (7.103), we obtain

$$s^2 Y(s) + 3sY(s) + 2Y(s) = 4sX(s) + X(s)$$

Dividing both sides of the above expression by s^2 and rearranging, we get

$$Y(s) = -3s^{-1}Y(s) - 2s^{-2}Y(s) + 4s^{-1}X(s) + s^{-2}X(s)$$

from which (noting that s^{-k} corresponds to integration of k times) the simulation diagram in Fig. 7-13 can be drawn. Choosing the outputs of integrators as state variables as shown in Fig. 7-13, we get

$$\begin{aligned} \dot{q}_1(t) &= -3q_1(t) + q_2(t) + 4x(t) \\ \dot{q}_2(t) &= -2q_1(t) + x(t) \\ y(t) &= q_1(t) \end{aligned}$$

In matrix form

$$\begin{aligned} \dot{\mathbf{q}}(t) &= \begin{bmatrix} -3 & 1 \\ -2 & 0 \end{bmatrix} + \begin{bmatrix} 4 \\ 1 \end{bmatrix} x(t) \\ y(t) &= \begin{bmatrix} 1 & 0 \end{bmatrix} \mathbf{q}(t) \end{aligned} \tag{7.104}$$

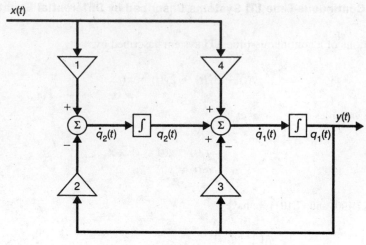

Fig. 7-13

7.16. Find state equations of a continuous-time LTI system with system function

$$H(s) = \frac{b_0 s^3 + b_1 s^2 + b_2 s + b_3}{s_3 + a_1 s^2 + a_2 s + a_3} \qquad (7.105)$$

From the definition of the system function [Eq.(3.37)]

$$H(s) = \frac{Y(s)}{X(s)} = \frac{b_0 s^3 + b_1 s^2 + b_2 s + b_3}{s^3 + a_1 s^2 + a_2 s + a_3}$$

we have

$$(s^3 + a_1 s^2 + a_2 s + a_3) Y(s) = (b_0 s^3 + b_1 s^2 + b_2 s + b_3) X(s)$$

Dividing both sides of the above expression by s^3 and rearranging, we get

$$Y(s) = -a_1 s^{-1} Y(s) - a_2 s^{-2} Y(s) - a_3 s^{-3} Y(s)$$
$$+ b_0 X(s) + b_1 s^{-1} X(s) + b_2 s^{-2} X(s) + b_3 s^{-3} X(s)$$

from which (noting that s^{-k} corresponds to integration of k times) the simulation diagram in Fig. 7-14 can be drawn. Choosing the outputs of integrators as state variables as shown in Fig. 7-14, we get

$$y(t) = q_1(t) + b_0 x(t)$$
$$\dot{q}_1(t) = -a_1 y(t) + q_2(t) + b_1 x(t)$$
$$= -a_1 q_1(t) + q_2(t) + (b_1 - a_1 b_0) x(t)$$
$$\dot{q}_2(t) = -a_2 y(t) + q_3(t) + b_2 x(t)$$
$$= -a_2 q_1(t) + q_3(t) + (b_2 - a_2 b_0) x(t)$$
$$\dot{q}_3(t) = -a_3 y(t) + b_3 x(t)$$
$$= -a_3 q_1(t) + (b_3 - a_3 b_0) x(t)$$

In matrix form

$$\dot{\mathbf{q}}(t) = \begin{bmatrix} -a_1 & 1 & 0 \\ -a_2 & 0 & 1 \\ -a_3 & 0 & 0 \end{bmatrix} \mathbf{q}(t) + \begin{bmatrix} b_1 - a_1 b_0 \\ b_2 - a_2 b_0 \\ b_3 - a_3 b_0 \end{bmatrix} x(t)$$

$$y(t) = \begin{bmatrix} 1 & 0 & 0 \end{bmatrix} \mathbf{q}(t) + b_0 x(t) \qquad (7.106)$$

As in the discrete-time case, the simulation of $H(s)$ shown in Fig. 7-14 is known as the canonical simulation of the first form, and Eq. (7.106) is known as the canonical state representation of the first form.

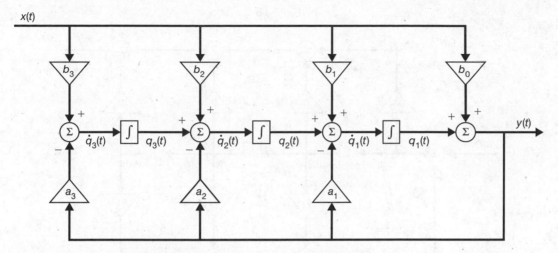

Fig. 7-14 Canonical simulation of the first form.

7.17. Redo Prob. 7.16 by expressing $H(s)$ as

$$H(s) = H_1(s)H_2(s)$$

where

$$H_1(s) = \frac{1}{s^3 + a_1 s^2 + a_2 s + a_3}$$

$$H_2(s) = b_0 s^3 + b_1 s^2 + b_2 s + b_3$$

Let

$$H_1(s) = \frac{W(s)}{X(s)} = \frac{1}{s^3 + a_1 s^2 + a_2 s + a_3}$$

$$H_2(s) = \frac{Y(s)}{W(s)} = b_0 s^3 + b_1 s^2 + b_2 s + b_3$$

(7.107)

Then we have

$$(s^3 + a_1 s^2 + a_2 s + a_3)W(s) = X(s)$$

$$Y(s) = (b_0 s^3 + b_1 s^2 + b_2 s + b_3)W(s)$$

Rearranging the above equations, we get

$$s^3 W(s) = -a_1 s^2 W(s) - a_2 s W(s) - a_3 W(s) + X(s)$$

$$Y(s) = b_0 s^3 W(s) + b_1 s^2 W(s) + b_2 s W(s) + b_3 W(s)$$

from which, noting the relation shown in Fig. 7-15, the simulation diagram in Fig. 7-16 can be drawn. Choosing the outputs of integrators as state variables as shown in Fig. 7-16, we have

$$\dot{v}_1(t) = v_2(t)$$
$$\dot{v}_2(t) = v_3(t)$$
$$\dot{v}_3(t) = -a_3 v_1(t) - a_2 v_2(t) - a_1 v_3(t) + x(t)$$
$$y(t) = b_3 v_1(t) + b_2 v_2(t) + b_1 v_3(t) + b_0 \dot{v}_3(t)$$
$$= (b_3 - a_3 b_0)v_1(t) + (b_2 - a_2 b_0)v_2(t)$$
$$+ (b_1 - a_1 b_0)v_3(t) + b_0 x(t)$$

(7.108)

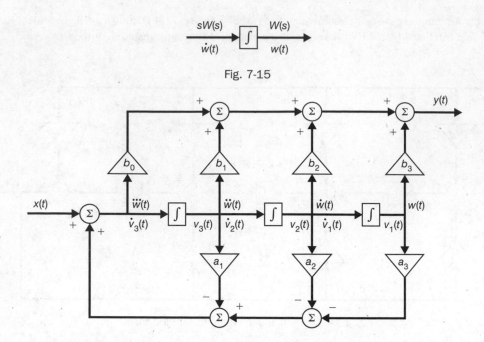

Fig. 7-15

Fig. 7-16 Canonical simulation of the second form.

In matrix form

$$\dot{\mathbf{v}}(t) = \begin{bmatrix} 0 & 1 & 0 \\ 0 & 0 & 1 \\ -a_3 & -a_2 & -a_1 \end{bmatrix} \mathbf{v}(t) + \begin{bmatrix} 0 \\ 0 \\ 1 \end{bmatrix} x(t) \tag{7.109}$$

$$y(t) = \begin{bmatrix} b_3 - a_3 b_0 & b_2 - a_2 b_0 & b_1 - a_1 b_0 \end{bmatrix} \mathbf{v}(t) + b_0 x(t)$$

As in the discrete-time case, the simulation of $H(s)$ shown in Fig. 7-16 is known as the canonical simulation of the second form, and Eq. (7.109) is known as the canonical state representation of the second form.

7.18. Consider a continuous-time LTI system with system function

$$H(s) = \frac{3s + 7}{(s + 1)(s + 2)(s + 5)} \tag{7.110}$$

Find a state representation of the system.

Rewrite $H(s)$ as

$$H(s) = \frac{3s + 7}{(s + 1)(s + 2)(s + 5)} = \frac{3s + 7}{s^3 + 8s^2 + 17s + 10} \tag{7.111}$$

Comparing Eq. (7.111) with Eq. (7.105) in Prob. 7.16, we see that

$$a_1 = 8 \qquad a_2 = 17 \qquad a_3 = 10 \qquad b_0 = b_1 = 0 \qquad b_2 = 3 \qquad b_3 = 7$$

Substituting these values into Eq. (7.106) in Prob. 7.16, we get

$$\dot{\mathbf{q}}(t) = \begin{bmatrix} -8 & 1 & 0 \\ -17 & 0 & 1 \\ -10 & 0 & 0 \end{bmatrix} \mathbf{q}(t) + \begin{bmatrix} 0 \\ 3 \\ 7 \end{bmatrix} x(t) \tag{7.112}$$

$$y(t) = \begin{bmatrix} 1 & 0 & 0 \end{bmatrix} \mathbf{q}(t)$$

7.19. Consider a continuous-time LTI system with system function

$$H(s) = \frac{3s+7}{(s+1)(s+2)(s+5)} \tag{7.113}$$

Find a state representation of the system such that its system matrix **A** is diagonal.

First we expand $H(s)$ in partial fractions as

$$H(s) = \frac{3s+7}{(s+1)(s+2)(s+5)} = \frac{1}{s+1} - \frac{\dfrac{1}{3}}{s+2} - \frac{\dfrac{2}{3}}{s+5}$$
$$= H_1(s) + H_2(s) + H_3(s)$$

where $\qquad H_1(s) = \dfrac{1}{s+1} \qquad H_2(s) = -\dfrac{\dfrac{1}{3}}{s+2} \qquad H_3(s) = -\dfrac{\dfrac{2}{3}}{s+5}$

Let

$$H_k(s) = \frac{\alpha_k}{s - p_k} = \frac{Y_k(s)}{X(s)} \tag{7.114}$$

Then

$$(s - p_k)Y_k(s) = \alpha_k X(s)$$

or

$$Y_k(s) = p_k s^{-1} Y_k(s) + \alpha_k s^{-1} X(s)$$

from which the simulation diagram in Fig. 7-17 can be drawn. Thus, $H(s) = H_1(s) + H_2(s) + H_3(s)$ can be simulated by the diagram in Fig. 7-18 obtained by parallel connection of three systems. Choosing the outputs of integrators as state variables as shown in Fig. 7-18, we get

$$\dot{q}_1(t) = -q_1(t) + x(t)$$

$$\dot{q}_2(t) = -2q_2(t) - \frac{1}{3}x(t)$$

$$\dot{q}_3(t) = -5q_3(t) - \frac{2}{3}x(t)$$

$$y(t) = q_1(t) + q_2(t) + q_3(t)$$

In matrix form

$$\dot{\mathbf{q}}(t) = \begin{bmatrix} -1 & 0 & 0 \\ 0 & -2 & 0 \\ 0 & 0 & -5 \end{bmatrix} \mathbf{q}(t) + \begin{bmatrix} 1 \\ -\dfrac{1}{3} \\ -\dfrac{2}{3} \end{bmatrix} x(t) \tag{7.115}$$

$$y(t) = \begin{bmatrix} 1 & 1 & 1 \end{bmatrix} \mathbf{q}(t)$$

Note that the system matrix **A** is a diagonal matrix whose diagonal elements consist of the poles of $H(s)$.

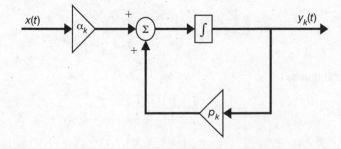

Fig. 7-17

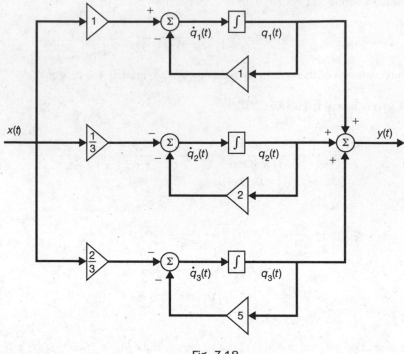

Fig. 7-18

Solutions of State Equations for Discrete-Time LTI Systems

7.20. Find $\mathbf{A}^n$ for

$$\mathbf{A} = \begin{bmatrix} 0 & 1 \\ -\dfrac{1}{8} & \dfrac{3}{4} \end{bmatrix}$$

by the Cayley-Hamilton theorem method.

First, we find the characteristic polynomial $c(\lambda)$ of $\mathbf{A}$.

$$c(\lambda) = |\lambda\mathbf{I} - \mathbf{A}| = \begin{bmatrix} \lambda & -1 \\ \dfrac{1}{8} & \lambda - \dfrac{3}{4} \end{bmatrix}$$

$$= \lambda^2 - \frac{3}{4}\lambda + \frac{1}{8} = \left(\lambda - \frac{1}{2}\right)\left(\lambda - \frac{1}{4}\right)$$

Thus, the eigenvalues of $\mathbf{A}$ are $\lambda_1 = \frac{1}{2}$ and $\lambda_2 = \frac{1}{4}$. Hence, by Eqs. (7.27) and (7.28) we have

$$\mathbf{A}^n = b_0\mathbf{I} + b_1\mathbf{A} = \begin{bmatrix} b_0 & b_1 \\ -\dfrac{1}{8}b_1 & b_0 + \dfrac{3}{4}b_1 \end{bmatrix}$$

and b_0 and b_1 are the solutions of

$$b_0 + b_1\left(\frac{1}{2}\right) = \left(\frac{1}{2}\right)^n$$

$$b_0 + b_1\left(\frac{1}{4}\right) = \left(\frac{1}{4}\right)^n$$

from which we get

$$b_0 = -\left(\frac{1}{2}\right)^n + 2\left(\frac{1}{4}\right)^n \qquad b_1 = 4\left(\frac{1}{2}\right)^n - 4\left(\frac{1}{4}\right)^n$$

Hence,

$$\mathbf{A}^n = \begin{bmatrix} -\left(\frac{1}{2}\right)^n + 2\left(\frac{1}{4}\right)^n & 4\left(\frac{1}{2}\right)^n - 4\left(\frac{1}{4}\right)^n \\[4mm] -\frac{1}{2}\left(\frac{1}{2}\right)^n + \frac{1}{2}\left(\frac{1}{4}\right)^n & 2\left(\frac{1}{2}\right)^n - \left(\frac{1}{4}\right)^n \end{bmatrix}$$

$$= \left(\frac{1}{2}\right)^n \begin{bmatrix} -1 & 4 \\ -\frac{1}{2} & 2 \end{bmatrix} + \left(\frac{1}{4}\right)^n \begin{bmatrix} 2 & -4 \\ \frac{1}{2} & -1 \end{bmatrix}$$

7.21. Repeat Prob. 7.20 using the diagonalization method.

Let $\mathbf{x}$ be an eigenvector of $\mathbf{A}$ associated with λ. Then

$$[\lambda \mathbf{I} - \mathbf{A}]\mathbf{x} = 0$$

For $\lambda = \lambda_1 = \frac{1}{2}$ we have

$$\begin{bmatrix} \frac{1}{2} & -1 \\ \frac{1}{8} & -\frac{1}{4} \end{bmatrix} \begin{bmatrix} x_1 \\ x_2 \end{bmatrix} = \begin{bmatrix} 0 \\ 0 \end{bmatrix}$$

The solutions of this system are given by $x_1 = 2x_2$. Thus, the eigenvectors associated with λ_1 are those vectors of the form

$$\mathbf{x}_1 = \alpha \begin{bmatrix} 2 \\ 1 \end{bmatrix} \qquad \alpha \neq 0$$

For $\lambda = \lambda_2 = \frac{1}{4}$ we have

$$\begin{bmatrix} \frac{1}{4} & -1 \\ \frac{1}{8} & -\frac{1}{2} \end{bmatrix} \begin{bmatrix} x_1 \\ x_2 \end{bmatrix} = \begin{bmatrix} 0 \\ 0 \end{bmatrix}$$

The solutions of this system are given by $x_1 = 4x_2$. Thus, the eigenvectors associated with λ_2 are those vectors of the form

$$\mathbf{x}_2 = \beta \begin{bmatrix} 4 \\ 1 \end{bmatrix} \qquad \beta \neq 0$$

Let $\alpha = \beta = 1$ in the above expressions and let

$$\mathbf{P} = \begin{bmatrix} \mathbf{x}_1 & \mathbf{x}_2 \end{bmatrix} = \begin{bmatrix} 2 & 4 \\ 1 & 1 \end{bmatrix}$$

Then

$$\mathbf{P}^{-1} = -\frac{1}{2} \begin{bmatrix} 1 & -4 \\ -1 & 2 \end{bmatrix} = \begin{bmatrix} -\frac{1}{2} & 2 \\ \frac{1}{2} & -1 \end{bmatrix}$$

and by Eq. (7.29) we obtain

$$
\mathbf{A}^n = \mathbf{P}\mathbf{\Lambda}^n\mathbf{P}^{-1} = \mathbf{P}\begin{bmatrix} \left(\dfrac{1}{2}\right)^n & 0 \\[2mm] 0 & \left(\dfrac{1}{4}\right)^n \end{bmatrix}\mathbf{P}^{-1} = \begin{bmatrix} 2 & 4 \\ 1 & 1 \end{bmatrix}\begin{bmatrix} \left(\dfrac{1}{2}\right)^n & 0 \\[2mm] 0 & \left(\dfrac{1}{4}\right)^n \end{bmatrix}\begin{bmatrix} -\dfrac{1}{2} & 2 \\[2mm] \dfrac{1}{2} & -1 \end{bmatrix}
$$

$$
= \begin{bmatrix} -\left(\dfrac{1}{2}\right)^n + 2\left(\dfrac{1}{4}\right)^n & 4\left(\dfrac{1}{2}\right)^n - 4\left(\dfrac{1}{4}\right)^n \\[3mm] -\dfrac{1}{2}\left(\dfrac{1}{2}\right)^n + \dfrac{1}{2}\left(\dfrac{1}{2}\right)^n & 2\left(\dfrac{1}{2}\right)^n - \left(\dfrac{1}{4}\right)^n \end{bmatrix}
$$

$$
= \left(\dfrac{1}{2}\right)^n\begin{bmatrix} -1 & 4 \\[2mm] -\dfrac{1}{2} & 2 \end{bmatrix} + \left(\dfrac{1}{4}\right)^n\begin{bmatrix} 2 & -4 \\[2mm] \dfrac{1}{2} & -1 \end{bmatrix}
$$

7.22. Repeat Prob. 7.20 using the spectral decomposition method.

Since all eigenvalues of **A** are distinct, by Eq. (7.33) we have

$$
\mathbf{E}_1 = \frac{1}{\lambda_1 - \lambda_2}(\mathbf{A} - \lambda_2\mathbf{I}) = \frac{1}{\dfrac{1}{2} - \dfrac{1}{4}}\left(\mathbf{A} - \dfrac{1}{4}\mathbf{I}\right) = 4\begin{bmatrix} -\dfrac{1}{4} & 1 \\[2mm] -\dfrac{1}{8} & \dfrac{1}{2} \end{bmatrix} = \begin{bmatrix} -1 & 4 \\[2mm] -\dfrac{1}{2} & 2 \end{bmatrix}
$$

$$
\mathbf{E}_2 = \frac{1}{\lambda_2 - \lambda_1}(\mathbf{A} - \lambda_1\mathbf{I}) = \frac{1}{\dfrac{1}{4} - \dfrac{1}{2}}\left(\mathbf{A} - \dfrac{1}{2}\mathbf{I}\right) = -4\begin{bmatrix} -\dfrac{1}{2} & 1 \\[2mm] -\dfrac{1}{8} & \dfrac{1}{4} \end{bmatrix} = \begin{bmatrix} 2 & -4 \\[2mm] \dfrac{1}{2} & -1 \end{bmatrix}
$$

Then, by Eq. (7.34) we obtain

$$
\mathbf{A}^n = \left(\dfrac{1}{2}\right)^n\mathbf{E}_1 + \left(\dfrac{1}{4}\right)^n\mathbf{E}_2 = \left(\dfrac{1}{2}\right)^n\begin{bmatrix} -1 & 4 \\[2mm] -\dfrac{1}{2} & 2 \end{bmatrix} + \left(\dfrac{1}{4}\right)^n\begin{bmatrix} 2 & -4 \\[2mm] \dfrac{1}{2} & -1 \end{bmatrix}
$$

$$
= \begin{bmatrix} -\left(\dfrac{1}{2}\right)^n + 2\left(\dfrac{1}{4}\right)^n & 4\left(\dfrac{1}{2}\right)^n - 4\left(\dfrac{1}{4}\right)^n \\[3mm] -\dfrac{1}{2}\left(\dfrac{1}{2}\right)^n + \dfrac{1}{2}\left(\dfrac{1}{4}\right)^n & 2\left(\dfrac{1}{2}\right)^n - \left(\dfrac{1}{4}\right)^n \end{bmatrix}
$$

7.23. Repeat Prob. 7.20 using the *z*-transform method.

First, we must find $(z\mathbf{I} - \mathbf{A})^{-1}$.

$$
(z\mathbf{I} - \mathbf{A})^{-1} = \begin{bmatrix} z & -1 \\[2mm] \dfrac{1}{8} & z - \dfrac{3}{4} \end{bmatrix}^{-1} = \frac{1}{\left(z - \dfrac{1}{2}\right)\left(z - \dfrac{1}{4}\right)}\begin{bmatrix} z - \dfrac{3}{4} & 1 \\[2mm] -\dfrac{1}{8} & z \end{bmatrix}
$$

$$
= \begin{bmatrix} \dfrac{z - \dfrac{3}{4}}{\left(z - \dfrac{1}{2}\right)\left(z - \dfrac{1}{4}\right)} & \dfrac{1}{\left(z - \dfrac{1}{2}\right)\left(z - \dfrac{1}{4}\right)} \\[6mm] \dfrac{-\dfrac{1}{8}}{\left(z - \dfrac{1}{2}\right)\left(z - \dfrac{1}{4}\right)} & \dfrac{z}{\left(z - \dfrac{1}{2}\right)\left(z - \dfrac{1}{4}\right)} \end{bmatrix}
$$

$$
= \begin{bmatrix} -\dfrac{1}{z-\dfrac{1}{2}} + 2\dfrac{1}{z-\dfrac{1}{4}} & 4\dfrac{1}{z-\dfrac{1}{2}} - 4\dfrac{1}{z-\dfrac{1}{4}} \\[4mm] -\dfrac{1}{2}\dfrac{1}{z-\dfrac{1}{2}} + \dfrac{1}{2}\dfrac{1}{z-\dfrac{1}{4}} & 2\dfrac{1}{z-\dfrac{1}{2}} - \dfrac{1}{z-\dfrac{1}{4}} \end{bmatrix}
$$

Then by Eq. (7.35) we obtain

$$
\mathbf{A}^n = \mathfrak{Z}_I^{-1}\{(z\mathbf{I}-\mathbf{A})^{-1}z\}
$$

$$
= \mathfrak{Z}_I^{-1}\begin{bmatrix} -\dfrac{z}{z-\dfrac{1}{2}} + 2\dfrac{z}{z-\dfrac{1}{4}} & 4\dfrac{z}{z-\dfrac{1}{2}} - 4\dfrac{z}{z-\dfrac{1}{4}} \\[4mm] -\dfrac{1}{2}\dfrac{z}{z-\dfrac{1}{2}} + \dfrac{1}{2}\dfrac{z}{z-\dfrac{1}{4}} & 2\dfrac{z}{z-\dfrac{1}{2}} - \dfrac{z}{z-\dfrac{1}{4}} \end{bmatrix}
$$

$$
= \begin{bmatrix} -\left(\dfrac{1}{2}\right)^n + 2\left(\dfrac{1}{4}\right)^n & 4\left(\dfrac{1}{2}\right)^n - 4\left(\dfrac{1}{4}\right)^n \\[4mm] -\dfrac{1}{2}\left(\dfrac{1}{2}\right)^n + \dfrac{1}{2}\left(\dfrac{1}{4}\right)^n & 2\left(\dfrac{1}{2}\right)^n - \left(\dfrac{1}{4}\right)^n \end{bmatrix}
$$

$$
= \left(\dfrac{1}{2}\right)^n \begin{bmatrix} -1 & 4 \\[2mm] -\dfrac{1}{2} & 2 \end{bmatrix} + \left(\dfrac{1}{4}\right)^n \begin{bmatrix} 2 & -4 \\[2mm] \dfrac{1}{2} & -1 \end{bmatrix}
$$

From the above results we note that when the eigenvalues of **A** are all distinct, the spectral decomposition method is computationally the most efficient method of evaluating $\mathbf{A}^n$.

7.24. Find $\mathbf{A}^n$ for

$$
\mathbf{A} = \begin{bmatrix} 0 & 1 \\[2mm] -\dfrac{1}{3} & \dfrac{4}{3} \end{bmatrix}
$$

The characteristic polynomial $c(\lambda)$ of **A** is

$$
c(\lambda) = |\lambda\mathbf{I}-\mathbf{A}| = \begin{vmatrix} \lambda & -1 \\[2mm] \dfrac{1}{3} & \lambda-\dfrac{4}{3} \end{vmatrix}
$$

$$
= \lambda^2 - \dfrac{4}{3}\lambda + \dfrac{1}{3} = (\lambda-1)\left(\lambda-\dfrac{1}{3}\right)
$$

Thus, the eigenvalues of **A** are $\lambda_1 = 1$ and $\lambda_2 = \frac{1}{3}$, and by Eq. (7.33) we have

$$
\mathbf{E}_1 = \dfrac{1}{\lambda_1-\lambda_2}(\mathbf{A}-\lambda_2\mathbf{I}) = \dfrac{1}{1-\dfrac{1}{3}}\left(\mathbf{A}-\dfrac{1}{3}\mathbf{I}\right) = \dfrac{3}{2}\begin{bmatrix} -\dfrac{1}{3} & 1 \\[2mm] -\dfrac{1}{3} & 1 \end{bmatrix} = \begin{bmatrix} -\dfrac{1}{2} & \dfrac{3}{2} \\[2mm] -\dfrac{1}{2} & \dfrac{3}{2} \end{bmatrix}
$$

$$
\mathbf{E}_2 = \dfrac{1}{\lambda_2-\lambda_1}(\mathbf{A}-\lambda_1\mathbf{I}) = \dfrac{1}{\dfrac{1}{3}-1}(\mathbf{A}-\mathbf{I}) = -\dfrac{3}{2}\begin{bmatrix} -1 & 1 \\[2mm] -\dfrac{1}{3} & \dfrac{1}{3} \end{bmatrix} = \begin{bmatrix} \dfrac{3}{2} & -\dfrac{3}{2} \\[2mm] \dfrac{1}{2} & -\dfrac{1}{2} \end{bmatrix}
$$

Thus, by Eq. (7.34) we obtain

$$\mathbf{A}^n = (1)^n \mathbf{E}_1 + \left(\frac{1}{3}\right)^n \mathbf{E}_2 = \begin{bmatrix} -\frac{1}{2} & \frac{3}{2} \\ -\frac{1}{2} & \frac{3}{2} \end{bmatrix} + \left(\frac{1}{3}\right)^n \begin{bmatrix} \frac{3}{2} & -\frac{3}{2} \\ \frac{1}{2} & -\frac{1}{2} \end{bmatrix}$$

$$= \begin{bmatrix} -\frac{1}{2} + \frac{3}{2}\left(\frac{1}{3}\right)^n & \frac{3}{2} - \frac{3}{2}\left(\frac{1}{3}\right)^n \\ -\frac{1}{2} + \frac{1}{2}\left(\frac{1}{3}\right)^n & \frac{3}{2} - \frac{1}{2}\left(\frac{1}{3}\right)^n \end{bmatrix}$$

7.25. Find $\mathbf{A}^n$ for

$$\mathbf{A} = \begin{bmatrix} 2 & 1 \\ 0 & 2 \end{bmatrix}$$

The characteristic polynomial $c(\lambda)$ of $\mathbf{A}$ is

$$c(\lambda) = |\lambda\mathbf{I} - \mathbf{A}| = \begin{vmatrix} \lambda - 2 & 1 \\ 0 & \lambda - 2 \end{vmatrix} = (\lambda - 2)^2$$

Thus, the eigenvalues of $\mathbf{A}$ are $\lambda_1 = \lambda_2 = 2$. We use the Cayley-Hamilton theorem to evaluate $\mathbf{A}^n$. By Eq. (7.27) we have

$$\mathbf{A}^n = b_0\mathbf{I} + b_1\mathbf{A} = \begin{bmatrix} b_0 + 2b_1 & b_1 \\ 0 & b_0 + 2b_1 \end{bmatrix}$$

where b_0 and b_1 are determined by setting $\lambda = 2$ in the following equations [App. A, Eqs. (A.59) and (A.60)]:

$$b_0 + b_1\lambda = \lambda^n$$
$$b_1 = n\lambda^{n-1}$$

Thus,

$$b_0 + 2b_1 = 2^n$$
$$b_1 = n2^{n-1}$$

from which we get

$$b_0 = (1-n)2^n \qquad b_1 = n2^{n-1}$$

and

$$\mathbf{A}^n = \begin{bmatrix} 2^n & n2^{n-1} \\ 0 & 2^n \end{bmatrix}$$

7.26. Consider the matrix $\mathbf{A}$ in Prob. 7.25. Let $\mathbf{A}$ be decomposed as

$$\mathbf{A} = \begin{bmatrix} 2 & 1 \\ 0 & 2 \end{bmatrix} = \begin{bmatrix} 2 & 0 \\ 0 & 2 \end{bmatrix} + \begin{bmatrix} 0 & 1 \\ 0 & 0 \end{bmatrix} = \mathbf{D} + \mathbf{N}$$

where

$$\mathbf{D} = \begin{bmatrix} 2 & 0 \\ 0 & 2 \end{bmatrix} \qquad \text{and} \qquad \mathbf{N} = \begin{bmatrix} 0 & 1 \\ 0 & 0 \end{bmatrix}$$

(a)   Show that $\mathbf{N}^2 = \mathbf{0}$.

(b)   Show that $\mathbf{D}$ and $\mathbf{N}$ commute, that is, $\mathbf{DN} = \mathbf{ND}$.

(c)   Using the results from parts (a) and (b), find $\mathbf{A}^n$.

(a) By simple multiplication we see that

$$\mathbf{N}^2 = \begin{bmatrix} 0 & 1 \\ 0 & 0 \end{bmatrix}\begin{bmatrix} 0 & 1 \\ 0 & 0 \end{bmatrix} = \begin{bmatrix} 0 & 0 \\ 0 & 0 \end{bmatrix} = \mathbf{0}$$

(b) Since the diagonal matrix $\mathbf{D}$ can be expressed as $2\mathbf{I}$, we have

$$\mathbf{DN} = 2\mathbf{IN} = 2\mathbf{N} = 2\mathbf{NI} = \mathbf{N}(2\mathbf{I}) = \mathbf{ND}$$

that is, $\mathbf{D}$ and $\mathbf{N}$ commute.

(c) Using the binomial expansion and the result from part(b), we can write

$$(\mathbf{D} + \mathbf{N})^n = \mathbf{D}^n + n\mathbf{D}^{n-1}\mathbf{N} + \frac{n(n-1)}{2!}\mathbf{D}^{n-2}\mathbf{N}^2 + \cdots + \mathbf{N}^n$$

Since $\mathbf{N}^2 = 0$, then $\mathbf{N}^k = 0$ for $k \geq 2$, and we have

$$\mathbf{A}^n = (\mathbf{D} + \mathbf{N})^n = \mathbf{D}^n + n\mathbf{D}^{n-1}\mathbf{N}$$

Thus [see App. A, Eq. (A.43)],

$$\mathbf{A}^n = \begin{bmatrix} 2 & 0 \\ 0 & 2 \end{bmatrix}^n + n\begin{bmatrix} 2 & 0 \\ 0 & 2 \end{bmatrix}^{n-1}\begin{bmatrix} 0 & 1 \\ 0 & 0 \end{bmatrix}$$

$$= \begin{bmatrix} 2^n & 0 \\ 0 & 2^n \end{bmatrix} + n\begin{bmatrix} 2^{n-1} & 0 \\ 0 & 2^{n-1} \end{bmatrix}\begin{bmatrix} 0 & 1 \\ 0 & 0 \end{bmatrix}$$

$$= \begin{bmatrix} 2^n & 0 \\ 0 & 2^n \end{bmatrix} + n\begin{bmatrix} 0 & 2^{n-1} \\ 0 & 0 \end{bmatrix} = \begin{bmatrix} 2^n & n2^{n-1} \\ 0 & 2^n \end{bmatrix}$$

which is the same result obtained in Prob. 7.25.

Note that a square matrix $\mathbf{N}$ is called *nilpotent of index r* if $\mathbf{N}^{r-1} \neq 0$ and $\mathbf{N}^r = 0$.

7.27. The *minimal polynomial $m(\lambda)$* of $\mathbf{A}$ is the polynomial of lowest order having 1 as its leading coefficient such that $m(\mathbf{A}) = \mathbf{0}$. Consider the matrix

$$\mathbf{A} = \begin{bmatrix} 2 & 0 & 0 \\ 0 & -2 & 1 \\ 0 & 4 & 1 \end{bmatrix}$$

(a) Find the minimal polynomial $m(\lambda)$ of $\mathbf{A}$.

(b) Using the result from part (a), find $\mathbf{A}^n$.

(a) The characteristic polynomial $c(\lambda)$ of $\mathbf{A}$ is

$$c(\lambda) = |\lambda\mathbf{I} - \mathbf{A}| = \begin{vmatrix} \lambda - 2 & 0 & 0 \\ 0 & \lambda + 2 & -1 \\ 0 & -4 & \lambda - 1 \end{vmatrix} = (\lambda + 3)(\lambda - 2)^2$$

Thus, the eigenvalues of $\mathbf{A}$ are $\lambda_1 = -3$ and $\lambda_2 = \lambda_3 = 2$. Consider

$$m(\lambda) = (\lambda + 3)(\lambda - 2) = \lambda^2 + \lambda - 6$$

Now

$$m(\mathbf{A}) = \mathbf{A}^2 + \mathbf{A} - 6\mathbf{I} = \begin{bmatrix} 2 & 0 & 0 \\ 0 & -2 & 1 \\ 0 & 4 & 1 \end{bmatrix}^2 + \begin{bmatrix} 2 & 0 & 0 \\ 0 & -2 & 1 \\ 0 & 4 & 1 \end{bmatrix} - 6\begin{bmatrix} 1 & 0 & 0 \\ 0 & 1 & 0 \\ 0 & 0 & 1 \end{bmatrix}$$

$$= \begin{bmatrix} 4 & 0 & 0 \\ 0 & 8 & -1 \\ 0 & -4 & 5 \end{bmatrix} + \begin{bmatrix} 2 & 0 & 0 \\ 0 & -2 & 1 \\ 0 & 4 & 1 \end{bmatrix} - \begin{bmatrix} 6 & 0 & 0 \\ 0 & 6 & 0 \\ 0 & 0 & 6 \end{bmatrix} = \begin{bmatrix} 0 & 0 & 0 \\ 0 & 0 & 0 \\ 0 & 0 & 0 \end{bmatrix} = \mathbf{0}$$

Thus, the minimal polynomial of $\mathbf{A}$ is

$$m(\lambda) = (\lambda + 3)(\lambda - 2) = \lambda^2 + \lambda - 6$$

(b) From the result from part (a) we see that $\mathbf{A}^n$ can be expressed as a linear combination of $\mathbf{I}$ and $\mathbf{A}$ only, even though the order of $\mathbf{A}$ is 3. Thus, similar to the result from the Cayley-Hamilton theorem, we have

$$\mathbf{A}^n = b_0\mathbf{I} + b_1\mathbf{A} = \begin{bmatrix} b_0 & & b_1 \\ -\dfrac{1}{8}b_1 & & b_0 + \dfrac{3}{4}b_1 \end{bmatrix}$$

where b_0 and b_1 are determined by setting $\lambda = -3$ and $\lambda = 2$ in the equation

$$b_0 + b_1\lambda = \lambda^n$$

Thus,

$$b_0 - 3b_1 = (-3)^n$$
$$b_0 + 2b_1 = 2^n$$

from which we get

$$b_0 = \frac{2}{5}(-3)^n + \frac{3}{5}(2)^n \qquad b_1 = -\frac{1}{5}(-3)^n + \frac{1}{5}(2)^n$$

and

$$\mathbf{A}^n = \begin{bmatrix} (2)^n & 0 & 0 \\ 0 & \dfrac{4}{5}(-3)^n + \dfrac{1}{5}(2)^n & -\dfrac{1}{5}(-3)^n + \dfrac{1}{5}(2)^n \\ 0 & -\dfrac{4}{5}(-3)^n + \dfrac{4}{5}(2)^n & \dfrac{1}{5}(-3)^n + \dfrac{4}{5}(2)^n \end{bmatrix}$$

$$= (-3)^n \begin{bmatrix} 0 & 0 & 0 \\ 0 & \dfrac{4}{5} & -\dfrac{1}{5} \\ 0 & -\dfrac{4}{5} & \dfrac{1}{5} \end{bmatrix} + (2)^n \begin{bmatrix} 1 & 0 & 0 \\ 0 & \dfrac{1}{5} & \dfrac{1}{5} \\ 0 & \dfrac{4}{5} & \dfrac{4}{5} \end{bmatrix}$$

7.28. Using the spectral decomposition method, evaluate $\mathbf{A}^n$ for matrix $\mathbf{A}$ in Prob. 7.27.

Since the minimal polynomial of $\mathbf{A}$ is

$$m(\lambda) = (\lambda + 3)(\lambda - 2) = (\lambda - \lambda_1)(\lambda - \lambda_2)$$

which contains only simple factors, we can apply the spectral decomposition method to evaluate $\mathbf{A}^n$. Thus, by Eq. (7.33) we have

$$\mathbf{E}_1 = \frac{1}{\lambda_1 - \lambda_2}(\mathbf{A} - \lambda_2 \mathbf{I}) = \frac{1}{-3-2}(\mathbf{A} - 2\mathbf{I})$$

$$= -\frac{1}{5}\begin{bmatrix} 0 & 0 & 0 \\ 0 & -4 & 1 \\ 0 & 4 & -1 \end{bmatrix} = \begin{bmatrix} 0 & 0 & 0 \\ 0 & \dfrac{4}{5} & -\dfrac{1}{5} \\ 0 & -\dfrac{4}{5} & \dfrac{1}{5} \end{bmatrix}$$

$$\mathbf{E}_2 = \frac{1}{\lambda_2 - \lambda_1}(\mathbf{A} - \lambda_1 \mathbf{I}) = \frac{1}{2-(-3)}(\mathbf{A} + 3\mathbf{I})$$

$$= \frac{1}{5}\begin{bmatrix} 5 & 0 & 0 \\ 0 & 1 & 1 \\ 0 & 4 & 4 \end{bmatrix} = \begin{bmatrix} 1 & 0 & 0 \\ 0 & \dfrac{1}{5} & \dfrac{1}{5} \\ 0 & \dfrac{4}{5} & \dfrac{4}{5} \end{bmatrix}$$

Thus, by Eq. (7.34) we get

$$\mathbf{A}^n = (-3)^n \mathbf{E}_1 + (2)^n \mathbf{E}_2$$

$$= (-3)^n \begin{bmatrix} 0 & 0 & 0 \\ 0 & \dfrac{4}{5} & -\dfrac{1}{5} \\ 0 & -\dfrac{4}{5} & \dfrac{1}{5} \end{bmatrix} + (2)^n \begin{bmatrix} 1 & 0 & 0 \\ 0 & \dfrac{1}{5} & \dfrac{1}{5} \\ 0 & \dfrac{4}{5} & \dfrac{4}{5} \end{bmatrix}$$

$$= \begin{bmatrix} (2)^n & 0 & 0 \\ 0 & \dfrac{4}{5}(-3)^n + \dfrac{1}{5}(2)^n & -\dfrac{1}{5}(-3)^n + \dfrac{1}{5}(2)^n \\ 0 & -\dfrac{4}{5}(-3)^n + \dfrac{4}{5}(2)^n & \dfrac{1}{5}(-3)^n + \dfrac{4}{5}(2)^n \end{bmatrix}$$

which is the same result obtained in Prob. 7.27(*b*).

7.29. Consider the discrete-time system in Prob. 7.7. Assume that the system is initially relaxed.

(*a*)   Using the state space representation, find the unit step response of the system.

(*b*)   Find the system function $H(z)$.

(*a*)   From the result of Prob. 7.7 we have

$$\mathbf{q}[n+1] = \mathbf{A}\mathbf{q}[n] + \mathbf{b}x[n]$$
$$y[n] = \mathbf{c}\mathbf{q}[n] + dx[n]$$

where

$$\mathbf{A} = \begin{bmatrix} 0 & 1 \\ -\dfrac{1}{8} & \dfrac{3}{4} \end{bmatrix} \qquad \mathbf{b} = \begin{bmatrix} 0 \\ 1 \end{bmatrix} \qquad \mathbf{c} = \begin{bmatrix} -\dfrac{1}{8} & \dfrac{3}{4} \end{bmatrix} \qquad d = 1$$

Setting $\mathbf{q}[0] = \mathbf{0}$ and $x[n] = u[n]$ in Eq. (7.25), the unit step response $s[n]$ is given by

$$s[n] = \sum_{k=0}^{n-1} \mathbf{c}\mathbf{A}^{n-1-k}\mathbf{b}u[k] + du[n] \tag{7.116}$$

Now, from Prob. 7.20 we have

$$\mathbf{A}^n = \left(\frac{1}{2}\right)^n \begin{bmatrix} -1 & 4 \\ -\frac{1}{2} & 2 \end{bmatrix} + \left(\frac{1}{4}\right)^n \begin{bmatrix} 2 & -4 \\ \frac{1}{2} & -1 \end{bmatrix}$$

and

$$\mathbf{cA}^{n-1-k}\mathbf{b} = \begin{bmatrix} -\frac{1}{8} & \frac{3}{4} \end{bmatrix} \left\{ \left(\frac{1}{2}\right)^{n-1-k} \begin{bmatrix} -1 & 4 \\ -\frac{1}{2} & 2 \end{bmatrix} + \left(\frac{1}{4}\right)^{n-1-k} \begin{bmatrix} 2 & -4 \\ \frac{1}{2} & -1 \end{bmatrix} \right\} \begin{bmatrix} 0 \\ 1 \end{bmatrix}$$

$$= \left(\frac{1}{2}\right)^{n-1-k} \begin{bmatrix} -\frac{1}{8} & \frac{3}{4} \end{bmatrix} \begin{bmatrix} -1 & 4 \\ -\frac{1}{2} & 2 \end{bmatrix} \begin{bmatrix} 0 \\ 1 \end{bmatrix}$$

$$+ \left(\frac{1}{4}\right)^{n-1-k} \begin{bmatrix} -\frac{1}{8} & \frac{3}{4} \end{bmatrix} \begin{bmatrix} 2 & -4 \\ \frac{1}{2} & -1 \end{bmatrix} \begin{bmatrix} 0 \\ 1 \end{bmatrix}$$

$$= \left(\frac{1}{2}\right)^{n-1-k} - \frac{1}{4}\left(\frac{1}{4}\right)^{n-1-k} = 2\left(\frac{1}{2}\right)^{n-k} - \left(\frac{1}{4}\right)^{n-k}$$

Thus,

$$s[n] = \sum_{k=0}^{n-1} \left[2\left(\frac{1}{2}\right)^{n-k} - \left(\frac{1}{4}\right)^{n-k} \right] + 1$$

$$= 2\left(\frac{1}{2}\right)^n \sum_{k=0}^{n-1} 2^k - \left(\frac{1}{4}\right)^n \sum_{k=0}^{n-1} 4^k + 1$$

$$= 2\left(\frac{1}{2}\right)^n \left(\frac{1-2^n}{1-2}\right) - \left(\frac{1}{4}\right)^n \left(\frac{1-4^n}{1-4}\right) + 1$$

$$= -2\left(\frac{1}{2}\right)^n + 2 + \frac{1}{3}\left(\frac{1}{4}\right)^n - \frac{1}{3} + 1$$

$$= \frac{8}{3} - 2\left(\frac{1}{2}\right)^n + \frac{1}{3}\left(\frac{1}{4}\right)^n \qquad n \geq 0$$

which is the same result obtained in Prob. 4.32(*c*)

(*b*) By Eq. (7.44) the system function $H(z)$ is given by

$$H(z) = \mathbf{c}(z\mathbf{I} - \mathbf{A})^{-1}\mathbf{b} + d$$

Now

$$(z\mathbf{I} - \mathbf{A})^{-1} = \begin{bmatrix} z & -1 \\ \frac{1}{8} & z-\frac{3}{4} \end{bmatrix}^{-1} = \frac{1}{\left(z-\frac{1}{2}\right)\left(z-\frac{1}{4}\right)} \begin{bmatrix} z-\frac{3}{4} & 1 \\ -\frac{1}{8} & z \end{bmatrix}$$

Thus,

$$H(z) = \frac{1}{\left(z-\frac{1}{2}\right)\left(z-\frac{1}{4}\right)} \begin{bmatrix} -\frac{1}{8}, & \frac{3}{4} \end{bmatrix} \begin{bmatrix} z-\frac{3}{4} & 1 \\ -\frac{1}{8} & z \end{bmatrix} \begin{bmatrix} 0 \\ 1 \end{bmatrix} + 1$$

$$= \frac{-\frac{1}{8}+\frac{3}{4}z}{\left(z-\frac{1}{2}\right)\left(z-\frac{1}{4}\right)} + 1 = \frac{z^2}{\left(z-\frac{1}{2}\right)\left(z-\frac{1}{4}\right)}$$

which is the same result obtained in Prob. 4.32(*a*).

7.30. Consider the discrete-time LTI system described by

$$\mathbf{q}[n+1] = \mathbf{A}\mathbf{q}[n] + \mathbf{b}x[n]$$
$$y[n] = \mathbf{c}\mathbf{q}[n] + dx[n]$$

(a) Show that the unit impulse response $h[n]$ of the system is given by

$$h[n] = \begin{cases} d & n=0 \\ \mathbf{c}\mathbf{A}^{n-1}\mathbf{b} & n>0 \\ 0 & n<0 \end{cases} \tag{7.117}$$

(b) Using Eq. (7.117), find the unit impulse response $h[n]$ of the system in Prob. 7.29.

(a) By setting $\mathbf{q}[0] = \mathbf{0}$, $x[k], = \delta[k]$, and $x[n] = \delta[n]$ in Eq. (7.25), we obtain

$$h[n] = \sum_{k=0}^{n-1} \mathbf{c}\mathbf{A}^{n-1-k}\mathbf{b}\delta[k] + d\delta[n] \tag{7.118}$$

Note that the sum in Eq. (7.118) has no terms for $n = 0$ and that the first term is $\mathbf{c}\mathbf{A}^{n-1}\mathbf{b}$ for $n > 0$. The second term on the right-hand side of Eq. (7.118) is equal to d for $n = 0$ and zero otherwise. Thus, we conclude that

$$h[n] = \begin{cases} d & n=0 \\ \mathbf{c}\mathbf{A}^{n-1}\mathbf{b} & n>0 \\ 0 & n<0 \end{cases}$$

(b) From the result of Prob. 7.29 we have

$$\mathbf{A} = \begin{bmatrix} 0 & 1 \\ -\dfrac{1}{8} & \dfrac{3}{4} \end{bmatrix} \qquad \mathbf{b} = \begin{bmatrix} 0 \\ 1 \end{bmatrix} \qquad \mathbf{c} = \begin{bmatrix} -\dfrac{1}{8} & \dfrac{3}{4} \end{bmatrix} \qquad d = 1$$

and

$$\mathbf{c}\mathbf{A}^{n-1}\mathbf{b} = \left(\frac{1}{2}\right)^{n-1} - \frac{1}{4}\left(\frac{1}{4}\right)^{n-1} \qquad n \geq 1$$

Thus, by Eq. (7.117) $h[n]$ is

$$h[n] = \begin{cases} 1 & n=0 \\ \left(\dfrac{1}{2}\right)^{n-1} - \left(\dfrac{1}{4}\right)^{n-1} & n \geq 1 \\ 0 & n<0 \end{cases}$$

which is the same result obtained in Prob. 4.32(b).

7.31. Use the state space method to solve the difference equation [Prob. 4.38(b)]

$$3y[n] - 4y[n-1] + y[n-2] = x[n] \tag{7.119}$$

with $x[n] = (\frac{1}{2})^n u[n]$ and $y[-1] = 1$, $y[-2] = 2$.

Rewriting Eq. (7.119), we have

$$y[n] - \frac{4}{3}y[n-1] + \frac{1}{3}y[n-2] = \frac{1}{3}x[n]$$

Let $q_1[n] = y[n-2]$ and $q_2[n] = y[n-1]$. Then

$$q_1[n+1] = q_2[n]$$

$$q_2[n+1] = -\frac{1}{3}q_1[n] + \frac{4}{3}q_2[n] + \frac{1}{3}x[n]$$

$$y[n] = -\frac{1}{3}q_1[n] + \frac{4}{3}q_2[n] + \frac{1}{3}x[n]$$

In matrix form

$$\mathbf{q}_1[n+1] = \mathbf{A}\mathbf{q}[n] + \mathbf{b}x[n]$$

$$y[n] = \mathbf{c}\mathbf{q}[n] + dx[n]$$

where

$$\mathbf{A} = \begin{bmatrix} 0 & 1 \\ -\dfrac{1}{3} & \dfrac{4}{3} \end{bmatrix} \qquad \mathbf{b} = \begin{bmatrix} 0 \\ \dfrac{1}{3} \end{bmatrix} \qquad \mathbf{c} = \begin{bmatrix} -\dfrac{1}{3} & \dfrac{4}{3} \end{bmatrix} \qquad d = \frac{1}{3}$$

and

$$\mathbf{q}[0] = \begin{bmatrix} q_1[0] \\ q_2[0] \end{bmatrix} = \begin{bmatrix} y[-2] \\ y[-1] \end{bmatrix} = \begin{bmatrix} 2 \\ 1 \end{bmatrix}$$

Then, by Eq. (7.25)

$$y[n] = \mathbf{c}\mathbf{A}^n\mathbf{q}[0] + \sum_{k=0}^{n-1} \mathbf{c}\mathbf{A}^{n-1-k}\mathbf{b}x[k] + dx[n] \qquad n > 0$$

Now from the result of Prob. 7.24 we have

$$\mathbf{A}^n = \begin{bmatrix} 0 & 1 \\ -\dfrac{1}{3} & \dfrac{4}{3} \end{bmatrix}^n = \begin{bmatrix} -\dfrac{1}{2} & \dfrac{3}{2} \\ -\dfrac{1}{2} & \dfrac{3}{2} \end{bmatrix} + \left(\dfrac{1}{3}\right)^n \begin{bmatrix} \dfrac{3}{2} & -\dfrac{3}{2} \\ \dfrac{1}{2} & -\dfrac{1}{2} \end{bmatrix}$$

and

$$\mathbf{c}\mathbf{A}^n\mathbf{q}[0] = \begin{bmatrix} -\dfrac{1}{3} & \dfrac{4}{3} \end{bmatrix} \left\{ \begin{bmatrix} -\dfrac{1}{2} & \dfrac{3}{2} \\ -\dfrac{1}{2} & \dfrac{3}{2} \end{bmatrix} + \left(\dfrac{1}{3}\right)^n \begin{bmatrix} \dfrac{3}{2} & -\dfrac{3}{2} \\ \dfrac{1}{2} & -\dfrac{1}{2} \end{bmatrix} \right\} \begin{bmatrix} 2 \\ 1 \end{bmatrix}$$

$$= \frac{1}{2} + \frac{1}{6}\left(\frac{1}{3}\right)^n$$

$$\mathbf{c}\mathbf{A}^{n-1-k}\mathbf{b} = \begin{bmatrix} -\dfrac{1}{3} & \dfrac{4}{3} \end{bmatrix} \left\{ \begin{bmatrix} -\dfrac{1}{2} & \dfrac{3}{2} \\ -\dfrac{1}{2} & \dfrac{3}{2} \end{bmatrix} + \left(\dfrac{1}{3}\right)^{n-1-k} \begin{bmatrix} \dfrac{3}{2} & -\dfrac{3}{2} \\ \dfrac{1}{2} & -\dfrac{1}{2} \end{bmatrix} \right\} \begin{bmatrix} 0 \\ \dfrac{1}{3} \end{bmatrix}$$

$$= \frac{1}{2} - \frac{1}{18}\left(\frac{1}{3}\right)^{n-1-k} = \frac{1}{2} - \frac{1}{2}\left(\frac{1}{3}\right)^{n+1-k}$$

Thus,

$$y[n] = \frac{1}{2} + \frac{1}{6}\left(\frac{1}{3}\right)^n + \sum_{k=0}^{n-1}\left[\frac{1}{2} - \frac{1}{2}\left(\frac{1}{3}\right)^{n+1-k}\right]\left(\frac{1}{2}\right)^k + \frac{1}{3}\left(\frac{1}{2}\right)^n$$

$$= \frac{1}{2} + \frac{1}{6}\left(\frac{1}{3}\right)^n + \frac{1}{2}\sum_{k=0}^{n-1}\left(\frac{1}{2}\right)^k - \frac{1}{2}\left(\frac{1}{3}\right)^{n+1}\sum_{k=0}^{n-1}\left(\frac{3}{2}\right)^k + \frac{1}{3}\left(\frac{1}{2}\right)^n$$

$$= \frac{1}{2} + \frac{1}{6}\left(\frac{1}{3}\right)^n + \frac{1}{2}\left[\frac{1-\left(\frac{1}{2}\right)^n}{1-\frac{1}{2}}\right] - \frac{1}{2}\left(\frac{1}{3}\right)^{n+1}\left[\frac{1-\left(\frac{3}{2}\right)^n}{1-\frac{3}{2}}\right] + \frac{1}{3}\left(\frac{1}{2}\right)^n$$

$$= \frac{1}{2} + \frac{1}{6}\left(\frac{1}{3}\right)^n + 1 - \left(\frac{1}{2}\right)^n + \frac{1}{3}\left(\frac{1}{3}\right)^n - \frac{1}{3}\left(\frac{1}{2}\right)^n + \frac{1}{3}\left(\frac{1}{2}\right)^n$$

$$= \frac{3}{2} - \left(\frac{1}{2}\right)^n + \frac{1}{2}\left(\frac{1}{3}\right)^n \qquad n > 0$$

which is the same result obtained in Prob. 4.38(*b*).

7.32. Consider the discrete-time **LTI** system shown in Fig. 7-19.

(*a*) Is the system asymptotically stable?

(*b*) Find the system function $H(z)$.

(*c*) Is the system **BIBO** stable?

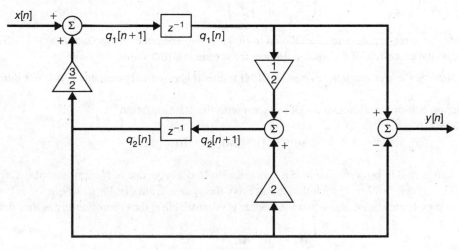

Fig. 7-19

(*a*) From Fig. 7-19 and choosing the state variables $q_1[n]$ and $q_2[n]$ as shown, we obtain

$$q_1[n+1] = \frac{3}{2}q_2[n] + x[n]$$

$$q_2[n+1] = -\frac{1}{2}q_1[n] + 2q_2[n]$$

$$y[n] = q_1[n] - q_2[n]$$

In matrix form

$$\mathbf{q}[n+1] = \mathbf{A}\mathbf{q}[n] + \mathbf{b}x[n]$$

$$y[n] = \mathbf{c}\mathbf{q}[n]$$

where
$$\mathbf{A} = \begin{bmatrix} 0 & \dfrac{3}{2} \\ -\dfrac{1}{2} & 2 \end{bmatrix} \qquad \mathbf{b} = \begin{bmatrix} 1 \\ 0 \end{bmatrix} \qquad \mathbf{c} = \begin{bmatrix} 1 & -1 \end{bmatrix}$$

Now
$$c(\lambda) = |\lambda \mathbf{I} - \mathbf{A}| = \begin{vmatrix} \lambda & -\dfrac{3}{2} \\ \dfrac{1}{2} & \lambda - 2 \end{vmatrix} = \lambda(\lambda - 2) + \frac{3}{4} = \left(\lambda - \frac{1}{2}\right)\left(\lambda - \frac{3}{2}\right)$$

Thus, the eigenvalues of $\mathbf{A}$ are $\lambda_1 = \frac{1}{2}$ and $\lambda_2 = \frac{3}{2}$. Since $|\lambda_2| > 1$, the system is not asymptotically stable.

(b) By Eq. (7.44) the system function $H(z)$ is given by

$$H(z) = \mathbf{c}(z\mathbf{I} - \mathbf{A})^{-1}\mathbf{b} = \begin{bmatrix} 1 & -1 \end{bmatrix} \begin{bmatrix} z & -\dfrac{3}{2} \\ \dfrac{1}{2} & z - 2 \end{bmatrix}^{-1} \begin{bmatrix} 1 \\ 0 \end{bmatrix}$$

$$= \frac{1}{\left(z - \dfrac{1}{2}\right)\left(z - \dfrac{3}{2}\right)} \begin{bmatrix} 1 & -1 \end{bmatrix} \begin{bmatrix} z - 2 & \dfrac{3}{2} \\ -\dfrac{1}{2} & z \end{bmatrix} \begin{bmatrix} 1 \\ 0 \end{bmatrix}$$

$$= \frac{z - \dfrac{3}{2}}{\left(z - \dfrac{1}{2}\right)\left(z - \dfrac{3}{2}\right)} = \frac{1}{z - \dfrac{1}{2}}$$

(c) Note that there is pole-zero cancellation in $H(z)$ at $z = \frac{3}{2}$. Thus, the only pole of $H(z)$ is $\frac{1}{2}$, which lies inside the unit circle of the z-plane. Hence, the system is BIBO stable.

Note that even though the system is BIBO stable, it is essentially unstable if it is not initially relaxed.

7.33. Consider an Nth-order discrete-time LTI system with the state equation

$$\mathbf{q}[n + 1] = \mathbf{A}\mathbf{q}[n] + \mathbf{b}x[n]$$

The system is said to be *controllable* if it is possible to find a sequence of N input samples $x[n_0]$, $x[n_0 + 1], \ldots, x[n_0 + N - 1]$ such that it will drive the system from $\mathbf{q}[n_0] = \mathbf{q}_0$ to $\mathbf{q}[n_0 + N] = \mathbf{q}_1$ and $\mathbf{q}_0$ and $\mathbf{q}_1$ are any finite states. Show that the system is controllable if the *controllability matrix* defined by

$$\mathbf{M}_c = [\mathbf{b} \quad \mathbf{A}\mathbf{b} \quad \cdots \quad \mathbf{A}^{N-1}\mathbf{b}] \tag{7.120}$$

has rank N.

We assume that $n_0 = 0$ and $\mathbf{q}[0] = \mathbf{0}$. Then, by Eq. (7.23) we have

$$\mathbf{q}[N] = \sum_{k=0}^{N-1} \mathbf{A}^{N-1-k}\, \mathbf{b}x[k] \tag{7.121}$$

which can be rewritten as

$$\mathbf{q}[N] = [\mathbf{b} \quad \mathbf{A}\mathbf{b} \quad \cdots \quad \mathbf{A}^{N-1}\mathbf{b}] \begin{bmatrix} x[N-1] \\ x[N-1] \\ \vdots \\ x[0] \end{bmatrix} \tag{7.122}$$

Thus, if $\mathbf{q}[N]$ is to be an arbitrary N-dimensional vector and also to have a nonzero input sequence, as required for controllability, the coefficient matrix in Eq. (7.122) must be nonsingular; that is, the matrix

$$\mathbf{M}_c = [\mathbf{b} \quad \mathbf{Ab} \quad \cdots \quad \mathbf{A}^{N-1}\mathbf{b}]$$

must have rank N.

7.34. Consider an Nth-order discrete-time LTI system with state space representation

$$\mathbf{q}[n + 1] = \mathbf{Aq}[n] + \mathbf{b}x[n]$$
$$y[n] = \mathbf{cq}[n]$$

The system is said to be *observable* if, starting at an arbitrary time index n_0, it is possible to determine the state $\mathbf{q}[n_0] = \mathbf{q}_0$ from the output sequence $y[n_0], y[n_0 + 1], \ldots, y[n_0 + N - 1]$. Show that the system is observable if the *observability matrix* defined by

$$\mathbf{M}_o = \begin{bmatrix} \mathbf{c} \\ \mathbf{cA} \\ \vdots \\ \mathbf{cA}^{N-1} \end{bmatrix} \tag{7.123}$$

has rank N.

We assume that $n_0 = 0$ and $x[n] = 0$. Then, by Eq. (7.25) the output $y[n]$ for $n = 0, 1, \ldots, N - 1$, with $x[n] = 0$, is given by

$$y[n] = \mathbf{cA}^n\mathbf{q}[0] \qquad n = 0, 1, \ldots, N - 1 \tag{7.124}$$

or

$$y[0] = \mathbf{cq}[0]$$
$$y[1] = \mathbf{cAq}[0] \tag{7.125}$$
$$\vdots$$
$$y[N - 1] = \mathbf{cA}^{N-1}\mathbf{q}[0]$$

Rewriting Eq. (7.125) as a matrix equation, we get

$$\begin{bmatrix} y[0] \\ y[1] \\ \vdots \\ y[N-1] \end{bmatrix} = \begin{bmatrix} \mathbf{c} \\ \mathbf{cA} \\ \vdots \\ \mathbf{cA}^{N-1} \end{bmatrix} \mathbf{q}[0] \tag{7.126}$$

Thus, to find a unique solution for $\mathbf{q}[0]$, the coefficient matrix of Eq. (7.126) must be nonsingular; that is, the matrix

$$\mathbf{M}_o = \begin{bmatrix} \mathbf{c} \\ \mathbf{cA} \\ \vdots \\ \mathbf{cA}^{N-1} \end{bmatrix}$$

must have rank N.

7.35. Consider the system in Prob. 7.7.

    (*a*)   Is the system controllable?

    (*b*)   Is the system observable?

    (*c*)   Find the system function $H(z)$.

(a)   From the result of Prob. 7.7 we have

$$\mathbf{A} = \begin{bmatrix} 0 & 1 \\ -\dfrac{1}{8} & \dfrac{3}{4} \end{bmatrix} \qquad \mathbf{b} = \begin{bmatrix} 0 \\ 1 \end{bmatrix} \qquad \mathbf{c} = \begin{bmatrix} -\dfrac{1}{8} & \dfrac{3}{4} \end{bmatrix} \qquad d = 1$$

Now
$$\mathbf{Ab} = \begin{bmatrix} 0 & 1 \\ -\dfrac{1}{8} & \dfrac{3}{4} \end{bmatrix} \begin{bmatrix} 0 \\ 1 \end{bmatrix} = \begin{bmatrix} 1 \\ \dfrac{3}{4} \end{bmatrix}$$

and by Eq. (7.120) the controllability matrix is

$$\mathbf{M}_c = \begin{bmatrix} \mathbf{b} & \mathbf{Ab} \end{bmatrix} = \begin{bmatrix} 0 & 1 \\ 1 & \dfrac{3}{4} \end{bmatrix}$$

and $|\mathbf{M}_c| = -1 \neq 0$. Thus, its rank is 2, and hence the system is controllable.

(b)   Similarly,

$$\mathbf{cA} = \begin{bmatrix} -\dfrac{1}{8} & \dfrac{3}{4} \end{bmatrix} \begin{bmatrix} 0 & 1 \\ -\dfrac{1}{8} & \dfrac{3}{4} \end{bmatrix} = \begin{bmatrix} -\dfrac{3}{32} & \dfrac{7}{16} \end{bmatrix}$$

and by Eq. (7.123) the observability matrix is

$$\mathbf{M}_o = \begin{bmatrix} \mathbf{c} \\ \mathbf{cA} \end{bmatrix} = \begin{bmatrix} -\dfrac{1}{8} & \dfrac{3}{4} \\ -\dfrac{3}{32} & \dfrac{7}{16} \end{bmatrix}$$

and $|\mathbf{M}_o| = -\dfrac{1}{64} \neq 0$. Thus, its rank is 2, and hence the system is observable.

(c)   By Eq. (7.44) the system function $H(z)$ is given by

$$H(z) = \mathbf{c}(z\mathbf{I} - \mathbf{A})^{-1}\mathbf{b} + d = \begin{bmatrix} -\dfrac{1}{8} & \dfrac{3}{4} \end{bmatrix} \begin{bmatrix} z & -1 \\ \dfrac{1}{8} & z - \dfrac{3}{4} \end{bmatrix}^{-1} \begin{bmatrix} 0 \\ 1 \end{bmatrix} + 1$$

$$= \dfrac{1}{\left(z - \dfrac{1}{2}\right)\left(z - \dfrac{1}{4}\right)} \begin{bmatrix} -\dfrac{1}{8} & \dfrac{3}{4} \end{bmatrix} \begin{bmatrix} z - \dfrac{3}{4} & 1 \\ -\dfrac{1}{8} & z \end{bmatrix} \begin{bmatrix} 0 \\ 1 \end{bmatrix} + 1$$

$$= \dfrac{\dfrac{3}{4}z - \dfrac{1}{8}}{\left(z - \dfrac{1}{2}\right)\left(z - \dfrac{1}{4}\right)} + 1 = \dfrac{z^2}{\left(z - \dfrac{1}{2}\right)\left(z - \dfrac{1}{4}\right)}$$

$$= \dfrac{1}{1 - \dfrac{3}{4}z^{-1} + \dfrac{1}{8}z^{-2}}$$

7.36. Consider the system in Prob. 7.7. Assume that

$$\mathbf{q}[0] = \begin{bmatrix} 0 \\ 1 \end{bmatrix}$$

Find $x[0]$ and $x[1]$ such that $\mathbf{q}[2] = \mathbf{0}$.

From Eq. (7.23) we have

$$\mathbf{q}[2] = \mathbf{A}^2\mathbf{q}[0] + \mathbf{Ab}x[0] + \mathbf{b}x[1] = \mathbf{A}^2\mathbf{q}[0] + \begin{bmatrix} \mathbf{b} & \mathbf{Ab} \end{bmatrix} \begin{bmatrix} x[1] \\ x[0] \end{bmatrix}$$

Thus,

$$\begin{bmatrix} 0 \\ 0 \end{bmatrix} = \begin{bmatrix} 0 & 1 \\ -\dfrac{1}{8} & \dfrac{3}{4} \end{bmatrix}^2 \begin{bmatrix} 0 \\ 1 \end{bmatrix} + \begin{bmatrix} 0 & 1 \\ 1 & \dfrac{3}{4} \end{bmatrix} \begin{bmatrix} x[1] \\ x[0] \end{bmatrix}$$

$$= \begin{bmatrix} \dfrac{3}{4} \\ \dfrac{7}{16} \end{bmatrix} + \begin{bmatrix} x[0] \\ x[1] + \dfrac{3}{4}x[0] \end{bmatrix}$$

from which we obtain $x[0] = -\dfrac{3}{4}$ and $x[1] = \dfrac{1}{8}$.

7.37. Consider the system in Prob. 7.7. We observe $y[0] = 1$ and $y[1] = 0$ with $x[0] = x[1] = 0$. Find the initial state $\mathbf{q}[0]$.

Using Eq. (7.125), we have

$$\begin{bmatrix} y[0] \\ y[1] \end{bmatrix} = \begin{bmatrix} \mathbf{c} \\ \mathbf{cA} \end{bmatrix} \mathbf{q}[0]$$

Thus,

$$\begin{bmatrix} 1 \\ 0 \end{bmatrix} = \begin{bmatrix} -\dfrac{1}{8} & \dfrac{3}{4} \\ -\dfrac{3}{32} & \dfrac{7}{16} \end{bmatrix} \begin{bmatrix} q_1[0] \\ q_2[0] \end{bmatrix}$$

Solving for $q_1[0]$ and $q_2[0]$, we obtain

$$\mathbf{q}[0] = \begin{bmatrix} q_1[0] \\ q_2[0] \end{bmatrix} = \begin{bmatrix} -\dfrac{1}{8} & \dfrac{3}{4} \\ -\dfrac{3}{32} & \dfrac{7}{16} \end{bmatrix}^{-1} \begin{bmatrix} 1 \\ 0 \end{bmatrix} = \begin{bmatrix} 28 \\ 6 \end{bmatrix}$$

7.38. Consider the system in Prob. 7.32.

(a) Is the system controllable?

(b) Is the system observable?

(a) From the result of Prob. 7.32 we have

$$\mathbf{A} = \begin{bmatrix} 0 & \dfrac{3}{2} \\ -\dfrac{1}{2} & 2 \end{bmatrix} \qquad \mathbf{b} = \begin{bmatrix} 1 \\ 0 \end{bmatrix} \qquad \mathbf{c} = \begin{bmatrix} 1 & -1 \end{bmatrix}$$

Now

$$\mathbf{Ab} = \begin{bmatrix} 0 & \dfrac{3}{2} \\ -\dfrac{1}{2} & 2 \end{bmatrix} \begin{bmatrix} 1 \\ 0 \end{bmatrix} = \begin{bmatrix} 0 \\ -\dfrac{1}{2} \end{bmatrix}$$

and by Eq. (7.120) the controllability matrix is

$$\mathbf{M}_c = \begin{bmatrix} \mathbf{b} & \mathbf{Ab} \end{bmatrix} = \begin{bmatrix} 1 & 0 \\ 0 & -\dfrac{1}{2} \end{bmatrix}$$

and $|\mathbf{M}_c| = -\dfrac{1}{2} \neq 0$. Thus, its rank is 2, and hence the system is controllable.

(*b*) Similarly,

$$cA = \begin{bmatrix} 1 & -1 \end{bmatrix} \begin{bmatrix} 0 & \dfrac{3}{2} \\ -\dfrac{1}{2} & 2 \end{bmatrix} = \begin{bmatrix} \dfrac{1}{2} & -\dfrac{1}{2} \end{bmatrix}$$

and by Eq. (7.123) the observability matrix is

$$M_o = \begin{bmatrix} c \\ cA \end{bmatrix} = \begin{bmatrix} 1 & -1 \\ \dfrac{1}{2} & -\dfrac{1}{2} \end{bmatrix}$$

and $|M_o| = 0$. Thus, its rank is less than 2, and hence the system is not observable.

Note from the result from Prob. 7.32(*b*) that the system function $H(z)$ has pole-zero cancellation. If $H(z)$ has pole-zero cancellation, then the system cannot be both controllable and observable.

Solutions of State Equations for Continuous-Time LTI Systems

7.39. Find e^{At} for

$$A = \begin{bmatrix} 0 & 1 \\ -6 & -5 \end{bmatrix}$$

using the Cayley-Hamilton theorem method.

First, we find the characteristic polynomial $c(\lambda)$ of A.

$$c(\lambda) = |\lambda I - A| = \begin{vmatrix} \lambda & -1 \\ 6 & \lambda + 5 \end{vmatrix}$$
$$= \lambda^2 + 5\lambda + 6 = (\lambda + 2)(\lambda + 3)$$

Thus, the eigenvalues of A are $\lambda_1 = -2$ and $\lambda_2 = -3$. Hence, by Eqs. (7.66) and (7.67) we have

$$e^{At} = b_0 I + b_1 A = \begin{bmatrix} b_0 & b_1 \\ -6b_1 & b_0 - 5b_1 \end{bmatrix}$$

and b_0 and b_1 are the solutions of

$$b_0 - 2b_1 = e^{-2t}$$
$$b_0 - 3b_1 = e^{-3t}$$

from which we get

$$b_0 = 3e^{-2t} - 2e^{-3t} \qquad b_1 = e^{-2t} - e^{-3t}$$

Hence,

$$e^{At} = \begin{bmatrix} 3e^{-2t} - 2e^{-3t} & e^{-2t} - e^{-3t} \\ -6e^{-2t} + 6e^{-3t} & -2e^{-2t} + 3e^{-3t} \end{bmatrix}$$
$$= e^{-2t} \begin{bmatrix} 3 & 1 \\ -6 & -2 \end{bmatrix} + e^{-3t} \begin{bmatrix} -2 & -1 \\ 6 & 3 \end{bmatrix}$$

7.40. Repeat Prob. 7.39 using the diagonalization method.

Let **x** be an eigenvector of **A** associated with λ. Then

$$[\lambda\mathbf{I} - \mathbf{A}]\mathbf{x} = \mathbf{0}$$

For λ = λ₁ = −2 we have

$$\begin{bmatrix} -2 & -1 \\ 6 & 3 \end{bmatrix}\begin{bmatrix} x_1 \\ x_2 \end{bmatrix} = \begin{bmatrix} 0 \\ 0 \end{bmatrix}$$

The solutions of this system are given by $x_2 = -2x_1$. Thus, the eigenvectors associated with λ₁ are those vectors of the form

$$\mathbf{x}_1 = \alpha\begin{bmatrix} 1 \\ -2 \end{bmatrix} \qquad \text{with } \alpha \neq 0$$

For λ = λ₂ = −3 we have

$$\begin{bmatrix} -3 & -1 \\ 6 & 2 \end{bmatrix}\begin{bmatrix} x_1 \\ x_2 \end{bmatrix} = \begin{bmatrix} 0 \\ 0 \end{bmatrix}$$

The solutions of this system are given by $x_2 = -3x_1$. Thus, the eigenvectors associated with λ₂ are those vectors of the form

$$\mathbf{x}_2 = \beta\begin{bmatrix} 1 \\ -3 \end{bmatrix} \qquad \text{with } \beta \neq 0$$

Let α = β = 1 in the above expressions and let

$$\mathbf{P} = \begin{bmatrix} \mathbf{x}_1 & \mathbf{x}_2 \end{bmatrix} = \begin{bmatrix} 1 & 1 \\ -2 & -3 \end{bmatrix}$$

Then

$$\mathbf{P}^{-1} = -\begin{bmatrix} -3 & -1 \\ 2 & 1 \end{bmatrix} = \begin{bmatrix} 3 & 1 \\ -2 & -1 \end{bmatrix}$$

and by Eq. (7.68) we obtain

$$e^{\mathbf{A}t} = \begin{bmatrix} 1 & 1 \\ -2 & -3 \end{bmatrix}\begin{bmatrix} e^{-2t} & 0 \\ 0 & e^{-3t} \end{bmatrix}\begin{bmatrix} 3 & 1 \\ -2 & -1 \end{bmatrix} = \begin{bmatrix} 3e^{-2t} - 2e^{-3t} & e^{-2t} - e^{-3t} \\ -6e^{-2t} + 6e^{-3t} & -2e^{-2t} + 3e^{-3t} \end{bmatrix}$$

$$= e^{-2t}\begin{bmatrix} 3 & 1 \\ -6 & -2 \end{bmatrix} + e^{-3t}\begin{bmatrix} -2 & -1 \\ 6 & 3 \end{bmatrix}$$

7.41. Repeat Prob. 7.39 using the spectral decomposition method.

Since all eigenvalues of **A** are distinct, by Eq. (7.33) we have

$$\mathbf{E}_1 = \frac{1}{\lambda_1 - \lambda_2}(\mathbf{A} - \lambda_2\mathbf{I}) = \mathbf{A} + 3\mathbf{I} = \begin{bmatrix} 3 & 1 \\ -6 & -2 \end{bmatrix}$$

$$\mathbf{E}_2 = \frac{1}{\lambda_2 - \lambda_1}(\mathbf{A} - \lambda_1\mathbf{I}) = -(\mathbf{A} + 2\mathbf{I}) = \begin{bmatrix} -2 & -1 \\ 6 & 3 \end{bmatrix}$$

Then by Eq. (7.70) we obtain

$$e^{\mathbf{A}t} = e^{-2t}\mathbf{E}_1 + e^{-3t}\mathbf{E}_2 = e^{-2t}\begin{bmatrix} 3 & 1 \\ -6 & -2 \end{bmatrix} + e^{-3t}\begin{bmatrix} -2 & -1 \\ 6 & 3 \end{bmatrix}$$

$$= \begin{bmatrix} 3e^{-2t} - 2e^{-3t} & e^{-2t} - e^{-3t} \\ -6e^{-2t} + 6e^{-3t} & -2e^{-2t} + 3e^{-3t} \end{bmatrix}$$

7.42. Repeat Prob. 7.39 using the Laplace transform method.

First, we must find $(s\mathbf{I} - \mathbf{A})^{-1}$.

$$(s\mathbf{I} - \mathbf{A})^{-1} = \begin{bmatrix} s & -1 \\ 6 & s+5 \end{bmatrix}^{-1} = \frac{1}{(s+2)(s+3)}\begin{bmatrix} s+5 & 1 \\ -6 & s \end{bmatrix}$$

$$= \begin{bmatrix} \dfrac{s+5}{(s+2)(s+3)} & \dfrac{1}{(s+2)(s+3)} \\ -\dfrac{6}{(s+2)(s+3)} & \dfrac{s}{(s+2)(s+3)} \end{bmatrix}$$

$$= \begin{bmatrix} \dfrac{3}{s+2} - \dfrac{2}{s+3} & \dfrac{1}{s+2} - \dfrac{1}{s+3} \\ -\dfrac{6}{s+2} + \dfrac{6}{s+3} & -\dfrac{2}{s+2} + \dfrac{3}{s+3} \end{bmatrix}$$

Then, by Eq. (7.71) we obtain

$$e^{\mathbf{A}t} = \mathscr{L}^{-1}\left\{(s\mathbf{I} - \mathbf{A})^{-1}\right\} = \begin{bmatrix} 3e^{-2t} - 2e^{-3t} & e^{-2t} - e^{-3t} \\ -6e^{-2t} + 6e^{-3t} & -2e^{-2t} + 3e^{-3t} \end{bmatrix}$$

Again we note that when the eigenvalues of $\mathbf{A}$ are all distinct, the spectral decomposition method is computationally the most efficient method of evaluating $e^{\mathbf{A}t}$.

7.43. Find $e^{\mathbf{A}t}$ for

$$\mathbf{A} = \begin{bmatrix} -2 & 1 \\ 1 & -2 \end{bmatrix}$$

The characteristic polynomial $c(\lambda)$ of $\mathbf{A}$ is

$$c(\lambda) = |\lambda\mathbf{I} - \mathbf{A}| = \begin{vmatrix} \lambda+2 & -1 \\ -1 & \lambda+2 \end{vmatrix}$$

$$= \lambda^2 + 4\lambda + 3 = (\lambda+1)(\lambda+3)$$

Thus, the eigenvalues of $\mathbf{A}$ are $\lambda_1 = -1$ and $\lambda_2 = -3$. Since all eigenvalues of $\mathbf{A}$ are distinct, by Eq. (7.33) we have

$$\mathbf{E}_1 = -\frac{1}{2}(\mathbf{A} + 3\mathbf{I}) = \frac{1}{2}\begin{bmatrix} 1 & 1 \\ 1 & 1 \end{bmatrix} = \begin{bmatrix} \dfrac{1}{2} & \dfrac{1}{2} \\ \dfrac{1}{2} & \dfrac{1}{2} \end{bmatrix}$$

$$\mathbf{E}_2 = -\frac{1}{2}(\mathbf{A} + \mathbf{I}) = -\frac{1}{2}\begin{bmatrix} -1 & 1 \\ 1 & -1 \end{bmatrix} = \begin{bmatrix} \dfrac{1}{2} & -\dfrac{1}{2} \\ -\dfrac{1}{2} & \dfrac{1}{2} \end{bmatrix}$$

Then, by Eq. (7.70) we obtain

$$e^{\mathbf{A}t} = e^{-t}\begin{bmatrix} \dfrac{1}{2} & \dfrac{1}{2} \\[2mm] \dfrac{1}{2} & \dfrac{1}{2} \end{bmatrix} + e^{-3t}\begin{bmatrix} \dfrac{1}{2} & -\dfrac{1}{2} \\[2mm] -\dfrac{1}{2} & \dfrac{1}{2} \end{bmatrix}$$

$$= \begin{bmatrix} \dfrac{1}{2}e^{-t} + \dfrac{1}{2}e^{-3t} & \dfrac{1}{2}e^{-t} - \dfrac{1}{2}e^{-3t} \\[2mm] \dfrac{1}{2}e^{-t} - \dfrac{1}{2}e^{-3t} & \dfrac{1}{2}e^{-t} + \dfrac{1}{2}e^{-3t} \end{bmatrix}$$

7.44. Given matrix

$$\mathbf{A} = \begin{bmatrix} 0 & -2 & 1 \\ 0 & 0 & 3 \\ 0 & 0 & 0 \end{bmatrix}$$

(a)   Show that $\mathbf{A}$ is nilpotent of index 3.

(b)   Using the result from part (a) find $e^{\mathbf{A}t}$.

(a)   By direct multiplication we have

$$\mathbf{A}^2 = \mathbf{A}\mathbf{A} = \begin{bmatrix} 0 & -2 & 1 \\ 0 & 0 & 3 \\ 0 & 0 & 0 \end{bmatrix}\begin{bmatrix} 0 & -2 & 1 \\ 0 & 0 & 3 \\ 0 & 0 & 0 \end{bmatrix} = \begin{bmatrix} 0 & 0 & -6 \\ 0 & 0 & 0 \\ 0 & 0 & 0 \end{bmatrix}$$

$$\mathbf{A}^3 = \mathbf{A}^2\mathbf{A} = \begin{bmatrix} 0 & 0 & -6 \\ 0 & 0 & 0 \\ 0 & 0 & 0 \end{bmatrix}\begin{bmatrix} 0 & -2 & 1 \\ 0 & 0 & 3 \\ 0 & 0 & 0 \end{bmatrix} = \begin{bmatrix} 0 & 0 & 0 \\ 0 & 0 & 0 \\ 0 & 0 & 0 \end{bmatrix}$$

Thus, $\mathbf{A}$ is nilpotent of index 3.

(b)   By definition (7.53) and the result from part (a)

$$e^{\mathbf{A}t} = \mathbf{I} + t\mathbf{A} + \frac{t^2}{2!}\mathbf{A}^2 + \frac{t^3}{3!}\mathbf{A}^3 + \cdots = \mathbf{I} + t\mathbf{A} + \frac{t^2}{2}\mathbf{A}^2$$

$$= \begin{bmatrix} 1 & 0 & 0 \\ 0 & 1 & 0 \\ 0 & 0 & 1 \end{bmatrix} + t\begin{bmatrix} 0 & -2 & 1 \\ 0 & 0 & 3 \\ 0 & 0 & 0 \end{bmatrix} + \frac{t^2}{2}\begin{bmatrix} 0 & 0 & -6 \\ 0 & 0 & 0 \\ 0 & 0 & 0 \end{bmatrix} = \begin{bmatrix} 1 & -2t & t-3t^2 \\ 0 & 1 & 3t \\ 0 & 0 & 1 \end{bmatrix}$$

7.45. Find $e^{\mathbf{A}t}$ for matrix $\mathbf{A}$ in Prob. 7.44 using the Cayley-Hamilton theorem method.

First, we find the characteristic polynomial $c(\lambda)$ of $\mathbf{A}$.

$$c(\lambda) = |\lambda\mathbf{I} - \mathbf{A}| = \begin{bmatrix} \lambda & 2 & -1 \\ 0 & \lambda & -3 \\ 0 & 0 & \lambda \end{bmatrix} = \lambda^3$$

Thus, $\lambda = 0$ is the eigenvalues of $\mathbf{A}$ with multiplicity 3. By Eq. (7.66) we have

$$e^{\mathbf{A}t} = b_0\mathbf{I} + b_1\mathbf{A} + b_2\mathbf{A}^2$$

where b_0, b_1, and b_2 are determined by setting $\lambda = 0$ in the following equations [App. A, Eqs. (A.59) and (A.60)]:

$$b_0 + b_1\lambda + b_2\lambda^2 = e^{\lambda t}$$
$$b_1 + 2b_2\lambda = te^{\lambda t}$$
$$2b_2 = t^2 e^{\lambda t}$$

Thus,

$$b_0 = 1 \qquad b_1 = t \qquad b_2 = \frac{t^2}{2}$$

Hence,

$$e^{\mathbf{A}t} = \mathbf{I} + t\mathbf{A} + \frac{t^2}{2}\mathbf{A}^2$$

which is the same result obtained in Prob. 7.44(*b*).

7.46. Show that

$$e^{\mathbf{A}+\mathbf{B}} = e^{\mathbf{A}}e^{\mathbf{B}}$$

provided **A** and **B** commute; that is, $\mathbf{AB} = \mathbf{BA}$.

By Eq. (7.53)

$$e^{\mathbf{A}}e^{\mathbf{B}} = \left(\sum_{k=0}^{\infty} \frac{1}{k!}\mathbf{A}^k\right)\left(\sum_{m=0}^{\infty} \frac{1}{m!}\mathbf{B}^m\right)$$

$$= \left(\mathbf{I} + \mathbf{A} + \frac{1}{2!}\mathbf{A}^2 + \cdots\right)\left(\mathbf{I} + \mathbf{B} + \frac{1}{2!}\mathbf{B}^2 + \cdots\right)$$

$$= \mathbf{I} + \mathbf{A} + \mathbf{B} + \frac{1}{2!}\mathbf{A}^2 + \mathbf{AB} + \frac{1}{2!}\mathbf{B}^2 + \cdots$$

$$e^{\mathbf{A}+\mathbf{B}} = \mathbf{I} + (\mathbf{A}+\mathbf{B}) + \frac{1}{2!}(\mathbf{A}+\mathbf{B})^2 + \cdots$$

$$= \mathbf{I} + \mathbf{A} + \mathbf{B} + \frac{1}{2!}\mathbf{A}^2 + \frac{1}{2}\mathbf{AB} + \frac{1}{2}\mathbf{BA} + \frac{1}{2!}\mathbf{B}^2 + \cdots$$

and

$$e^{\mathbf{A}}e^{\mathbf{B}} - e^{\mathbf{A}+\mathbf{B}} = \frac{1}{2}(\mathbf{AB} - \mathbf{BA}) + \cdots$$

Thus, if $\mathbf{AB} = \mathbf{BA}$, then

$$e^{\mathbf{A}+\mathbf{B}} = e^{\mathbf{A}}e^{\mathbf{B}}$$

7.47. Consider the matrix

$$\mathbf{A} = \begin{bmatrix} 2 & 1 & 0 \\ 0 & 2 & 1 \\ 0 & 0 & 2 \end{bmatrix}$$

Now we decompose **A** as

$$\mathbf{A} = \mathbf{\Lambda} + \mathbf{N}$$

where

$$\mathbf{\Lambda} = \begin{bmatrix} 2 & 0 & 0 \\ 0 & 2 & 0 \\ 0 & 0 & 2 \end{bmatrix} \qquad \text{and} \qquad \mathbf{N} = \begin{bmatrix} 0 & 1 & 0 \\ 0 & 0 & 1 \\ 0 & 0 & 0 \end{bmatrix}$$

(a) Show that the matrix $\mathbf{N}$ is nilpotent of index 3.

(b) Show that $\mathbf{\Lambda}$ and $\mathbf{N}$ commute; that is, $\mathbf{\Lambda N} = \mathbf{N \Lambda}$.

(c) Using the results from parts (a) and (b), find $e^{\mathbf{\Lambda}t}$.

(a) By direct multiplication we have

$$\mathbf{N}^2 = \mathbf{NN} = \begin{bmatrix} 0 & 1 & 0 \\ 0 & 0 & 1 \\ 0 & 0 & 0 \end{bmatrix}\begin{bmatrix} 0 & 1 & 0 \\ 0 & 0 & 1 \\ 0 & 0 & 0 \end{bmatrix} = \begin{bmatrix} 0 & 0 & 1 \\ 0 & 0 & 0 \\ 0 & 0 & 0 \end{bmatrix}$$

$$\mathbf{N}^3 = \mathbf{N}^2\mathbf{N} = \begin{bmatrix} 0 & 0 & 1 \\ 0 & 0 & 0 \\ 0 & 0 & 0 \end{bmatrix}\begin{bmatrix} 0 & 1 & 0 \\ 0 & 0 & 1 \\ 0 & 0 & 0 \end{bmatrix} = \begin{bmatrix} 0 & 0 & 0 \\ 0 & 0 & 0 \\ 0 & 0 & 0 \end{bmatrix}$$

Thus, $\mathbf{N}$ is nilpotent of index 3.

(b) Since the diagonal matrix $\mathbf{\Lambda}$ can be expressed as $2\mathbf{I}$, we have

$$\mathbf{\Lambda N} = 2\mathbf{IN} = 2\mathbf{N} = 2\mathbf{NI} = \mathbf{N}(2\mathbf{I}) = \mathbf{N\Lambda}$$

that is, $\mathbf{\Lambda}$ and $\mathbf{N}$ commute.

(c) Since $\mathbf{\Lambda}$ and $\mathbf{N}$ commute, then, by the result from Prob. 7.46

$$e^{\mathbf{\Lambda}t} = e^{(\mathbf{\Lambda} + \mathbf{N})t} = e^{\mathbf{\Lambda}t}e^{\mathbf{N}t}$$

Now [see App. A, Eq. (A.49)]

$$e^{\mathbf{\Lambda}t} = \begin{bmatrix} e^{2t} & 0 & 0 \\ 0 & e^{2t} & 0 \\ 0 & 0 & e^{2t} \end{bmatrix} = e^{2t}\begin{bmatrix} 1 & 0 & 0 \\ 0 & 1 & 0 \\ 0 & 0 & 1 \end{bmatrix} = e^{2t}\mathbf{I}$$

and using similar justification as in Prob. 7.44(b), we have

$$e^{\mathbf{N}t} = \mathbf{I} + t\mathbf{N} + \frac{t^2}{2!}\mathbf{N}^2$$

$$= \begin{bmatrix} 1 & 0 & 0 \\ 0 & 1 & 0 \\ 0 & 0 & 1 \end{bmatrix} + \begin{bmatrix} 0 & t & 0 \\ 0 & 0 & t \\ 0 & 0 & 0 \end{bmatrix} + \begin{bmatrix} 0 & 0 & \dfrac{t^2}{2} \\ 0 & 0 & 0 \\ 0 & 0 & 0 \end{bmatrix} = \begin{bmatrix} 1 & t & \dfrac{t^2}{2} \\ 0 & 1 & t \\ 0 & 0 & 1 \end{bmatrix}$$

Thus,

$$e^{\mathbf{\Lambda}t} = e^{\mathbf{\Lambda}t}e^{\mathbf{N}t} = e^{2t}\mathbf{I}e^{\mathbf{N}t} = e^{2t}e^{\mathbf{N}t} = e^{2t}\begin{bmatrix} 1 & t & \dfrac{t^2}{2} \\ 0 & 1 & t \\ 0 & 0 & 1 \end{bmatrix}$$

7.48. Using the state variables method, solve the second-order linear differential equation

$$y''(t) + 5y'(t) + 6y(t) = x(t) \tag{7.127}$$

with the initial conditions $y(0) = 2$, $y'(0) = 1$, and $x(t) = e^{-t}u(t)$ (Prob. 3.38).

Let the state variables $q_1(t)$ and $q_2(t)$ be

$$q_1(t) = y(t) \qquad q_2(t) = y'(t)$$

Then the state space representation of Eq. (7.127) is given by [Eq. (7.19)]

$$\dot{\mathbf{q}}(t) = \mathbf{A}\mathbf{q}(t) + \mathbf{b}x(t)$$
$$y(t) = \mathbf{c}\mathbf{q}(t)$$

with
$$\mathbf{A} = \begin{bmatrix} 0 & 1 \\ -6 & -5 \end{bmatrix} \quad \mathbf{b} = \begin{bmatrix} 0 \\ 1 \end{bmatrix} \quad \mathbf{c} = \begin{bmatrix} 1 & 0 \end{bmatrix} \quad \mathbf{q}[0] = \begin{bmatrix} q_1[0] \\ q_2[0] \end{bmatrix} = \begin{bmatrix} 2 \\ 1 \end{bmatrix}$$

Thus, by Eq. (7.65)

$$y(t) = \mathbf{c}e^{\mathbf{A}t}\mathbf{q}(0) + \int_0^t \mathbf{c}e^{\mathbf{A}(t-\tau)}\mathbf{b}x(\tau)\,d\tau$$

with $d = 0$. Now, from the result of Prob. 7.39,

$$e^{\mathbf{A}t} = e^{-2t}\begin{bmatrix} 3 & 1 \\ -6 & -2 \end{bmatrix} + e^{-3t}\begin{bmatrix} -2 & -1 \\ 6 & 3 \end{bmatrix}$$

and
$$\mathbf{c}e^{\mathbf{A}t}\mathbf{q}(0) = \begin{bmatrix} 1 & 0 \end{bmatrix}\left\{ e^{-2t}\begin{bmatrix} 3 & 1 \\ -6 & -2 \end{bmatrix} + e^{-3t}\begin{bmatrix} -2 & -1 \\ 6 & 3 \end{bmatrix} \right\}\begin{bmatrix} 2 \\ 1 \end{bmatrix}$$

$$= 7e^{-2t} - 5e^{-3t}$$

$$\mathbf{c}e^{\mathbf{A}(t-\tau)}\mathbf{b} = \begin{bmatrix} 1 & 0 \end{bmatrix}\left\{ e^{-2(t-\tau)}\begin{bmatrix} 3 & 1 \\ -6 & -2 \end{bmatrix} + e^{-3(t-\tau)}\begin{bmatrix} -2 & -1 \\ 6 & 3 \end{bmatrix} \right\}\begin{bmatrix} 0 \\ 1 \end{bmatrix}$$

$$= e^{-2(t-\tau)} - e^{-3(t-\tau)}$$

Thus,

$$y(t) = 7e^{-2t} - 5e^{-3t} + \int_0^t (e^{-2(t-\tau)} - e^{-3(t-\tau)})e^{-\tau}\,d\tau$$

$$= 7e^{-2t} - 5e^{-3t} + e^{-2t}\int_0^t e^{\tau}\,d\tau - e^{-3t}\int_0^t e^{2\tau}\,d\tau$$

$$= \frac{1}{2}e^{-t} + 6e^{-2t} - \frac{9}{2}e^{-3t} \qquad t > 0$$

which is the same result obtained in Prob. 3.38.

7.49. Consider the network shown in Fig. 7-20. The initial voltages across the capacitors C_1 and C_2 are $\frac{1}{2}$ V and 1 V, respectively. Using the state variable method, find the voltages across these capacitors for $t > 0$. Assume that $R_1 = R_2 = R_3 = 1\ \Omega$ and $C_1 = C_2 = 1$ F.

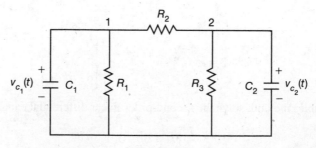

Fig. 7-20

Let the state variables $q_1(t)$ and $q_2(t)$ be

$$q_1(t) = v_{C_1}(t) \qquad q_2(t) = v_{C_2}(t)$$

Applying Kirchhoff's current law at nodes 1 and 2, we get

$$C_1 \dot{q}_1(t) + \frac{q_1(t)}{R_1} + \frac{q_1(t) - q_2(t)}{R_2} = 0$$

$$C_2 \dot{q}_2(t) + \frac{q_2(t)}{R_3} + \frac{q_2(t) - q_1(t)}{R_2} = 0$$

Substituting the values of R_1, R_2, R_3, C_1, and C_2 and rearranging, we obtain

$$\dot{q}_1(t) = -2q_1(t) + q_2(t)$$

$$\dot{q}_2(t) = q_1(t) - 2q_2(t)$$

In matrix form

$$\dot{\mathbf{q}}(t) = \mathbf{A}\mathbf{q}(t)$$

with
$$\mathbf{A} = \begin{bmatrix} -2 & 1 \\ 1 & -2 \end{bmatrix} \quad \text{and} \quad \mathbf{q}(0) = \begin{bmatrix} \frac{1}{2} \\ 1 \end{bmatrix}$$

Then, by Eq. (7.63) with $x(t) = 0$ and using the result from Prob. 7.43, we get

$$\mathbf{q}(t) = e^{\mathbf{A}t}\mathbf{q}(0) = \left\{ e^{-t} \begin{bmatrix} \frac{1}{2} & \frac{1}{2} \\ \frac{1}{2} & \frac{1}{2} \end{bmatrix} + e^{-3t} \begin{bmatrix} \frac{1}{2} & -\frac{1}{2} \\ -\frac{1}{2} & \frac{1}{2} \end{bmatrix} \right\} \begin{bmatrix} \frac{1}{2} \\ 1 \end{bmatrix}$$

$$= \begin{bmatrix} \dfrac{3}{4}e^{-t} - \dfrac{1}{4}e^{-3t} \\ \dfrac{3}{4}e^{-t} + \dfrac{1}{4}e^{-3t} \end{bmatrix}$$

Thus,

$$v_{C_1}(t) = \frac{3}{4}e^{-t} - \frac{1}{4}e^{-3t} \quad \text{and} \quad v_{C_2}(t) = \frac{3}{4}e^{-t} + \frac{1}{4}e^{-3t}$$

7.50. Consider the continuous-time LTI system shown in Fig. 7-21.

(a) Is the system asymptotically stable?

(b) Find the system function $H(s)$.

(c) Is the system BIBO stable?

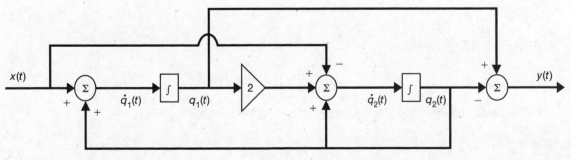

Fig. 7-21

(a)   From Fig. 7-21 and choosing the state variables $q_1(t)$ and $q_2(t)$ as shown, we obtain

$$\dot{q}_1(t) = q_2(t) + x(t)$$
$$\dot{q}_2(t) = 2q_1(t) + q_2(t) - x(t)$$
$$y(t) = q_1(t) - q_2(t)$$

In matrix form

$$\dot{\mathbf{q}}(t) = \mathbf{A}\mathbf{q}(t) + \mathbf{b}x(t)$$
$$y(t) = \mathbf{c}\mathbf{q}(t)$$

where

$$\mathbf{A} = \begin{bmatrix} 0 & 1 \\ 2 & 1 \end{bmatrix} \qquad \mathbf{b} = \begin{bmatrix} 1 \\ -1 \end{bmatrix} \qquad \mathbf{c} = \begin{bmatrix} 1 & -1 \end{bmatrix}$$

Now

$$c(\lambda) = |\lambda\mathbf{I} - \mathbf{A}| = \begin{bmatrix} \lambda & -1 \\ -2 & \lambda - 1 \end{bmatrix} = \lambda^2 - \lambda - 2 = (\lambda + 1)(\lambda - 2)$$

Thus, the eigenvalues of $\mathbf{A}$ are $\lambda_1 = -1$ and $\lambda_2 = 2$. Since $\text{Re}\{\lambda_2\} > 0$, the system is not asymptotically stable.

(b)   By Eq. (7.52) the system function $H(s)$ is given by

$$H(s) = \mathbf{c}(s\mathbf{I} - \mathbf{A})^{-1}\mathbf{b} = \begin{bmatrix} 1 & -1 \end{bmatrix}\begin{bmatrix} s & -1 \\ -2 & s-1 \end{bmatrix}^{-1}\begin{bmatrix} 1 \\ -1 \end{bmatrix}$$

$$= \frac{1}{(s+1)(s-2)}\begin{bmatrix} 1 & -1 \end{bmatrix}\begin{bmatrix} s-1 & 1 \\ 2 & s \end{bmatrix}\begin{bmatrix} 1 \\ -1 \end{bmatrix}$$

$$= \frac{2(s-2)}{(s+1)(s-2)} = \frac{2}{s+1}$$

(c)   Note that there is pole-zero cancellation in $H(s)$ at $s = 2$. Thus, the only pole of $H(s)$ is -1, which is located in the left-hand side of the s-plane. Hence, the system is BIBO stable.

Again, it is noted that the system is essentially unstable if the system is not initially relaxed.

7.51. Consider an Nth-order continuous-time LTI system with state equation

$$\dot{\mathbf{q}}(t) = \mathbf{A}\mathbf{q}(t) + \mathbf{b}x(t)$$

The system is said to be *controllable* if it is possible to find an input $x(t)$ which will drive the system from $\mathbf{q}(t_0) = \mathbf{q}_0$ to $\mathbf{q}(t_1) = \mathbf{q}_1$ in a specified finite time and $\mathbf{q}_0$ and $\mathbf{q}_1$ are any finite state vectors. Show that the system is controllable if the *controllability matrix* defined by

$$\mathbf{M}_c = [\mathbf{b} \quad \mathbf{A}\mathbf{b} \quad \cdots \quad \mathbf{A}^{N-1}\mathbf{b}] \tag{7.128}$$

has rank N.

We assume that $t_0 = 0$ and $\mathbf{q}[0] = \mathbf{0}$. Then, by Eq. (7.63) we have

$$\mathbf{q}_1 = \mathbf{q}(t_1) = e^{\mathbf{A}t_1}\int_0^{t_1} e^{-\mathbf{A}\tau}\mathbf{b}x(\tau)\,d\tau \tag{7.129}$$

Now, by the Cayley-Hamilton theorem we can express $e^{-\mathbf{A}\tau}$ as

$$e^{-\mathbf{A}\tau} = \sum_{k=0}^{N-1} \alpha_k(\tau)\mathbf{A}^k \tag{7.130}$$

Substituting Eq. (7.130) into Eq. (7.129) and rearranging, we get

$$\mathbf{q}_1 = e^{\mathbf{A}t_1} = \left[\sum_{k=0}^{N-1} \mathbf{A}^k \mathbf{b} \int_0^{t_1} \alpha_k(\tau) x(\tau) d\tau \right] \tag{7.131}$$

Let

$$\int_0^{t_1} \alpha_k(\tau) x(\tau) d\tau = \beta_k$$

Then Eq. (7.131) can be rewritten as

$$e^{-\mathbf{A}t_1} \mathbf{q}_1 = \sum_{k=0}^{N-1} \mathbf{A}^k \mathbf{b} \beta_k$$

or

$$e^{-\mathbf{A}t_1} \mathbf{q}_1 = \begin{bmatrix} \mathbf{b} & \mathbf{Ab} & \cdots & \mathbf{A}^{N-1}\mathbf{b} \end{bmatrix} \begin{bmatrix} \beta_0 \\ \beta_1 \\ \vdots \\ \beta_{N-1} \end{bmatrix} \tag{7.132}$$

For any given state $\mathbf{q}_1$ we can determine from Eq. (7.132) unique β_k's ($k = 0, 1, \ldots, N-1$), and hence $x(t)$, if the coefficients matrix of Eq. (7.132) is nonsingular, that is, the matrix

$$\mathbf{M}_c = \begin{bmatrix} \mathbf{b} & \mathbf{Ab} & \cdots & \mathbf{A}^{N-1}\mathbf{b} \end{bmatrix}$$

has rank N.

7.52. Consider an Nth-order continuous-time LTI system with state space representation

$$\dot{\mathbf{q}}(t) = \mathbf{Aq}(t) + \mathbf{b}x(t)$$
$$y(t) = \mathbf{cq}(t)$$

The system is said to be *observable* if any initial state $\mathbf{q}(t_0)$ can be determined by examining the system output $y(t)$ over some finite period of time from t_0 to t_1. Show that the system is observable if the *observability matrix* defined by

$$\mathbf{M}_o = \begin{bmatrix} \mathbf{c} \\ \mathbf{cA} \\ \vdots \\ \mathbf{cA}^{N-1} \end{bmatrix} \tag{7.133}$$

has rank N.

We prove this by contradiction. Suppose that the rank of $\mathbf{M}_o$ is less than N. Then there exists an initial state $\mathbf{q}[0] = \mathbf{q}_0 \neq \mathbf{0}$ such that

$$\mathbf{M}_o \mathbf{q}_0 = \mathbf{0}$$

or

$$\mathbf{cq}_0 = \mathbf{cAq}_0 = \cdots = \mathbf{cA}^{N-1}\mathbf{q}_0 = 0 \tag{7.134}$$

Now from Eq. (7.65), for $x(t) = 0$ and $t_0 = 0$,

$$y(t) = \mathbf{c}e^{\mathbf{A}t}\mathbf{q}_0 \tag{7.135}$$

However, by the Cayley-Hamilton theorem, $e^{\mathbf{A}t}$ can be expressed as

$$e^{\mathbf{A}t} = \sum_{k=0}^{N-1} \alpha_k(t)\mathbf{A}^k \tag{7.136}$$

Substituting Eq. (7.136) into Eq. (7.135), we get

$$y(t) = \sum_{k=0}^{N-1} \alpha_k(t)\mathbf{c}\mathbf{A}^k\mathbf{q}_0 = 0 \tag{7.137}$$

in view of Eq. (7.134). Thus, $\mathbf{q}_0$ is indistinguishable from the zero state, and hence, the system is not observable. Therefore, if the system is to be observable, then $\mathbf{M}_o$ must have rank N.

7.53. Consider the system in Prob. 7.50.

(*a*) Is the system controllable?

(*b*) Is the system observable?

(*a*) From the result from Prob. 7.50 we have

$$\mathbf{A} = \begin{bmatrix} 0 & 1 \\ 2 & 1 \end{bmatrix} \qquad \mathbf{b} = \begin{bmatrix} 1 \\ -1 \end{bmatrix} \qquad \mathbf{c} = \begin{bmatrix} 1 & -1 \end{bmatrix}$$

Now

$$\mathbf{Ab} = \begin{bmatrix} 0 & 1 \\ 2 & 1 \end{bmatrix}\begin{bmatrix} 1 \\ -1 \end{bmatrix} = \begin{bmatrix} -1 \\ 1 \end{bmatrix}$$

and by Eq. (7.128) the controllability matrix is

$$\mathbf{M}_c = \begin{bmatrix} \mathbf{b} & \mathbf{Ab} \end{bmatrix} = \begin{bmatrix} 1 & -1 \\ -1 & 1 \end{bmatrix}$$

and $|\mathbf{M}_c| = 0$. Thus, it has a rank less than 2, and hence, the system is not controllable.

(*b*) Similarly,

$$\mathbf{cA} = \begin{bmatrix} 1 & -1 \end{bmatrix}\begin{bmatrix} 0 & 1 \\ 2 & 1 \end{bmatrix} = \begin{bmatrix} -2 & 0 \end{bmatrix}$$

and by Eq. (7.133) the observability matrix is

$$\mathbf{M}_o = \begin{bmatrix} \mathbf{c} \\ \mathbf{cA} \end{bmatrix} = \begin{bmatrix} 1 & -1 \\ -2 & 0 \end{bmatrix}$$

and $|\mathbf{M}_0| = -2 \neq 0$. Thus, its rank is 2, and hence, the system is observable.

Note from the result from Prob. 7.50(*b*) that the system function $H(s)$ has pole-zero cancellation. As in
the discrete-time case, if $H(s)$ has pole-zero cancellation, then the system cannot be both controllable and observable.

7.54. Consider the system shown in Fig. 7-22.

(*a*) Is the system controllable?

(*b*) Is the system observable?

(*c*) Find the system function $H(s)$.

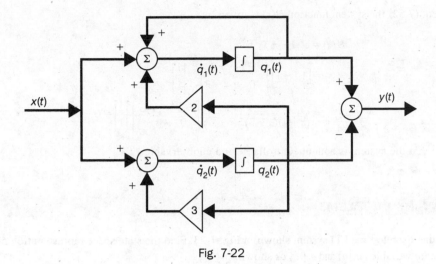

Fig. 7-22

(*a*) From Fig. 7-22 and choosing the state variables $q_1(t)$ and $q_2(t)$ as shown, we have

$$\dot{q}_1(t) = q_1(t) + 2q_2(t) + x(t)$$
$$\dot{q}_2(t) = 3q_2(t) + x(t)$$
$$y(t) = q_1(t) - q_2(t)$$

In matrix form

$$\dot{\mathbf{q}}_1(t) = \mathbf{A}\mathbf{q}(t) + \mathbf{b}x(t)$$
$$y(t) = \mathbf{c}\mathbf{q}(t)$$

where

$$\mathbf{A} = \begin{bmatrix} 1 & 2 \\ 0 & 3 \end{bmatrix} \qquad \mathbf{b} = \begin{bmatrix} 1 \\ 1 \end{bmatrix} \qquad \mathbf{c} = \begin{bmatrix} 1 & -1 \end{bmatrix}$$

Now

$$\mathbf{A}\mathbf{b} = \begin{bmatrix} 1 & 2 \\ 0 & 3 \end{bmatrix}\begin{bmatrix} 1 \\ 1 \end{bmatrix} = \begin{bmatrix} 3 \\ 3 \end{bmatrix}$$

and by Eq. (7.128) the controllability matrix is

$$\mathbf{M}_c = \begin{bmatrix} \mathbf{b} & \mathbf{A}\mathbf{b} \end{bmatrix} = \begin{bmatrix} 1 & 3 \\ 1 & 3 \end{bmatrix}$$

and $|\mathbf{M}_c| = 0$. Thus, its rank is less than 2, and hence, the system is not controllable.

(*b*) Similarly,

$$\mathbf{c}\mathbf{A} = \begin{bmatrix} 1 & -1 \end{bmatrix}\begin{bmatrix} 1 & 2 \\ 0 & 3 \end{bmatrix} = \begin{bmatrix} 1 & -1 \end{bmatrix}$$

and by Eq. (7.133) the observability matrix is

$$\mathbf{M}_o = \begin{bmatrix} \mathbf{c} \\ \mathbf{c}\mathbf{A} \end{bmatrix} = \begin{bmatrix} 1 & -1 \\ 1 & -1 \end{bmatrix}$$

and $|\mathbf{M}_o| = 0$. Thus, its rank is less than 2, and hence, the system is not observable.

(*c*) By Eq. (7.52) the system function $H(s)$ is given by

$$H(s) = \mathbf{c}(s\mathbf{I} - \mathbf{A})^{-1}\mathbf{b}$$

$$= \begin{bmatrix} 1 & -1 \end{bmatrix} \begin{bmatrix} s-1 & -2 \\ 0 & s-3 \end{bmatrix}^{-1} \begin{bmatrix} 1 \\ 1 \end{bmatrix}$$

$$= \frac{1}{(s-1)(s-3)} \begin{bmatrix} 1 & -1 \end{bmatrix} \begin{bmatrix} s-3 & 2 \\ 0 & s-1 \end{bmatrix} \begin{bmatrix} 1 \\ 1 \end{bmatrix} = 0$$

Note that the system is both uncontrollable and unobservable.

SUPPLEMENTARY PROBLEMS

7.55. Consider the discrete-time LTI system shown in Fig. 7-23. Find the state space representation of the system with the state variables $q_1[n]$ and $q_2[n]$ as shown.

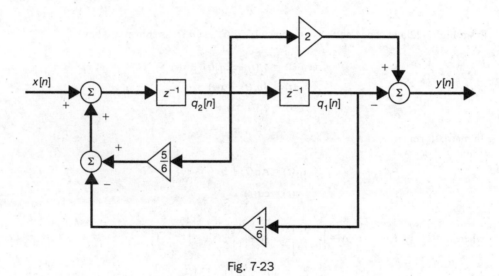

Fig. 7-23

7.56. Consider the discrete-time LTI system shown in Fig. 7-24. Find the state space representation of the system with the state variables $q_1[n]$ and $q_2[n]$ as shown.

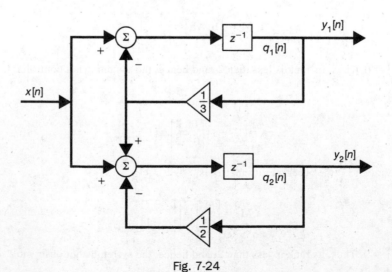

Fig. 7-24

7.57. Consider the discrete-time LTI system shown in Fig. 7-25.

    (*a*) Find the state space representation of the system with the state variables $q_1[n]$ and $q_2[n]$ as shown.

    (*b*) Find the system function $H(z)$.

    (*c*) Find the difference equation relating $x[n]$ and $y[n]$.

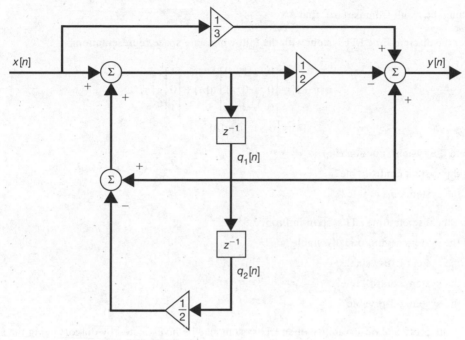

Fig. 7-25

7.58. A discrete-time LTI system is specified by the difference equation

$$y[n] + y[n-1] - 6y[n-2] = 2\,x[n-1] + x[n-2]$$

Write the two canonical forms of state representation for the system.

7.59. Find $\mathbf{A}^n$ for

$$\mathbf{A} = \begin{bmatrix} 0 & 1 \\ -\dfrac{1}{6} & \dfrac{5}{6} \end{bmatrix}$$

    (*a*) Using the Cayley-Hamilton theorem method.

    (*b*) Using the diagonalization method.

7.60. Find $\mathbf{A}^n$ for

$$\mathbf{A} = \begin{bmatrix} 3 & 0 & 0 \\ 0 & -2 & 1 \\ 0 & 4 & 1 \end{bmatrix}$$

    (*a*) Using the spectral decomposition method.

    (*b*) Using the z-transform method.

7.61. Given a matrix

$$\mathbf{A} = \begin{bmatrix} -1 & 2 & 2 \\ 2 & -1 & 2 \\ 2 & 2 & -1 \end{bmatrix}$$

    (a) Find the minimal polynomial $m(\lambda)$ of $\mathbf{A}$.

    (b) Using the result from part (a), find $\mathbf{A}^n$.

7.62. Consider the discrete-time LTI system with the following state space representation:

$$\mathbf{q}[n+1] = \begin{bmatrix} 0 & 1 & 0 \\ 0 & 0 & 1 \\ 0 & -1 & 2 \end{bmatrix} \mathbf{q}[n] + \begin{bmatrix} 1 \\ 0 \\ 1 \end{bmatrix} x[n]$$

$$y[n] = \begin{bmatrix} 0 & 1 & 0 \end{bmatrix} \mathbf{q}[n]$$

    (a) Find the system function $H(z)$.

    (b) Is the system controllable?

    (c) Is the system observable?

7.63. Consider the discrete-time LTI system in Prob. 7.55.

    (a) Is the system asymptotically stable?

    (b) Is the system BIBO stable?

    (c) Is the system controllable?

    (d) Is the system observable?

7.64. The controllability and observability of an LTI system may be investigated by diagonalizing the system matrix $\mathbf{A}$. A system with a state space representation

$$\mathbf{v}[n+1] = \mathbf{\Lambda}\mathbf{v}[n] + \hat{\mathbf{b}}x[n]$$

$$y[n] = \hat{\mathbf{c}}\mathbf{v}[n]$$

(where $\mathbf{\Lambda}$ is a diagonal matrix) is controllable if the vector $\hat{\mathbf{b}}$ has no zero elements, and it is observable if the vector $\hat{\mathbf{c}}$ has no zero elements. Consider the discrete-time LTI system in Prob. 7.55.

    (a) Let $\mathbf{v}[n] = \mathbf{T}\mathbf{q}[n]$. Find the matrix $\mathbf{T}$ such that the new state space representation will have a diagonal system matrix.

    (b) Write the new state space representation of the system.

    (c) Using the result from part (b), investigate the controllability and observability of the system.

 7.65. Consider the network shown in Fig. 7-26. Find a state space representation for the network with the state variables $q_1(t) = i_L(t)$, $q_2(t) = v_C(t)$ and outputs $y_1(t), = i_1(t), y_2(t) = v_C(t)$, assuming $R_1 = R_2 = 1\ \Omega, L = 1\ H$, and $C = 1\ F$.

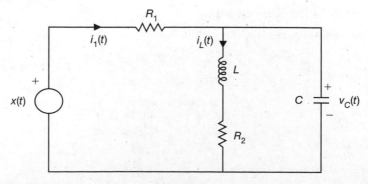

Fig. 7-26

7.66. Consider the continuous-time LTI system shown in Fig. 7-27

(a) Find the state space representation of the system with the state variables $q_1(t)$ and $q_2(t)$ as shown.

(b) For what values of α will the system be asymptotically stable?

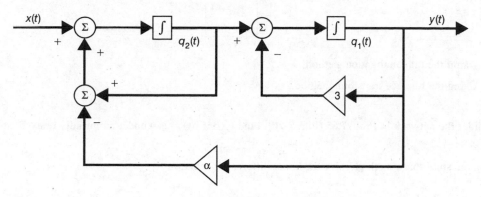

Fig. 7-27

7.67. A continuous-time LTI system is described by

$$H(s) = \frac{3s^2 - 1}{s^3 + 3s^2 - s - 2}$$

Write the two canonical forms of state representation for the system.

7.68. Consider the continuous-time LTI system shown in Fig. 7-28.

(a) Find the state space representation of the system with the state variables $q_1(t)$ and $q_2(t)$ as shown.

(b) Is the system asymptotically stable?

(c) Find the system function $H(s)$.

(d) Is the system BIBO stable?

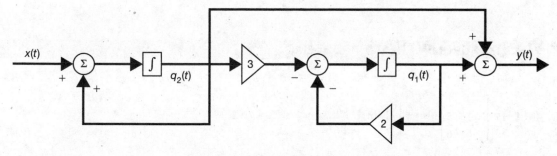

Fig. 7-28

7.69. Find $e^{\mathbf{A}t}$ for

$$\mathbf{A} = \begin{bmatrix} -1 & 1 \\ -1 & -1 \end{bmatrix}$$

(a) Using the Cayley-Hamilton theorem method.

(b) Using the spectral decomposition method.

7.70. Consider the matrix $\mathbf{A}$ in Prob. 7.69. Find $e^{-\mathbf{A}t}$ and show that $e^{-\mathbf{A}t} = [e^{\mathbf{A}t}]^{-1}$.

7.71. Find $e^{\mathbf{A}t}$ for

$$\mathbf{A} = \begin{bmatrix} 0 & 1 \\ -2 & -3 \end{bmatrix}$$

    (*a*) Using the diagonalization method.

    (*b*) Using the Laplace transform method.

7.72. Consider the network in Prob. 7.65 (Fig. 7-26). Find $v_C(t)$ if $x(t) = u(t)$ under an initially relaxed condition.

7.73. Using the state space method, solve the linear differential equation

$$y''(t) + 3y'(t) + 2y(t) = 0$$

with the initial conditions $y(0) = 0, y'(0) = 1$.

7.74. As in the discrete-time case, controllability and observability of a continuous-time LTI system may be investigated by diagonalizing the system matrix $\mathbf{A}$. A system with state space representation

$$\dot{\mathbf{v}}(t) = \mathbf{\Lambda}\mathbf{v}(t) + \hat{\mathbf{b}}x(y)$$
$$y(t) = \hat{\mathbf{c}}\mathbf{v}(t)$$

where $\mathbf{\Lambda}$ is a diagonal matrix, is controllable if the vector $\hat{\mathbf{b}}$ has no zero elements, and is observable if the vector $\hat{\mathbf{c}}$ has no zero elements. Consider the continuous-time system in Prob. 7.50.

    (*a*) Find a new state space representation of the system by diagonalizing the system matrix $\mathbf{A}$.

    (*b*) Is the system controllable?

    (*c*) Is the system observable?

ANSWERS TO SUPPLEMENTARY PROBLEMS

7.55. $\mathbf{q}[n+1] = \begin{bmatrix} 0 & 1 \\ -\dfrac{1}{6} & \dfrac{5}{6} \end{bmatrix} \mathbf{q}[n] + \begin{bmatrix} 0 \\ 1 \end{bmatrix} x[n]$

$y[n] = \begin{bmatrix} -1 & 2 \end{bmatrix} \mathbf{q}[n]$

7.56. $\mathbf{q}[n+1] = \begin{bmatrix} -\dfrac{1}{3} & 0 \\ \dfrac{1}{3} & -\dfrac{1}{2} \end{bmatrix} \mathbf{q}[n] + \begin{bmatrix} 1 \\ 1 \end{bmatrix} x[n]$

$y[n] = \begin{bmatrix} 1 & 0 \\ 0 & 1 \end{bmatrix} \mathbf{q}[n]$

7.57. (a) $\mathbf{q}[n+1] = \begin{bmatrix} 1 & -\dfrac{1}{2} \\ 1 & 0 \end{bmatrix} \mathbf{q}[n] + \begin{bmatrix} 1 \\ 0 \end{bmatrix} x[n]$

$$y[n] = \begin{bmatrix} \dfrac{1}{2} & \dfrac{1}{4} \end{bmatrix} \mathbf{q}[n] - \dfrac{1}{6} x[n]$$

(b) $H(z) = -\dfrac{1}{6} \dfrac{z^2 - 4z - 1}{z^2 - z + \dfrac{1}{2}}$

(c) $y[n] - y[n-1] + \dfrac{1}{2} y[n-2] = -\dfrac{1}{6} x[n] + \dfrac{2}{3} x[n-1] + \dfrac{1}{6} x[n-2]$

7.58. (1) $\mathbf{q}[n+1] = \begin{bmatrix} -1 & 1 \\ 6 & 0 \end{bmatrix} \mathbf{q}[n] + \begin{bmatrix} 2 \\ 1 \end{bmatrix} x[n]$

$$y[n] = \begin{bmatrix} 1 & 0 \end{bmatrix} \mathbf{q}[n]$$

(2) $\mathbf{v}[n+1] = \begin{bmatrix} 0 & 1 \\ 6 & -1 \end{bmatrix} \mathbf{v}[n] + \begin{bmatrix} 0 \\ 1 \end{bmatrix} x[n]$

$$y[n] = \begin{bmatrix} 1 & 2 \end{bmatrix} \mathbf{v}[n]$$

7.59. $\mathbf{A}^n = \begin{bmatrix} -2\left(\dfrac{1}{2}\right)^n + 3\left(\dfrac{1}{3}\right)^n & 6\left(\dfrac{1}{2}\right)^n - 6\left(\dfrac{1}{3}\right)^n \\ -\left(\dfrac{1}{2}\right)^n + \left(\dfrac{1}{3}\right)^n & 3\left(\dfrac{1}{2}\right)^n - 2\left(\dfrac{1}{3}\right)^n \end{bmatrix}$

7.60. $\mathbf{A}^n = \begin{bmatrix} (3)^n & 0 & 0 \\ 0 & \dfrac{1}{5}(2)^n + \dfrac{4}{5}(-3)^n & \dfrac{1}{5}(2)^n - \dfrac{1}{5}(-3)^n \\ 0 & \dfrac{4}{5}(2)^n - \dfrac{4}{5}(-3)^n & \dfrac{4}{5}(2)^n - \dfrac{1}{5}(-3)^n \end{bmatrix}$

7.61. (a) $m(\lambda) = (\lambda - 3)(\lambda + 3) = \lambda^2 - 9$

(b) $\mathbf{A}^n = \dfrac{1}{3} \begin{bmatrix} 3^n + 2(-3)^n & 3^n - (-3)^n & 3^n - (-3)^n \\ 3^n - (-3)^n & 3^n + 2(-3)^n & 3^n - (-3)^n \\ 3^n - (-3)^n & 3^n - (-3)^n & 3^n + 2(-3)^n \end{bmatrix}$

7.62. (a) $H(z) = \dfrac{1}{(z-1)^2}$

(b) The system is controllable.

(c) The system is not observable.

7.63. (a) The system is asymptotically stable.

(b) The system is BIBO stable.

(c) The system is controllable.

(d) The system is not observable.

7.64. *(a)* $\mathbf{T} = \begin{bmatrix} 1 & -2 \\ -1 & 3 \end{bmatrix}$

     (b) $\mathbf{v}[n+1] = \begin{bmatrix} \dfrac{1}{3} & 0 \\ 0 & \dfrac{1}{2} \end{bmatrix} \mathbf{v}[n] + \begin{bmatrix} -2 \\ 3 \end{bmatrix} x[n]$

         $y[n] = \begin{bmatrix} -1 & 0 \end{bmatrix} \mathbf{v}[n]$

     (c) The system is controllable but not observable.

7.65. $\dot{\mathbf{q}}(t) = \begin{bmatrix} -1 & 1 \\ -1 & -1 \end{bmatrix} \mathbf{q}(t) + \begin{bmatrix} 0 \\ 1 \end{bmatrix} x(t)$

     $\mathbf{y}(t) = \begin{bmatrix} 0 & -1 \\ 0 & 1 \end{bmatrix} \mathbf{q}(t) + \begin{bmatrix} 1 \\ 0 \end{bmatrix} x(t)$

7.66. *(a)* $\dot{\mathbf{q}}(t) = \begin{bmatrix} -3 & 1 \\ -\alpha & 1 \end{bmatrix} \mathbf{q}(t) + \begin{bmatrix} 0 \\ 1 \end{bmatrix} x(t)$

         $y(t) = \begin{bmatrix} 1 & 0 \end{bmatrix} \mathbf{q}(t)$

     (b) $\alpha \geq 4$

7.67. (1) $\dot{\mathbf{q}}(t) = \begin{bmatrix} -3 & 1 & 0 \\ 1 & 0 & 1 \\ 2 & 0 & 0 \end{bmatrix} \mathbf{q}(t) + \begin{bmatrix} 3 \\ 0 \\ -1 \end{bmatrix} x(t)$

         $y(t) = \begin{bmatrix} 1 & 0 & 0 \end{bmatrix} \mathbf{q}(t)$

     (2) $\dot{\mathbf{v}}(t) = \begin{bmatrix} 0 & 1 & 0 \\ 0 & 0 & 1 \\ 2 & 1 & -3 \end{bmatrix} \mathbf{v}(t) + \begin{bmatrix} 0 \\ 0 \\ 1 \end{bmatrix} x(t)$

         $y(t) = \begin{bmatrix} -1 & 0 & 3 \end{bmatrix} \mathbf{v}(t)$

7.68. *(a)* $\dot{\mathbf{q}}(t) = \begin{bmatrix} -2 & -3 \\ 0 & 1 \end{bmatrix} \mathbf{q}(t) + \begin{bmatrix} 0 \\ 1 \end{bmatrix} x(t)$

         $y(t) = \begin{bmatrix} 1 & 1 \end{bmatrix} \mathbf{q}(t)$

     (b) The system is not asymptotically stable.

     (c) $H(s) = \dfrac{1}{s+2}$

     (d) The system is BIBO stable.

7.69. $e^{\mathbf{A}t} = e^{-t} \begin{bmatrix} \cos t & \sin t \\ -\sin t & \cos t \end{bmatrix}$

7.70. $e^{-\mathbf{A}t} = e^{t} \begin{bmatrix} \cos t & -\sin t \\ \sin t & \cos t \end{bmatrix}$

7.71. $e^{\mathbf{A}t} = \begin{bmatrix} 2e^{-t} - e^{-2t} & e^{-t} - e^{-2t} \\ -2e^{-t} + 2e^{-2t} & -e^{-t} + 2e^{-2t} \end{bmatrix}$

7.72. $v_C(t) = \dfrac{1}{2}(1 + e^{-t}\sin t - e^{-t}\cos t),\ t > 0$

7.73. $y(t) = e^{-t} - e^{-2t},\ t > 0$

7.74. (a) $\dot{\mathbf{v}}(t) = \begin{bmatrix} -1 & 0 \\ 0 & 2 \end{bmatrix}\mathbf{v}(t) + \begin{bmatrix} 1 \\ 0 \end{bmatrix}x(t)$

$y(t) = \begin{bmatrix} 2 & -1 \end{bmatrix}\mathbf{v}(t)$

(b) The system is not controllable.

(c) The system is observable.

CHAPTER 8

Random Signals

8.1 Introduction

Random signals, as mentioned in Chap. 1, are those signals that take random values at any given time and must be characterized statistically. However, when observed over a long period, a random signal may exhibit certain regularities that can be described in terms of probabilities and statistical averages. The probabilistic model used to describe random signals is called a random (or stochastic) process.

8.2 Random Processes

A. Definition:

Consider a random experiment with outcomes λ and a sample space S. If to every outcome $\lambda \in S$ we assign a real-valued time function $X(t, \lambda)$, we create a *random* (or *stochastic*) *process*. A random process $X(t, \lambda)$ is therefore a function of two parameters, the time t and the outcome λ. For a specific λ, say, λ_i, we have a single time function $X(t, \lambda_i) = x_i(t)$. This time function is called a *sample function* or a *realization of the process*. The totality of all sample functions is called an *ensemble*. For a specific time t_j, $X(t_j, \lambda) = X_j$ denotes a random variable. For fixed $t(= t_j)$ and fixed $\lambda(= \lambda_i)$, $X(t_j, \lambda_i) = x_i(t_j)$ is a number.

Thus, a random process is sometimes defined as a family of random variables indexed by the parameter $t \in T$, where T is called the *index set*.

Fig. 8-1 illustrates the concepts of the sample space of the random experiment, outcomes of the experiment, associated sample functions, and random variables resulting from taking two measurements of the sample functions.

In the following we use the notation $X(t)$ to represent $X(t, \lambda)$.

EXAMPLE 8.1 Consider a random experiment of flipping a coin. The sample space is $S = \{H, T\}$ where H denotes the outcome that "head" appears and T denotes the outcome that "tail" appears. Let

$$X(t, H) = x_1(t) = \sin \omega_1 t$$
$$X(t, T) = x_2(t) = \sin \omega_2 t$$

where ω_1 and ω_2 are some fixed numbers. Then $X(t)$ is a random signal with $x_1(t)$ and $x_2(t)$ as sample functions. Note that $x_1(t)$ and $x_2(t)$ are deterministic signals. Randomness of $X(t)$ comes from the outcomes of flipping a coin.

EXAMPLE 8.2 Consider a random experiment of flipping a coin repeatedly and observing the sequence of outcomes. Then $S = \{\lambda_i, i = 1, 2, \ldots\}$, where $\lambda_i = H$ or T.

Let
$$X(t, \lambda_i) = \sin (\Omega_i\, t), \quad (i - 1)\, T \leq t \leq iT$$

where $\Omega_i = \omega_1$ if $\lambda_i = H$ and $\Omega_i = \omega_2$ if $\lambda_i = T$.

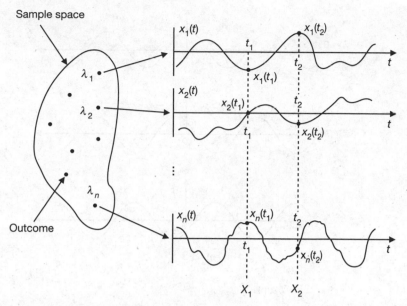

Fig. 8-1 Random process.

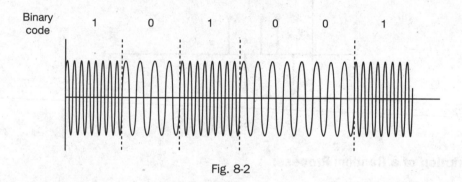

Fig. 8-2

One realization (or sample function) of the random signal $X(t)$ is shown in Fig. 8-2. This kind of random signal is the sort of signal that might be produced by a frequency shift keying (FSK) modem where the frequencies are determined by random sequence of data bits 1 or 0 (by replacing $H = 1$ and $T = 0$).

EXAMPLE 8.3   Often a random signal $X(t)$ is specified in terms of random variables.

$$X(t) = a \cos(\omega_0 t + \Theta)$$

where a and ω_0 are fixed amplitude and frequency and Θ is a random variable (r.v.) uniformly distributed over $[0, 2\pi]$; that is, r.v. Θ is defined by $\Theta(\lambda) = \lambda$ for each λ in $S = [0, 2\pi]$. That is,

$$X(t, \lambda) = a \cos(\omega_0 t + \lambda) \text{ for } 0 \le \lambda \le 2\pi$$

The ensemble of $X(t, \lambda)$ is the set of cosine functions that have the same amplitude and frequency, but whose phase angle are functions of uniform r.v. over $S = [0, 2\pi]$. Some sampling functions of $X(t, \lambda)$ are plotted in Fig. 8-3.

EXAMPLE 8.4   Let $X_1, X_2, \ldots$ be independent r.v. with

$$P\{X_n = 1\} = P\{X_n = -1\} = \frac{1}{2} \text{ for each } n. \text{ Let } X(n) = \{X_n, n \ge 0\} \text{ with } X_0 = 0.$$

Then $X(n)$ is a discrete-time random sequence. A sample sequence of $X(n)$ is shown in Fig. 8-4.

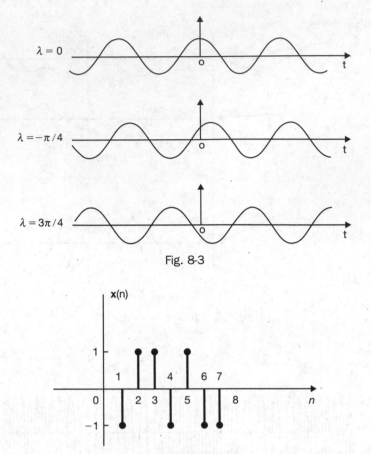

Fig. 8-3

Fig. 8-4

B.   Description of a Random Process:

In a random process $\{X(t), t \in T\}$, the index set T is called the *parameter set* of the random process. The values assumed by $X(t)$ are called *states*, and the set of all possible values forms the *state space E* of the random process. If the index set T of a random process is discrete, then the process is called a *discrete-parameter* (or *discrete-time*) process. A discrete-parameter process is also called a *random sequence* and is denoted by $\{X_n, n = 1, 2, ...\}$. If T is continuous, then we have a *continuous-parameter* (or *continuous-time*) process. If the state space E of a random process is discrete, then the process is called a *discrete-state* process, often referred to as a *chain*. In this case, the state space E is often assumed to be $\{0, 1, 2, ...\}$. If the state space E is continuous, then we have a *continuous-state* process.

A complex random process $X(t)$ is defined by

$$X(t) = X_1(t) + jX_2(t)$$

where $X_1(t)$ and $X_2(t)$ are (real) random processes and $j = \sqrt{-1}$. Throughout this book, all random processes are real random processes unless specified otherwise.

8.3   Statistics of Random Processes

A.   Probabilistic Expressions:

Consider a random process $X(t)$. For a particular time t_1, $X(t_1) = X_1$ is a random variable, and its distribution function $F_X(x_1; t_1)$ is defined as

$$F_X(x_1; t_1) = P\{X(t_1) \leq x_1\} \tag{8.1}$$

where x_1 is any real number.

And $F_X(x_1; t_1)$ is called the *first-order distribution* of $X(t)$. The corresponding first-order density function is obtained by

$$f_X(x_1; t_1) = \frac{\partial F_X(x_1; t_1)}{\partial x_1} \tag{8.2}$$

Similarly, given t_1 and t_2, $X(t_1) = X_1$ and $X(t_2) = X_2$ represent two random variables. Their joint distribution is called the *second-order distribution* and is given by

$$F_X(x_1, x_2; t_1, t_2) = P\{X(t_1) \le x_1, X(t_2) \le x_2\} \tag{8.3}$$

where x_1 and x_2 are any real numbers.

The corresponding second-order density function is obtained by

$$f_X(x_1, x_2; t_1, t_2) = \frac{\partial^2 F_X(x_1, x_2; t_1, t_2)}{\partial x_1 \partial x_2} \tag{8.4}$$

In a similar manner, for n random variables $X(t_i) = X_i (i = 1, \ldots, n)$, the *nth-order distribution* is

$$F_X(x_1, \ldots, x_n; t_1, \ldots, t_n) = P\{X(t_1) \le x_1, \ldots, X(t_n) \le x_n\} \tag{8.5}$$

The corresponding nth-order density function is

$$f_X(x_1, \ldots, x_n; t_1, \ldots, t_n) = \frac{\partial^n F_X(x_1, \ldots, x_n; t_1, \ldots, t_n)}{\partial x_1 \ldots \partial x_n} \tag{8.6}$$

In a similar manner, we can define a joint distribution between two random processes $X(t)$ and $Y(t)$. The joint distribution for $X(t_1)$ and $Y(t_2)$ is defined by

$$F_{XY}(x_1, y_2; t_1, t_2) = P\{X(t_1) \le x_1, Y(t_2) \le y_2\} \tag{8.7}$$

and corresponding joint density function by

$$f_{XY}(x_1, y_2; t_1, t_2) = \frac{\partial^2}{\partial x_1 \partial y_2} F_{XY}(x_1, y_2; t_1, t_2) \tag{8.8}$$

The joint nth-order distribution for $X(t)$ and $Y(t)$ is defined by

$$F_{XY}(x_1, \ldots, x_n; y_1, \ldots, y_n; t_1, \ldots, t_n)$$
$$= P\{X(t_1) \le x_1, \ldots, X(t_{n1}) \le x_n; Y(t_1) \le y_1, \ldots, Y(t_n) \le y_n\} \tag{8.9}$$

and the corresponding nth-order density function by

$$f_{XY}(x_1, \ldots, x_n; y_1, \ldots, y_n; t_1, \ldots, t_n)$$
$$= \frac{\partial^{2n}}{\partial x_1 \cdots \partial x_n \partial y_1 \cdots \partial y_n} F_{XY}(x_1, \ldots, x_n; y_1, \ldots, y_n; t_1, \ldots, t_n) \tag{8.10}$$

B. Statistical Averages:

As in the case of random variables, random processes are often described by using *statistical averages* (or *ensemble averages*).

The *mean* of $X(t)$ is defined by

$$\mu_X(t) = E[X(t)] = \int_{-\infty}^{\infty} x f_X(x; t) \, dx \tag{8.11}$$

where $X(t)$ is treated as a random variable for a fixed value of t.

For discrete time processes, we use the following notation:

$$\mu_X(n) = E[X(n)] = \sum_n x_n p_X(x_n) \tag{8.12}$$

where $p_X(x_n) = P(X = x_n)$.

The *autocorrelation* of $X(t)$ is defined by

$$R_{XX}(t_1, t_2) = E[X(t_1)X(t_2)]$$

$$= \int_{-\infty}^{\infty} \int_{-\infty}^{\infty} x_1 x_2 f_X(x_1, x_2; t_1, t_2) dx_1 dx_2 \tag{8.13}$$

The autocorrelation describes the relationship (correlation) between two samples of $X(t)$. In order to see how the correlation between two samples depends on how far apart the samples are spaced, the autocorrelation function is often expressed as

$$R_{XX}(t, t+\tau) = E[X(t)X(t+\tau)] \tag{8.14}$$

Note that

$$R_{XX}(t_1, t_2) = E[X(t_1)X(t_2)] = E[X(t_2)X(t_1)] = R_{XX}(t_2, t_1) \tag{8.15}$$

and

$$R_{XX}(t, t) = E[X^2(t)] \tag{8.16}$$

The *autocovariance* of $X(t)$ is defined by

$$C_{XX}(t_1, t_2) = E\{[X(t_1) - \mu_X(t_1)][X(t_2) - \mu_X(t_2)]\}$$

$$= R_{XX}(t_1, t_2) - \mu_X(t_1)\mu_X(t_2) \tag{8.17}$$

It is clear that if $\mu_X(t) = 0$, then $C_{XX}(t_1, t_2) = R_{XX}(t_1, t_2)$.

Note that $C_{XX}(t_1, t_2)$ and $R_{XX}(t_1, t_2)$ are deterministic functions of t_1 and t_2.

The variance of $X(t)$ is given by

$$\sigma_X^2(t) = \text{Var}[X(t)] = E\{[X(t) - \mu_X(t)]^2\} = C_{XX}(t, t) \tag{8.18}$$

If $X(t)$ is a complex random process, then the autocorrelation and autocovariance of $X(t)$ are defined by

$$R_{XX}(t_1, t_2) = E[X(t_1)X^*(t_2)] \tag{8.19}$$

$$C_{XX}(t_1, t_2) = E\{[X(t_1) - \mu_X(t_1)][X(t_2) - \mu_X(t_2)]^*\} \tag{8.20}$$

where $*$ denotes the complex conjugate.

In a similar manner, for discrete-time random processes (or random sequences), $X(n)$, the autocorrelation and autocovariance of $X(n)$ are defined by

$$R_{XX}(n_1, n_2) = E[X(n_1)X(n_2)] \tag{8.21}$$

$$C_{XX}(n_1, n_2) = E\{[X(n_1) - \mu_X(n_1)][X(n_2) - \mu_X(n_2)]\} \tag{8.22}$$

and

$$R_{XX}(n_1, n_2) = R_{XX}(n_2, n_1) \tag{8.23}$$

$$R_{XX}(n, n) = E[X^2(n)] \tag{8.24}$$

If $\mu_X(n) = 0$, then $C_{XX}(n_1, n_2) = R_{XX}(n_1, n_2)$.

For two different random signals $X(t)$ and $Y(t)$, we have the following definitions. The *cross-correlation* of $X(t)$ and $Y(t)$ is defined by

$$R_{XY}(t_1, t_2) = E[X(t_1)Y(t_2)]$$

$$= \int_{-\infty}^{\infty} \int_{-\infty}^{\infty} x_1 y_2 f_{XY}(x_1, y_2; t_1, t_2) dx_1 dy_2 \tag{8.25}$$

The *cross-covariance* of $X(t)$ and $Y(t)$ is defined by

$$C_{XY}(t_1, t_2) = E\{[X(t_1) - \mu_X(t_1)][Y(t_2) - \mu_Y(t_2)]\}$$
$$= R_{XY}(t_1, t_2) - \mu_X(t_1)\mu_Y(t_2) \tag{8.26}$$

Some Properties of X(t) and Y(t):

Two random processes $X(t)$ and $Y(t)$ are *independent* if for all t_1 and t_2,

$$F_{XY}(x, y; t_1, t_2) = F_X(x; t_1)F_Y(y; t_2) \tag{8.27}$$

They are *uncorrelated* if for all t_1 and t_2

$$C_{XY}(t_1, t_2) = R_{XY}(t_1, t_2) - \mu_X(t_1)\mu_Y(t_2) = 0 \tag{8.28}$$

or

$$R_{XY}(t_1, t_2) = \mu_X(t_1)\mu_Y(t_2) \tag{8.29}$$

They are *orthogonal* if for all t_1 and t_2

$$R_{XY}(t_1, t_2) = 0 \tag{8.30}$$

By changing t_1 and t_2 by n_1 and n_2, respectively, similar definitions can be obtained for two different random sequences $X(n)$ and $Y(n)$.

C. Stationarity:

1. Strict-Sense Stationary:

A random process $X(t)$ is called *strict-sense stationary* (SSS) if its statistics are invariant to a shift of origin. In other words, the process $X(t)$ is SSS if

$$f_X(x_1, \ldots, x_n; t_1, \ldots, t_n) = f_X(x_1, \ldots, x_n; t_1 + c, \ldots, t_n + c) \tag{8.31}$$

for any c.

From Eq. (8.31) it follows that $f_X(x_1; t_1) = f_X(x_1; t_1 + c)$ for any c. Hence, the first-order density of a stationary $X(t)$ is independent of t:

$$f_X(x_1; t) = f_X(x_1) \tag{8.32}$$

Similarly, $f_X(x_1, x_2; t_1, t_2) = f_X(x_1, x_2; t_1 + c, t_2 + c)$ for any c. Setting $c = -t_1$, we obtain

$$f_X(x_1, x_2; t_1, t_2) = f_X(x_1, x_2; t_2 - t_1) \tag{8.33}$$

which indicates that if $X(t)$ is SSS, the joint density of the random variables $X(t)$ and $X(t + \tau)$ is independent of t and depends only on the time difference τ.

2. Wide-Sense Stationary:

A random process $X(t)$ is called *wide-sense stationary* (WSS) if its mean is constant

$$E[X(t)] = \mu_X \tag{8.34}$$

and its autocorrelation depends only on the time difference τ

$$E[X(t)X(t + \tau)] = R_{XX}(\tau) \tag{8.35}$$

From Eqs. (8.17) and (8.35) it follows that the autocovariance of a WSS process also depends only on the time difference τ:

$$C_{XX}(\tau) = R_{XX}(\tau) - \mu_X^2 \tag{8.36}$$

Setting $\tau = 0$ in Eq. (8.35), we obtain

$$E[X^2(t)] = R_{XX}(0) \tag{8.37}$$

Thus, the average power of a WSS process is independent of t and equals $R_{XX}(0)$.

Similarly, a discrete-time random process $X(n)$ is WSS if

$$E[X(n)] = \mu_X = \text{constant} \tag{8.38}$$

and

$$E[X(n)X(n + k)] = R_{XX}(k) \tag{8.39}$$

Then

$$C_{XX}(k) = R_{XX}(k) - \mu_X^2 \tag{8.40}$$

Setting $k = 0$ in Eq. (8.39) we have

$$E[X^2(n)] = R_{XX}(0) \tag{8.41}$$

Note that an SSS process is WSS but a WSS process is not necessarily SSS.

Two processes $X(t)$ and $Y(t)$ are called *jointly wide-sense stationary* (jointly WSS) if each is WSS and their cross-correlation depends only on the time difference τ:

$$R_{XY}(t, t + \tau) = E[X(t)Y(t + \tau)] = R_{XY}(\tau) \tag{8.42}$$

From Eq. (8.42) it follows that the cross-covariance of jointly WSS $X(t)$ and $Y(t)$ also depends only on the time difference τ:

$$C_{XY}(\tau) = R_{XY}(\tau)\mu_X\mu_Y \tag{8.43}$$

Similar to Eqs. (8.27) to (8.30), two jointly WSS random process $X(t)$ and $Y(t)$ are independent if for all x and y

$$f_{XY}(x, y) = f_X(x)f_Y(y) \tag{8.44}$$

They are uncorrelated if for all τ

$$C_{XY}(\tau) = R_{XY}(\tau) - \mu_X\mu_Y = 0 \tag{8.45}$$

or

$$R_{XY}(\tau) = \mu_X\mu_Y \tag{8.46}$$

They are orthogonal if for all τ

$$R_{XY}(\tau) = 0 \tag{8.47}$$

Similarly, two random sequences $X(n)$ and $Y(n)$ are jointly WSS if each is WSS and their cross-correlation depends only on the time difference k:

$$R_{XY}(n, n + k) = E[X(n)Y(n + k)] = R_{XY}(k) \tag{8.48}$$

Then the cross-covariance of jointly WSS $X(n)$ and $Y(n)$ is

$$C_{XY}(k) = R_{XY}(k) - \mu_X\mu_Y \tag{8.49}$$

They are uncorrelated if for all k

$$C_{XY}(k) = 0 \tag{8.50}$$

or

$$R_{XY}(k) = \mu_X\mu_Y \tag{8.51}$$

They are orthogonal if for all k

$$R_{XY}(k) = 0 \tag{8.52}$$

D. Time Averages and Ergodicity:

The *time-averaged mean* of a sample function $x(t)$ of a random process $X(t)$ is defined as

$$\bar{x} = \langle x(t) \rangle = \lim_{T \to \infty} \frac{1}{T} \int_{-T/2}^{T/2} x(t) \, dt \tag{8.53}$$

where the symbol $\langle \cdot \rangle$ denotes *time-averaging*.

Similarly, the *time-averaged autocorrelation* of the sample function $x(t)$ is defined as

$$\bar{R}_{XX}(\tau) = \langle x(t)x(t + \tau) \rangle = \lim_{T \to \infty} \frac{1}{T} \int_{-T/2}^{T/2} x(t)x(t + \tau) \, dt \tag{8.54}$$

Note that $\bar{x}$ and $\bar{R}_{XX}(\tau)$ are random variables; their values depend on which sample function of $X(t)$ is used in the time-averaging evaluations.

If $X(t)$ is stationary, then by taking the expected value on both sides of Eqs. (7.20) and (7.21), we obtain

$$E[\bar{x}] = \lim_{T \to \infty} \frac{1}{T} \int_{-T/2}^{T/2} E[x(t)] \, dt = \mu_X \tag{8.55}$$

which indicates that the expected value of the time-averaged mean is equal to the ensemble mean, and

$$E[\bar{R}_{XX}(\tau)] = \lim_{T \to \infty} \frac{1}{T} \int_{-T/2}^{T/2} E[x(t)x(t + \tau)] \, dt = R_{XX}(\tau) \tag{8.56}$$

which also indicates that the expected value of the time-averaged autocorrelation is equal to the ensemble autocorrelation.

A random process $X(t)$ is said to be *ergodic* if time averages are the same for all sample functions and equal to the corresponding ensemble averages. Thus, in an ergodic process, all its statistics can be obtained by observing a single sample function $x(t) = X(t, \lambda)$ (λ fixed) of the process.

A stationary process $X(t)$ is called *ergodic* in the *mean* if

$$\bar{x} = \langle x(t) \rangle = E[X(t)] = \mu_X \tag{8.57}$$

Similarly, a stationary process $X(t)$ is called *ergodic in the autocorrelation* if

$$\bar{R}_{XX}(\tau) = \langle x(t)x(t + \tau) \rangle = E[X(t)X(t + \tau)] = R_{XX}(\tau) \tag{8.58}$$

The time-averaged mean of a sample sequence $x(n)$ of a random sequence $X(n)$ is defined as

$$\bar{x} = \langle x(n) \rangle = \lim_{N \to \infty} \frac{1}{2N + 1} \sum_{n=-N}^{N} x(n) \tag{8.59}$$

Similarly, the time-average autocorrelation of the sample sequence $x(n)$ is defined as

$$\bar{R}_{XX}(k) = \langle x(n)x(n + k) \rangle = \lim_{N \to \infty} \frac{1}{2N + 1} \sum_{n=-N}^{N} x(n)x(n + k) \tag{8.60}$$

If $X(n)$ is stationary, then

$$E[\bar{x}] = \lim_{N \to \infty} \frac{1}{2N + 1} \sum_{n=-N}^{N} E[x(n)] = \mu_X \tag{8.61}$$

and

$$E[\bar{R}_{XX}(k)] = \lim_{N \to \infty} \frac{1}{2N + 1} \sum_{n=-N}^{N} E[x(n)x(n + k)] = R_{XX}(k) \tag{8.62}$$

Thus, $X(n)$ is also ergodic in the mean and autocorrelation if

$$\bar{x} = \langle x(n) \rangle = E[X(n)] = \mu_X \tag{8.63}$$

$$\bar{R}_{XX}(k) = \langle x(n)x(n+k) \rangle = E[x(n)x(n+k)] = R_{XX}(k) \tag{8.64}$$

Testing for the ergodicity of a random process is usually very difficult. A reasonable assumption in the random analysis of most random signals is that the random waveforms are ergodic in the mean and in the autocorrelation. Fundamental electrical engineering parameters, such as dc value, root-mean-square (rms) value, and average power can be related to the statistical averages of an ergodic random process. They are summarized in the following:

1. $\bar{x} = \langle x(t) \rangle$ is equal to the dc level of the signal.
2. $[\bar{x}]^2 = \langle x(t) \rangle^2$ is equal to the normalized power in the dc component.
3. $\bar{R}_{XX}(0) = \langle x^2(t) \rangle$ is equal to the total average normalized power.
4. $\bar{\sigma}_X^2 = \langle x^2(t) \rangle - \langle x(t) \rangle^2$ is equal to the average normalized power in the time-varying or ac component of the signal.
5. $\bar{\sigma}_X$ is equal to the rms value of the ac component of the signal.

8.4   Gaussian Random Process:

Consider a random process $X(t)$, and define n random variables $X(t_1), \ldots, X(t_n)$ corresponding to n time instants $t_1, \ldots, t_n$. Let $\mathbf{X}$ be a *random vector* ($n \times 1$ matrix) defined by

$$\mathbf{X} = \begin{bmatrix} X(t_1) \\ \vdots \\ X(t_n) \end{bmatrix} \tag{8.65}$$

Let $\mathbf{x}$ be an n-dimensional vector ($n \times 1$ matrix) defined by

$$\mathbf{x} = \begin{bmatrix} x_1 \\ \vdots \\ x_n \end{bmatrix} \tag{8.66}$$

so that the event $\{X(t_1) \leq x_1, \ldots, X(t_n) \leq x_n\}$ is written $\{\mathbf{X} \leq \mathbf{x}\}$. Then $X(t)$ is called a *Gaussian* (or normal) process if $\mathbf{X}$ has a jointly multivariate Gaussian density function for every finite set of $\{t_i\}$ and every n.

The multivariate Gaussian density function is given by

$$f_{\mathbf{X}}(\mathbf{x}) = \frac{1}{(2\pi)^{n/2} \left| \det \mathbf{C} \right|^{1/2}} \exp\left[-\frac{1}{2}(\mathbf{x} - \boldsymbol{\mu})^T \mathbf{C}^{-1} (\mathbf{x} - \boldsymbol{\mu}) \right] \tag{8.67}$$

where T denotes the "transpose," $\boldsymbol{\mu}$ is the *vector means*, and $\mathbf{C}$ is the *covariance matrix*, given by

$$\boldsymbol{\mu} = E[\mathbf{X}] = \begin{bmatrix} \mu_1 \\ \vdots \\ \mu_n \end{bmatrix} = \begin{bmatrix} E[X(t_1)] \\ \vdots \\ E[X(t_n)] \end{bmatrix} \tag{8.68}$$

$$\mathbf{C} = \begin{bmatrix} C_{11} & \cdots & C_{1n} \\ \cdots & \cdots & \cdots \\ C_{n1} & \cdots & C_{nn} \end{bmatrix} \tag{8.69}$$

where

$$C_{ij} = C_{XX}(t_i, t_j) = R_{XX}(t_i, t_j) - \mu_i \mu_j \tag{8.70}$$

which is the covariance of $X(t_i)$ and $X(t_j)$, and det $\mathbf{C}$ is the determinant of the matrix $\mathbf{C}$.

Alternate Definition:

A random process $X(t)$ is a Gaussian process if for any integers n and any subset $\{t_1, \ldots, t_n\}$ of T, and any real coefficients $a_k (1 \le k \le n)$, the r.v.

$$\sum_{k=1}^{n} a_k X(t_k) = a_1 X(t_1) + a_2 X(t_2) + \cdots a_n X(t_n) \tag{8.71}$$

is a Gaussian r.v..

Some of the important properties of a Gaussian process are as follows:

1. A Gaussian process $X(t)$ is completely specified by the set of means

$$\mu_i = E[X(t_i)] \qquad i = 1, \ldots, n$$

and the set of autocorrelations

$$R_{XX}(t_i, t_j) = E[X(t_i)X(t_j)] \qquad i, j = 1, \ldots, n$$

2. If the set of random variables $X(t_i)$, $i = 1, \ldots, n$, is uncorrelated, that is,

$$C_{ij} = 0 \qquad i \ne j$$

then $X(t_i)$ are independent.
3. If a Gaussian process $X(t)$ is WSS, then $X(t)$ is *SSS*.
4. If the input process $X(t)$ of a linear system is Gaussian, then the output process $Y(t)$ is also Gaussian.

SOLVED PROBLEMS

8.1 Consider a random process $X(t)$ defined by

$$X(t) = Y \cos \omega t \qquad t \ge 0 \tag{8.72}$$

where ω is a constant and Y is a uniform r.v. over $(0, 1)$.

(a) Describe $X(t)$.

(b) Sketch a few typical sample functions of $X(t)$.

(a) The random process $X(t)$ is a continuous-parameter (or time), continuous-state random process. The state space is $E = \{x: -1 < x < 1\}$ and the index parameter set is $T = \{t: t \ge 0\}$.

(b) Three sample functions of $X(t)$ are sketched in Fig. 8-5.

8.2. Consider a random signal $X(t)$ given by

$$X(t) = \sum_{k=-\infty}^{\infty} A_k p(t - kT_b - T_d) \tag{8.73}$$

where $\{A_k\}$ is a sequence of independent r.v.'s with $P[A_k = A] = P[A_k = -A] = \frac{1}{2}$, $p(t)$ is a unit amplitude pulse of duration T_b, and T_d is a r.v. uniformly distributed over $[0, T_b]$.

(a) Describe $X(t)$.

(b) Sketch a sample function of $X(t)$.

(a) The random signal $X(t)$ is a continuous time, discrete-state random process. The state space is $(A, -A)$, and the index parameter set is $T = (t; -\infty < t < \infty)$. $X(t)$ is known as a *random binary signal*.

(b) A sample function of $X(t)$ is sketched in Fig. 8-6.

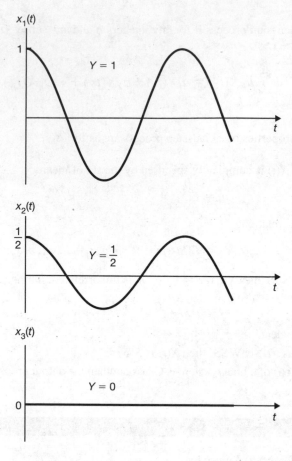

Fig. 8-5

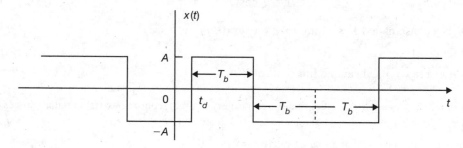

Fig. 8-6   Random binary signal.

8.3. Consider a random process $\{X(t); t \geq 0\}$, where $X(t)$ represents the total number of "events" that have occurred in the interval $(0, t)$.

(a) Describe $X(t)$.

(b) Sketch a sample function of $X(t)$.

(a) From definition, $X(t)$ must satisfy the following conditions:

1. $X(t) \geq 0$ and $X(0) = 0$

2. $X(t)$ is integer valued.

3. $X(t_1) \leq X(t_2)$ if $t_1 < t_2$

4. $X(t_2) - X(t_1)$ equals the number of events that have occurred on the interval (t_1, t_2).

Thus, $X(t)$ is a continuous-time discrete state random process.

Note that $X(t)$ is known as a *counting* process. A counting process $X(t)$ is said to possess *independent increments* if the number of events which occur in disjoint intervals are independent.

(*b*) A sample function of $X(t)$ is sketched in Fig. 8-7.

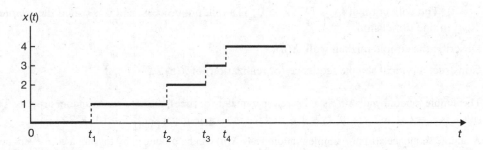

Fig. 8-7 A sampling function of a counting process.

8.4. Let W_1, W_2, ... be independent identically distributed (i.i.d.) zero-mean Gaussian r.v.'s. Let

$$X_n = \sum_{k=1}^{n} W_k = W_1 + W_2 + \cdots + W_n \qquad n = 1, 2, \ldots \tag{8.74}$$

with $X_0 = 0$. The collection of r.v.'s $X(n) = \{X_n, n \geq 0\}$ is a random process.

(*a*) Describe $X(n)$.

(*b*) Sketch a sample function of $X(n)$.

(*a*) The random signal $X(n)$ is a discrete time, continuous-state random process. The state space is $E = (-\infty, \infty)$ and the index parameter set is $T = \{0, 1, 2, \ldots\}$.

(*b*) A sample function of $X(t)$ is sketched in Fig. 8-8.

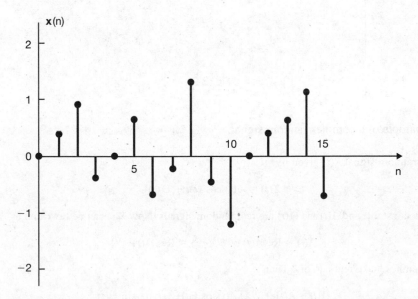

Fig. 8-8

8.5. Let $Z_1, Z_2, \ldots$ be independent identically distributed r.v.'s with $P(Z_n = 1) = p$ and $P(Z_n = -1) = q = 1 - p$ for all n. Let

$$X_n = \sum_{i=1}^{n} Z_i \qquad n = 1, 2, \ldots \tag{8.75}$$

and $X_0 = 0$. The collection of r.v.'s $\{X_n, n \geq 0\}$ is a random process, and it is called the *simple random walk* $X(n)$ in one dimension.

(*a*) Describe the simple random walk $X(n)$.

(*b*) Construct a typical sample sequence (or realization) of $X(n)$.

(*a*) The simple random walk $X(n)$ is a discrete-parameter (or time), discrete-state random process. The state space is $E = \{\ldots, -2, -1, 0, 1, 2, \ldots\}$, and the index parameter set is $T = \{0, 1, 2, \ldots\}$.

(*b*) A sample sequence $x(n)$ of a simple random walk $X(n)$ can be produced by tossing a coin every second and letting $x(n)$ increase by unity if a head appears and decrease by unity if a tail appears. Thus, for instance,

n	0	1	2	3	4	5	6	7	8	9	10	$\cdots$
Coin tossing		H	T	T	H	H	H	T	H	H	T	$\cdots$
$x(n)$	0	1	0	-1	0	1	2	1	2	3	2	$\cdots$

The sample sequence $x(n)$ obtained above is plotted in Fig. 8-9. The simple random walk $X(n)$ specified in this problem is said to be *unrestricted* because there are no bounds on the possible values of X_n.

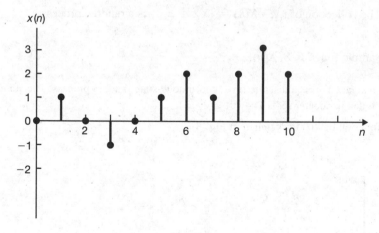

Fig. 8-9

8.6. Give an example of a complex random signal.

Consider a random signal $X(t)$ given by

$$X(t) = A(t) \cos [\omega t + \Theta(t)] \tag{8.76}$$

where ω is a constant, and $A(t)$ and $\Theta(t)$ are real random signals. Now $X(t)$ can be rewritten as

$$X(t) = \text{Re}\{A(t) e^{j\Theta(t)} e^{j\omega t}\} = \text{Re}\{Y(t) e^{j\omega t}\} \tag{8.77}$$

where Re denotes "take real part of." Then

$$Y(t) = A(t)e^{j\Theta(t)} = A(t)\cos \Theta(t) + j A(t)\sin \Theta(t) \tag{8.78}$$

is a complex random signal.

Statistics of Random Processes

8.7. Let a random signal $X(t)$ be specified by

$$X(t) = t - Y \tag{8.79}$$

where Y is an exponential r.v. with pdf

$$f_Y(y) = \begin{cases} e^{-y}, & y \geq 0 \\ 0, & y < 0 \end{cases}$$

Find the first-order cdf of $X(t)$, $F_X(x; t)$.

$$\begin{aligned} F_X(x; t) &= P\{X(t) \leq x\} = P\{t - Y \leq x\} \\ &= P\{Y \leq t - X\} \\ &= \int_{t-x}^{\infty} f_Y(y)\, dy = \int_{t-x}^{\infty} e^{-y} dy = e^{-(t-x)} \quad t \geq x \end{aligned}$$

Next, if $x > t$, then $t - x < 0$ and $Y \geq 0$, and

$$\begin{aligned} F_X(x; t) &= P\{Y \leq t - X\} = P\{Y \geq 0\} \\ &= \int_0^{\infty} f_Y(y)\, dy = \int_0^{\infty} e^{-y} dy = 1 \quad t < x \end{aligned}$$

Thus,

$$F_X(x; t) = \begin{cases} e^{-(t-x)}, & t \geq x \\ 1, & t < x \end{cases} \tag{8.80}$$

8.8. A discrete-time random sequence $X(n)$ is defined by $X(n) = A^n (n \geq 0)$, where A is a uniform r.v. over $(0, 1)$. Find the mean $\mu_X(n)$ and autocorrelation $R_{XX}(n, m)$ of $X(n)$.

The pdf of A is given by

$$f_A(a) = \begin{cases} 1, & 0 < a < 1 \\ 0, & \text{otherwise} \end{cases}$$

Then

$$\mu_X(n) = E[X(n)] = E[A^n] = \int_0^1 a^n\, da = \frac{1}{n+1} \tag{8.81}$$

and

$$R_{XX}(n, m) = E[X(n)X(m)] = E[A^{n+m}] = \int_0^1 a^{n+m} da = \frac{1}{n+m+1} \tag{8.82}$$

8.9. Show that

$$R_{XX}(t, t) \geq 0 \tag{8.83}$$

From definition (8.13)

$$R_{XX}(t, t) = E[X(t)X(t)] = E[X^2(t)] \geq 0$$

since $E[Y^2] \geq 0$, for any r.v. Y.

8.10. Show that

$$\left| R_{XX}(t_1, t_2) \right| \leq \frac{R_{XX}(t_1, t_1) + R_{XX}(t_2, t_2)}{2} \tag{8.84}$$

Since $E[Y^2] \geq 0$, for any r.v. Y, we have

$$0 \leq E[(X(t_1) + X(t_2))^2] = R_{XX}(t_1, t_1) + 2R_{XX}(t_1, t_2) + R_{XX}(t_2, t_2) \qquad (8.85)$$

$$0 \leq E[(X(t_1) - X(t_2))^2] = R_{XX}(t_1, t_1) - 2R_{XX}(t_1, t_2) + R_{XX}(t_2, t_2) \qquad (8.86)$$

From Eqs. (8.85) and (8.86) we have

$$-R_{XX}(t_1, t_2) \leq \frac{R_{XX}(t_1, t_1) + R_{XX}(t_2, t_2)}{2}$$

$$R_{XX}(t_1, t_2) \leq \frac{R_{XX}(t_1, t_1) + R_{XX}(t_2, t_2)}{2}$$

which imply that

$$\left| R_{XX}(t_1, t_2) \right| \leq \frac{R_{XX}(t_1, t_1) + R_{XX}(t_2, t_2)}{2}$$

8.11. Consider a random signal given by

$$X(t) = A \cos \omega_0 t \qquad (8.87)$$

where ω_0 is a constant and A is an uniform r.v. over [0,1]. Find the mean $\mu_X(t)$ and autocorrelation $R_{XX}(t_1, t_2)$ of $X(t)$.

The pdf of A is given by

$$f_A(a) = \begin{cases} 1, & 0 < a < 1 \\ 0, & \text{otherwise} \end{cases}$$

$$\mu_X(t) = E[X(t)] = E[A \cos \omega_0 t] = E[A] \cos \omega_0 t = \frac{1}{2} \cos \omega_0 t \qquad (8.88)$$

since

$$E[A] = \int_0^1 a f_A(a)\, da = \int_0^1 a\, da = \frac{1}{2}.$$

$$R_{XX}(t_1, t_2) = E[X(t_1) X(t_2)] = E[A^2 \cos \omega_0 t_1 \cos \omega_0 t_2]$$
$$= E[A^2] \cos \omega_0 t_1 \cos \omega_0 t_2 = \frac{1}{3} \cos \omega_0 t_1 \cos \omega_0 t_2 \qquad (8.89)$$

since

$$E[A^2] = \int_0^1 a^2 f_A(a)\, da = \int_0^1 a^2\, da = \frac{1}{3}.$$

8.12. A random sequence $X(n)$ is defined as

$$X(n) = An + B \qquad (8.90)$$

where A and B are independent zero mean Gaussian r.v.'s of variance σ_A^2 and σ_B^2, respectively.

(a) Find the mean $\mu_X(n)$ and autocorrelation $R_{XX}(n, m)$ of $X(n)$.

(b) Find $E[X^2(n)]$.

(a)

$$\mu_X(n) = E[X(n)] = E[An + B] = E[A]n + E[B] = 0 \qquad (8.91)$$

since $E[A] = E[B] = 0$.

$$R_{XX}(n, m) = E[X(n)X(m)] = E[(An + B)(Am + B)]$$
$$= E[A^2]nm + E[AB](n + m) + E[B^2] \qquad (8.92)$$
$$= \sigma_A^2 nm + \sigma_B^2$$

since $E[AB] = E[BA] = E[A]E[B] = 0$.

(b)   Setting $m = n$ in Eq. (8.92), we obtain

$$E[X^2(n)] = \sigma_A^2 n^2 + \sigma_B^2 \tag{8.93}$$

8.13. A counting process $X(t)$ of Prob. 8.3 is said to be a Poisson process with rate (or intensity) $\lambda \ (> 0)$ if

1. $X(0) = 0$.

2. $X(t)$ has independent increments.

3.   The number of events in any interval of length t is Poisson distributed with mean λt; that is, for all $s, t > 0$,

$$P[X(t + s) - X(s) = n] = e^{-\lambda t} \frac{(\lambda t)^n}{n!} \qquad n = 0, 1, 2, \ldots \tag{8.94}$$

(a)   Find the mean $\mu_X(t)$ and $E[X^2(t)]$.

(b)   Find the autocorrelation $R_{XX}(t_1, t_2)$ of $X(t)$.

(a)   Setting $s = 0$ in Eq. (8.94) and using condition 1, we have

$$P[X(t) = n] = e^{-\lambda t} \frac{(\lambda t)^n}{n!} \qquad n = 0, 1, 2, \ldots \tag{8.95}$$

Thus,

$$\mu_X(t) = E[X(t)] = \sum_{n=1}^{\infty} n P[X(t) = n] = \sum_{n=1}^{\infty} n e^{-\lambda t} \frac{(\lambda t)^n}{n!} \tag{8.96}$$

Now, the Taylor expansion of $e^{\lambda t}$ is given by

$$e^{\lambda t} = \sum_{n=0}^{\infty} \frac{(\lambda t)^n}{n!}$$

Differentiating twice with respect to λt, we obtain

$$e^{\lambda t} = \sum_{n=0}^{\infty} n \frac{(\lambda t)^{n-1}}{n!} = \frac{1}{\lambda t} \sum_{n=1}^{\infty} n \frac{(\lambda t)^n}{n!} \tag{8.97}$$

$$e^{\lambda t} = \sum_{n=0}^{\infty} n(n-1) \frac{(\lambda t)^{n-2}}{n!} = \frac{1}{(\lambda t)^2} \sum_{n=1}^{\infty} n^2 \frac{(\lambda t)^n}{n!} - \frac{1}{(\lambda t)^2} \sum_{n=1}^{\infty} n \frac{(\lambda t)^n}{n!} \tag{8.98}$$

Using Eqs. (9.97) and (9.98), we obtain

$$\mu_X(t) = E[X(t)] = e^{-\lambda t} \sum_{n=1}^{\infty} n \frac{(\lambda t)^n}{n!} = e^{-\lambda t} e^{\lambda t} (\lambda t) = \lambda t \tag{8.99}$$

and

$$E[X^2(t)] = \sum_{n=1}^{\infty} n^2 P[X(t) = n] = e^{-\lambda t} \sum_{n=1}^{\infty} n^2 \frac{(\lambda t)^n}{n!} \tag{8.100}$$

$$= e^{-\lambda t} \left[(\lambda t)^2 e^{\lambda t} + (\lambda t) e^{\lambda t} \right] = (\lambda t)^2 + (\lambda t)$$

(b)   Next, let $t_1 < t_2$, the r.v.'s $X(t_1)$ and $X(t_2 - t_1)$ are independent since the intervals $(0, t_1)$ and (t_1, t_2) are non-overlapping, and they are Poisson distributed with mean λt_1 and $\lambda(t_2 - t_1)$, respectively. Thus,

$$E\{X(t_1)[X(t_2) - X(t_1)]\} = E[X(t_1)]E[X(t_2) - X(t_1)] = \lambda t_1 \lambda (t_2 - t_1)) \tag{8.101}$$

Now using identity

$$X(t_1)\,X(t_2) = X(t_1)\,[X(t_1) + X(t_2) - X(t_1)] = [X^2(t_1)] + X(t_1)\,[X(t_2) - X(t_1)]$$

we have

$$R_{XX}(t_1, t_2) = E[X(t_1)\,X(t_2)] = E[X^2(t_1)] + E\{X(t_1)[X(t_2) - X(t_1)]\}$$
$$= \lambda t_1 + \lambda^2 t_1^2 + \lambda t_1\,\lambda(t_2 - t_1) = \lambda t_1 + \lambda^2 t_1 t_2 \qquad t_1 \le t_2 \tag{8.102}$$

Interchanging t_1 and t_2, we have

$$R_{XX}(t_1, t_2) = \lambda t_2 + \lambda^2 t_1 t_2 \qquad t_1 \ge t_2 \tag{8.103}$$

Thus, combining Eqs. (102) and (8.103), we obtain

$$R_{XX}(t_1, t_2) = \lambda \min (t_1, t_2) + \lambda^2 t_1 t_2 \tag{8.104}$$

8.14. Let $X(t)$ and $Y(t)$ be defined by

$$X(t) = A\cos \omega t + B\sin \omega t \tag{8.105}$$
$$Y(t) = B\cos \omega t - A\sin \omega t \tag{8.106}$$

where ω is constant and A and B are independent random variables both having zero mean and variance σ^2. Find the cross-correlation of $X(t)$ and $Y(t)$.

The cross-correlation of $X(t)$ and $Y(t)$ is

$$R_{XY}(t_1, t_2) = E[X(t_1)Y(t_2)]$$
$$= E[(A\cos \omega t_1 + B\sin \omega t_1)(B\cos \omega t_2 - A\sin \omega t_2)]$$
$$= E[AB](\cos \omega t_1 \cos \omega t_2 - \sin \omega t_1 \sin \omega t_2)$$
$$- E[A^2]\cos \omega t_1 \sin \omega t_2 + E[B^2]\sin \omega t_1 \cos \omega t_2$$

Since $\qquad E[AB] = E[A]E[B] = 0 \qquad E[A^2] = E[B^2] = \sigma^2$

we have $\qquad R_{XY}(t_1, t_2) = \sigma^2(\sin \omega t_1 \cos \omega t_2 - \cos \omega t_1 \sin \omega t_2)$
$$= \sigma^2 \sin \omega(t_1 - t_2)$$

or $\qquad R_{XY}(\tau) = -\sigma^2 \sin \omega\tau$

where $\tau = t_2 - t_1$. $\tag{8.107}$

8.15. Consider a random process $X(t)$ given by

$$X(t) = A\cos (\omega t + \Theta) \tag{8.108}$$

where A and ω are constants and Θ is a uniform random variable over $[-\pi, \pi]$. Show that $X(t)$ is WSS.

From Eq. (B.57) (Appendix B), we have

$$f_\Theta(\theta) = \begin{cases} \dfrac{1}{2\pi} & -\pi \le \theta \le \pi \\ 0 & \text{otherwise} \end{cases}$$

Thus, $\qquad \mu_X(t) = E[X(t)] = \int_{-\infty}^{\infty} A\cos (\omega t + \theta)f_\Theta(\theta)\,d\theta$
$$= \frac{A}{2\pi}\int_{-\pi}^{\pi} \cos(\omega t + \theta)\,d\theta = 0 \tag{8.109}$$

$$R_{XX}(t, t + \tau) = E[X(t)X(t + \tau)]$$

$$= \frac{A^2}{2\pi} \int_{-\pi}^{\pi} \cos(\omega t + \theta)\cos[\omega(t + \tau) + \theta]d\theta$$

$$= \frac{A^2}{2\pi} \int_{-\pi}^{\pi} \frac{1}{2}[\cos \omega t + \cos(2\omega t + 2\theta + \omega \tau)]d\theta$$

$$= \frac{A^2}{2} \cos \omega \tau \tag{8.110}$$

Since the mean of $X(t)$ is a constant and the autocorrelation of $X(t)$ is a function of time difference only, we conclude that $X(t)$ is WSS.

Note that $R_{XX}(\tau)$ is periodic with the period $T_0 = 2\pi/\omega$. A WSS random process is called *periodic* if its autocorrelation is periodic.

8.16. Consider a random process $X(t)$ given by

$$X(t) = A\cos(\omega t + \theta) \tag{8.111}$$

where ω and θ are constants and A is a random variable. Determine whether $X(t)$ is WSS.

$$\mu_X(t) = E[X(t)] = E[A\cos(\omega t + \theta)]$$
$$= \cos(\omega t + \theta)E[A] \tag{8.112}$$

which indicates that the mean of $X(t)$ is not constant unless $E[A] = 0$.

$$R_{XX}(t, t + \tau) = E[X(t)X(t + \tau)]$$
$$= E[(A^2 \cos(\omega t + \theta)\cos[\omega(t + \tau) + \theta)]]$$
$$= \frac{1}{2}[\cos \omega \tau + \cos(2\omega t + 2\theta + \omega \tau)]E[A^2] \tag{8.113}$$

Thus, we see that the autocorrelation of $X(t)$ is not a function of the time difference τ only, and the process $X(t)$ is not WSS.

8.17. Consider a random process $X(t)$ given by

$$X(t) = A\cos \omega t + B\sin \omega t \tag{8.114}$$

where ω is constant and A and B are random variables.

(a) Show that the condition

$$E[A] = E[B] = 0 \tag{8.115}$$

is necessary for $X(t)$ to be stationary.

(b) Show that $X(t)$ is WSS if and only if the random variables A and B are uncorrelated with equal variance; that is,

$$E[AB] = 0 \tag{8.116}$$

and

$$E[A^2] = E[B^2] = \sigma^2 \tag{8.117}$$

(a) $\mu_X(t) = E[X(t)] = E[A]\cos \omega t + E[B]\sin \omega t$ must be independent of t for $X(t)$ to be stationary. This is possible only if $\mu_X(t) = 0$; that is,

$$E[A] = E[B] = 0$$

(*b*) If $X(t)$ is WSS, then from Eq. (8.37)

$$E[X^2(0)] = E\left[X^2\left(\frac{\pi}{2\omega}\right)\right] = R_{XX}(0) = \sigma_X^2$$

But

$$X(0) = A \quad \text{and} \quad X\left(\frac{\pi}{2\omega}\right) = B$$

Thus,

$$E[A^2] = E[B^2] = \sigma_X^2 = \sigma^2$$

Using the preceding result, we obtain

$$
\begin{aligned}
R_{XX}(t, t+\tau) &= E[X(t)X(t+\tau)] \\
&= E[(A\cos\omega t + B\sin\omega t)[A\cos\omega(t+\tau) + B\sin\omega(t+\tau)]] \\
&= \sigma^2\cos\omega t + E[AB]\sin(2\omega t + \omega\tau)
\end{aligned}
\tag{8.118}
$$

which will be a function of τ only if $E[AB] = 0$.

Conversely, if $E[AB] = 0$ and $E[A^2] = E[B^2] = \sigma^2$, then from the result of part (*a*) and Eq. (8.118), we have

$$\mu_X(t) = 0$$
$$R_{XX}(t, t+\tau) = \sigma^2\cos\omega\tau = R_{XX}(\tau)$$

Hence, $X(t)$ is WSS.

8.18. A random process $X(t)$ is said to be *covariance-stationary* if the covariance of $X(t)$ depends only on the time difference $\tau = t_2 - t_1$; that is,

$$C_{XX}(t, t+\tau) = C_{XX}(\tau) \tag{8.119}$$

Let $X(t)$ be given by

$$X(t) = (A + 1)\cos t + B\sin t$$

where A and B are independent random variables for which

$$E[A] = E[B] = 0 \quad \text{and} \quad E[A^2] = E[B^2] = 1$$

Show that $X(t)$ is not WSS, but it is covariance-stationary.

$$
\begin{aligned}
\mu_X(t) = E[X(t)] &= E[(A+1)\cos t + B\sin t)] \\
&= E[A+1]\cos t + E[B]\sin t \\
&= \cos t
\end{aligned}
$$

which depends on t. Thus, $X(t)$ cannot be WSS.

$$
\begin{aligned}
R_{XX}(t_1, t_2) &= E[X(t_1)X(t_2)] \\
&= E[[(A+1)\cos t_1 + B\sin t_1][(A+1)\cos t_2 + B\sin t_2]] \\
&= E[(A+1)^2]\cos t_1\cos t_2 + E[B^2]\sin t_1\sin t_2 \\
&\quad + E[(A+1)B](\cos t_1\sin t_2 + \sin t_1\cos t_2)
\end{aligned}
$$

Now

$$E[(A+1)^2] = E[A^2 + 2A + 1] = E[A^2] + 2E[A] + 1 = 2$$
$$E[(A+1)B] = E[AB] + E[B] = E[A]E[B] + E[B] = 0$$
$$E[B^2] = 1$$

Substituting these values into the expression of $R_{XX}(t_1, t_2)$, we obtain

$$R_{XX}(t_1, t_2) = 2 \cos t_1 \cos t_2 + \sin t_1 \sin t_2$$
$$= \cos(t_2 - t_1) + \cos t_1 \cos t_2$$

From Eq. (8.17), we have

$$C_{XX}(t_1, t_2) = R_{XX}(t_1, t_2) - \mu_X(t_1)\,\mu_X(t_2)$$
$$= \cos(t_2 - t_1) + \cos t_1 \cos t_2 - \cos t_1 \cos t_2$$
$$= \cos(t_2 - t_1)$$

Thus, $X(t)$ is covariance-stationary.

8.19. Show that if a random process $X(t)$ is WSS, then it must also be covariance stationary.

If $X(t)$ is WSS, then

$$E[X(t)] = \mu \text{ (constant)} \qquad \text{for all } t$$
$$R_{XX}(t, t + \tau) = R_{XX}(\tau) \qquad \text{for all } t$$

Now

$$C_{XX}(t, t + \tau) = \text{Cov}[X(t)X(t + \tau)] = R_{XX}(t, t + \tau) - E[X(t)]\,E[X(t + \tau)]$$
$$= R_{XX}(\tau) - \mu^2$$

which indicates that $C_{XX}(t, t + \tau)$ depends only on τ. Thus, $X(t)$ is covariance stationary.

8.20. Show that if $X(t)$ is WSS, then

$$E[[X(t + \tau) - X(t)]^2] = 2[R_{XX}(0) - R_{XX}(\tau)] \tag{8.120}$$

where $R_{XX}(\tau)$ is the autocorrelation of $X(t)$.

Using the linearity of E (the expectation operator) and Eqs. (8.35) and (8.37), we have

$$E\Big[[X(t + \tau) - X(t)]^2\Big] = E\Big[X^2(t + \tau) - 2X(t + \tau)X(t) + X^2(t)\Big]$$
$$= E\Big[X^2(t + \tau)\Big] - 2E[X(t + \tau)X(t)] + E\Big[X^2(t)\Big]$$
$$= R_{XX}(0) - 2R_{XX}(\tau) + R_{XX}(0)$$
$$= 2[R_{XX}(0) - R_{XX}(\tau)]$$

8.21. Let $X(t) = A \cos(\omega t + \Theta)$, where ω is constant and both A and Θ are r.v.'s with pdf $f_A(a)$ and $f_\Theta(\theta)$, respectively. Find the conditions that $X(t)$ is WSS.

$$\mu_X(t) = E[X(t)] = E[A \cos(\omega t + \Theta)] = \iint a \cos(\omega t + \theta)\, f_{A\Theta}(a, \theta)\, d\theta\, da \tag{8.121}$$

The first condition for the double integral to be independent of t is for A and Θ to be statistically independent. Then

$$\mu_X(t) = E[A \cos(\omega t + \Theta)] = \iint a \cos(\omega t + \theta)\, f_A(a)f_\Theta(\theta)\, d\theta\, da \tag{8.122}$$

The second condition is for Θ to be uniformly distributed over $[0, 2\pi]$. Then we have $\mu_X(t) = 0$ since $\frac{1}{2\pi} \int_0^{2\pi} \cos(\omega t + \theta)\, d\theta = 0$.

Next,

$$R_{XX}(t_1, t_2) = E[X(t_1 X(t_2)] = E[A^2 \cos(\omega t_1 + \Theta) \cos(\omega t_2 + \Theta)] \tag{8.123}$$

Since A and Θ are independent, we have

$$R_{XX}(t_1, t_2) = \frac{1}{2} E[A^2] E\{\cos \omega (t_2 - t_1) + \cos[\omega(t_2 + t_1) + 2\Theta]\} \tag{8.124}$$

and $E[\cos \omega(t_2 + t_1) + 2\Theta] = 0$ since Θ is uniformly distributed over $[0, 2\pi]$.

Thus,

$$R_{XX}(t_1, t_2) = \frac{1}{2} E[A^2] \cos \omega (t_2 - t_1) = \frac{1}{2} E[A^2] \cos \omega \tau \tag{8.125}$$

So, we conclude that $X(t)$ is WSS if A and Θ are independent, and Θ is uniformly distributed over $[0, 2\pi]$.

8.22. Let $Z(t) = X(t) + Y(t)$, where random processes $X(t)$ and $Y(t)$ are independent and WSS. Is $Z(t)$ WSS?

$$\mu_Z(t) = E[Z(t)] = E[X(t) + Y(t)] = \mu_X + \mu_Y = \text{constant} \tag{8.126}$$

$$\begin{aligned} R_{ZZ}(t, t + \tau) &= E[Z(t) Z(t + \tau)] = E\{[X(t) + Y(t)][X(t + \tau) + Y(t + \tau)]\} \\ &= E[X(t)X(t + \tau)] + E[Y(t)Y(t + \tau)] + E[X(t)Y(t + \tau)] + E[Y(t)X(t + \tau)] \\ &= R_{XX}(\tau) + R_{YY}(\tau) + E[X(t)]E[Y(t + \tau)] + E[Y(t)]E[X(t + \tau)] \\ &= R_{XX}(\tau) + R_{YY}(\tau) + 2\mu_X \mu_Y \end{aligned} \tag{8.127}$$

Since the mean of $Z(t)$ is constant and its autocorrelation depends only on τ, $Z(t)$ is WSS.

8.23. Let $Z(t) = X(t) + Y(t)$, where random processes $X(t)$ and $Y(t)$ are jointly WSS. Show that if $X(t)$ and $Y(t)$ are orthogonal, then

$$R_{ZZ}(\tau) = R_{XX}(\tau) + R_{YY}(\tau) \tag{8.128}$$

$$\begin{aligned} R_{ZZ}(t, t + \tau) &= E[Z(t) Z(t + \tau)] = E\{[X(t) + Y(t)][X(t + \tau) + Y(t + \tau)]\} \\ &= E[X(t)X(t + \tau)] + E[Y(t)Y(t + \tau)] + E[X(t)Y(t + \tau)] + E[Y(t)X(t + \tau)] \\ &= R_{XX}(\tau) + R_{YY}(\tau) + R_{XY}(\tau) + R_{YX}(\tau) \end{aligned} \tag{8.129}$$

Since $X(t)$ and $Y(t)$ are orthogonal, then $R_{XY}(\tau) = 0$, and we have

$$R_{ZZ}(\tau) = R_{XX}(\tau) + R_{YY}(\tau)$$

8.24. A random signal $X(t)$ is defined as $X(t) = At + B$, where A and B are independent r.v.'s with both zero mean and unit variance. Is $X(t)$ WSS?

$$\mu_X(t) = E[X(t)] = E[At + B] = E[A] t + E[B] = 0$$

since $E[A] = E[B] = 0$.

$$\begin{aligned} R_{XX}(t_1, t_2) &= E[X(t_1)X(t_2)] = E[(At_1 + B)(At_2 + B)] \\ &= E[A^2 t_1 t_2 + A Bt_1 + B At_2 + B^2] \\ &= E[A^2] t_1 t_2 + E[AB] t_1 + E[BA] t_2 + E[B^2] \\ &= 1 + t_1 t_2 \end{aligned} \tag{8.130}$$

since $E[A^2] = E[B^2] = 1$ and $E[AB] = E[BA] = E[A] E[B] = 0$.

Since $R_{XX}(t_1, t_2)$ is not the function of $|t_2 - t_1|$, $X(t)$ is not WSS.

8.25. Let $X(n) = \{X_n, n \geq 0\}$ be a random sequence of iid r.v.'s with mean 0 and variance 1. Show that $X(n)$ is WSS.

$$\mu_X(n) = E[X(n)] = E[X_n] = 0 \text{ constant}$$

$$R_{XX}(n, n+k) = E\big[X(n)X(n+k)\big] = E\big[X_n X_{n+k}\big] = \begin{cases} E(X_n)E(X_{n+k}) = 0, & k \neq 0 \\ E(X_n^2) = 1, & k = 0 \end{cases} \qquad (8.131)$$

which depends only on k. Thus $X(n)$ is WSS.

8.26. A random signal $X(t)$ is defined as $X(t) = A$, where A is a r.v. uniformly distributed over $[0, 1]$. Is $X(t)$ ergodic in the mean?

$$\mu_X(t) = E[X(t)] = E[A] = \int_0^1 a\, da = \frac{1}{2}$$

$$\bar{x} = \langle x(t) \rangle = \frac{1}{2T} \int_{-T}^{T} x(t)\, dt = A \text{ as } T \to \infty$$

Since $\bar{x} \neq \mu_X(t)$, $X(t)$ is not ergodic in the mean.

8.27. Show that the process $X(t)$ defined in Eq. (8.108) (Prob. 8.15) is ergodic in both the mean and the autocorrelation.

From Eq. (8.53), we have

$$\bar{x} = \langle x(t) \rangle = \lim_{T \to \infty} \frac{1}{T} \int_{-T/2}^{T/2} A\cos(\omega t + \theta)\, dt$$

$$= \frac{A}{T_0} \int_{-T_0/2}^{T_0/2} \cos(\omega t + \theta)\, dt = 0 \qquad (8.132)$$

where $T_0 = 2\pi/\omega$.

From Eq. (8.54), we have

$$\bar{R}_{XX}(\tau) = \langle x(t)x(t+\tau) \rangle$$

$$= \lim_{T \to \infty} \frac{1}{T} \int_{-T/2}^{T/2} A^2 \cos(\omega t + \theta)\cos[\omega(t+\tau) + \theta]\, dt$$

$$= \frac{A^2}{T_0} \int_{-T_0/2}^{T_0/2} \frac{1}{2}[\cos \omega\tau + \cos(2\omega t + 2\theta + \omega\tau)]\, dt$$

$$= \frac{A^2}{2} \cos \omega\tau \qquad (8.133)$$

Thus, we have

$$\mu_X(t) = E[X(t)] = \langle x(t) \rangle = \bar{x}$$

$$R_{XX}(\tau) = E[X(t)X(t+\tau)] = \langle x(t)X(t+\tau) \rangle = \bar{R}_{XX}(\tau)$$

Hence, by definitions (8.57) and (8.58), we conclude that $X(t)$ is ergodic in both the mean and the autocorrelation.

8.28. Consider a random process $Y(t)$ defined by

$$Y(t) = \int_0^t X(\tau)\, d\tau \qquad (8.134)$$

where $X(t)$ is given by

$$X(t) = A\cos \omega t \qquad (8.135)$$

where ω is constant and $A = N[0; \sigma^2]$.

(a) Determine the pdf of $Y(t)$ at $t = t_k$.

(b) Is $Y(t)$ WSS?

(a)
$$Y(t_k) = \int_0^{t_k} A \cos \omega \tau \, d\tau = \frac{\sin \omega t_k}{\omega} A \tag{8.136}$$

Then from the result of Example B.10 (Appendix B) we see that $Y(t_k)$ is a Gaussian random variable with

$$E[Y(t_k)] = \frac{\sin \omega t_k}{\omega} E[A] = 0 \tag{8.137}$$

and
$$\sigma_Y^2 = \mathrm{var}[Y(t_k)] = \left(\frac{\sin \omega t_k}{\omega}\right)^2 \sigma^2 \tag{8.138}$$

Hence, by Eq. (B.53), the pdf of $Y(t_k)$ is

$$f_Y(y) = \frac{1}{\sqrt{2\pi}\sigma_Y} e^{-y^2/(2\sigma_Y^2)} \tag{8.139}$$

(b)   From Eqs. (8.137) and (8.138), the mean and variance of $Y(t)$ depend on time $t(t_k)$, so $Y(t)$ is not WSS.

8.29. Show that if a Gaussian random process is WSS, then it is SSS.

If the Gaussian process $X(t)$ is WSS, then

$$\mu_i = E[X(t_i)] = \mu(= \text{constant}) \qquad \text{for all } t_i$$

and
$$R_{XX}(t_i, t_j) = R_{XX}(t_j - t_i)$$

Therefore, in the expression for the joint probability density of Eq. (8.67) and Eqs. (8.68), (8.69), and (8.70).

$$\mu_1 = \mu_2 = \cdots = \mu_n = \mu \to E[X(t_i)] = E[X(t_i + c)]$$
$$C_{ij} = C_{XX}(t_i, t_j) = R_{XX}(t_i, t_j) - \mu_i \mu_j$$
$$= R_{XX}(t_j - t_i) - \mu^2 = C_{XX}(t_i + c, t_j + c)$$

for any c. It then follows that

$$f_{\mathbf{X}(t_i)}(\mathbf{x}) = f_{\mathbf{X}(t_i + c)}(\mathbf{x})$$

for any c. Therefore, $X(t)$ is SSS by Eq. (8.31)

8.30. Let **X** be an n-dimensional Gaussian random vector [Eq. (8.65)] with independent components. Show that the multivariate Gaussian joint density function is given by

$$f_{\mathbf{X}}(\mathbf{x}) = \frac{1}{(2\pi)^{n/2} \prod_{i=1}^{n} \sigma_i} \exp\left[-\frac{1}{2} \sum_{i=1}^{n}\left(\frac{x_i - \mu_i}{\sigma_i}\right)^2\right] \tag{8.140}$$

where $\mu_i = E[X_i]$ and $\sigma_i^2 = \mathrm{var}(X_i)$.

The multivariate Gaussian density function is given by Eq. (8.67). Since $X_i = X(t_i)$ are independent, we have

$$C_{ij} = \begin{cases} \sigma_i^2 & i = j \\ 0 & i \neq j \end{cases} \tag{8.141}$$

Thus, from Eq. (8.69) the covariance matrix **C** becomes

$$\mathbf{C} = \begin{bmatrix} \sigma_1^2 & 0 & \cdots & 0 \\ 0 & \sigma_2^2 & \cdots & 0 \\ \cdots & \cdots & \cdots & \cdots \\ 0 & 0 & \cdots & \sigma_n^2 \end{bmatrix} \tag{8.142}$$

It therefore follows that

$$|\det \mathbf{C}|^{1/2} = \sigma_1 \sigma_2 \cdots \sigma_n = \prod_{i=1}^{n} \sigma_1 \tag{8.143}$$

and

$$\mathbf{C}^{-1} = \begin{bmatrix} \dfrac{1}{\sigma_1^2} & 0 & \cdots & 0 \\ 0 & \dfrac{1}{\sigma_2^2} & \cdots & 0 \\ \cdots & \cdots & \cdots & \cdots \\ 0 & 0 & \cdots & \dfrac{1}{\sigma_n^2} \end{bmatrix} \tag{8.144}$$

Then we can write

$$(\mathbf{x} - \boldsymbol{\mu})^T \mathbf{C}^{-1} (\mathbf{x} - \boldsymbol{\mu}) = \sum_{i=1}^{n} \left(\frac{x_i - \mu_i}{\sigma_i} \right)^2 \tag{8.145}$$

Substituting Eqs. (8.143) and (8.145) into Eq. (8.67), we obtain Eq. (8.140).

SUPPLEMENTARY PROBLEMS

8.31. Consider a random process $X(t)$ defined by

$$X(t) = \cos \Omega t$$

where Ω is a random variable uniformly distributed over $[0, \omega_0]$. Determine whether $X(t)$ is stationary.

8.32. Consider the random process $X(t)$ defined by

$$X(t) = A\cos \omega t$$

where ω is a constant and A is a random variable uniformly distributed over $[0, 1]$. Find the autocorrelation and autocovariance of $X(t)$.

8.33. Let $X(t)$ be a WSS random process with autocorrelation

$$R_{XX}(\tau) = A e^{-\alpha|\tau|}$$

Find the second moment of the random variable $Y = X(5) - X(2)$.

8.34. A random signal $X(t)$ is given by $X(t) = A(t) \cos (\omega t + \Theta)$, where $A(t)$ is a zero mean WSS random signal with autocorrelation $R_{AA}(\tau)$, and Θ is a r.v. uniformly distributed over $[0, 2\pi]$ and independent of $A(t)$. The total average power of $A(t)$ is 1 watt.

  (*a*) Show that $X(t)$ is WSS.

  (*b*) Find the total average power of $X(t)$.

8.35. A random signal $X(t)$ is given by $X(t) = A + B \cos (\omega t + \Theta)$, where A, B, and Θ are independent r.v.'s uniformly distributed over $[0, 1]$, $[0, 2]$ and $[0, 2\pi]$, respectively. Find the mean and the autocorrelation of $X(t)$.

8.36. Let $X(t)$ be a WSS random process with mean μ_X. Let $Y(t) = a + X(t)$. Is $Y(t)$ WSS?

8.37. Let $X(t)$ and $Y(t)$ be defined by

$$X(t) = A + Bt, \, Y(t) = B + At$$

where A and B are independent r.v.'s with zero means and variance σ_A^2 and σ_B^2, respectively. Find the autocorrelations and cross-correlation of $X(t)$ and $Y(t)$.

8.38. Let $Z(t) = X(t)Y(t)$, where $X(t)$ and $Y(t)$ are independent and WSS. Is $Z(t)$ WSS?

8.39. Two random signals $X(t)$ and $Y(t)$ are given by

$$X(t) = A \cos \omega t + B \sin \omega t, \quad Y(t) = B \cos \omega t - A \sin \omega t$$

where ω is a constant, and A and B are independent r.v.'s with zero mean and same variance σ^2. Find the cross-correlation function of $X(t)$ and $Y(t)$.

ANSWERS TO SUPPLEMENTARY PROBLEMS

8.31. Nonstationary.

Hint: Examine specific sample functions of $X(t)$ for different frequencies, say, $\Omega = \pi/2$, π, and 2π.

8.32. $R_{XX}(t_1, t_2) = \dfrac{1}{3} \cos t_1 \cos t_2$

$C_{XX}(t_1, t_2) = \dfrac{1}{12} \cos t_1 \cos t_2$

8.33. $2A(1 - e^{-3\alpha})$

8.34. (a) Yes.   (b) ½ watts.

8.35. $\mu_X(t) = \dfrac{1}{2}$, $R_{XX}(t, t+\tau) = \dfrac{1}{3} + \dfrac{1}{3} \cos \omega \tau$

8.36. Yes.

8.37. $R_{XX}(t_1, t_2) = \sigma_A^2 + \sigma_B^2 t_1 t_2, \; R_{YY}(t_1, t_2) = \sigma_B^2 + \sigma_A^2 t_1 t_2$

$R_{XY}(t_1, t_2) = \sigma_A^2 t_2 + \sigma_B^2 t_1$

8.38. Yes.

8.39. $R_{XY}(t_1, t_2) = \sigma^2 \sin \omega(t_1 - t_2)$

CHAPTER 9

Power Spectral Density and Random Signals in Linear System

9.1 Introduction

In this chapter, the notion of power spectral density for a random signal is introduced. This concept enables us to study wide-sense stationary random signals in the frequency domain and define a white-noise process. The response of a linear system to random signal is then studied.

9.2 Correlations and Power Spectral Densities

In the following, we assume that all random processes are WSS.

A. Autocorrelation $R_{XX}(\tau)$:

The autocorrelation of $X(t)$ is [Eq. (8.35)]

$$R_{XX}(\tau) = E[X(t)X(t + \tau)] \tag{9.1}$$

Properties of $R_{XX}(\tau)$:

1. $R_{XX}(-\tau) = R_{XX}(\tau)$ (9.2)
2. $|R_{XX}(\tau)| \le R_{XX}(0)$ (9.3)
3. $R_{XX}(0) = E[X^2(t)]$ (9.4)

Property 3 [Eq. (9.4)] is easily obtained by setting $\tau = 0$ in Eq. (9.1). If we assume that $X(t)$ is a voltage waveform across a $1-\Omega$ resistor, then $E[X^2(t)]$ is the average value of power delivered to the $1-\Omega$ resistor by $X(t)$. Thus, $E[X^2(t)]$ is often called the average power of $X(t)$. Properties 1 and 2 are verified in Prob. (9.1).

    In case of a discrete-time random process $X(n)$, the autocorrelation function of $X(n)$ is defined by [Eq. (8.39)]

$$R_{XX}(k) = E[X(n)X(n+ k)] \tag{9.5}$$

Various properties of $R_{XX}(k)$ similar to those of $R_{XX}(\tau)$ can be obtained by replacing τ by k in Eqs. (9.2) to (9.4).

B. Cross-Correlation $R_{XY}(\tau)$:

The cross-correlation of $X(t)$ and $Y(t)$ is [Eq. (8.42)]

$$R_{XY}(\tau) = E[X(t)Y(t + \tau)] \tag{9.6}$$

Properties of $R_{XY}(\tau)$:

1. $R_{XY}(-\tau) = R_{YX}(\tau)$ (9.7)

2. $\left| R_{XY}(\tau) \right| \leq \sqrt{R_{XX}(0)R_{YY}(0)}$ (9.8)

3. $\left| R_{XY}(\tau) \right| \leq \dfrac{1}{2}[R_{XX}(0) + R_{YY}(0)]$ (9.9)

These properties are verified in Prob. 9.2.

Similarly, the cross-correlation function of two discrete-time jointly WSS random sequences $X(n)$ and $Y(n)$ is defined by

$$R_{XY}(k) = E[X(n)Y(n + k)]$$ (9.10)

And various properties of $R_{XY}(k)$ similar to those of $R_{XY}(\tau)$ can be obtained by replacing τ by k in Eqs. (9.7) to (9.9).

C.   Power Spectral Density or Power Spectrum:

Let $R_{XX}(\tau)$ be the autocorrelation of $X(t)$. Then the *power spectral density* (or *power spectrum*) of $X(t)$ is defined by the Fourier transform of $R_{XX}(\tau)$ as

$$S_{XX}(\omega) = \int_{-\infty}^{\infty} R_{XX}(\tau)e^{-j\omega\tau}\,d\tau$$ (9.11)

Thus, $R_{XX}(\tau) = \dfrac{1}{2\pi}\int_{-\infty}^{\infty} S_{XX}(\omega)e^{j\omega\tau}\,d\omega$ (9.12)

Equations (9.11) and (9.12) are known as the *Wiener-Khinchin relations*.

Properties of $S_{XX}(\omega)$:

1. $S_{XX}(\omega)$ is real and $S_{XX}(\omega) \geq 0$ (9.13)
2. $S_{XX}(-\omega) = S_{XX}(\omega)$ (9.14)

3. $\dfrac{1}{2\pi}\int_{-\infty}^{\infty} S_{XX}(\omega)\,d\omega = R_{XX}(0) = E[X^2(t)]$ (9.15)

Similarly, the power spectral density $S_{XX}(\Omega)$ of a discrete-time random process $X(n)$ is defined as the Fourier transform of $R_{XX}(k)$:

$$S_{XX}(\Omega) = \sum_{k=-\infty}^{\infty} R_{XX}(k)e^{-j\Omega k}$$ (9.16)

Thus, taking the inverse Fourier transform of $S_{XX}(\Omega)$, we obtain

$$R_{XX}(k) = \dfrac{1}{2\pi}\int_{-\pi}^{\pi} S_{XX}(\Omega)e^{j\Omega k}\,d\Omega$$ (9.17)

Properties of $S_X(\Omega)$:

1. $S_{XX}(\Omega + 2\pi) = S_{XX}(\Omega)$ (9.18)
2. $S_{XX}(\Omega)$ is real and $S_{XX}(\Omega) \geq 0$.   (9.19)
3. $S_{XX}(-\Omega) = S_{XX}(\Omega)$ (9.20)

4. $E[X^2(n)] = R_{XX}(0) = \dfrac{1}{2\pi}\int_{-\pi}^{\pi} S_{XX}(\Omega)\,d\Omega$ (9.21)

Note that property 1 [Eq. (9.18)] follows from the fact that $e^{-j\Omega k}$ is periodic with period 2π. Hence it is sufficient to define $S_{XX}(\Omega)$ only in the range $(-\pi, \pi)$.

D. Cross-Power Spectral Densities:

The *cross-power spectral density* (or *cross-power spectrum*) $S_{XY}(\omega)$ of two continuous-time random processes $X(t)$ and $Y(t)$ is defined as the Fourier transform of $R_{XY}(\tau)$:

$$S_{XY}(\omega) = \int_{-\infty}^{\infty} R_{XY}(\tau)e^{-j\omega\tau}\,d\tau \tag{9.22}$$

Thus, taking the inverse Fourier transform of $S_{XY}(\omega)$, we get

$$R_{XY}(\tau) = \frac{1}{2\pi}\int_{-\infty}^{\infty} S_{XY}(\omega)e^{j\omega\tau}\,d\omega \tag{9.23}$$

Properties of $S_{XY}(\omega)$:

Unlike $S_{XX}(\omega)$, which is a real-valued function of ω, $S_{XY}(\omega)$, in general, is a complex-valued function.

1. $S_{XY}(\omega) = S_{YX}(-\omega)$ $\tag{9.24}$
2. $S_{XY}(-\omega) = S_{XY}^*(\omega)$ $\tag{9.25}$

Similarly, the cross-power spectral density $S_{XY}(\Omega)$ of two discrete-time random processes $X(n)$ and $Y(n)$ is defined as the Fourier transform of $R_{XY}(k)$:

$$S_{XY}(\Omega) = \sum_{k=-\infty}^{\infty} R_{XY}(k)e^{-j\Omega k} \tag{9.26}$$

Thus, taking the inverse Fourier transform of $S_{XY}(\Omega)$, we get

$$R_{XY}(k) = \frac{1}{2\pi}\int_{-\pi}^{\pi} S_{XY}(\Omega)e^{j\Omega k}\,d\Omega \tag{9.27}$$

Properties of $S_{XY}(\Omega)$:

Unlike $S_{XX}(\Omega)$, which is a real-valued function of Ω, $S_{XY}(\Omega)$, in general, is a complex-valued function.

1. $S_{XY}(\Omega + 2\pi) = S_{XY}(\Omega)$ $\tag{9.28}$
2. $S_{XY}(\Omega) = S_{YX}(-\Omega)$ $\tag{9.29}$
3. $S_{XY}(-\Omega) = S_{XY}^*(\Omega)$ $\tag{9.30}$

9.3 White Noise

A random process $X(t)$ is called *white noise* if [Fig. 9-1(a)]

$$S_{XX}(\omega) = \frac{\eta}{2} \tag{9.31}$$

Taking the inverse Fourier transform of Eq. (9.31), we have

$$R_{XX}(\tau) = \frac{\eta}{2}\delta(\tau) \tag{9.32}$$

which is illustrated in Fig. 9-1(b). It is usually assumed that the mean of white noise is zero.

Similarly, a zero-mean discrete-time random sequence $X(n)$ is called a discrete-time white noise if

$$S_{XX}(\Omega) = \sigma^2 \tag{9.33}$$

Again the power spectral density of $X(n)$ is constant. Note that $S_{XX}(\Omega + 2\pi) = S_{XX}(\Omega)$ and the average power of $X(n)$ is $\sigma^2 = \text{Var}[X(n)]$, which is constant. Taking the discrete-time inverse Fourier transform of Eq. (9.33), we have

$$R_{XX}(k) = \sigma^2\,\delta(k) \tag{9.34}$$

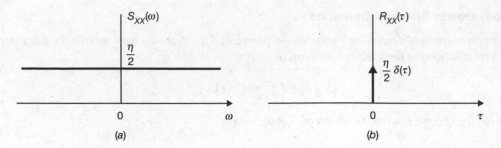

Fig. 9-1 White noise.

Band-Limited White Noise:

A random process $X(t)$ is called *band-limited white noise* if

$$S_{XX}(\omega) = \begin{cases} \dfrac{\eta}{2} & |\omega| \le \omega_B \\ 0 & |\omega| > \omega_B \end{cases} \tag{9.35}$$

Then

$$R_{XX}(\tau) = \frac{1}{2\pi} \int_{-\omega_B}^{\omega_B} \frac{\eta}{2} e^{j\omega\tau} \, d\omega = \frac{\eta\omega_B}{2\pi} \frac{\sin \omega_B \tau}{\omega_B \tau} \tag{9.36}$$

And $S_{XX}(\omega)$ and $R_{XX}(\tau)$ of band-limited white noise are shown in Fig. 9-2.

Note that the term *white* or *band-limited white* refers to the spectral shape of the process $X(t)$ only, and these terms do not imply that the distribution associated with $X(t)$ is Gaussian.

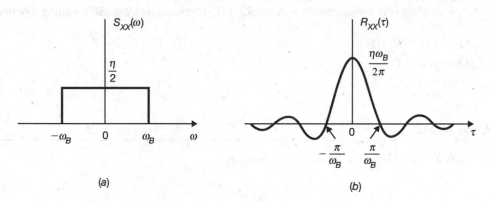

Fig. 9-2 Band-limited white noise.

Narrowband Random Process:

Suppose that $X(t)$ is a WSS process with zero mean and its power spectral density $S_{XX}(\omega)$ is nonzero only in some narrow frequency band of width $2W$ that is very small compared to a center frequency ω_c. Then the process $X(t)$ is called a *narrowband random process*.

In many communication systems, a narrowband process (or noise) is produced when white noise (or broadband noise) is passed through a narrowband linear filter. When a sample function of the narrowband process is viewed on an oscilloscope, the observed waveform appears as a sinusoid of random amplitude and phase. For this reason, the narrowband noise $X(t)$ is conveniently represented by the expression

$$X(t) = V(t) \cos [\omega_c t + \phi(t)] \tag{9.37}$$

9.4   Response of Linear System to Random Input

A.   Linear System:

As we discussed in Chap. 1 (Sec. 1.5), a system is a mathematical model for a physical process that relates the input (or excitation) signal x to the output (or response) y, and the system is viewed as a transformation (or mapping) of x into y. This transformation is represented by the operator $\mathbf{T}$ as (Eq. (1.60))

$$y = \mathbf{T}x \tag{9.38}$$

For a continuous-time linear time-invariant (LTI) system, Eq. (9.38) can be expressed as Eq. (2.60)

$$y(t) = \int_{-\infty}^{\infty} h(\alpha) x(t - \alpha) \, d\alpha = h(t) * x(t) \tag{9.39}$$

where $h(t)$ is the *impulse response* of a continuous-time LTI system. For a discrete-time LTI system, Eq. (9.38) can be expressed as (Eq. (2.45))

$$y(n) = \sum_{i=-\infty}^{\infty} h(i) x(n-i) = h(n) * x(n) \tag{9.40}$$

where $h(n)$ is the *impulse response* (or *unit sample response*) of a discrete-time LTI system.

B.   Response of a Continuous-Time Linear System to Random Input:

When the input to a continuous-time linear system represented by Eq. (9.38) is a random process $\{X(t), t \in T_x\}$, then the output will also be a random process $\{Y(t), t \in T_y\}$; that is,

$$\mathbf{T}\{X(t), t \in T_x\} = \{Y(t), t \in T_y\} \tag{9.41}$$

For any input sample function $x_i(t)$, the corresponding output sample function is

$$y_i(t) = \mathbf{T}\{x_i(t)\} \tag{9.42}$$

If the system is LTI, then by Eq. (9.39), we can write

$$Y(t) = \int_{-\infty}^{\infty} h(\alpha) X(t - \alpha) \, d\alpha = h(t) * X(t) \tag{9.43}$$

Note that Eq. (9.43) is a stochastic integral. Then

$$\mu_Y(t) = E[Y(t)] = E\left[\int_{-\infty}^{\infty} h(\alpha) X(t - \alpha) \, d\alpha\right]$$

$$= \int_{-\infty}^{\infty} h(\alpha) E[X(t - \alpha)] \, d\alpha$$

$$= \int_{-\infty}^{\infty} h(\alpha) \mu_X(t - \alpha) \, d\alpha = h(t) * \mu_X(t) \tag{9.44}$$

$$R_{YY}(t_1, t_2) = E[Y(t_1) Y(t_2)]$$

$$= E\left[\int_{-\infty}^{\infty} \int_{-\infty}^{\infty} h(\alpha) X(t_1 - \alpha) h(\beta) X(t_2 - \beta) \, d\alpha d\beta\right]$$

$$= \int_{-\infty}^{\infty} \int_{-\infty}^{\infty} h(\alpha) h(\beta) E[X(t_1 - \alpha) X(t_2 - \beta)] \, d\alpha d\beta$$

$$= \int_{-\infty}^{\infty} \int_{-\infty}^{\infty} h(\alpha) h(\beta) R_{XX}(t_1 - \alpha, t_2 - \beta)] \, d\alpha d\beta \tag{9.45}$$

If the input $X(t)$ is WSS, then from Eq. (9.43) we have

$$E[Y(t)] = \int_{-\infty}^{\infty} h(\alpha) \mu_X \, d\alpha = \mu_X \int_{-\infty}^{\infty} h(\alpha) \, d\alpha = \mu_X H(0) \tag{9.46}$$

where $H(0)$ is the frequency response of the linear system at $\omega = 0$. Thus, the mean of the output is a constant.

The autocorrelation of the output given in Eq. (9.45) becomes

$$R_{YY}(t_1, t_2) = \int_{-\infty}^{\infty} \int_{-\infty}^{\infty} h(\alpha) h(\beta) R_{XX}(t_2 - t_1 + \alpha - \beta)\, d\alpha d\beta \tag{9.47}$$

which indicates that $R_{YY}(t_1, t_2)$ is a function of the time difference $\tau = t_2 - t_1$. Hence,

$$R_{YY}(\tau) = \int_{-\infty}^{\infty} \int_{-\infty}^{\infty} h(\alpha) h(\beta) R_{XX}(\tau + \alpha - \beta)\, d\alpha d\beta \tag{9.48}$$

Thus, we conclude that if the input $X(t)$ is WSS, the output $Y(t)$ is also WSS.

The cross-correlation function between input $X(t)$ and $Y(t)$ is given by

$$\begin{aligned} R_{XY}(t_1, t_2) &= E\big[X(t_1) Y(t_2)\big] \\ &= E\Big[X(t_1) \int_{-\infty}^{\infty} h(\alpha) X(t_2 - \alpha)\, d\alpha\Big] \\ &= \int_{-\infty}^{\infty} h(\alpha) E\big[X(t_1) X(t_2 - \alpha)\big]\, d\alpha \\ &= \int_{-\infty}^{\infty} h(\alpha) R_{XX}(t_1, t_2 - \alpha)\, d\alpha \end{aligned} \tag{9.49}$$

When input $X(t)$ is WSS, Eq. (9.49) becomes

$$R_{XY}(t_1, t_2) = \int_{-\infty}^{\infty} h(\alpha) R_{XX}(t_2 - t_1 - \alpha)\, d\alpha \tag{9.50}$$

which indicates that $R_{XY}(t_1, t_2)$ is a function of the time difference $\tau = t_2 - t_1$. Hence

$$R_{XY}(\tau) = \int_{-\infty}^{\infty} h(\alpha) R_{XX}(\tau - \alpha)\, d\alpha = h(\tau) * R_{XX}(\tau) \tag{9.51}$$

Thus, we conclude that if the input $X(t)$ to an LTI system is WSS, the output $Y(t)$ is also WSS. Moreover, the input $X(t)$ and output $Y(t)$ are jointly WSS.

In a similar manner, it can be shown that (Prob. 9.11)

$$R_{YY}(\tau) = \int_{-\infty}^{\infty} h(-\alpha) R_{XY}(\tau - \alpha)\, d\alpha = h(-\tau) * R_{XY}(\tau) \tag{9.52}$$

Substituting Eq. (9.51) into Eq. (9.52), we have

$$R_{YY}(\tau) = h(-\tau) * h(\tau) * R_{XX}(\tau) \tag{9.53}$$

Now taking Fourier transforms of Eq. (9.51), (9.52), and (9.53) and using convolution property of Fourier transform [Eq. (5.58)], we obtain

$$S_{XY}(\omega) = H(\omega) S_{XX}(\omega) \tag{9.54}$$

$$S_{YY}(\omega) = H^*(\omega) S_{XY}(\omega) \tag{9.55}$$

$$S_{YY}(\omega) = H^*(\omega) H(\omega) S_{XX}(\omega) = |H(\omega)|^2 S_{XX}(\omega) \tag{9.56}$$

The schematic of these relations is shown in Fig. 9-3.

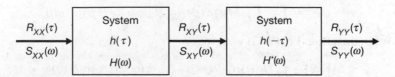

Fig. 9-3

Equation (9.56) indicates the important result that the power spectral density of the output is the product of the power spectral density of the input and the magnitude squared of the frequency response of the system.

When the autocorrelation of the output $R_{YY}(\tau)$ is desired, it is easier to determine the power spectral density $S_{YY}(\omega)$ and then to evaluate the inverse Fourier transform (Prob. 9.13). Thus,

$$R_{YY}(\tau) = \frac{1}{2\pi} \int_{-\infty}^{\infty} S_{YY}(\omega) e^{j\omega\tau} d\omega$$

$$= \frac{1}{2\pi} \int_{-\infty}^{\infty} |H(\omega)|^2 S_{XX}(\omega) e^{j\omega\tau} d\omega \tag{9.57}$$

By Eq. (9.4), the average power in the output $Y(t)$ is

$$E[Y^2(t)] = R_{YY}(0) = \frac{1}{2\pi} \int_{-\infty}^{\infty} |H(\omega)|^2 S_{XX}(\omega) d\omega \tag{9.58}$$

C. Response of a Discrete-Time LTI System to Random Input:

When the input to a discrete-time LTI system is a discrete-time random sequence $X(n)$, then by Eq. (2.39), the output $Y(n)$ is

$$Y(n) = \sum_{i=-\infty}^{\infty} h(i) X(n-i) \tag{9.59}$$

The autocorrelation function of $Y(n)$ is given by (Prob. 9.22)

$$R_{YY}(n,m) = \sum_{i=-\infty}^{\infty} \sum_{l=-\infty}^{\infty} h(i)h(l) R_{XX}(n-i, m-l) \tag{9.60}$$

The cross-correlation function of $X(n)$ and $Y(n)$ is given by (Prob. 9.23)

$$R_{XY}(n,m) = E[X(n)Y(m)] = \sum_{i=-\infty}^{\infty} h(l) R_{XX}(n, m-l) \tag{9.61}$$

When $X(n)$ is WSS, then from Eq. (9.59)

$$\mu_Y(n) = E[Y(n)] = \mu_X \sum_{i=-\infty}^{\infty} h(i) = \mu_X H(0) \tag{9.62}$$

where $H(0) = H(\Omega)\big|_{\Omega=0}$ and $H(\Omega)$ is the frequency response of the system defined by the Fourier transform of $H(n)$.

The autocorrelation function of $Y(n)$ is, from Eq. (9.60)

$$R_{YY}(n,m) = \sum_{i=-\infty}^{\infty} \sum_{l=-\infty}^{\infty} h(i)h(l) R_{XX}(m-n+i-l) \tag{9.63}$$

Setting $m = n + k$, we get

$$R_{YY}(n, n+k) = \sum_{i=-\infty}^{\infty} \sum_{l=-\infty}^{\infty} h(i)h(l) R_{XX}(k+i-l) = R_{YY}(k) \tag{9.64}$$

Similarly, from Eq. (9.61), we obtain

$$R_{XY}(k) = \sum_{l=-\infty}^{\infty} h(l) R_{XX}(k-l) = h(k) * R_{XX}(k) \tag{9.65}$$

and (Prob. 9.24)

$$R_{YY}(k) = \sum_{l=-\infty}^{\infty} h(-l) R_{XY}(k-l) = h(-k) * R_{XY}(k) \tag{9.66}$$

Substituting Eq. (9.65) into Eq. (9.66), we obtain

$$R_{YY}(k) = h(-k) * h(k) * R_{XX}(k) \tag{9.67}$$

Now taking Fourier transforms of Eq. (9.65), (9.66) and (9.67), we obtain

$$S_{XY}(\Omega) = H(\Omega) S_{XX}(\Omega) \tag{9.68}$$

$$S_{YY}(\Omega) = H^*(\Omega) S_{XY}(\Omega) \tag{9.69}$$

$$S_{YY}(\Omega) = H^*(\Omega)H(\Omega)S_{XX}(\Omega) = |H(\Omega)|^2 S_{XX}(\Omega) \tag{9.70}$$

Similarly, when the autocorrelation function of the output $R_{YY}(k)$ is desired, it is easier to determine the power spectral density $S_{YY}(\Omega)$ and then take the inverse Fourier transform. Thus,

$$R_{YY}(k) = \frac{1}{2\pi}\int_{-\pi}^{\pi} S_{YY}(\Omega)e^{j\Omega k}\, d\Omega = \frac{1}{2\pi}\int_{-\pi}^{\pi} |H(\Omega)|^2 S_{XX}(\Omega)e^{j\Omega k}\, d\Omega \tag{9.71}$$

By Eq. (9.21), the average power in the output $Y(n)$ is

$$E\big[Y^2(n)\big] = R_{YY}(0) = \frac{1}{2\pi}\int_{-\pi}^{\pi} |H(\Omega)|^2 S_{XX}(\Omega)\, d\Omega \tag{9.72}$$

SOLVED PROBLEMS

Correlations and Power Spectral Densities

9.1. Let $X(t)$ be a WSS random process. Verify Eqs. (9.2) and (9.3); that is,

    (a) $R_{XX}(-\tau) = R_{XX}(\tau)$

    (b) $|R_{XX}(\tau)| \le R_{XX}(0)$

    (a) From Eq. (8.35)

$$R_{XX}(\tau) = E[X(t)X(t+\tau)]$$

        Setting $t + \tau = t'$, we have

$$R_{XX}(\tau) = E[X(t'-\tau)X(t')]$$
$$= E[X(t')X(t'-\tau)] = R_{XX}(-\tau)$$

    (b) $E[[X(t) \pm X(t+\tau)]^2] \ge 0$

        or $E[X^2(t) \pm 2X(t)X(t+\tau) + X^2(t+\tau)] \ge 0$

        or $E[X^2(t)] \pm 2E[X(t)X(t+\tau)] + E[X^2(t+\tau)] \ge 0$

        or $2R_{XX}(0) \pm 2R_{XX}(\tau) \ge 0$

        Hence, $R_{XX}(0) \ge |R_{XX}(\tau)|$

9.2. Let $X(t)$ and $Y(t)$ be WSS random processes. Verify Eqs. (9.7) and (9.8); that is,

    (a) $R_{XY}(-\tau) = R_{YX}(\tau)$

    (b) $|R_{XY}(\tau)| \le \sqrt{R_{XX}(0)R_{YY}(0)}$

(a) By Eq. (8.42)

$$R_{XY}(-\tau) = E[X(t)Y(t-\tau)]$$

Setting $t - \tau = t'$, we obtain

$$R_{XY}(-\tau) = E[X(t'+\tau)Y(t')] = E[Y(t')X(t'+\tau)] = R_{YX}(\tau)$$

(b) From the Cauchy-Schwarz inequality Eq. (B.129) (Appendix B), it follows that

$$\{E[X(t)Y(t+\tau)]\}^2 \le E[X^2(t)]E[Y^2(t+\tau)]$$

or
$$[R_{XY}(\tau)]^2 \le R_{XX}(0)R_{YY}(0)$$

Thus,
$$|R_{XY}(\tau)| \le \sqrt{R_{XX}(0)R_{YY}(0)}$$

9.3. Show that the power spectrum of a (real) random process $X(t)$ is real, and verify Eq. (9.14); that is,

$$S_{XX}(-\omega) = S_{XX}(\omega)$$

From Eq. (9.11) and by expanding the exponential, we have

$$S_{XX}(\omega) = \int_{-\infty}^{\infty} R_{XX}(\tau)e^{-j\omega\tau}\,d\tau$$
$$= \int_{-\infty}^{\infty} R_{XX}(\tau)(\cos\omega\tau - j\sin\omega\tau)\,d\tau \qquad (9.73)$$
$$= \int_{-\infty}^{\infty} R_{XX}(\tau)\cos\omega\tau\,d\tau - j\int_{-\infty}^{\infty} R_{XX}(\tau)\sin\omega\tau\,d\tau$$

Since $R_{XX}(-\tau) = R_{XX}(\tau)$ [Eq. (9.2)] imaginary term in Eq. (9.73) then vanishes and we obtain

$$S_{XX}(\omega) = \int_{-\infty}^{\infty} R_{XX}(\tau)\cos\omega\tau\,d\tau \qquad (9.74)$$

which indicates that $S_{XX}(\omega)$ is real.

Since the cosine is an even function of its arguments, that is, $\cos(-\omega\tau) = \cos\omega\tau$, it follows that

$$S_{XX}(-\omega) = S_{XX}(\omega)$$

which indicates that the power spectrum of $X(t)$ is an even function of frequency.

9.4. Let $X(t)$ and $Y(t)$ be both zero-mean and WSS random processes. Consider the random process $Z(t)$ defined by

$$Z(t) = X(t) + Y(t) \qquad (9.75)$$

(a) Determine the autocorrelation and the power spectrum of $Z(t)$ if $X(t)$ and $Y(t)$ are jointly WSS.

(b) Repeat part (a) if $X(t)$ and $Y(t)$ are orthogonal.

(c) Show that if $X(t)$ and $Y(t)$ are orthogonal, then the mean square of $Z(t)$ is equal to the sum of the mean squares of $X(t)$ and $Y(t)$.

(a) The autocorrelation of $Z(t)$ is given by

$$\begin{aligned} R_{ZZ}(t_1, t_2) &= E[Z(t_1)Z(t_2)] \\ &= E[[X(t_1)+Y(t_1)][X(t_2)+Y(t_2)]] \\ &= E[X(t_1)X(t_2)] + E[X(t_1)Y(t_2)] \\ &\quad + E[Y(t_1)X(t_2)] + E[Y(t_1)Y(t_2)] \\ &= R_{XX}(t_1,t_2) + R_{XY}(t_1,t_2) + R_{YX}(t_1,t_2) + R_{YY}(t_1,t_2) \end{aligned} \qquad (9.76)$$

If $X(t)$ and $Y(t)$ are jointly WSS, then we have

$$R_{ZZ}(\tau) = R_{XX}(\tau) + R_{XY}(\tau) + R_{YX}(\tau) + R_{YY}(\tau) \tag{9.77}$$

where $\tau = t_2 - t_1$.

Taking the Fourier transform of both sides of Eq. (9.77), we obtain

$$S_{ZZ}(\omega) = S_{XX}(\omega) + S_{XY}(\omega) + S_{YX}(\omega) + S_{YY}(\omega) \tag{9.78}$$

(b) If $X(t)$ and $Y(t)$ are orthogonal [Eq. (8.47)],

$$R_{XY}(\tau) = R_{YX}(\tau) = 0$$

Then Eqs. (9.77) and (9.78) become

$$R_{ZZ}(\tau) = R_{XX}(\tau) + R_{YY}(\tau) \tag{9.79}$$

and
$$S_{ZZ}(\omega) = S_{XX}(\omega) + S_{YY}(\omega) \tag{9.80}$$

(c) From Eqs. (9.79) and (8.37)

$$R_{ZZ}(0) = R_{XX}(0) + R_{YY}(0)$$

or
$$E[Z^2(t)] = E[X^2(t)] + E[Y^2(t)] \tag{9.81}$$

which indicates that the mean square of $Z(t)$ is equal to the sum of the mean squares of $X(t)$ and $Y(t)$.

9.5. Two random processes $X(t)$ and $(Y(t)$ are given by

$$X(t) = A\cos(\omega t + \Theta) \tag{9.82}$$
$$Y(t) = A\sin(\omega t + \Theta) \tag{9.83}$$

where A and ω are constants and Θ is a uniform random variable over $[0, 2\pi]$. Find the cross-correlation of $X(t)$ and $Y(t)$, and verify Eq. (9.7).

From Eq. (8.42), the cross-correlation of $X(t)$ and $Y(t)$ is

$$\begin{aligned}
R_{XY}(t, t + \tau) &= E[X(t)Y(t + \tau)] \\
&= E[A^2 \cos(\omega t + \Theta)\sin[\omega(t + \tau) + \Theta]] \\
&= \frac{A^2}{2} E[\sin(2\omega t + \omega\tau + 2\Theta) - \sin(-\omega\tau)] \\
&= \frac{A^2}{2}\sin\omega\tau = R_{XY}(\tau)
\end{aligned} \tag{9.84}$$

Similarly,
$$\begin{aligned}
R_{YX}(t, t + \tau) &= E[Y(t)X(t + \tau)] \\
&= E[A^2 \sin(\omega t + \Theta)\cos[\omega(t + \tau) + \Theta]] \\
&= \frac{A^2}{2} E[\sin(2\omega t + \omega\tau + 2\Theta) + \sin(-\omega\tau)] \\
&= -\frac{A^2}{2}\sin\omega\tau = R_{YX}(\tau)
\end{aligned} \tag{9.85}$$

From Eqs. (9.84) and (9.85)

$$R_{XY}(-\tau) = \frac{A^2}{2}\sin\omega(-\tau) = -\frac{A^2}{2}\sin\omega\tau = R_{YX}(\tau)$$

which verifies Eq. (9.7).

9.6. A class of modulated random signal $Y(t)$ is defined by

$$Y(t) = AX(t)\cos(\omega_c t + \Theta) \tag{9.86}$$

where $X(t)$ is the random message signal and $A\cos(\omega_c t + \Theta)$ is the carrier. The random message signal $X(t)$ is a zero-mean stationary random process with autocorrelation $R_{XX}(\tau)$ and power spectrum $S_{XX}(\omega)$. The carrier amplitude A and the frequency ω_c are constants, and phase Θ is a random variable uniformly distributed over $[0, 2\pi]$. Assuming that $X(t)$ and Θ are independent, find the mean, autocorrelation, and power spectrum of $Y(t)$.

$$\mu_Y(t) = E[Y(t)] = E[AX(t)\cos(\omega_c t + \Theta)]$$
$$= AE[X(t)]E[\cos(\omega_c t + \Theta)] = 0$$

since $X(t)$ and Θ are independent and $E[X(t)] = 0$.

$$R_{YY}(t, t+\tau) = E[Y(t)Y(t+\tau)]$$
$$= E[A^2 X(t)X(t+\tau)\cos(\omega_c t + \Theta)\cos[\omega_c(t+\tau)+\Theta]]$$
$$= \frac{A^2}{2}E[X(t)X(t+\tau)]E[\cos\omega_c\tau + \cos(2\omega_c t + \omega_c\tau + 2\Theta)]$$
$$= \frac{A^2}{2}R_{XX}(\tau)\cos\omega_c\tau = R_{YY}(\tau) \tag{9.87}$$

Since the mean of $Y(t)$ is a constant and the autocorrelation of $Y(t)$ depends only on the time difference τ, $Y(t)$ is WSS. Thus,

$$S_{YY}(\omega) = \mathcal{F}[R_{YY}(\tau)] = \frac{A^2}{2}\mathcal{F}[R_{XX}(\tau)\cos\omega_c\tau]$$

By Eqs. (9.11) and (5.144)

$$\mathcal{F}[R_{XX}(\tau)] = S_{XX}(\omega)$$
$$\mathcal{F}(\cos\omega_c\tau) = \pi\delta(\omega - \omega_c) + \pi\delta(\omega + \omega_c)$$

Then, using the frequency convolution theorem (5.59) and Eq. (2.59), we obtain

$$S_{YY}(\omega) = \frac{A^2}{4\pi}S_{XX}(\omega) * [\pi\delta(\omega - \omega_c) + \pi\delta(\omega + \omega_c)]$$
$$= \frac{A^2}{4}[S_{XX}(\omega - \omega_c) + S_{XX}(\omega + \omega_c)] \tag{9.88}$$

9.7. Consider a random process $X(t)$ that assumes the values $\pm A$ with equal probability. A typical sample function of $X(t)$ is shown in Fig. 9-4. The average number of polarity switches (zero crossings) per unit time is α. The probability of having exactly k crossings in time τ is given by the Poisson distribution [Eq. (B.48)]

$$P(Z = k) = e^{-\alpha\tau}\frac{(\alpha\tau)^k}{k!} \tag{9.89}$$

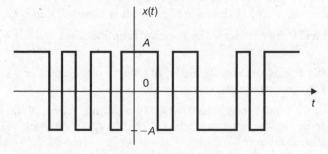

Fig. 9-4 Telegraph signal.

where Z is the random variable representing the number of zero crossing. The process $X(t)$ is known as the *telegraph* signal. Find the autocorrelation and the power spectrum of $X(t)$.

If τ is any positive time interval, then

$$
\begin{aligned}
R_{XX}(t, t+\tau) &= E[X(t)X(t+\tau)] \\
&= A^2 P[\, X(t) \text{ and } X(t+\tau) \text{ have same signs}] \\
&\quad + (-A^2)P[X(t) \text{ and } X(t+\tau) \text{ have different signs}] \\
&= A^2 P[Z \text{ even in } (t, t+\tau)] - A^2 P[Z \text{ odd in } (t, t+\tau)] \\
&= A^2 \sum_{k\,\text{even}} e^{-\alpha\tau}\frac{(\alpha\tau)^k}{k!} - A^2 \sum_{k\,\text{odd}} e^{-\alpha\tau}\frac{(\alpha\tau)^k}{k!} \\
&= A^2 e^{-\alpha\tau} \sum_{k=0}^{\infty} \frac{(\alpha\tau)^k}{k!}(-1)^k \\
&= A^2 e^{-\alpha\tau} \sum_{k=0}^{\infty} \frac{(-\alpha\tau)^k}{k!} = A^2 e^{-\alpha\tau}e^{-\alpha\tau} = A^2 e^{-2\alpha\tau}
\end{aligned}
\tag{9.90}
$$

which indicates that the autocorrelation depends only on the time difference τ. By Eq. (9.2), the complete solution that includes $\tau < 0$ is given by

$$
R_{XX}(\tau) = A^2 e^{-2\alpha|\tau|}
\tag{9.91}
$$

which is sketched in Fig. 9-5(a).

Taking the Fourier transform of both sides of Eq. (9.91), we see that the power spectrum of $X(t)$ is [Eq. (5.138)]

$$
S_{XX}(\omega) = A^2 \frac{4\alpha}{\omega^2 + (2\alpha)^2}
\tag{9.92}
$$

which is sketched in Fig. 9-5(b).

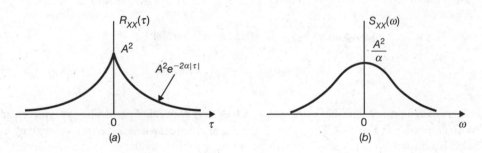

Fig. 9-5

9.8. Consider a random binary process $X(t)$ consisting of a random sequence of binary symbols 1 and 0. A typical sample function of $X(t)$ is shown in Fig. 9-6. It is assumed that

1. The symbols 1 and 0 are represented by pulses of amplitude $+A$ and $-A$ V, respectively, and duration $T_b s$.

2. The two symbols 1 and 0 are equally likely, and the presence of a 1 or 0 in any one interval is independent of the presence in all other intervals.

3. The pulse sequence is not synchronized, so that the starting time t_d of the first pulse after $t = 0$ is equally likely to be anywhere between 0 to T_b. That is, t_d is the sample value of a random variable T_d uniformly distributed over $[0, T_b]$.

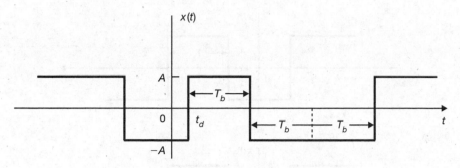

Fig. 9-6 Random binary signal.

Find the autocorrelation and power spectrum of $X(t)$.

The random binary process $X(t)$ can be represented by

$$X(t) = \sum_{k=-\infty}^{\infty} A_k p(t - kT_b - T_d) \tag{9.93}$$

where $\{A_k\}$ is a sequence of independent random variables with $P[A_k = A] = P[A_k = -A] = \frac{1}{2}$, $p(t)$ is a unit amplitude pulse of duration T_b, and T_d is a random variable uniformly distributed over $[0, T_b]$.

$$\mu_X(t) = E[X(t)] = E[A_k] = \frac{1}{2}A + \frac{1}{2}(-A) = 0 \tag{9.94}$$

Let $t_2 > t_1$. When $t_2 - t_1 > T_b$, then t_1 and t_2 must fall in different pulse intervals [Fig. 9-7(a)] and the random variables $X(t_1)$ and $X(t_2)$ are therefore independent. We thus have

$$R_{XX}(t_1, t_2) = E[X(t_1)X(t_2)] = E[X(t_1)]E[X(t_2)] = 0 \tag{9.95}$$

When $t_2 - t_1 < T_b$, then depending on the value of T_d, t_1 and t_2 may or may not be in the same pulse interval [Fig. 9-7(b) and (c)]. If we let B denote the random event "t_1 and t_2 are in adjacent pulse intervals," then we have

$$R_{XX}(t_1, t_2) = E[X(t_1)X(t_2)|B]P(B) + E[X(t_1)X(t_2)|\bar{B}]P(\bar{B})$$

Now
$$E[X(t_1)X(t_2)|B] = E[X(t_1)]E[X(t_2)] = 0$$

$$E[X(t_1)X(t_2)|\bar{B}] = A^2$$

Since $P(B)$ will be the same when t_1 and t_2 fall in any time range of length T_b, it suffices to consider the case $0 < t < T_b$, as shown in Fig. 9-7(b). From Fig. 9-7 (b);

$$P(B) = P(t_1 < T_d < t_2)$$
$$= \int_{t_1}^{t_2} f_{T_d}(t_d)\, dt_d = \int_{t_1}^{t_2} \frac{1}{T_b}\, dt_d = \frac{t_2 - t_1}{T_b}$$

From Eq. (B.4), we have

$$P(\bar{B}) = 1 - P(B) = 1 - \frac{t_2 - t_1}{T_b}$$

Thus,
$$R_{XX}(t_1, t_2) = A^2\left(1 - \frac{t_2 - t_1}{T_b}\right) = R_{XX}(\tau) \tag{9.96}$$

where $\tau = t_2 - t_1$.

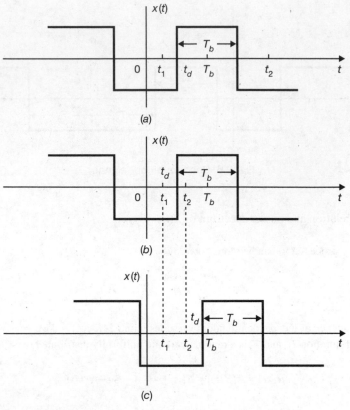

Fig. 9-7

Since $R_{XX}(-\tau) = R_{XX}(\tau)$, we conclude that

$$R_{XX}(\tau) = \begin{cases} A^2\left(1 - \dfrac{|\tau|}{T_b}\right) & |\tau| \le T_b \\ 0 & |\tau| > T_b \end{cases} \tag{9.97}$$

which is plotted in Fig. 9-8(a).

From Eqs. (9.94) and (9.97), we see that $X(t)$ is WSS. Thus, from Eq. (9.11), the power spectrum of $X(t)$ is (Prob. 5.67)

$$S_{XX}(\omega) = A^2 T_b \left[\frac{\sin(\omega T_b/2)}{\omega T_b/2} \right]^2 \tag{9.98}$$

which is plotted in Fig. 9-8(b).

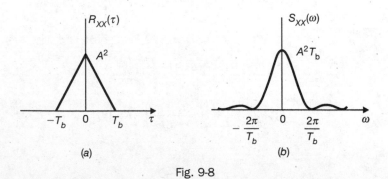

Fig. 9-8

Response of Linear System to Random Input

9.9. A WSS random process $X(t)$ is applied to the input of an LTI system with impulse response $h(t) = 3e^{-2t} u(t)$. Find the mean value of the output $Y(t)$ of the system if $E[X(t)] = 2$.

By Eq. (5.45), the frequency response $H(\omega)$ of the system is

$$H(\omega) = \mathcal{F}[h(t)] = 3\frac{1}{j\omega + 2}$$

Then, by Eq. (9.46), the mean value of $Y(t)$ is

$$\mu_Y(t) = E[Y(t)] = \mu_X H(0) = 2\left(\frac{3}{2}\right) = 3$$

9.10. Let $Y(t)$ be the output of an LTI system with impulse response $h(t)$, when $X(t)$ is applied as input. Show that

(a) $\quad R_{XY}(t_1, t_2) = \int_{-\infty}^{\infty} h(\beta) R_{XX}(t_1, t_2 - \beta)\, d\beta \qquad\qquad\qquad (9.99)$

(b) $\quad R_{YY}(t_1, t_2) = \int_{-\infty}^{\infty} h(\alpha) R_{XY}(t_1, -\alpha, t_2)\, d\alpha \qquad\qquad\quad (9.100)$

(a) Using Eq. (9.43), we have

$$R_{XY}(t_1, t_2) = E[X(t_1) Y(t_2)]$$
$$= E\left[X(t_1)\int_{-\infty}^{\infty} h(\beta) X(t_2 - \beta)\, d\beta\right]$$
$$= \int_{-\infty}^{\infty} h(\beta) E[X(t_1) X(t_2 - \beta)]\, d\beta$$
$$= \int_{-\infty}^{\infty} h(\beta) R_{XX}(t_1, t_2 - \beta)\, d\beta$$

(b) Similarly,

$$R_{YY}(t_1, t_2) = E[Y(t_1) Y(t_2)]$$
$$= E\left[\int_{-\infty}^{\infty} h(\alpha) X(t_1 - \alpha)\, d\alpha\, Y(t_2)\right]$$
$$= \int_{-\infty}^{\infty} h(\alpha) E[X(t_1 - \alpha)] Y(t_2)\, d\alpha$$
$$= \int_{-\infty}^{\infty} h(\alpha) R_{XY}(t_1 - \alpha, t_2)\, d\alpha$$

9.11. Let $X(t)$ be a WSS random input process to an LTI system with impulse response $h(t)$, and let $Y(t)$ be the corresponding output process. Show that

(a) $\quad R_{XY}(\tau) = h(\tau) * R_{XX}(\tau) \qquad\qquad\qquad\qquad\qquad\qquad (9.101)$

(b) $\quad R_{YY}(\tau) = h(-\tau) * R_{XY}(\tau) \qquad\qquad\qquad\qquad\qquad\quad (9.102)$

(c) $\quad S_{XY}(\omega) = H(\omega) S_{XX}(\omega) \qquad\qquad\qquad\qquad\qquad\quad (9.103)$

(d) $\quad S_{YY}(\omega) = H^*(\omega) S_{XY}(\omega) \qquad\qquad\qquad\qquad\qquad\quad (9.104)$

where $*$ denotes the convolution and $H^*(\omega)$ is the complex conjugate of $H(\omega)$.

(a) If $X(t)$ is WSS, then Eq. (9.99) of Prob. 9.10 becomes

$$R_{XY}(t_1, t_2) = \int_{-\infty}^{\infty} h(\beta) R_{XX}(t_2 - t_1 - \beta)\, d\beta \qquad\qquad (9.105)$$

which indicates that $R_{XY}(t_1, t_2)$ is a function of the time difference $\tau = t_2 - t_1$ only. Hence, Eq. (9.105) yields

$$R_{XY}(\tau) = \int_{-\infty}^{\infty} h(\beta) R_{XX}(\tau - \beta)\, d\beta = h(\tau) * R_{XX}(\tau)$$

(b) Similarly, if $X(t)$ is WSS, then Eq. (9.100) becomes

$$R_{YY}(t_1, t_2) = \int_{-\infty}^{\infty} h(\alpha) R_{XY}(t_2 - t_1 + \alpha)\, d\alpha$$

or

$$R_{YY}(\tau) = \int_{-\infty}^{\infty} h(\alpha) R_{XY}(\tau + \alpha)\, d\alpha = h(-\tau) * R_{XY}(\tau)$$

(c) Taking the Fourier transform of both sides of Eq. (9.101) and using Eqs. (9.22) and (5.58), we obtain

$$S_{XY}(\omega) = H(\omega)\, S_{XX}(\omega)$$

(d) Similarly, taking the Fourier transform of both sides of Eq. (9.102) and using Eqs. (9.11), (5.58), and (5.53), we obtain

$$S_{YY}(\omega) = H^*(\omega) S_{XY}(\omega)$$

Note that by combining Eqs. (9.103) and (9.104), we obtain Eq. (9.56); that is,

$$S_{YY}(\omega) = H^*(\omega) H(\omega) S_{XX}(\omega) = |H(\omega)|^2 S_{XX}(\omega)$$

9.12. Let $X(t)$ and $Y(t)$ be the wide-sense stationary random input process and random output process, respectively, of a quadrature phase-shifting filter ($-\pi/2$ rad phase shifter of Prob. 5.48). Show that

(a) $R_{XX}(\tau) = R_{YY}(\tau)$ (9.106)

(b) $R_{XY}(\tau) = \hat{R}_{XX}(\tau)$ (9.107)

where $\hat{R}_{XX}(\tau)$ is the Hilbert transform of $R_{XX}(\tau)$.

(a) The Hilbert transform $\hat{X}(t)$ of $X(t)$ was defined in Prob. 5.48 as the output of a quadrature phase-shifting filter with

$$h(t) = \frac{1}{\pi t} \quad H(\omega) = -j\,\text{sgn}(\omega)$$

Since $|H(\omega)|^2 = 1$, we conclude that if $X(t)$ is a WSS random signal, then $Y(t) = \hat{X}(t)$ and by Eq. (9.56)

$$S_{YY}(\omega) = |H(\omega)|^2 S_{XX}(\omega) = S_{XX}(\omega)$$

Hence, $$R_{YY}(\tau) = \mathscr{F}^{-1}[S_{YY}(\omega)] = \mathscr{F}^{-1}[S_{XX}(\omega)] = R_{XX}(\tau)$$

(b) Using Eqs. (9.101) and (5.174), we have

$$R_{XY}(\tau) = h(\tau) * R_{XX}(\tau) = \frac{1}{\pi t} * R_{XX}(\tau) = \hat{R}_{XX}(\tau)$$

9.13. A WSS random process $X(t)$ with autocorrelation

$$R_{XX}(\tau) = A e^{-a|\tau|}$$

where A and a are real positive constants, is applied to the input of an LTI system with impulse response

$$h(t) = e^{-bt} u(t)$$

where b is a real positive constant. Find the autocorrelation of the output $Y(t)$ of the system.

Using Eq. (5.45), we see that the frequency response $H(\omega)$ of the system is

$$H(\omega) = \mathscr{F}[h(t)] = \frac{1}{j\omega + b}$$

So $$|H(\omega)|^2 = \frac{1}{\omega^2 + b^2}$$

Using Eq. (5.138), we see that the power spectral density of $X(t)$ is

$$S_{XX}(\omega) = \mathscr{F}[R_{XX}(\tau)] = A\frac{2a}{\omega^2 + a^2}$$

By Eq. (9.56), the power spectral density of $Y(t)$ is

$$\begin{aligned}
S_{YY}(\omega) &= |H(\omega)|^2 S_{XX}(\omega) \\
&= \left(\frac{1}{\omega^2 + b^2}\right)\left(\frac{2aA}{\omega^2 + a^2}\right) \\
&= \frac{aA}{(a^2 - b^2)b}\left(\frac{2b}{\omega^2 + b^2}\right) - \frac{A}{a^2 - b^2}\left(\frac{2a}{\omega^2 + a^2}\right)
\end{aligned}$$

Taking the inverse Fourier transform of both sides of the above equation and using Eq. (5.139), we obtain

$$R_{YY}(\tau) = \frac{aA}{(a^2 - b^2)b}e^{-b|\tau|} - \frac{A}{a^2 - b^2}e^{-a|\tau|}$$

9.14. Verify Eq. (9.13); that is, the power spectrum of any WSS process $X(t)$ is real and

$$S_{XX}(\omega) \geq 0$$

for every ω.

The realness of the power spectrum of $X(t)$ was shown in Prob. 9.3. Consider an ideal bandpass filter with frequency response (Fig. 9-9)

$$H(\omega) = \begin{cases} 1 & \omega_1 \leq |\omega| \leq \omega_2 \\ 0 & \text{otherwise} \end{cases}$$

with a random process $X(t)$ as its input. From Eq. (9.56) it follows that the power spectrum $S_{YY}(\omega)$ of the resulting output $Y(t)$ equals

$$S_{YY}(\omega) = \begin{cases} S_{XX}(\omega) & \omega_1 \leq |\omega| \leq \omega_2 \\ 0 & \text{otherwise} \end{cases}$$

Hence, from Eq. (9.58), we have

$$E[Y^2(t)] = \frac{1}{2\pi}\int_{-\infty}^{\infty} S_{YY}(\omega)\, d\omega = 2\left(\frac{1}{2\pi}\right)\int_{\omega_1}^{\omega_2} S_{XX}(\omega)\, d\omega \geq 0 \tag{9.108}$$

which indicates that the area of $S_{XX}(\omega)$ in any interval of ω is nonnegative. This is possible only if $S_{XX}(\omega) \geq 0$ for every ω.

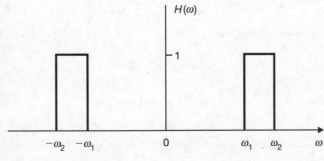

Fig. 9-9

9.15. Consider a WSS process $X(t)$ with autocorrelation $R_{XX}(\tau)$ and power spectrum $S_{XX}(\omega)$. Let $X'(t) = dX(t)/dt$. Show that

(a)　$R_{XX'}(\tau) = \dfrac{dR_{XX}(\tau)}{d\tau}$ 　　　　　　　　　　　　　　　(9.109)

(b)　$R_{X'X'}(\tau) = -\dfrac{d^2 R_{XX}(\tau)}{d\tau^2}$ 　　　　　　　　　　　　(9.110)

(c)　$S_{X'X'}(\omega) = \omega^2 S_{XX}(\omega)$ 　　　　　　　　　　　　　　(9.111)

A system with frequency response $H(\omega) = j\omega$ is a differentiator (Fig. 9-10). Thus, if $X(t)$ is its input, then its output is $Y(t) = X'(t)$ [see Eq. (5.55)].

(a)　From Eq. (9.103)

$$S_{XX'}(\omega) = H(\omega)S_{XX}(\omega) = j\omega S_{XX}(\omega)$$

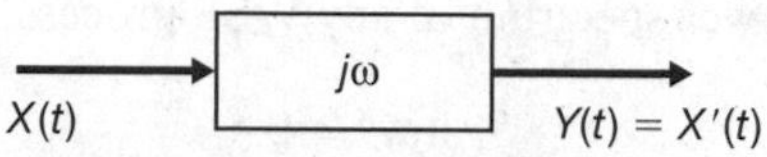

Fig. 9-10 Differentiator.

Taking the inverse Fourier transform of both sides, we obtain

$$R_{XX'}(\tau) = \frac{dR_{XX}(\tau)}{d\tau}$$

(b)　From Eq. (9.104)

$$S_{X'X}(\omega) = H^*(\omega)S_{XX'}(\omega) = -j\omega S_{XX'}(\omega)$$

Again taking the inverse Fourier transform of both sides and using the result of part (a), we have

$$R_{X'X}(\tau) = -\frac{dR_{XX'}(\tau)}{d\tau} = -\frac{d^2 R_{XX}(\tau)}{d\tau^2}$$

(c)　From Eq. (9.56)

$$S_{X'X'}(\omega) = |H(\omega)|^2 S_{XX}(\omega) = |j\omega|^2 S_{XX}(\omega) = \omega^2 S_{XX}(\omega)$$

9.16. Suppose that the input to the differentiator of Fig. 9-10 is the zero-mean random telegraph signal of Prob. 9.7.

(a)　Determine the power spectrum of the differentiator output and plot it.

(b)　Determine the mean-square value of the differentiator output.

(a)　From Eq. (9.92) of Prob. 9.7

$$S_{XX}(\omega) = A^2 \frac{4\alpha}{\omega^2 + (2\alpha)^2}$$

For the differentiator $H(\omega) = j\omega$, and from Eq. (9.56), we have

$$S_{YY}(\omega) = |H(\omega)|^2 S_{XX}(\omega) = A^2 \frac{4\alpha\omega^2}{\omega^2 + (2\alpha)^2} \tag{9.112}$$

which is plotted in Fig. 9-11 .

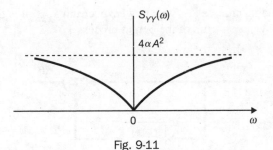

Fig. 9-11

(b)   From Eq. (9.58) or Fig. 9-11

$$E[Y^2(t)] = \frac{1}{2\pi}\int_{-\infty}^{\infty} S_{YY}(\omega)\,d\omega = \infty$$

9.17. Suppose the random telegraph signal of Prob. 9.7 is the input to an ideal bandpass filter with unit gain and narrow bandwidth $W_B(= 2\pi B)(\ll \omega_c)$ centered at $\omega_c = 2\alpha$. Find the dc component and the average power of the output.

From Eqs. (9.56) and (9.92) and Fig. 9-5 (b), the resulting output power spectrum

$$S_{YY}(\omega) - |H(\omega)|^2 S_{XX}(\omega)$$

is shown in Fig. 9-12. Since $H(0) = 0$, from Eq. (9.46) we see that

$$\mu_Y = H(0)\mu_X = 0$$

Hence, the dc component of the output is zero.

From Eq. (9.92) (Prob. 9.7)

$$S_{XX}(\pm \omega_c) = A^2 \frac{4\alpha}{(2\alpha)^2 + (2\alpha)^2} = \frac{A^2}{(2\alpha)}$$

Since $W_B \ll \omega_c$,

$$S_{YY}(\omega) \approx \begin{cases} \dfrac{A^2}{2\alpha} & |\omega - \omega_c| < \dfrac{W_B}{2} \\ 0 & \text{otherwise} \end{cases}$$

The average output power is

$$E[Y^2(t)] = \frac{1}{2\pi}\int_{-\infty}^{\infty} S_{YY}(\omega)\,d\omega$$

$$\approx \frac{1}{2\pi}(2W_B)\left(\frac{A^2}{2\alpha}\right) = \frac{A^2 W_B}{2\pi\alpha} = \frac{A^2 B}{\alpha} \qquad (9.113)$$

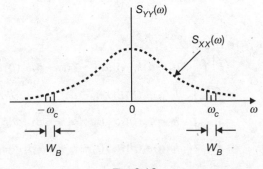

Fig. 9-12

9.18. Suppose that a WSS random process $X(t)$ with power spectrum $S_{XX}(\omega)$ is the input to the filter shown in Fig. 9-13. Find the power spectrum of the output process $Y(t)$.

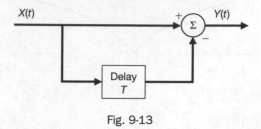

Fig. 9-13

From Fig. 9-13, $Y(t)$ can be expressed as

$$Y(t) = X(t) - X(t - T) \tag{9.114}$$

From Eq. (2.1) the impulse response of the filter is

$$h(t) = \delta(t) - \delta(t - T)$$

and by Eqs. (5.140) and (5.50) the frequency response of the filter is

$$H(\omega) = 1 - e^{-j\omega T}$$

Thus, by Eq. (9.56) the output power spectrum is

$$\begin{aligned}
S_{YY}(\omega) &= |H(\omega)|^2 S_{XX}(\omega) = \left|1 - e^{-j\omega T}\right|^2 S_{XX}(\omega) \\
&= \left[(1 - \cos \omega T)^2 + \sin^2 \omega T\right] S_{XX}(\omega) \\
&= 2(1 - \cos \omega T) S_{XX}(\omega)
\end{aligned} \tag{9.115}$$

9.19. Suppose that $X(t)$ is the input to an LTI system with impulse response $h_1(t)$ and that $Y(t)$ is the input to another LTI system with impulse response $h_2(t)$. It is assumed that $X(t)$ and $Y(t)$ are jointly wide-sense stationary. Let $V(t)$ and $Z(t)$ denote the random process at the respective system outputs (Fig. 9-14). Find the cross-correlation and cross spectral density of $V(t)$ and $Z(t)$ in terms of the cross-correlation and cross spectral density of $X(t)$ and $Y(t)$.

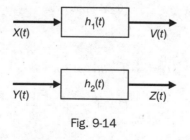

Fig. 9-14

Using Eq. (9.43), we have

$$\begin{aligned}
R_{VZ}(t_1, t_2) &= E[V(t_1)Z(t_2)] \\
&= E\left[\int_{-\infty}^{\infty} X(t_1 - \alpha)h_1(\alpha)\,d\alpha \int_{-\infty}^{\infty} Y(t_2 - \beta)h_2(\beta)\,d\beta\right] \\
&= \int_{-\infty}^{\infty}\int_{-\infty}^{\infty} h_1(\alpha)h_2(\beta) E[X(t_1 - \alpha)Y(t_2 - \beta)]\,d\alpha\,d\beta \tag{9.116} \\
&= \int_{-\infty}^{\infty}\int_{-\infty}^{\infty} h_1(\alpha)h_2(\beta) R_{XY}(t_1 - \alpha, t_2 - \beta)\,d\alpha\,d\beta \\
&= \int_{-\infty}^{\infty}\int_{-\infty}^{\infty} h_1(\alpha)h_2(\beta) R_{XY}(t_2 - t_1 + \alpha - \beta)\,d\alpha\,d\beta
\end{aligned}$$

since $X(t)$ and $Y(t)$ are jointly WSS.

Equation (9.116) indicates that $R_{VZ}(t_1, t_2)$ depends only on the time difference $\tau = t_2 - t_1$. Thus,

$$R_{VZ}(\tau) = \int_{-\infty}^{\infty} \int_{-\infty}^{\infty} h_1(\alpha) h_2(\beta) R_{XY}(\tau + \alpha - \beta) \, d\alpha \, d\beta \tag{9.117}$$

Taking the Fourier transform of both sides of Eq. (9.117), we obtain

$$S_{VZ}(\omega) = \int_{-\infty}^{\infty} R_{VZ}(\tau) e^{-j\omega\tau} d\tau$$

$$= \int_{-\infty}^{\infty} \int_{-\infty}^{\infty} \int_{-\infty}^{\infty} h_1(\alpha) h_2(\beta) R_{XY}(\tau + \alpha - \beta) e^{-j\omega\tau} d\alpha \, d\beta \, d\tau$$

Let $\tau + \alpha - \beta = \lambda$, or equivalently $\tau = \lambda - \alpha + \beta$. Then

$$S_{VZ}(\omega) = \int_{-\infty}^{\infty} \int_{-\infty}^{\infty} \int_{-\infty}^{\infty} h_1(\alpha) h_2(\beta) R_{XY}(\lambda) e^{-j\omega(\lambda - \alpha + \beta)} d\alpha \, d\beta \, d\lambda$$

$$= \int_{-\infty}^{\infty} h_1(\alpha) e^{j\omega\alpha} d\alpha \int_{-\infty}^{\infty} h_2(\beta) e^{-j\omega\beta} d\beta \int_{-\infty}^{\infty} R_{XY}(\lambda) e^{-j\omega\lambda} d\lambda$$

$$= H_1(-\omega) H_2(\omega) S_{XY}(\omega)$$

$$= H_1^*(\omega) H_2(\omega) S_{XY}(\omega) \tag{9.118}$$

where $H_1(\omega)$ and $H_2(\omega)$ are the frequency responses of the respective systems in Fig. 9-14.

9.20. The input $X(t)$ to the *RC* filter shown in Fig. 9-15 is a white-noise process.

(*a*) Determine the power spectrum of the output process $Y(t)$.

(*b*) Determine the autocorrelation and the mean-square value of $Y(t)$.

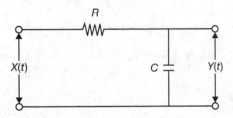

Fig. 9-15 *RC* filter.

From Eq. (5.91) the frequency response of the *RC* filter is

$$H(\omega) = \frac{1}{1 + j\omega RC}$$

(*a*) From Eqs. (9.31) and (9.56)

$$S_{XX}(\omega) = \frac{\eta}{2}$$

$$S_{YY}(\omega) = |H(\omega)|^2 S_{XX}(\omega) = \frac{1}{1 + (\omega RC)^2} \frac{\eta}{2} \tag{9.119}$$

(*b*) Rewriting Eq. (9.119) as

$$S_{YY}(\omega) = \frac{\eta}{2} \frac{1}{2RC} \frac{2[1/(RC)]}{\omega^2 + [1/(RC)]^2}$$

and using the Fourier transform pair Eq. (5.138), we obtain

$$R_{YY}(\tau) = \frac{\eta}{2} \frac{1}{2RC} e^{-|\tau|/(RC)} \tag{9.120}$$

Finally, from Eq. (9.120)

$$E[Y^2(t)] = R_{YY}(0) = \frac{\eta}{4RC} \tag{9.121}$$

9.21. The input $X(t)$ to an ideal bandpass filter having the frequency response characteristic shown in Fig. 9-16 is a white-noise process. Determine the total noise power at the output of the filter.

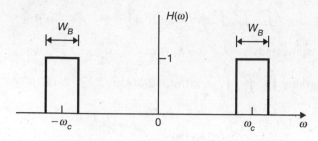

Fig. 9-16

$$S_{XX}(\omega) = \frac{\eta}{2}$$

$$S_{YY}(\omega) = |H(\omega)|^2 S_{XX}(\omega) = \frac{\eta}{2}|H(\omega)|^2$$

The total noise power at the output of the filter is

$$E[Y^2(t)] = \frac{1}{2\pi}\int_{-\infty}^{\infty} S_{YY}(\omega)\,d\omega = \frac{1}{2\pi}\frac{\eta}{2}\int_{-\infty}^{\infty} |H(\omega)|^2\,d\omega$$

$$= \frac{\eta}{2}\frac{1}{2\pi}(2W_B) = \eta B \tag{9.122}$$

where $B = W_B/(2\pi)$ (in Hz).

9.22. Verify Eq. (9.60); that is

$$R_{YY}(n,m) = \sum_{i=-\infty}^{\infty}\sum_{l=-\infty}^{\infty} h(i)h(l)\,R_{XX}(n-i,m-l)$$

From Eq. (9.59) we have

$$R_{YY}(n,m) = E[Y(n)Y(m)] = E\left[\sum_{i=-\infty}^{\infty}\sum_{l=-\infty}^{\infty} h(i)h(l)X(n-i)X(m-l)\right]$$

$$= \sum_{i=-\infty}^{\infty}\sum_{l=-\infty}^{\infty} h(i)h(l)E[X(n-i)X(m-l)]$$

$$= \sum_{i=-\infty}^{\infty}\sum_{l=-\infty}^{\infty} h(i)h(l)R_{XX}(n-i,m-l)$$

9.23. Verify Eq. (9.61); that is

$$R_{XY}(n,m) = \sum_{l=-\infty}^{\infty} h(l)R_{XX}(n,m-l)$$

From Eq. (9.59), we have

$$R_{XY}(n, m) = E[X(n)Y(m)] = E\left[X(n)\sum_{l=-\infty}^{\infty} h(l)X(m-l)\right]$$

$$= \sum_{l=-\infty}^{\infty} h(l)E[X(n)X(m-l)]$$

$$= \sum_{l=-\infty}^{\infty} h(l)R_{XX}(n, m-l)$$

9.24. Verify Eq. (9.66); that is

$$R_{YY}(k) = h(-k) * R_{XY}(k)$$

From Eq. (9.64), and using Eq. (9.65), we obtain

$$R_{YY}(k) = \sum_{i=-\infty}^{\infty} \sum_{l=-\infty}^{\infty} h(i)h(l)R_{XX}(k+i-l)$$

$$= \sum_{i=-\infty}^{\infty} h(i)R_{XY}(k+i)$$

$$= \sum_{l=-\infty}^{\infty} h(-l)R_{XY}(k-l) = h(-k) * R_{XY}(k)$$

9.25. The output $Y(n)$ of a discrete-time system is related to the input $X(n)$ by

$$Y(n) = X(n) - X(n-1) \tag{9.123}$$

If the input is a zero-mean discrete-time white noise with power spectral density σ^2, find $E[Y^2(t)]$.

The impulse response $h(n)$ of the system is given by

$$h(n) = \delta(n) - \delta(n-1)$$

Taking the discrete-time Fourier transform of $h(n)$, the frequency response $H(\Omega)$ of the system is given by

$$H(\Omega) = 1 - e^{-j\Omega}$$

Then

$$|H(\Omega)|^2 = |1 - e^{-j\Omega}|^2 = (1 - \cos\Omega)^2 + \sin^2\Omega = 2(1 - \cos\Omega) \tag{9.124}$$

Since $S_{XX}(\Omega) = \sigma^2$, and by Eq. (9.70), we have

$$S_{YY}(\Omega) = S_{XX}(\Omega)|H(\Omega)|^2 = 2\sigma^2(1 - \cos\Omega) \tag{9.125}$$

Thus, by Eq. (9.72) we obtain

$$E[Y^2(t)] = R_{YY}(0) = \frac{1}{2\pi}\int_{-\pi}^{\pi} S_{YY}(\Omega)\,d\Omega$$

$$= \frac{2\sigma^2}{2\pi}\int_{-\pi}^{\pi}(1 - \cos\Omega)\,d\Omega = 2\sigma^2 \tag{9.126}$$

9.26. The discrete-time system shown in Fig. 9-17 consists of one unit delay element and one scalar multiplier $(a < 1)$. The input $X(n)$ is discrete-time white noise with average power σ^2. Find the spectral density and average power of the output $Y(n)$.

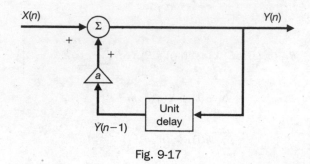

Fig. 9-17

From Fig. 9-17, $Y(n)$ and $X(n)$ are related by

$$Y(n) = aY(n-1) + X(n) \tag{9.127}$$

The impulse response $h(n)$ of the system is defined by

$$h(n) = ah(n-1) + \delta(n) \tag{9.128}$$

Solving Eq. (9.128), we obtain

$$h(n) = a^n u(n) \tag{9.129}$$

where $u(n)$ is the unit step sequence defined by

$$u(n) = \begin{cases} 1 & n \ge 0 \\ 0 & n < 0 \end{cases}$$

Taking the Fourier transform of Eq. (9.129), we obtain

$$H(\Omega) = \sum_{n=0}^{\infty} a^n e^{-j\Omega n} = \frac{1}{1 - ae^{-j\Omega}} \qquad a < 1, |\Omega| < \pi$$

Now, by Eq. (9.34),

$$S_{XX}(\Omega) = \sigma^2 \qquad |\Omega| < \pi$$

and by Eq. (9.70) the power spectral density of $Y(n)$ is

$$S_{YY}(\Omega) = |H(\Omega)|^2 S_{XX}(\Omega) = H(\Omega)H(-\Omega)S_{XX}(\Omega)$$

$$= \frac{\sigma^2}{(1 - ae^{-j\Omega})(1 - ae^{j\Omega})}$$

$$= \frac{\sigma^2}{1 + a^2 - 2a\cos\Omega} \qquad |\Omega| < \pi \tag{9.130}$$

Taking the inverse Fourier transform of Eq. (9.130), we obtain

$$R_{YY}(k) = \frac{\sigma^2}{1 - a^2} a^{|k|}$$

Thus, by Eq. (9.72) the average power of $Y(n)$ is

$$E[Y^2(n)] = R_{YY}(0) = \frac{\sigma^2}{1 - a^2}$$

SUPPLEMENTARY PROBLEMS

9.27. A sample function of a random telegraph signal $X(t)$ is shown in Fig. 9-18. This signal makes independent random shifts between two equally likely values, A and 0. The number of shifts per unit time is governed by the Poisson distribution with parameter α.

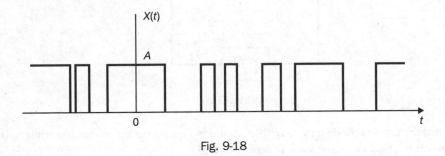

Fig. 9-18

(a) Find the autocorrelation and the power spectrum of $X(t)$.

(b) Find the rms value of $X(t)$.

9.28. Suppose that $X(t)$ is a Gaussian process with

$$\mu_X = 2 \qquad R_{XX}(\tau) = 5e^{-0.2|\tau|}$$

Find the probability that $X(4) \leq 1$.

9.29. The output of a filter is given by

$$Y(t) = X(t + T) - X(t - T)$$

where $X(t)$ is a WSS process with power spectrum $S_{XX}(\omega)$ and T is a constant. Find the power spectrum of $Y(t)$.

9.30. Let $\hat{X}(t)$ is the Hilbert transform of a WSS process $X(t)$. Show that

$$R_{X\hat{X}}(0) = E[X(t)\hat{X}(t)] = 0$$

9.31. A WSS random process $X(t)$ is applied to the input of an LTI system with impulse response $h(t) = 3e^{-2t}u(t)$. Find the mean value of $Y(t)$ of the system if $E[X(t)] = 2$.

9.32. The input $X(t)$ to the RC filter shown in Fig. 9-19 is a white noise specified by Eq. (9.31). Find the mean-square value of $Y(t)$.

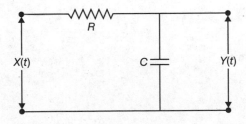

Fig. 9-19 *RC* filter.

9.33. The input $X(t)$ to a differentiator is the random telegraph signal of Prob. 9.7.

(a) Determine the power spectral density of the differentiator output.

(b) Find the mean-square value of the differentiator output.

9.34. Suppose that the input to the filter shown in Fig. 9-20 is a white noise specified by Eq. (9.51). Find the power spectral density of $Y(t)$.

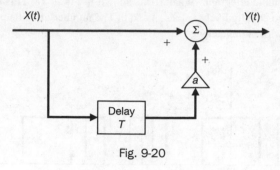

$$X(t) \quad\quad\quad\quad\quad Y(t)$$

Fig. 9-20

9.35. Suppose that the input to the discrete-time filter shown in Fig. 9-21 is a discrete-time white noise with average power σ^2. Find the power spectral density of $Y(n)$.

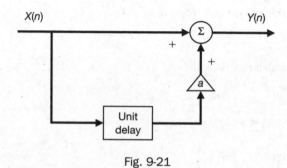

Fig. 9-21

ANSWERS TO SUPPLEMENTARY PROBLEMS

9.27. (a) $R_{XX}(\tau) = \dfrac{A^2}{4}(1 + e^{-2\alpha|\tau|})$; $S_{XX}(\omega) = \dfrac{A^2}{2}\pi\delta(\omega) + A^2\dfrac{4\alpha}{\omega^2 + (2\alpha)^2}$

(b) $\dfrac{A}{2}$

9.28. 0.159

9.29. $S_{YY}(\omega) = 4\sin^2\omega T S_{XX}(\omega)$

9.30. *Hint:* Use relation (b) of Prob. 9.12 and definition (5.174).

9.31. *Hint:* Use Eq. (9.46).

3

9.32. *Hint:* Use Eqs. (9.57) and (9.58).

$\eta/(4RC)$

9.33. (a) $S_Y(\omega) = \dfrac{4\alpha\omega^2}{\omega^2 + 4\alpha^2}$

(b) $E[Y^2(t)] = \infty$

9.34. $S_Y(\omega) = \sigma^2 \rightarrow \dfrac{\eta}{2}(1 + a^2 + 2a\cos\omega T)$

9.35. $S_Y(\Omega) = \sigma^2(1 + a^2 + 2a\cos\Omega)$

APPENDIX A

Review of Matrix Theory

A.1 Matrix Notation and Operations

A. Definitions:

1. An $m \times n$ matrix $\mathbf{A}$ is a rectangular array of elements having m rows and n columns and is denoted as

$$\mathbf{A} = \begin{bmatrix} a_{11} & a_{12} & \cdots & a_{1n} \\ a_{21} & a_{22} & \cdots & a_{2n} \\ \vdots & \vdots & \ddots & \vdots \\ a_{m1} & a_{m2} & \cdots & a_{mn} \end{bmatrix} = [a_{ij}]_{m \times n} \tag{A.1}$$

When $m = n$, $\mathbf{A}$ is called a *square matrix of order n*.

2. A $1 \times n$ matrix is called an n-dimensional *row vector*:

$$[a_{11} \quad a_{12} \quad \cdots \quad a_{1n}] \tag{A.2}$$

An $m \times 1$ matrix is called an m-dimensional *column vector*:

$$\begin{bmatrix} a_{11} \\ a_{21} \\ \vdots \\ a_{m1} \end{bmatrix} \tag{A.3}$$

3. A *zero* matrix $\mathbf{0}$ is a matrix having all its elements zero.

4. A *diagonal* matrix $\mathbf{D}$ is a square matrix in which all elements not on the main diagonal are zero:

$$\mathbf{D} = \begin{bmatrix} d_1 & 0 & \cdots & 0 \\ 0 & d_2 & \cdots & 0 \\ \vdots & \vdots & \ddots & \vdots \\ 0 & 0 & \cdots & d_n \end{bmatrix} \tag{A.4}$$

Sometimes the diagonal matrix $\mathbf{D}$ in Eq. (A.4) is expressed as

$$\mathbf{D} = \text{diag}(d_1 \quad d_2 \quad \cdots \quad d_n) \tag{A.5}$$

5. The *identity* (or *unit*) matrix **I** is a diagonal matrix with all of its diagonal elements equal to 1.

$$\mathbf{I} = \begin{bmatrix} 1 & 0 & \cdots & 0 \\ 0 & 1 & \cdots & 0 \\ \vdots & \vdots & \ddots & \vdots \\ 0 & 0 & \cdots & 1 \end{bmatrix} \tag{A.6}$$

B. Operations:

Let $\mathbf{A} = [a_{ij}]_{m \times n}$, $\mathbf{B} = [b_{ij}]_{m \times n}$, and $\mathbf{C} = [c_{ij}]_{m \times n}$.

a. Equality of Two Matrices:

$$\mathbf{A} = \mathbf{B} \Rightarrow a_{ij} = b_{ij} \tag{A.7}$$

b. Addition:

$$\mathbf{C} = \mathbf{A} + \mathbf{B} \Rightarrow c_{ij} = a_{ij} + b_{ij} \tag{A.8}$$

c. Multiplication by a Scalar:

$$\mathbf{B} = \alpha \mathbf{A} \Rightarrow b_{ij} = \alpha a_{ij} \tag{A.9}$$

If $\alpha = -1$, then $\mathbf{B} = -\mathbf{A}$ is called the *negative* of **A**.

EXAMPLE A.1 Let

$$\mathbf{A} = \begin{bmatrix} 1 & 2 & 3 \\ -1 & 0 & 4 \end{bmatrix} \qquad \mathbf{B} = \begin{bmatrix} 2 & 0 & -1 \\ 4 & 1 & -2 \end{bmatrix}$$

Then

$$\mathbf{A} + \mathbf{B} = \begin{bmatrix} 1+2 & 2+0 & 3-1 \\ -1+4 & 0+1 & 4-2 \end{bmatrix} = \begin{bmatrix} 3 & 2 & 2 \\ 3 & 1 & 2 \end{bmatrix}$$

$$-\mathbf{B} = (-1)\mathbf{B} = \begin{bmatrix} -2 & 0 & 1 \\ -4 & -1 & 2 \end{bmatrix}$$

$$\mathbf{A} - \mathbf{B} = \begin{bmatrix} 1-2 & 2-0 & 3+1 \\ -1-4 & 0-1 & 4+2 \end{bmatrix} = \begin{bmatrix} -1 & 2 & 4 \\ -5 & -1 & 6 \end{bmatrix}$$

Notes:

1. $\mathbf{A} = \mathbf{B}$ and $\mathbf{B} = \mathbf{C} \Rightarrow \mathbf{A} = \mathbf{C}$
2. $\mathbf{A} + \mathbf{B} = \mathbf{B} + \mathbf{A}$
3. $(\mathbf{A} + \mathbf{B}) + \mathbf{C} = \mathbf{A} + (\mathbf{B} + \mathbf{C})$
4. $\mathbf{A} + \mathbf{0} = \mathbf{0} + \mathbf{A} = \mathbf{A}$ $\qquad\qquad$ (A.10)
5. $\mathbf{A} - \mathbf{A} = \mathbf{A} + (-\mathbf{A}) = \mathbf{0}$
6. $(\alpha + \beta)\mathbf{A} = \alpha\mathbf{A} + \alpha\mathbf{B}$
7. $\alpha(\mathbf{A} + \mathbf{B}) = \alpha\mathbf{A} + \alpha\mathbf{B}$
8. $\alpha(\beta\mathbf{A}) = (\alpha\beta)\mathbf{A} = \beta(\alpha\mathbf{A})$

d. Multiplication:

Let $\mathbf{A} = [a_{ij}]_{m \times n}$, $\mathbf{B} = [b_{ij}]_{n \times p}$, and $\mathbf{C} = [c_{ij}]_{m \times p}$.

$$\mathbf{C} = \mathbf{AB} \Rightarrow c_{ij} = \sum_{k=1}^{n} a_{ik} b_{kj} \tag{A.11}$$

The matrix product **AB** is defined only when the number of columns of **A** is equal to the number of rows of **B**. In this case **A** and **B** are said to be *conformable*.

EXAMPLE A.2 Let

$$\mathbf{A} = \begin{bmatrix} 0 & -1 \\ 1 & 2 \\ 2 & -3 \end{bmatrix} \qquad \mathbf{B} = \begin{bmatrix} 1 & 2 \\ 3 & -1 \end{bmatrix}$$

Then

$$\mathbf{AB} = \begin{bmatrix} 0 & -1 \\ 1 & 2 \\ 2 & -3 \end{bmatrix} \begin{bmatrix} 1 & 2 \\ 3 & -1 \end{bmatrix} = \begin{bmatrix} 0(1) + (-1)\,3 & 0(2) + (-1)\,(-1) \\ 1(1) + 2(3) & 1(2) + 2(-1) \\ 2(1) + (-3)\,3 & 2(2) + (-3)(-1) \end{bmatrix} = \begin{bmatrix} -3 & 1 \\ 7 & 0 \\ -7 & 7 \end{bmatrix}$$

but **BA** is not defined.

Furthermore, even if both **AB** and **BA** are defined, in general

$$\mathbf{AB} \neq \mathbf{BA} \tag{A.12}$$

EXAMPLE A.3 Let

$$\mathbf{A} = \begin{bmatrix} 0 & -1 \\ 1 & 2 \end{bmatrix} \qquad \mathbf{B} = \begin{bmatrix} 1 & 2 \\ 3 & -1 \end{bmatrix}$$

Then

$$\mathbf{AB} = \begin{bmatrix} 0 & -1 \\ 1 & 2 \end{bmatrix} \begin{bmatrix} 1 & 2 \\ 3 & -1 \end{bmatrix} = \begin{bmatrix} -3 & 1 \\ 7 & 0 \end{bmatrix}$$

$$\mathbf{BA} = \begin{bmatrix} 1 & 2 \\ 3 & -1 \end{bmatrix} \begin{bmatrix} 0 & -1 \\ 1 & 2 \end{bmatrix} = \begin{bmatrix} 2 & 3 \\ -1 & -5 \end{bmatrix} \neq \mathbf{AB}$$

An example of the case where **AB** = **BA** follows.

EXAMPLE A.4 Let

$$\mathbf{A} = \begin{bmatrix} 1 & 0 \\ 0 & 3 \end{bmatrix} \qquad \mathbf{B} = \begin{bmatrix} 2 & 0 \\ 0 & 4 \end{bmatrix}$$

Then

$$\mathbf{AB} = \mathbf{BA} = \begin{bmatrix} 2 & 0 \\ 0 & 12 \end{bmatrix}$$

Notes:

1. $\mathbf{A0} = \mathbf{0A} = \mathbf{0}$
2. $\mathbf{AI} = \mathbf{IA} = \mathbf{A}$
3. $(\mathbf{A} + \mathbf{B})\mathbf{C} = \mathbf{AC} + \mathbf{BC}$
4. $\mathbf{A}(\mathbf{B} + \mathbf{C}) = \mathbf{AB} + \mathbf{AC}$
5. $(\mathbf{AB})\mathbf{C} = \mathbf{A}(\mathbf{BC}) = \mathbf{ABC}$
6. $\alpha\,(\mathbf{AB}) = (\alpha\mathbf{A})\mathbf{B} = \mathbf{A}(\alpha\mathbf{B})$

$$\tag{A.13}$$

It is important to note that **AB** = **0** does not necessarily imply **A** = **0** or **B** = **0**.

EXAMPLE A.5 Let

$$\mathbf{A} = \begin{bmatrix} 2 & 1 \\ 2 & 1 \end{bmatrix} \quad \mathbf{B} = \begin{bmatrix} 1 & 1 \\ -2 & -2 \end{bmatrix}$$

Then

$$\mathbf{AB} = \begin{bmatrix} 2 & 1 \\ 2 & 1 \end{bmatrix} \begin{bmatrix} 1 & 1 \\ -2 & -2 \end{bmatrix} = \begin{bmatrix} 0 & 0 \\ 0 & 0 \end{bmatrix} = \mathbf{0}$$

A.2 Transpose and Inverse

A. Transpose:

Let $\mathbf{A}$ be an $n \times m$ matrix. The *transpose* of $\mathbf{A}$, denoted by $\mathbf{A}^T$, is an $m \times n$ matrix formed by interchanging the rows and columns of $\mathbf{A}$.

$$\mathbf{B} = \mathbf{A}^T \Rightarrow b_{ij} = a_{ji} \tag{A.14}$$

EXAMPLE A.6

$$\mathbf{A} = \begin{bmatrix} 1 & 2 & 3 \\ -1 & 0 & 4 \end{bmatrix} \quad \mathbf{A}^T = \begin{bmatrix} 1 & 2 & 3 \\ -1 & 0 & 4 \end{bmatrix}^T = \begin{bmatrix} 1 & -1 \\ 2 & 0 \\ 3 & 4 \end{bmatrix}$$

If $\mathbf{A}^T = \mathbf{A}$, then $\mathbf{A}$ is said to be *symmetric*, and if $\mathbf{A}^T = -\mathbf{A}$, then $\mathbf{A}$ is said to be *skew-symmetric*.

EXAMPLE A.7 Let

$$\mathbf{A} = \begin{bmatrix} 1 & 2 & 3 \\ 2 & 4 & -1 \\ 3 & -1 & 5 \end{bmatrix} \quad \mathbf{B} = \begin{bmatrix} 0 & 1 & -2 \\ -1 & 0 & 3 \\ 2 & -3 & 0 \end{bmatrix}$$

Then $\mathbf{A}$ is a symmetric matrix and $\mathbf{B}$ is a skew-symmetric matrix.

Note that if a matrix is skew-symmetric, then its diagonal elements are all zero.

Notes:

1. $(\mathbf{A}^T)^T = \mathbf{A}$
2. $(\mathbf{A} + \mathbf{B})^T = \mathbf{A}^T + \mathbf{B}^T$
3. $(\alpha\mathbf{A})^T = \alpha\mathbf{A}^T$
4. $(\mathbf{AB})^T = \mathbf{B}^T\mathbf{A}^T$

$$\tag{A.15}$$

B. Inverses:

A matrix $\mathbf{A}$ is said to be *invertible* if there exists a matrix $\mathbf{B}$ such that

$$\mathbf{BA} = \mathbf{AB} = \mathbf{I} \tag{A.16a}$$

The matrix $\mathbf{B}$ is called the *inverse* of $\mathbf{A}$ and is denoted by $\mathbf{A}^{-1}$. Thus,

$$\mathbf{A}^{-1}\mathbf{A} = \mathbf{AA}^{-1} = \mathbf{I} \tag{A.16b}$$

EXAMPLE A.8

$$\begin{bmatrix} 2 & 1 \\ 1 & 1 \end{bmatrix}\begin{bmatrix} 1 & -1 \\ -1 & 2 \end{bmatrix} = \begin{bmatrix} 1 & -1 \\ -1 & 2 \end{bmatrix}\begin{bmatrix} 2 & 1 \\ 1 & 1 \end{bmatrix} = \begin{bmatrix} 1 & 0 \\ 0 & 1 \end{bmatrix}$$

Thus,

$$\begin{bmatrix} 2 & 1 \\ 1 & 1 \end{bmatrix}^{-1} = \begin{bmatrix} 1 & -1 \\ -1 & 2 \end{bmatrix}$$

Notes:

1. $(\mathbf{A}^{-1})^{-1} = \mathbf{A}$
2. $(\mathbf{A}^{-1})^{T} = (\mathbf{A}^{T})^{-1}$ (A.17)
3. $(\alpha\mathbf{A})^{-1} = \frac{1}{\alpha}\mathbf{A}^{-1}$
4. $(\mathbf{AB})^{-1} = \mathbf{B}^{-1}\mathbf{A}^{-1}$

Note that if $\mathbf{A}$ is invertible, then $\mathbf{AB} = \mathbf{0}$ implies that $\mathbf{B} = \mathbf{0}$ since

$$\mathbf{A}^{-1}\mathbf{AB} = \mathbf{IB} = \mathbf{B} = \mathbf{A}^{-1}\mathbf{0} = \mathbf{0}$$

A.3   Linear Independence and Rank

A.   Linear independence:

Let $\mathbf{A} = [\mathbf{a}_1 \quad \mathbf{a}_2 \quad \cdots \quad \mathbf{a}_n]$, where $\mathbf{a}_i$ denotes the ith column vector of $\mathbf{A}$. A set of column vectors $\mathbf{a}_i (i = 1, 2, \ldots, n)$ is said to be *linearly dependent* if there exist numbers α_i ($i = 1, 2, \ldots, n$) not all zero such that

$$\alpha_1\mathbf{a}_1 + \alpha_2\mathbf{a}_2 + \cdots + \alpha_n\mathbf{a}_n = \mathbf{0} \tag{A.18}$$

If Eq. (A.18) holds only for all $\alpha_i = 0$, then the set is said to be *linearly independent*.

EXAMPLE A.9   Let

$$\mathbf{a}_1 = \begin{bmatrix} 1 \\ -1 \\ 0 \end{bmatrix} \qquad \mathbf{a}_2 = \begin{bmatrix} 2 \\ 1 \\ -1 \end{bmatrix} \qquad \mathbf{a}_3 = \begin{bmatrix} 4 \\ 5 \\ -3 \end{bmatrix}$$

Since $2\mathbf{a}_1 + (-3)\mathbf{a}_2 + \mathbf{a}_3 = \mathbf{0}$, $\mathbf{a}_1$, $\mathbf{a}_2$, and $\mathbf{a}_3$ are linearly dependent. Let

$$\mathbf{d}_1 = \begin{bmatrix} 1 \\ 0 \\ 0 \end{bmatrix} \qquad \mathbf{d}_2 = \begin{bmatrix} 0 \\ 1 \\ 0 \end{bmatrix} \qquad \mathbf{d}_3 = \begin{bmatrix} 0 \\ 0 \\ 1 \end{bmatrix}$$

Then

$$\alpha_1\mathbf{d}_1 + \alpha_2\mathbf{d}_2 + \alpha_3\mathbf{d}_3 = \begin{bmatrix} \alpha_1 \\ \alpha_2 \\ \alpha_3 \end{bmatrix} = \begin{bmatrix} 0 \\ 0 \\ 0 \end{bmatrix}$$

implies that $\alpha_1 = \alpha_2 = \alpha_3 = 0$. Thus, $\mathbf{d}_1$, $\mathbf{d}_2$, and $\mathbf{d}_3$ are linearly independent.

B.　Rank of a Matrix:

The number of linearly independent column vectors in a matrix $\mathbf{A}$ is called the *column rank* of $\mathbf{A}$, and the number of linearly independent row vectors in a matrix $\mathbf{A}$ is called the *row rank* of $\mathbf{A}$. It can be shown that

$$\text{Rank of } \mathbf{A} = \text{column rank of } \mathbf{A} = \text{row rank of } \mathbf{A} \tag{A.19}$$

Note:

If the rank of an $N \times N$ matrix $\mathbf{A}$ is N, then $\mathbf{A}$ is invertible and $\mathbf{A}^{-1}$ exists.

A.4　Determinants

A.　Definitions:

Let $\mathbf{A} = [a_{ij}]$ be a square matrix of order N. We associate with $\mathbf{A}$ a certain number called its *determinant*, denoted by $\det \mathbf{A}$ or $|\mathbf{A}|$. Let $\mathbf{M}_{ij}$ be the square matrix of order $(N-1)$ obtained from $\mathbf{A}$ by deleting the ith row and jth column. The number A_{ij} defined by

$$A_{ij} = (-1)^{i+j}|\mathbf{M}_{ij}| \tag{A.20}$$

is called the *cofactor* of a_{ij}. Then $\det \mathbf{A}$ is obtained by

$$\det \mathbf{A} = |\mathbf{A}| = \sum_{k=1}^{N} a_{ik}A_{ik} \qquad i = 1, 2, \ldots, N \tag{A.21a}$$

or

$$\det \mathbf{A} = |\mathbf{A}| = \sum_{k=1}^{N} a_{kj}A_{kj} \qquad j = 1, 2, \ldots, N \tag{A.21b}$$

Equation (A.21a) is known as the *Laplace expansion* of $|\mathbf{A}|$ along the ith row, and Eq. (A.21b) the Laplace expansion of $|\mathbf{A}|$ along the jth column.

EXAMPLE A.10　For a 1×1 matrix,

$$\mathbf{A} = [a_{11}] \rightarrow |\mathbf{A}| = a_{11} \tag{A.22}$$

For a 2×2 matrix,

$$\mathbf{A} = \begin{bmatrix} a_{11} & a_{12} \\ a_{21} & a_{22} \end{bmatrix} \rightarrow |\mathbf{A}| = \begin{vmatrix} a_{11} & a_{12} \\ a_{21} & a_{22} \end{vmatrix} = a_{11}a_{22} - a_{12}a_{21} \tag{A.23}$$

For a 3×3 matrix,

$$\mathbf{A} = \begin{bmatrix} a_{11} & a_{12} & a_{13} \\ a_{21} & a_{22} & a_{23} \\ a_{31} & a_{32} & a_{33} \end{bmatrix}$$

Using Eqs. (A.21a) and (A.23), we obtain

$$|\mathbf{A}| = \begin{vmatrix} a_{11} & a_{12} & a_{13} \\ a_{21} & a_{22} & a_{23} \\ a_{31} & a_{32} & a_{33} \end{vmatrix} = a_{11}\begin{vmatrix} a_{22} & a_{23} \\ a_{32} & a_{33} \end{vmatrix} - a_{12}\begin{vmatrix} a_{21} & a_{23} \\ a_{31} & a_{33} \end{vmatrix} + a_{13}\begin{vmatrix} a_{21} & a_{22} \\ a_{31} & a_{32} \end{vmatrix}$$

$$= a_{11}a_{22}a_{33} + a_{12}a_{23}a_{31} + a_{13}a_{21}a_{32} - a_{11}a_{23}a_{32} - a_{12}a_{21}a_{33} - a_{13}a_{22}a_{31} \tag{A.24}$$

B. Determinant Rank of a Matrix:

The *determinant rank* of a matrix $\mathbf{A}$ is defined as the order of the largest square submatrix $\mathbf{M}$ of $\mathbf{A}$ such that det $\mathbf{M} \neq 0$. It can be shown that the rank of $\mathbf{A}$ is equal to the determinant rank of $\mathbf{A}$.

EXAMPLE A.11　Let

$$\mathbf{A} = \begin{bmatrix} 1 & 2 & 4 \\ -1 & 1 & 5 \\ 0 & -1 & -3 \end{bmatrix}$$

Note that $|\mathbf{A}| = 0$. One of the largest submatrices whose determinant is not equal to zero is

$$\begin{bmatrix} 1 & 2 \\ -1 & 1 \end{bmatrix}$$

Hence the rank of the matrix $\mathbf{A}$ is 2. (See Example A.9.)

C. Inverse of a Matrix:

Using determinants, the inverse of an $N \times N$ matrix $\mathbf{A}$ can be computed as

$$\mathbf{A}^{-1} = \frac{1}{\det \mathbf{A}} \operatorname{adj} \mathbf{A} \tag{A.25}$$

and

$$\operatorname{adj} \mathbf{A} = [A_{ij}]^T = \begin{bmatrix} A_{11} & A_{21} & \cdots & A_{N1} \\ A_{12} & A_{22} & \cdots & A_{N2} \\ \vdots & \vdots & \ddots & \vdots \\ A_{1N} & A_{2N} & \cdots & A_{NN} \end{bmatrix} \tag{A.26}$$

where A_{ij} is the cofactor of a_{ij} defined in Eq. (A.20) and "adj" stands for the *adjugate* (or *adjoint*). Formula (A.25) is used mainly for $N = 2$ and $N = 3$.

EXAMPLE A.12　Let

$$\mathbf{A} = \begin{bmatrix} 1 & 0 & -3 \\ 1 & 2 & 0 \\ 3 & -1 & -2 \end{bmatrix}$$

Then

$$|\mathbf{A}| = 1 \begin{vmatrix} 2 & 0 \\ -1 & -2 \end{vmatrix} - 3 \begin{vmatrix} 1 & 2 \\ 3 & -1 \end{vmatrix} = -4 - 3(-7) = 17$$

$$\operatorname{adj} \mathbf{A} = \begin{bmatrix} \begin{vmatrix} 2 & 0 \\ -1 & -2 \end{vmatrix} & -\begin{vmatrix} 0 & -3 \\ -1 & -2 \end{vmatrix} & \begin{vmatrix} 0 & -3 \\ 2 & 0 \end{vmatrix} \\ -\begin{vmatrix} 1 & 0 \\ 3 & -2 \end{vmatrix} & \begin{vmatrix} 1 & -3 \\ 3 & -2 \end{vmatrix} & -\begin{vmatrix} 1 & -3 \\ 1 & 0 \end{vmatrix} \\ \begin{vmatrix} 1 & 2 \\ 3 & -1 \end{vmatrix} & -\begin{vmatrix} 1 & 0 \\ 3 & -1 \end{vmatrix} & \begin{vmatrix} 1 & 0 \\ 1 & 2 \end{vmatrix} \end{bmatrix} = \begin{bmatrix} -4 & 3 & 6 \\ 2 & 7 & -3 \\ -7 & 1 & 2 \end{bmatrix}$$

Thus,

$$\mathbf{A}^{-1} = \frac{1}{17} \begin{bmatrix} -4 & 3 & 6 \\ 2 & 7 & -3 \\ -7 & 1 & 2 \end{bmatrix}$$

For a 2×2 matrix,

$$\begin{bmatrix} a_{11} & a_{12} \\ a_{21} & a_{22} \end{bmatrix}^{-1} = \frac{1}{a_{11}a_{22} - a_{12}a_{21}} \begin{bmatrix} a_{22} & -a_{12} \\ -a_{21} & a_{11} \end{bmatrix} \tag{A.27}$$

From Eq. (A.25) we see that if det $\mathbf{A} = 0$, then $\mathbf{A}^{-1}$ does not exist. The matrix $\mathbf{A}$ is called *singular* if det $\mathbf{A} = 0$, and *nonsingular* if det $\mathbf{A} \neq 0$. Thus, if a matrix is nonsingular, then it is invertible and $\mathbf{A}^{-1}$ exists.

A.5   Eigenvalues and Eigenvectors

A.   Definitions:

Let $\mathbf{A}$ be an $N \times N$ matrix. If

$$\mathbf{A}\mathbf{x} = \lambda\mathbf{x} \tag{A.28}$$

for some scalar λ and nonzero column vector $\mathbf{x}$, then λ is called an *eigenvalue* (or *characteristic value*) of $\mathbf{A}$ and $\mathbf{x}$ is called an *eigenvector* associated with λ.

B.   Characteristic Equation:

Equation (A.28) can be rewritten as

$$(\lambda\mathbf{I} - \mathbf{A})\mathbf{x} = \mathbf{0} \tag{A.29}$$

where $\mathbf{I}$ is the identity matrix of Nth order. Equation (A.29) will have a nonzero eigenvector $\mathbf{x}$ only if $\lambda\mathbf{I} - \mathbf{A}$ is singular, that is,

$$|\lambda\mathbf{I} - \mathbf{A}| = 0 \tag{A.30}$$

which is called the *characteristic equation* of $\mathbf{A}$. The polynomial $c(\lambda)$ defined by

$$c(\lambda) = |\lambda\mathbf{I} - \mathbf{A}| = \lambda^N + c_{N-1}\lambda^{N-1} + \cdots + c_1\lambda + c_0 \tag{A.31}$$

is called the *characteristic polynomial* of $\mathbf{A}$. Now if $\lambda_1, \lambda_2, \ldots, \lambda_i$ are distinct eigenvalues of $\mathbf{A}$, then we have

$$c(\lambda) = (\lambda - \lambda_1)^{m_1}(\lambda - \lambda_2)^{m_2} \cdots (\lambda - \lambda_i)^{m_i} \tag{A.32}$$

where $m_1 + m_2 + \cdots + m_i = N$ and m_i is called the *algebraic multiplicity* of λ_i.

Theorem A.1:

Let λ_k ($k = 1, 2, \ldots, i$) be the distinct eigenvalues of $\mathbf{A}$ and let $\mathbf{x}_k$ be the eigenvectors associated with the eigenvalues λ_k. Then the set of eigenvectors $\mathbf{x}_1, \mathbf{x}_2, \ldots, \mathbf{x}_i$ are linearly independent.

Proof   The proof is by contradiction. Suppose that $\mathbf{x}_1, \mathbf{x}_2, \ldots, \mathbf{x}_i$ are linearly dependent. Then there exists $\alpha_1, \alpha_2, \ldots, \alpha_i$ not all zero such that

$$\alpha_1\mathbf{x}_1 + \alpha_2\mathbf{x}_2 + \cdots + \alpha_i\mathbf{x}_i = \sum_{K=1}^{i} \alpha_k\mathbf{x}_k = \mathbf{0} \tag{A.33}$$

Assuming $\alpha_1 \neq 0$, then by Eq. (A.33) we have

$$(\lambda_2 \mathbf{I} - \mathbf{A})(\lambda_3 \mathbf{I} - \mathbf{A}) \cdots (\lambda_i \mathbf{I} - \mathbf{A}) \left[\sum_{K=1}^{i} \alpha_k \mathbf{x}_k \right] = \mathbf{0} \qquad (A.34)$$

Now by Eq. (A.28)

$$(\lambda_j \mathbf{I} - \mathbf{A})\mathbf{x}_k = (\lambda_j - \lambda_k)\mathbf{x}_k \qquad j \neq k$$

and
$$(\lambda_k \mathbf{I} - \mathbf{A})\mathbf{x}_k = \mathbf{0}$$

Then Eq. (A.34) can be written as

$$\alpha_1 (\lambda_2 - \lambda_1)(\lambda_3 - \lambda_1) \cdots (\lambda_i - \lambda_1)\mathbf{x}_1 = \mathbf{0} \qquad (A.35)$$

Since λ_k $(k = 1, 2, \ldots, i)$ are distinct, Eq. (A.35) implies that $\alpha_1 = 0$, which is a contradiction. Thus, the set of eigenvectors $\mathbf{x}_1, \mathbf{x}_2, \ldots, \mathbf{x}_i$ are linearly independent.

A.6 Diagonalization and Similarity Transformation

A. Diagonalization:

Suppose that all eigenvalues of an $N \times N$ matrix $\mathbf{A}$ are distinct. Let $\mathbf{x}_1, \mathbf{x}_2, \ldots, \mathbf{x}_N$ be eigenvectors associated with the eigenvalues $\lambda_1, \lambda_2, \ldots, \lambda_N$. Let

$$\mathbf{P} = [\mathbf{x}_1 \quad \mathbf{x}_2 \quad \cdots \quad \mathbf{x}_N] \qquad (A.36)$$

Then
$$\begin{aligned}
\mathbf{AP} &= \mathbf{A} \begin{bmatrix} \mathbf{x}_1 & \mathbf{x}_2 & \cdots & \mathbf{x}_N \end{bmatrix} \\
&= \begin{bmatrix} \mathbf{A}\mathbf{x}_1 & \mathbf{A}\mathbf{x}_2 & \cdots & \mathbf{A}\mathbf{x}_N \end{bmatrix} \\
&= \begin{bmatrix} \lambda_1\mathbf{x}_1 & \lambda_2\mathbf{x}_2 & \cdots & \lambda_N\mathbf{x}_N \end{bmatrix} \\
&= \begin{bmatrix} \mathbf{x}_1 & \mathbf{x}_2 & \cdots & \mathbf{x}_N \end{bmatrix} \begin{bmatrix} \lambda_1 & 0 & \cdots & 0 \\ 0 & \lambda_2 & \cdots & 0 \\ \vdots & \vdots & \ddots & \vdots \\ 0 & 0 & \cdots & \lambda_N \end{bmatrix} = \mathbf{P}\boldsymbol{\Lambda}
\end{aligned} \qquad (A.37)$$

where
$$\boldsymbol{\Lambda} = \begin{bmatrix} \lambda_1 & 0 & \cdots & 0 \\ 0 & \lambda_2 & \cdots & 0 \\ \vdots & \vdots & \ddots & \vdots \\ 0 & 0 & \cdots & \lambda_N \end{bmatrix} \qquad (A.38)$$

By Theorem A.1, $\mathbf{P}$ has N linearly independent column vectors. Thus, $\mathbf{P}$ is nonsingular and $\mathbf{P}^{-1}$ exists, and hence

$$\mathbf{P}^{-1}\mathbf{AP} = \boldsymbol{\Lambda} = \begin{bmatrix} \lambda_1 & 0 & \cdots & 0 \\ 0 & \lambda_2 & \cdots & 0 \\ \vdots & \vdots & \ddots & \vdots \\ 0 & 0 & \cdots & \lambda_N \end{bmatrix} \qquad (A.39)$$

We call $\mathbf{P}$ the *diagonalization* matrix or *eigenvector* matrix, and $\boldsymbol{\Lambda}$ the *eigenvalue* matrix.

Notes:

1. A sufficient (but not necessary) condition that an $N \times N$ matrix **A** be diagonalizable is that **A** has N distinct eigenvalues.
2. If **A** does not have N independent eigenvectors, then **A** is not diagonalizable.
3. The diagonalization matrix **P** is not unique. Reordering the columns of **P** or multiplying them by nonzero scalars will produce a new diagonalization matrix.

B. Similarity Transformation:

Let **A** and **B** be two square matrices of the same order. If there exists a nonsingular matrix **Q** such that

$$\mathbf{B} = \mathbf{Q}^{-1}\mathbf{A}\mathbf{Q} \tag{A.40}$$

then we say that **B** is *similar* to **A** and Eq. (A.40) is called the *similarity transformation*.

Notes:

1. If **B** is similar to **A**, then **A** is similar to **B**.
2. If **A** is similar to **B** and **B** is similar to **C**, then **A** is similar to **C**.
3. If **A** and **B** are similar, then **A** and **B** have the same eigenvalues.
4. An $N \times N$ matrix **A** is similar to a diagonal matrix **D** if and only if there exist N linearly independent eigenvectors of **A**.

A.7 Functions of a Matrix

A. Powers of a Matrix:

We define powers of an $N \times N$ matrix **A** as

$$\mathbf{A}^n = \underbrace{\mathbf{A}\mathbf{A}\cdots\mathbf{A}}_{n}$$

$$\mathbf{A}^0 = \mathbf{I} \tag{A.41}$$

It can be easily verified by direct multiplication that if

$$\mathbf{D} = \begin{bmatrix} d_1 & 0 & \cdots & 0 \\ 0 & d_2 & \cdots & 0 \\ \vdots & \vdots & \ddots & \vdots \\ 0 & 0 & \cdots & d_N \end{bmatrix} \tag{A.42}$$

then

$$\mathbf{D}^n = \begin{bmatrix} d_1^n & 0 & \cdots & 0 \\ 0 & d_2^n & \cdots & 0 \\ \vdots & \vdots & \ddots & \vdots \\ 0 & 0 & \cdots & d_N^n \end{bmatrix} \tag{A.43}$$

Notes:

1. If the eigenvalues of **A** are $\lambda_1, \lambda_2, \ldots, \lambda_i$, then the eigenvalues of $\mathbf{A}^n$ are $\lambda_1^n, \lambda_2^n, \ldots, \lambda_i^n$.
2. Each eigenvector of **A** is still an eigenvector of $\mathbf{A}^n$.
3. If **P** diagonalizes **A**, that is,

$$\mathbf{P}^{-1}\mathbf{A}\mathbf{P} = \mathbf{\Lambda} = \begin{bmatrix} \lambda_1 & 0 & \cdots & 0 \\ 0 & \lambda_2 & \cdots & 0 \\ \vdots & \vdots & \ddots & \vdots \\ 0 & 0 & \cdots & \lambda_N \end{bmatrix} \tag{A.44}$$

then it also diagonalizes $\mathbf{A}^n$, that is,

$$\mathbf{P}^{-1}\mathbf{A}^n\mathbf{P} = \mathbf{\Lambda}^n = \begin{bmatrix} \lambda_1^n & 0 & \cdots & 0 \\ 0 & \lambda_2^n & \cdots & 0 \\ \vdots & \vdots & \ddots & \vdots \\ 0 & 0 & \cdots & \lambda_N^n \end{bmatrix} \tag{A.45}$$

since

$$(\mathbf{P}^{-1}\mathbf{A}\mathbf{P})(\mathbf{P}^{-1}\mathbf{A}\mathbf{P}) = \mathbf{P}^{-1}\mathbf{A}^2\mathbf{P} = \mathbf{\Lambda}^2$$

$$(\mathbf{P}^{-1}\mathbf{A}^2\mathbf{P})(\mathbf{P}^{-1}\mathbf{A}\mathbf{P}) = \mathbf{P}^{-1}\mathbf{A}^3\mathbf{P} = \mathbf{\Lambda}^3 \tag{A.46}$$

$$\vdots$$

B. Function of a Matrix:

Consider a function of λ defined by

$$f(\lambda) = a_0 + a_1\lambda + a_2\lambda^2 + \cdots = \sum_{k=0}^{\infty} a_k \lambda^k \tag{A.47}$$

With any such function we can associate a function of an $N \times N$ matrix $\mathbf{A}$:

$$f(\mathbf{A}) = a_0\mathbf{I} + a_1\mathbf{A} + a_2\mathbf{A}^2 + \cdots = \sum_{k=0}^{\infty} a_k \mathbf{A}^k \tag{A.48}$$

If $\mathbf{A}$ is a diagonal matrix $\mathbf{D}$ in Eq. (A.42), then using Eq. (A.43), we have

$$f(\mathbf{D}) = a_0\mathbf{I} + a_1\mathbf{D} + a_2\mathbf{D}^2 + \cdots = \sum_{k=0}^{\infty} a_k \mathbf{D}^k$$

$$= \begin{bmatrix} \displaystyle\sum_{k=0}^{\infty} a_k d_1^k & 0 & \cdots & 0 \\ 0 & \displaystyle\sum_{k=0}^{\infty} a_k d_2^k & \cdots & 0 \\ \vdots & \vdots & \ddots & \vdots \\ 0 & 0 & \cdots & \displaystyle\sum_{k=0}^{\infty} a_k d_N^k \end{bmatrix} = \begin{bmatrix} f(d_1) & 0 & \cdots & 0 \\ 0 & f(d_2) & \cdots & 0 \\ \vdots & \vdots & \ddots & \vdots \\ 0 & 0 & \cdots & f(d_N) \end{bmatrix} \tag{A.49}$$

If $\mathbf{P}$ diagonalizes $\mathbf{A}$, that is [Eq. (A.44)],

$$\mathbf{P}^{-1}\mathbf{A}\mathbf{P} = \mathbf{\Lambda}$$

then we have

$$\mathbf{A} = \mathbf{P}\mathbf{\Lambda}\mathbf{P}^{-1} \tag{A.50}$$

and

$$\mathbf{A}^2 = (\mathbf{P}\mathbf{\Lambda}\mathbf{P}^{-1})(\mathbf{P}\mathbf{\Lambda}\mathbf{P}^{-1}) = \mathbf{P}\mathbf{\Lambda}^2\mathbf{P}^{-1}$$

$$\mathbf{A}^3 = (\mathbf{P}\mathbf{\Lambda}^2\mathbf{P}^{-1})(\mathbf{P}\mathbf{\Lambda}\mathbf{P}^{-1}) = \mathbf{P}\mathbf{\Lambda}^3\mathbf{P}^{-1} \tag{A.51}$$

$$\vdots$$

Thus, we obtain

$$f(\mathbf{A}) = \mathbf{P}f(\mathbf{\Lambda})\mathbf{P}^{-1} \tag{A.52}$$

Replacing $\mathbf{D}$ by $\mathbf{\Lambda}$ in Eq. (A.49), we get

$$f(\mathbf{A}) = \mathbf{P} \begin{bmatrix} f(\lambda_1) & 0 & \cdots & 0 \\ 0 & f(\lambda_2) & \cdots & 0 \\ \vdots & \vdots & \ddots & \vdots \\ 0 & 0 & \cdots & f(\lambda_N) \end{bmatrix} \mathbf{P}^{-1} \tag{A.53}$$

where λ_k are the eigenvalues of $\mathbf{A}$.

C.   The Cayley-Hamilton Theorem:

Let the characteristic polynomial $c(\lambda)$ of an $N \times N$ matrix $\mathbf{A}$ be given by [Eq. (A.31)]

$$c(\lambda) = |\lambda\mathbf{I} - \mathbf{A}| = \lambda^N + c_{N-1}\lambda^{N-1} + \cdots + c_1\lambda + c_0$$

The *Cayley-Hamilton* theorem states that the matrix $\mathbf{A}$ satisfies its own characteristic equation; that is,

$$c(\mathbf{A}) = \mathbf{A}^N + c_{N-1}\mathbf{A}^{N-1} + \cdots + c_1\mathbf{A} + c_0\mathbf{I} = \mathbf{0} \tag{A.54}$$

EXAMPLE A.13   Let

$$\mathbf{A} = \begin{bmatrix} 2 & 1 \\ 0 & 3 \end{bmatrix}$$

Then, its characteristic polynomial is

$$c(\lambda) = |\lambda\mathbf{I} - \mathbf{A}| = \begin{vmatrix} \lambda - 2 & -1 \\ 0 & \lambda - 3 \end{vmatrix} = (\lambda - 2)(\lambda - 3) = \lambda^2 - 5\lambda + 6$$

and

$$c(\mathbf{A}) = \mathbf{A}^2 - 5\mathbf{A} + 6\mathbf{I} = \begin{bmatrix} 2 & 1 \\ 0 & 3 \end{bmatrix}^2 - 5\begin{bmatrix} 2 & 1 \\ 0 & 3 \end{bmatrix} + 6\begin{bmatrix} 1 & 0 \\ 0 & 1 \end{bmatrix}$$

$$= \begin{bmatrix} 4 & 5 \\ 0 & 9 \end{bmatrix} - \begin{bmatrix} 10 & 5 \\ 0 & 15 \end{bmatrix} + \begin{bmatrix} 6 & 0 \\ 0 & 6 \end{bmatrix}$$

$$= \begin{bmatrix} 0 & 0 \\ 0 & 0 \end{bmatrix} = \mathbf{0}$$

Rewriting Eq. (A.54), we have

$$\mathbf{A}^N = -c_0\mathbf{I} - c_1\mathbf{A} - \cdots - c_{N-1}\mathbf{A}^{N-1} \tag{A.55}$$

Multiplying through by $\mathbf{A}$ and then substituting the expression (A.55) for $\mathbf{A}^N$ on the right and rearranging, we get

$$\mathbf{A}^{N+1} = \alpha_0\mathbf{I} + \alpha_1\mathbf{A} + \cdots + \alpha_{N-1}\mathbf{A}^{N-1} \tag{A.56}$$

By continuing this process, we can express any positive integral power of $\mathbf{A}$ as a linear combination of $\mathbf{I}, \mathbf{A}, \ldots,$ $\mathbf{A}^{N-1}$. Thus, $f(\mathbf{A})$ defined by Eq. (A.48) can be represented by

$$f(\mathbf{A}) = b_0\mathbf{I} + b_1\mathbf{A} + \cdots + b_{N-1}\mathbf{A}^{N-1} = \sum_{m=0}^{N-1} b_m\mathbf{A}^m \tag{A.57}$$

In a similar manner, if λ is an eigenvalue of $\mathbf{A}$, then $f(\lambda)$ can also be expressed as

$$f(\lambda) = b_0 + b_1\lambda + \cdots + b_{N-1}\lambda^{N-1} = \sum_{m=0}^{N-1} b_m\lambda^m \tag{A.58}$$

Thus, if all eigenvalues of $\mathbf{A}$ are distinct, the coefficients b_m ($m = 0, 1, ..., N - 1$) can be determined by the following N equations:

$$f(\lambda_k) = b_0 + b_1\lambda_k + \cdots + b_{N-1}\lambda_k^{N-1} \quad k = 1, 2, ..., N \tag{A.59}$$

If all eigenvalues of $\mathbf{A}$ are not distinct, then Eq. (A.59) will not yield N equations. Assume that an eigenvalue λ_i has multiplicity r and all other eigenvalues are distinct. In this case differentiating both sides of Eq. (A.58) r times with respect to λ and setting $\lambda = \lambda_i$, we obtain r equations corresponding to λ_i:

$$\frac{d^{n-1}}{d\lambda^{n-1}} f(\lambda)\Big|_{\lambda=\lambda_i} = \frac{d^{n-1}}{d\lambda^{n-1}}\left(\sum_{m=0}^{N-1} b_m\lambda^m\right)\Big|_{\lambda=\lambda_i} \quad n = 1, 2, ..., r \tag{A.60}$$

Combining Eqs. (A.59) and (A.60), we can determine all coefficients b_m in Eq. (A.57).

D. Minimal Polynomial of A:

The *minimal* (or *minimum*) polynomial $m(\lambda)$ of an $N \times N$ matrix $\mathbf{A}$ is the polynomial of lowest degree having 1 as its leading coefficient such that $m(\mathbf{A}) = \mathbf{0}$. Since $\mathbf{A}$ satisfies its characteristic equation, the degree of $m(\lambda)$ is not greater than N.

EXAMPLE A.14 Let

$$\mathbf{A} = \begin{bmatrix} \alpha & 0 \\ 0 & \alpha \end{bmatrix}$$

The characteristic polynomial is

$$c(\lambda) = |\lambda\mathbf{I} - \mathbf{A}| = \begin{vmatrix} \lambda - \alpha & 0 \\ 0 & \lambda - \alpha \end{vmatrix} = (\lambda - \alpha)^2 = \lambda^2 - 2\alpha\lambda + \alpha^2$$

and the minimal polynomial is

$$m(\lambda) = \lambda - \alpha$$

since

$$m(\mathbf{A}) = \mathbf{A} - \alpha\mathbf{I} = \begin{bmatrix} \alpha & 0 \\ 0 & \alpha \end{bmatrix} - \alpha\begin{bmatrix} 1 & 0 \\ 0 & 1 \end{bmatrix} = \begin{bmatrix} 0 & 0 \\ 0 & 0 \end{bmatrix} = \mathbf{0}$$

Notes:

1. Every eigenvalue of $\mathbf{A}$ is a zero of $m(\lambda)$.
2. If all the eigenvalues of $\mathbf{A}$ are distinct, then $c(\lambda) = m(\lambda)$.
3. $c(\lambda)$ is divisible by $m(\lambda)$.
4. $m(\lambda)$ may be used in the same way as $c(\lambda)$ for the expression of higher powers of $\mathbf{A}$ in terms of a limited number of powers of $\mathbf{A}$.

It can be shown that $m(\lambda)$ can be determined by

$$m(\lambda) = \frac{c(\lambda)}{d(\lambda)} \tag{A.61}$$

where $d(\lambda)$ is the greatest common divisor (gcd) of all elements of $\text{adj}(\lambda\mathbf{I} - \mathbf{A})$.

EXAMPLE A.15 Let

$$\mathbf{A} = \begin{bmatrix} 5 & -6 & -6 \\ -1 & 4 & 2 \\ 3 & -6 & -4 \end{bmatrix}$$

Then

$$c(\lambda) = |\lambda\mathbf{I} - A| = \begin{bmatrix} \lambda - 5 & 6 & 6 \\ 1 & \lambda - 4 & -2 \\ -3 & 6 & \lambda + 4 \end{bmatrix}$$

$$= \lambda^3 - 5\lambda^2 + 8\lambda - 4 = (\lambda - 1)(\lambda - 2)^2$$

$$\text{adj}\big[\lambda\mathbf{I} - \mathbf{A}\big] = \begin{bmatrix} \begin{vmatrix} \lambda - 4 & -2 \\ 6 & \lambda + 4 \end{vmatrix} & -\begin{vmatrix} 6 & 6 \\ 6 & \lambda + 4 \end{vmatrix} & \begin{vmatrix} 6 & 6 \\ \lambda - 4 & -2 \end{vmatrix} \\ -\begin{vmatrix} 1 & -2 \\ -3 & \lambda + 4 \end{vmatrix} & \begin{vmatrix} \lambda - 5 & 6 \\ -3 & \lambda + 4 \end{vmatrix} & -\begin{vmatrix} \lambda - 5 & 6 \\ 1 & -2 \end{vmatrix} \\ \begin{vmatrix} 1 & \lambda - 4 \\ -3 & 6 \end{vmatrix} & -\begin{vmatrix} \lambda - 5 & 6 \\ -3 & 6 \end{vmatrix} & \begin{vmatrix} \lambda - 5 & 6 \\ 1 & \lambda - 4 \end{vmatrix} \end{bmatrix}$$

$$= \begin{bmatrix} (\lambda + 2)(\lambda - 2) & -6(\lambda - 2) & -6(\lambda - 2) \\ -(\lambda - 2) & (\lambda + 1)(\lambda - 2) & 2(\lambda - 2) \\ 3(\lambda - 2) & -6(\lambda - 2) & (\lambda - 2)(\lambda - 7) \end{bmatrix}$$

Thus, $d(\lambda) = \lambda - 2$ and

$$m(\lambda) = \frac{c(\lambda)}{d(\lambda)} = (\lambda - 1)(\lambda - 2) = \lambda^2 - 3\lambda + 2$$

and

$$m(\mathbf{A}) = (\mathbf{A} - \mathbf{I})(\mathbf{A} - 2\mathbf{I}) = \begin{bmatrix} 4 & -6 & -6 \\ -1 & 3 & 2 \\ 3 & -6 & -5 \end{bmatrix}\begin{bmatrix} 3 & -6 & -6 \\ -1 & 2 & 2 \\ 3 & -6 & -6 \end{bmatrix} = \begin{bmatrix} 0 & 0 & 0 \\ 0 & 0 & 0 \\ 0 & 0 & 0 \end{bmatrix}$$

E. Spectral Decomposition:

It can be shown that if the minimal polynomial $m(\lambda)$ of an $N \times N$ matrix $\mathbf{A}$ has the form

$$m(\lambda) = (\lambda - \lambda_1)(\lambda - \lambda_2) \cdots (\lambda - \lambda_i) \tag{A.62}$$

then $\mathbf{A}$ can be represented by

$$\mathbf{A} = \lambda_1\mathbf{E}_1 + \lambda_2\mathbf{E}_2 + \cdots + \lambda_i\mathbf{E}_i \tag{A.63}$$

where $\mathbf{E}_j$ $(j = 1, 2, \ldots, i)$ are called *constituent* matrices and have the following properties:

1. $\mathbf{I} = \mathbf{E}_1 + \mathbf{E}_2 + \cdots + \mathbf{E}_i$
2. $\mathbf{E}_m\mathbf{E}_k = \mathbf{0}, m \neq k$
3. $\mathbf{E}_k^2 = \mathbf{E}_k$
4. $\mathbf{A}\mathbf{E}_k = \mathbf{E}_k\mathbf{A} = \lambda_k\mathbf{E}_k$

$$\tag{A.64}$$

Any matrix $\mathbf{B}$ for which $\mathbf{B}^2 = \mathbf{B}$ is called *idempotent*. Thus, the constituent matrices $\mathbf{E}_j$ are idempotent matrices. The set of eigenvalues of $\mathbf{A}$ is called the *spectrum* of $\mathbf{A}$, and Eq. (A.63) is called the *spectral decomposition* of $\mathbf{A}$. Using the properties of Eq. (A.64), we have

$$\mathbf{A}^2 = \lambda_1^2\mathbf{E}_1 + \lambda_2^2\mathbf{E}_2 + \cdots + \lambda_i^2\mathbf{E}_i$$

$$\vdots$$

$$\mathbf{A}^n = \lambda_1^n\mathbf{E}_1 + \lambda_2^n\mathbf{E}_2 + \cdots + \lambda_i^n\mathbf{E}_i \tag{A.65}$$

and

$$f(\mathbf{A}) = f(\lambda_1)\mathbf{E}_1 + f(\lambda_2)\mathbf{E}_2 + \cdots + f(\lambda_i)\mathbf{E}_i \tag{A.66}$$

The constituent matrices $\mathbf{E}_j$ can be evaluated as follows. The partial-fraction expansion of

$$\frac{1}{m(\lambda)} = \frac{1}{(\lambda - \lambda_1)(\lambda - \lambda_2)\cdots(\lambda - \lambda_i)}$$

$$= \frac{k_1}{\lambda - \lambda_1} + \frac{k_2}{\lambda - \lambda_2} + \cdots + \frac{k_i}{\lambda - \lambda_i}$$

leads to

$$k_j = \frac{1}{\displaystyle\prod_{\substack{m=1 \\ m \neq j}}^{i} (\lambda_j - \lambda_m)}$$

Then

$$\frac{1}{m(\lambda)} = \frac{k_1 g_1(\lambda) + k_2 g_2(\lambda) + \cdots + k_i g_i(\lambda)}{(\lambda - \lambda_1)(\lambda - \lambda_2)\cdots(\lambda - \lambda_i)}$$

where

$$g_j(\lambda) = \prod_{\substack{m=1 \\ m \neq j}}^{i} (\lambda - \lambda_m)$$

Let $e_j(\lambda) = k_j g_j(\lambda)$. Then the constituent matrices $\mathbf{E}_j$ can be evaluated as

$$\mathbf{E}_j = e_j(\mathbf{A}) = \frac{\displaystyle\prod_{\substack{m=1 \\ m \neq j}}^{i} (\mathbf{A} - \lambda_m \mathbf{I})}{\displaystyle\prod_{\substack{m=1 \\ m \neq j}}^{i} (\lambda_j - \lambda_m)} \tag{A.67}$$

EXAMPLE A.16   Consider the matrix $\mathbf{A}$ in Example A.15:

$$\mathbf{A} = \begin{bmatrix} 5 & -6 & -6 \\ -1 & 4 & 2 \\ 3 & -6 & -4 \end{bmatrix}$$

From Example A.15, we have

$$m(\lambda) = (\lambda - 1)(\lambda - 2)$$

Then

$$\frac{1}{m(\lambda)} = \frac{1}{(\lambda - 1)(\lambda - 2)} = \frac{-1}{\lambda - 1} + \frac{1}{\lambda - 2}$$

and

$$e_1(\lambda) = -(\lambda - 2) \qquad e_2(\lambda) = \lambda - 1$$

Then

$$\mathbf{E}_1 = e_1(\mathbf{A}) = -(\mathbf{A} - 2\mathbf{I}) = \begin{bmatrix} -3 & 6 & 6 \\ 1 & -2 & -2 \\ -3 & 6 & 6 \end{bmatrix}$$

$$\mathbf{E}_2 = e_2(\mathbf{A}) = \mathbf{A} - \mathbf{I} = \begin{bmatrix} 4 & -6 & -6 \\ -1 & 3 & 2 \\ 3 & -6 & -5 \end{bmatrix}$$

$$\mathbf{A} = \lambda_1 \mathbf{E}_1 + \lambda_2 \mathbf{E}_2 = \mathbf{E}_1 + 2\mathbf{E}_2$$

$$= \begin{bmatrix} -3 & 6 & 6 \\ 1 & -2 & -2 \\ -3 & 6 & 6 \end{bmatrix} + 2\begin{bmatrix} 4 & -6 & -6 \\ -1 & 3 & 2 \\ 3 & -6 & -5 \end{bmatrix} = \begin{bmatrix} 5 & -6 & -6 \\ -1 & 4 & 2 \\ 3 & -6 & -4 \end{bmatrix}$$

A.8   Differentiation and Integration of Matrices

A.   Definitions:

The derivative of an $m \times n$ matrix $\mathbf{A}(t)$ is defined to be the $m \times n$ matrix, each element of which is the derivative of the corresponding element of $\mathbf{A}$; that is,

$$\frac{d}{dt}\mathbf{A}(t) = \left[\frac{d}{dt}a_{ij}(t)\right]_{m \times n}$$

$$= \begin{bmatrix} \dfrac{d}{dt}a_{11}(t) & \dfrac{d}{dt}a_{12}(t) & \cdots & \dfrac{d}{dt}a_{1n}(t) \\ \dfrac{d}{dt}a_{21}(t) & \dfrac{d}{dt}a_{22}(t) & \cdots & \dfrac{d}{dt}a_{2n}(t) \\ \vdots & \vdots & \ddots & \vdots \\ \dfrac{d}{dt}a_{m1}(t) & \dfrac{d}{dt}a_{m2}(t) & \cdots & \dfrac{d}{dt}a_{mn}(t) \end{bmatrix} \tag{A.68}$$

Similarly, the integral of an $m \times n$ matrix $\mathbf{A}(t)$ is defined to be

$$\int \mathbf{A}(t)\,dt = \left[\int a_{ij}(t)\,dt\right]_{m \times n}$$

$$= \begin{bmatrix} \int a_{11}(t)\,dt & \int a_{12}(t)\,dt & \cdots & \int a_{1n}(t)\,dt \\ \int a_{21}(t)\,dt & \int a_{22}(t)\,dt & \cdots & \int a_{2n}(t)\,dt \\ \vdots & \vdots & \ddots & \vdots \\ \int a_{m1}(t)\,dt & \int a_{m2}(t)\,dt & \cdots & \int a_{mn}(t)\,dt \end{bmatrix} \tag{A.69}$$

EXAMPLE A.17   Let

$$\mathbf{A} = \begin{bmatrix} t & t^2 \\ 1 & t^3 \end{bmatrix}$$

Then

$$\frac{d}{dt}\mathbf{A} = \begin{bmatrix} \dfrac{d}{dt}t & \dfrac{d}{dt}t^2 \\ \dfrac{d}{dt}1 & \dfrac{d}{dt}t^3 \end{bmatrix} = \begin{bmatrix} 1 & 2t \\ 0 & 3t^2 \end{bmatrix}$$

and

$$\int_0^1 \mathbf{A}\,dt = \begin{bmatrix} \int_0^1 t\,dt & \int_0^1 t^2\,dt \\ \int_0^1 1\,dt & \int_0^1 t^3\,dt \end{bmatrix} = \begin{bmatrix} \dfrac{1}{2} & \dfrac{1}{3} \\ 1 & \dfrac{1}{4} \end{bmatrix}$$

B.   Differentiation of the Product of Two Matrices:

If the matrices $\mathbf{A}(t)$ and $\mathbf{B}(t)$ can be differentiated with respect to t, then

$$\frac{d}{dt}[\mathbf{A}(t)\mathbf{B}(t)] = \frac{d\mathbf{A}(t)}{dt}\mathbf{B}(t) + \mathbf{A}(t)\frac{d\mathbf{B}(t)}{dt} \tag{A.70}$$

Review of Probability

B.1 Probability

A. Random Experiments:

In the study of probability, any process of observation is referred to as an *experiment*. The results of an observation are called the *outcomes* of the experiment. An experiment is called a *random experiment* if its outcome cannot be predicted. Typical examples of a random experiment are the roll of a die, the toss of a coin, drawing a card from a deck, or selecting a message signal for transmission from several messages.

B. Sample Space and Events:

The set of all possible outcomes of a random experiment is called the *sample space S*. An element in S is called a *sample point*. Each outcome of a random experiment corresponds to a sample point.

A set A is called a *subset* of B, denoted by $A \subset B$ if every element of A is also an element of B. Any subset of the sample space S is called an *event*. A sample point of S is often referred to as an *elementary event*. Note that the sample space S is the subset of itself, that is, $S \subset S$. Since S is the set of all possible outcomes, it is often called the *certain event*.

C. Algebra of Events:

1. The *complement* of event A, denoted $\bar{A}$, is the event containing all sample points in S but not in A.
2. The *union* of events A and B, denoted $A \cup B$, is the event containing all sample points in either A or B or both.
3. The *intersection* of events A and B, denoted $A \cap B$, is the event containing all sample points in both A and B.
4. The event containing no sample point is called the *null event*, denoted $\varnothing$. Thus $\varnothing$ corresponds to an impossible event.
5. Two events A and B are called *mutually exclusive* or *disjoint* if they contain no common sample point, that is, $A \cap B = \varnothing$.

By the preceding set of definitions, we obtain the following identities:

$$\bar{S} = \varnothing \quad \bar{\varnothing} = S$$
$$S \cup A = S \quad S \cap A = A$$
$$A \cup \bar{A} = S \quad A \cap \bar{A} = \varnothing \quad \bar{\bar{A}} = A$$

D. Venn Diagram:

A graphical representation that is very useful for illustrating set operation is the Venn diagram. For instance, in the three Venn diagrams shown in Fig. B-1, the shaded areas represent, respectively, the events $A \cup B$, $A \cap B$, and $\bar{A}$.

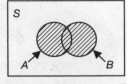
(a) Shaded region: $A \cup B$

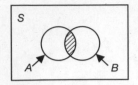
(b) Shaded region: $A \cap B$

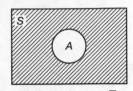
(c) Shaded region: $\overline{A}$

Fig. B-1

E. Probabilities of Events:

An assignment of real numbers to the events defined on S is known as the *probability measure*. In the *axiomatic* definition, the probability $P(A)$ of the event A is a real number assigned to A that satisfies the following three *axioms*:

Axiom 1:	$P(A) \geq 0$	(B.1)
Axiom 2:	$P(S) = 1$	(B.2)
Axiom 3:	$P(A \cup B) = P(A) + P(B) \quad$ if $A \cap B = \varnothing$	(B.3)

With the preceding axioms, the following useful properties of probability can be obtained.

1.	$P(\overline{A}) = 1 - P(A)$	(B.4)
2.	$P(\varnothing) = 0$	(B.5)
3.	$P(A) \leq P(B) \quad$ if $A \subset B$	(B.6)
4.	$P(A) \leq 1$	(B.7)
5.	$P(A \cup B) = P(A) + P(B) - P(A \cap B)$	(B.8)

Note that Property 4 can be easily derived from axiom 2 and property 3. Since $A \subset S$, we have

$$P(A) \leq P(S) = 1$$

Thus, combining with axiom 1, we obtain

$$0 \leq P(A) \leq 1 \tag{B.9}$$

Property 5 implies that

$$P(A \cup B) \leq P(A) + P(B) \tag{B.10}$$

since $P(A \cap B) \geq 0$ by axiom 1.

One can also define $P(A)$ intuitively, in terms of relative frequency. Suppose that a random experiment is repeated n times. If an event A occurs n_A times, then its probability $P(A)$ is defined as

$$P(A) = \lim_{n \to \infty} \frac{n_A}{n} \tag{B.11}$$

Note that this limit may not exist.

EXAMPLE B.1   Using the axioms of probability, prove Eq. (B.4).

$$S = A \cup \bar{A} \qquad \text{and} \qquad A \cap \bar{A} = \varnothing$$

Then the use of axioms 1 and 3 yields

$$P(S) = 1 = P(A) + P(\bar{A})$$

Thus
$$P(\bar{A}) = 1 - P(A)$$

EXAMPLE B.2   Verify Eq. (B.5).

$$A = A \cup \varnothing \qquad \text{and} \qquad A \cap \varnothing = \varnothing$$

Therefore, by axiom 3,

$$P(A) = P(A \cup \varnothing) = P(A) + P(\varnothing)$$

and we conclude that

$$P(\varnothing) = 0$$

EXAMPLE B.3   Verify Eq. (B.6).

Let $A \subset B$. Then from the Venn diagram shown in Fig. B-2, we see that

$$B = A \cup (B \cap \bar{A}) \quad \text{and} \quad A \cap (B \cap \bar{A}) = \varnothing$$

Hence, from axiom 3,

$$P(B) = P(A) + P(B \cap \bar{A}) \geq P(A)$$

because by axiom 1, $P(B \cap \bar{A}) \geq 0$.

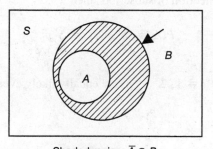

Shaded region: $\bar{A} \cap B$

Fig. B-2

EXAMPLE B.4   Verify Eq. (B.8).

From the Venn diagram of Fig. B-3, each of the sets $A \cup B$ and B can be expressed, respectively, as a union of mutually exclusive sets as follows:

$$A \cup B = A \cup (\bar{A} \cap B) \qquad \text{and} \qquad B = (A \cap B) \cup (\bar{A} \cap B)$$

Thus, by axiom 3,

$$P(A \cup B) = P(A) + P(\bar{A} \cap B) \tag{B.12}$$

and
$$P(B) = P(A \cap B) + P(\overline{A} \cap B) \tag{B.13}$$
From Eq. (B.13) we have

$$P(\overline{A} \cap B) = P(B) - P(A \cap B) \tag{B.14}$$

Substituting Eq. (B.14) into Eq. (B.12), we obtain

$$P(A \cup B) = P(A) + P(B) - P(A \cap B)$$

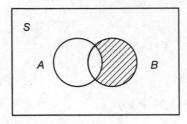
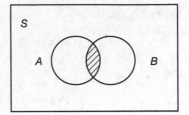

Shaded region: $\overline{A} \cap B$ Shaded region: $A \cap B$

Fig. B-3

F. Equally Likely Events:

Consider a finite sample space S with finite elements

$$S = \{\lambda_1, \lambda_2, ..., \lambda_n\}$$

where λ_i's are elementary events. Let $P(\lambda_i) = p_i$. Then

1. $0 \leq p_i \leq 1 \qquad i = 1, 2, ..., n$

2. $\displaystyle\sum_{i=1}^{n} p_i = p_1 + p_2 + \cdots + p_n = 1 \tag{B.15}$

3. If $A = \bigcup_{i \in I} \lambda_i$, where I is a collection of subscripts, then

$$P(A) = \sum_{\lambda_i \in A} p(\lambda_i) = \sum_{i \in I} p_i \tag{B.16}$$

When all elementary events λ_i $(i = 1, 2, ..., n)$ are equally likely events, that is

$$p_1 = p_2 = ... = p_n$$

then from Eq. (B.15), we have

$$p_i = \frac{1}{n} \quad i = 1, 2, ..., n \tag{B.17}$$

and
$$P(A) = \frac{n(A)}{n} \tag{B.18}$$

where $n(A)$ is the number of outcomes belonging to event A and n is the number of sample points in S.

G. Conditional Probability:

The *conditional probability* of an event A given the event B, denoted by $P(A|B)$, is defined as

$$P(A|B) = \frac{P(A \cap B)}{P(B)} \qquad P(B) > 0 \tag{B.19}$$

where $P(A \cap B)$ is the joint probability of A and B. Similarly,

$$P(B|A) = \frac{P(A \cap B)}{P(A)} \qquad P(A) > 0 \tag{B.20}$$

is the conditional probability of an event B given event A. From Eqs. (B.19) and (B.20) we have

$$P(A \cap B) = P(A|B)P(B) = P(B|A)P(A) \tag{B.21}$$

Equation (B.21) is often quite useful in computing the joint probability of events.

From Eq. (B.21) we can obtain the following *Bayes rule*:

$$P(A|B) = \frac{P(B|A) P(A)}{P(B)} \tag{B.22}$$

EXAMPLE B.5 Find $P(A|B)$ if (*a*) $A \cap B = \varnothing$, (*b*) $A \subset B$, and (*c*) $B \subset A$.

(*a*) If $A \cap B = \varnothing$, then $P(A \cap B) = P(\varnothing) = 0$. Thus,

$$P(A|B) = \frac{P(A \cap B)}{P(B)} = \frac{P(\varnothing)}{P(B)} = 0$$

(*b*) If $A \subset B$, then $A \cap B = A$ and

$$P(A|B) = \frac{P(A \cap B)}{P(B)} = \frac{P(A)}{P(B)}$$

(*c*) If $B \subset A$, then $A \cap B = B$ and

$$P(A|B) = \frac{P(A \cap B)}{P(B)} = \frac{P(B)}{P(B)} = 1$$

H. Independent Events:

Two events A and B are said to be *(statistically) independent* if

$$P(A|B) = P(A) \qquad \text{and} \qquad P(B|A) = P(B) \tag{B.23}$$

This, together with Eq. (B.21), implies that for two statistically independent events

$$P(A \cap B) = P(A)P(B) \tag{B.24}$$

We may also extend the definition of independence to more than two events. The events $A_1, A_2, \dots, A_n$ are independent if and only if for every subset $\{A_{i_1}, A_{i_2}, \dots, A_{i_k}\}$ $(2 \leq k \leq n)$ of these events,

$$P(A_{i_1} \cap A_{i_2} \cap \dots \cap A_{i_k}) = P(A_{i_1})P(A_{i_2}) \dots P(A_{i_k}) \tag{B.25}$$

I. Total Probability:

The events $A_1, A_2, \dots, A_n$ are called *mutually exclusive* and *exhaustive* if

$$\bigcup_{i=1}^{n} A_i = A_1 \cup A_2 \cup \dots \cup A_n = S \quad \text{and} \quad A_i \cap A_j = \varnothing \quad i \neq j \tag{B.26}$$

Let B be any event in S. Then

$$P(B) = \sum_{i=1}^{n} P(B \cap A_i) = \sum_{i=1}^{n} P(B|A_i) P(A_i) \tag{B.27}$$

which is known as the *total probability* of event B. Let $A = A_i$ in Eq. (B.22); using Eq. (B.27) we obtain

$$P(A_i|B) = \frac{P(B|A_i)P(A_i)}{\sum\limits_{i=1}^{n} P(B|A_i)P(A_i)} \tag{B.28}$$

Note that the terms on the right-hand side are all conditioned on events A_i, while that on the left is conditioned on B. Equation (B.28) is sometimes referred to as *Bayes' theorem*.

EXAMPLE B.6 Verify Eq. (B.27).

Since $B \cap S = B$ [and using Eq. (B.26)], we have

$$B = B \cap S = B \cap (A_1 \cup A_2 \cup \cdots \cup A_N)$$
$$= (B \cap A_1) \cup (B \cap A_2) \cup \cdots \cup (B \cap A_N)$$

Now the events $B \cap A_k$ ($k = 1, 2, \ldots, N$) are mutually exclusive, as seen from the Venn diagram of Fig. B-4. Then by axiom 3 of the probability definition and Eq. (B.21), we obtain

$$P(B) = P(B \cap S) = \sum_{k=1}^{N} P(B \cap A_k) = \sum_{k=1}^{N} P(B|A_k)P(A_k)$$

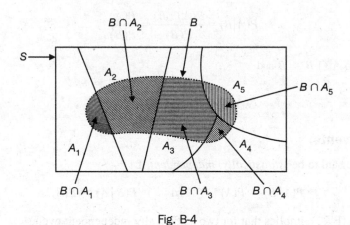

Fig. B-4

B.2 Random Variables

A. Random Variables:

Consider a random experiment with sample space S. A *random variable* $X(\lambda)$ is a single-valued real function that assigns a real number called the *value* of $X(\lambda)$ to each sample point λ of S. Often we use a single letter X for this function in place of $X(\lambda)$ and use r.v. to denote the random variable. A schematic diagram representing a r.v. is given in Fig. B-5.

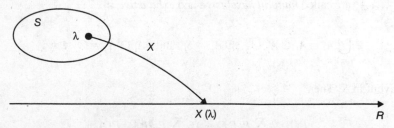

Fig. B-5 Random variable X as a function.

The sample space S is termed the *domain* of the r.v. X, and the collection of all numbers [values of $X(\lambda)$] is termed the *range* of the r.v. X. Thus, the range of X is a certain subset of the set of all real numbers and it is usually denoted by R_X. Note that two or more different sample points might give the same value of $X(\lambda)$, but two different numbers in the range cannot be assigned to the same sample point.

The r.v. X induces a probability measure on the real line as follows:

$$P(X = x) = P\{\lambda : X(\lambda) = x\}$$

$$P(X \le x) = P\{\lambda : X(\lambda) \le x\}$$

$$P(x_1 < X \le x_2) = P\{\lambda : x_1 < X(\lambda) \le x_2\}$$

If X can take on only a *countable* number of distinct values, then X is called a *discrete* random variable. If X can assume any values within one or more intervals on the real line, then X is called a *continuous* random variable. The number of telephone calls arriving at an office in a finite time is an example of a discrete random variable, and the exact time of arrival of a telephone call is an example of a continuous random variable.

B. Distribution Function:

The *distribution function* [or *cumulative distribution function* (cdf)] of X is the function defined by

$$F_X(x) = P(X \le x) \quad -\infty < x < \infty \tag{B.29}$$

Properties of $F_x(x)$:

1. $0 \le F_X(x) \le 1$ (B.30)
2. $F_X(x_1) \le F_X(x_2)$ if $x_1 < x_2$ (B.31)
3. $F_X(\infty) = 1$ (B.32)
4. $F_X(-\infty) = 0$ (B.33)

5. $F_X(a^+) = F_X(a) \quad a^+ = \lim_{0 < \varepsilon \to 0} a + \varepsilon$ (B.34)

From definition (B.29) we can compute other probabilities:

$$P(a < X \le b) = F_X(b) - F_X(a) \tag{B.35}$$

$$P(X > a) = 1 - F_X(a) \tag{B.36}$$

$$P(X < b) = F_X(b^-) \quad b^- = \lim_{0 < \varepsilon \to 0} b - \varepsilon \tag{B.37}$$

C. Discrete Random Variables and Probability Mass Functions:

Let X be a discrete r.v. with cdf $F_X(x)$. Then $F_X(x)$ is a staircase function (see Fig. B-6), and $F_X(x)$ changes values only in jumps (at most a countable number of them) and is constant between jumps.

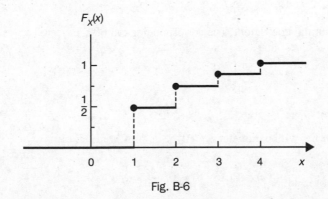

Fig. B-6

Suppose that the jumps in $F_X(x)$ of a discrete r.v. X occur at the points $x_1, x_2, \ldots$, where the sequence may be either finite or countably infinite, and we assume $x_i < x_j$ if $i < j$. Then

$$F_X(x_i) - F_X(x_{i-1}) = P(X \le x_i) - P(X \le x_{i-1}) = P(X = x_i) \tag{B.38}$$

Let

$$p_X(x) = P(X = x) \tag{B.39}$$

The function $p_X(x)$ is called the *probability mass function* (pmf) of the discrete r.v. X.

Properties of $p_x(x)$:

1. $0 \le p_X(x_i) \le 1 \qquad i = 1, 2, \ldots$ $\tag{B.40}$
2. $p_X(x) = 0 \quad$ if $x \ne x_i (i = 1, 2, \ldots)$ $\tag{B.41}$

3. $\displaystyle\sum_i p_X(x_i) = 1$ $\tag{B.42}$

The cdf $F_X(x)$ of a discrete r.v. X can be obtained by

$$F_X(x) = P(X \le x) = \sum_{x_i \le x} p_X(x_i) \tag{B.43}$$

D. Examples of Discrete Random Variables:

1. Bernoulli Distribution:

A r.v. X is called a *Bernoulli* r.v. with parameter p if its pmf is given by

$$p_X(k) = P(X = k) = p^k(1 - p)^{1-k} \qquad k = 0, 1 \tag{B.44}$$

where $0 \le p \le 1$. By Eq. (B.29), the cdf $F_X(x)$ of the Bernoulli r.v. X is given by

$$F_X(x) = \begin{cases} 0 & x < 0 \\ 1 - p & 0 \le x < 1 \\ 1 & x \ge 1 \end{cases} \tag{B.45}$$

2. Binomial Distribution:

A r.v. X is called a *binomial* r.v. with parameters (n, p) if its pmf is given by

$$p_X(k) = P(X = k) = \binom{n}{k} p^k(1 - p)^{n-k} \qquad k = 0, 1, \ldots, n \tag{B.46}$$

where $0 \le p \le 1$ and

$$\binom{n}{k} = \frac{n!}{k!(n - k)!}$$

which is known as the binomial coefficient. The corresponding cdf of X is

$$F_X(x) = \sum_{k=0}^{n} \binom{n}{k} p^k(1 - p)^{n-k} \qquad n \le x < n + 1 \tag{B.47}$$

3. Poisson Distribution:

A r.v. X is called a *Poisson* r.v. with parameter $\lambda \ (>0)$ if its pmf is given by

$$p_X(k) = P(X = k) = e^{-\lambda} \frac{\lambda^k}{k!} \qquad k = 0, 1, \ldots \tag{B.48}$$

The corresponding cdf of X is

$$F_X(x) = e^{-\lambda} \sum_{k=0}^{n} \frac{\lambda^k}{k!} \qquad n \le x < n+1 \tag{B.49}$$

E. Continuous Random Variables and Probability Density Functions:

Let X be a r.v. with cdf $F_X(x)$. Then $F_X(x)$ is continuous and also has a derivative $dF_X(x)/dx$ that exists everywhere except at possibly a finite number of points and is piecewise continuous. Thus, if X is a continuous r.v., then

$$P(X = x) = 0 \tag{B.50}$$

In most applications, the r.v. is either discrete or continuous. But if the cdf $F_X(x)$ of a r.v. X possesses both features of discrete and continuous r.v.'s, then the r.v. X is called the *mixed* r.v.

Let

$$f_X(x) = \frac{dF_X(x)}{dx} \tag{B.51}$$

The function $f_X(x)$ is called the *probability density function* (pdf) of the continuous r.v. X.

Properties of $f_X(x)$:

1. $f_X(x) \ge 0$ (B.52)
2. $\int_{-\infty}^{\infty} f_X(x)\,dx = 1$ (B.53)
3. $f_X(x)$ is piecewise continuous.
4. $P(a < X \le b) = \int_a^b f_X(x)\,dx$ (B.54)

The cdf $F_X(x)$ of a continuous r.v. X can be obtained by

$$F_X(x) = P(X \le x) = \int_{-\infty}^{x} f_X(\xi)\,d\xi \tag{B.55}$$

F. Examples of Continuous Random Variables:

1. Uniform Distribution:

A r.v. X is called a *uniform* r.v. over (a, b) if its pdf is given by

$$f_X(x) = \begin{cases} \dfrac{1}{b-a} & a < x < b \\ 0 & \text{otherwise} \end{cases} \tag{B.56}$$

The corresponding cdf of X is

$$F_X(x) = \begin{cases} 0 & x \le a \\ \dfrac{x-a}{b-a} & a < x < b \\ 1 & x \ge b \end{cases} \tag{B.57}$$

2. Exponential Distribution:

A r.v. X is called an *exponential* r.v. with parameter λ (> 0) if its pdf is given by

$$f_X(x) = \begin{cases} \lambda e^{-\lambda x} & x > 0 \\ 0 & x < 0 \end{cases} \tag{B.58}$$

The corresponding cdf of X is

$$F_X(x) = \begin{cases} 1 - e^{-\lambda x} & x \geq 0 \\ 0 & x < 0 \end{cases} \tag{B.59}$$

3. Normal (or Gaussian) Distribution:

A r.v. $X = N(\mu; \sigma^2)$ is called a *normal* (or *Gaussian*) r.v. if its pdf is given by

$$f_X(x) = \frac{1}{\sqrt{2\pi}\sigma} e^{-(x-\mu)^2/(2\sigma^2)} \tag{B.60}$$

The corresponding cdf of X is

$$F_X(x) = \frac{1}{\sqrt{2\pi}\sigma} \int_{-\infty}^{x} e^{-(\xi-\mu)^2/(2\sigma^2)} \, d\xi = \frac{1}{\sqrt{2\pi}} \int_{-\infty}^{(x-\mu)t} e^{-\xi^2/2} \, d\xi \tag{B.61}$$

B.3　Two-Dimensional Random Variables

A. Joint Distribution Function:

Let S be the sample space of a random experiment. Let X and Y be two r.v.'s defined on S. Then the pair (X, Y) is called a two-dimensional r.v. if each of X and Y associates a real number with every element of S. The *joint cumulative distribution* function (or joint cdf) of X and Y, denoted by $F_{XY}(x, y)$, is the function defined by

$$F_{XY}(x, y) = P(X \leq x, Y \leq y) \tag{B.62}$$

Two r.v.'s X and Y will be called *independent* if

$$F_{XY}(x, y) = F_X(x) F_Y(y) \tag{B.63}$$

for every value of x and y.

B. Marginal Distribution Function:

Since $\{X \leq \infty\}$ and $\{Y \leq \infty\}$ are certain events, we have

$$\{X \leq x, Y \leq \infty\} = \{X \leq x\} \qquad \{X \leq \infty, Y \leq y\} = \{Y \leq y\}$$

so that

$$F_{XY}(x, \infty) = F_X(x) \tag{B.64}$$

$$F_{XY}(\infty, y) = F_Y(y) \tag{B.65}$$

The cdf's $F_X(x)$ and $F_Y(y)$, when obtained by Eqs. (B.64) and (B.65), are referred to as the *marginal* cdf's of X and Y, respectively.

C. Joint Probability Mass Functions:

Let (X, Y) be a discrete two-dimensional r.v. and (X, Y) takes on the values (x_i, y_j) for a certain allowable set of integers i and j. Let

$$p_{XY}(x_i, y_j) = P(X = x_i, Y = y_j) \tag{B.66}$$

The function $p_{XY}(x_i, y_j)$ is called the *joint probability mass function* (joint pmf) of (X, Y).

Properties of $p_{XY}(x_i, y_j)$:

1. $\quad 0 \leq p_{XY}(x_i, y_j) \leq 1$ $\qquad$ (B.67)

2. $\quad \displaystyle\sum_{x_i} \sum_{y_j} p_{XY}(x_i, y_j) = 1$ $\qquad$ (B.68)

The joint cdf of a discrete two-dimensional r.v. (X, Y) is given by

$$F_{XY}(x, y) = \sum_{x_i \leq x} \sum_{y_j \leq y} p_{XY}(x_i, y_j) \qquad \text{(B.69)}$$

D. Marginal Probability Mass Functions:

Suppose that for a fixed value $X = x_i$, the r.v. Y can only take on the possible values y_j $(j = 1, 2, \ldots, n)$.

Then $\qquad\qquad\qquad\qquad p_X(x_i) = \displaystyle\sum_{y_j} p_{XY}(x_i, y_j)$ $\qquad$ (B.70)

Similarly, $\qquad\qquad\qquad p_Y(y_j) = \displaystyle\sum_{x_j} p_{XY}(x_i, y_j)$ $\qquad$ (B.71)

The pmf's $p_X(x_i)$ and $p_Y(y_j)$, when obtained by Eqs. (B.70) and (B.71), are referred to as the *marginal* pmf's of X and Y, respectively. If X and Y are independent r.v.'s, then

$$p_{XY}(x_i, y_j) = p_X(x_i) p_Y(y_j) \qquad \text{(B.72)}$$

E. Joint Probability Density Functions:

Let (X, Y) be a continuous two-dimensional r.v. with cdf $F_{XY}(x, y)$ and let

$$f_{XY}(x, y) = \frac{\partial^2 F_{XY}(x, y)}{\partial x \partial y} \qquad \text{(B.73)}$$

The function $f_{XY}(x, y)$ is called the *joint probability density function* (joint pdf) of (X, Y). By integrating Eq. (B.73), we have

$$F_{XY}(x, y) = \int_{-\infty}^{x} \int_{-\infty}^{y} f_{XY}(\xi, \eta) \, d\xi \, d\eta \qquad \text{(B.74)}$$

Properties of $f_{XY}(x, y)$:

1. $\quad f_{XY}(x, y) \geq 0$ $\qquad$ (B.75)
2. $\quad \int_{-\infty}^{\infty} \int_{-\infty}^{\infty} f_{XY}(x, y) \, dx \, dy = 1$ $\qquad$ (B.76)

F. Marginal Probability Density Functions:

By Eqs. (B.64), (B.65), and definition (B.51), we obtain

$$f_X(x) = \int_{-\infty}^{\infty} f_{XY}(x, y) \, dy \qquad \text{(B.77)}$$

$$f_Y(y) = \int_{-\infty}^{\infty} f_{XY}(x, y) \, dx \qquad \text{(B.78)}$$

The pdf's $f_X(x)$ and $f_Y(x)$, when obtained by Eqs. (B.77) and (B.78), are referred to as the *marginal* pdf's of X and Y, respectively. If X and Y are independent r.v.'s, then

$$f_{XY}(x, y) = f_X(x) f_Y(y) \qquad \text{(B.79)}$$

The conditional pdf of X given the event $\{Y = y\}$ is

$$f_{X|Y}(x|y) = \frac{f_{XY}(x, y)}{f_Y(y)} \quad f_Y(y) \neq 0 \tag{B.80}$$

where $f_Y(y)$ is the marginal pdf of Y.

B.4   Functions of Random Variables

A.   Random Variable $g(X)$:

Given a r.v. X and a function $g(x)$, the expression

$$Y = g(X) \tag{B.81}$$

defines a new r.v. Y. With y a given number, we denote D_y the subset of R_X (range of X) such that $g(x) \leq y$. Then

$$(Y \leq y) = [g(X) \leq y] = (X \in D_y)$$

where $(X \in D_y)$ is the event consisting of all outcomes λ such that the point $X(\lambda) \in D_y$. Hence,

$$F_Y(y) = P(Y \leq y) = P[g(X) \leq y] = P(X \in D_y) \tag{B.82}$$

If X is a continuous r.v. with pdf $f_X(x)$, then

$$F_Y(y) = \int_{D_y} f_X(x)\, dx \tag{B.83}$$

Determination of $f_Y(y)$ from $f_X(x)$:

Let X be a continuous r.v. with pdf $f_X(x)$. If the transformation $y = g(x)$ is one-to-one and has the inverse transformation

$$x = g^{-1}(y) = h(y) \tag{B.84}$$

then the pdf of Y is given by

$$f_Y(y) = f_X(x)\left|\frac{dx}{dy}\right| = f_X[h(y)]\left|\frac{dh(y)}{dy}\right| \tag{B.85}$$

Note that if $g(x)$ is a continuous monotonic increasing or decreasing function, then the transformation $y = g(x)$ is one-to-one. If the transformation $y = g(x)$ is not one-to-one, $f_Y(y)$ is obtained as follows: Denoting the real roots of $y = g(x)$ by x_k, that is,

$$y = g(x_1) = \ldots = g(x_k) = \ldots$$

then

$$f_Y(y) = \sum_k \frac{f_X(x_k)}{|g'(x_k)|} \tag{B.86}$$

where $g'(x)$ is the derivative of $g(x)$.

EXAMPLE B.7   Let $Y = aX + b$. Show that if $X = N(\mu; \sigma^2)$, then $Y = N(a\mu + b; a^2\sigma^2)$.

The equation $y = g(x) = ax + b$ has a single solution $x_1 = (y - b)/a$, and $g'(x) = a$. The range of y is $(-\infty, \infty)$. Hence, by Eq. (B.86)

$$f_Y(y) = \frac{1}{|a|} f_X\left(\frac{y - b}{a}\right) \tag{B.87}$$

Since $X = N(\mu; \sigma^2)$, by Eq. (B.60)

$$f_X(x) = \frac{1}{\sqrt{2\pi}\sigma} \exp\left[-\frac{1}{2\sigma^2}(x - \mu)^2\right] \tag{B.88}$$

Hence, by Eq. (B.87)

$$f_Y(y) = \frac{1}{\sqrt{2\pi}|a|\sigma} \exp\left[-\frac{1}{2\sigma^2}\left(\frac{y - b}{a} - \mu\right)^2\right]$$

$$= \frac{1}{\sqrt{2\pi}|a|\sigma} \exp\left[-\frac{1}{2a^2\sigma^2}(y - a\mu - b)^2\right] \tag{B.89}$$

which is the pdf of $N(a\mu + b; a^2\sigma^2)$. Thus, if $X = N(\mu; \sigma^2)$, then $Y = N(a\mu + b; a^2\sigma^2)$.

EXAMPLE B.8   Let $Y = X^2$. Find $f_Y(y)$ if $X = N(0; 1)$.

If $y < 0$, then the equation $y = x^2$ has no real solutions; hence, $f_Y(y) = 0$.
If $y > 0$, then $y = x^2$ has two solutions

$$x_1 = \sqrt{y} \quad x_2 = -\sqrt{y}$$

Now, $y = g(x) = x^2$ and $g'(x) = 2x$. Hence, by Eq. (B.86)

$$f_Y(y) = \frac{1}{2\sqrt{y}}\left[f_X\left(\sqrt{y}\right) + f_X\left(-\sqrt{y}\right)\right]u(y) \tag{B.90}$$

Since $X = N(0; 1)$ from Eq. (B.60), we have

$$f_X(x) = \frac{1}{\sqrt{2\pi}} e^{-x^2/2} \tag{B.91}$$

Since $f_X(x)$ is an even function from Eq. (B.90), we have

$$f_Y(y) = \frac{1}{\sqrt{y}} f_X\left(\sqrt{y}\right)u(y) = \frac{1}{\sqrt{2\pi y}} e^{-y/2}u(y) \tag{B.92}$$

B.   One Function of Two Random Variables:

Given two random variables X and Y and a function $g(x, y)$, the expression

$$Z = g(X, Y) \tag{B.93}$$

is a new random variable. With z a given number, we denote by D_z the region of the xy plane such that $g(x, y) \leq z$. Then

$$[Z \leq z] = \{g(X, Y) \leq z\} = \{(X, Y) \in D_z\}$$

where $\{(X, Y) \in D_z\}$ is the event consisting of all outcomes λ such that the point $\{X(\lambda), Y(\lambda)\}$ is in D_z. Hence,

$$F_Z(z) = P(Z \leq z) = P\{(X, Y) \in D_z\} \tag{B.94}$$

If X and Y are continuous r.v.'s with joint pdf $f_{XY}(x, y)$, then

$$f_Z(z) = \int\int_{D_Z} f_{XY}(x, y)\, dx\, dy \tag{B.95}$$

EXAMPLE B.9   Consider two r.v.'s X and Y with joint pdf $f_{XY}(x, y)$. Let $Z = X + Y$.

    (*a*)   Determine the pdf of Z.

    (*b*)   Determine the pdf of Z if X and Y are independent.

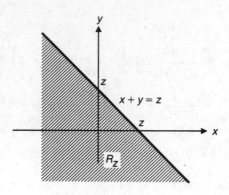

Fig. B-7

(*a*)   The range R_z of Z corresponding to the event $(Z \le z) = (X + Y \le z)$ is the set of points (x, y) which lie on and to the left of the line $z = x + y$ (Fig. B-7). Thus, we have

$$F_Z(z) = P(X + Y \le z) = \int_{-\infty}^{\infty} \left[\int_{-\infty}^{z-x} f_{XY}(x, y) dy \right] dx \qquad (B.96)$$

Then

$$f_Z(z) = \frac{d}{dz} F_Z(z) = \int_{-\infty}^{\infty} \left[\frac{d}{dz} \int_{-\infty}^{z-x} f_{XY}(x, y) dy \right] dx$$

$$= \int_{-\infty}^{\infty} f_{XY}(x, z - x) dx \qquad (B.97)$$

(*b*)   If X and Y are independent, then Eq. (B.97) reduces to

$$f_Z(z) = \int_{-\infty}^{\infty} f_X(x) f_Y(z - x) dx \qquad (B.98)$$

The integral on the right-hand side of Eq. (B.98) is known as a *convolution* of $f_X(z)$ and $f_Y(z)$. Since the convolution is commutative, Eq. (B.98) can also be written as

$$f_Z(z) = \int_{-\infty}^{\infty} f_Y(y) f_X(z - y) dy \qquad (B.99)$$

C.   Two Functions of Two Random Variables:

Given two r.v.'s. X and Y and two functions $g(x, y)$ and $h(x, y)$, the expression

$$Z = g(X, Y) \qquad W = h(X, Y) \qquad (B.100)$$

defines two new r.v.'s Z and W. With z and w two given numbers we denote D_{zw} the subset of R_{XY} [range of (X, Y)] such that $g(x, y) \le z$ and $h(x, y) \le w$. Then

$$(Z \le z, W \le w) = [g(x, y) \le z, h(x, y) \le w] = \{(X, Y) \in D_{zw}\}$$

where $\{(X, Y) \in D_{zw}\}$ is the event consisting of all outcomes λ such that the point $\{X(\lambda), Y(\lambda)\} \in D_{zw}$. Hence,

$$F_{ZW}(z, w) = P(Z \le z, W \le w) = P\{(X, Y) \in D_{zw}\} \qquad (B.101)$$

In the continuous case we have

$$f_{ZW}(z, w) = \int\limits_{D_{ZW}} \int f_{XY}(x, y)\, dx\, dy \tag{B.102}$$

Determination of $f_{ZW}(z, w)$ from $f_{XY}(x, y)$:
Let X and Y be two continuous r.v.'s with joint pdf $f_{XY}(x, y)$. If the transformation

$$z = g(x, y) \qquad w = h(x, y) \tag{B.103}$$

is one-to-one and has the inverse transformation

$$x = q(z, w) \qquad y = r(z, w) \tag{B.104}$$

then the joint pdf of Z and W is given by

$$f_{ZW}(z, w) = f_{XY}(x, y) \mid J(x, y) \mid^{-1} \tag{B.105}$$

where $x = q(z, w), y = r(z, w)$ and

$$J(x, y) = \begin{vmatrix} \dfrac{\partial g}{\partial x} & \dfrac{\partial g}{\partial y} \\[2mm] \dfrac{\partial h}{\partial x} & \dfrac{\partial h}{\partial y} \end{vmatrix} = \begin{vmatrix} \dfrac{\partial z}{\partial x} & \dfrac{\partial z}{\partial y} \\[2mm] \dfrac{\partial w}{\partial x} & \dfrac{\partial w}{\partial y} \end{vmatrix} \tag{B.106}$$

which is the Jacobian of the transformation (B.103).

EXAMPLE B.10 Consider the transformation

$$R = \sqrt{X^2 + Y^2} \qquad \Theta = \tan^{-1}\frac{Y}{X} \tag{B.107}$$

$$\bar{J}(x, y) = \begin{vmatrix} \dfrac{\partial x}{\partial r} & \dfrac{\partial x}{\partial \theta} \\[2mm] \dfrac{\partial y}{\partial r} & \dfrac{\partial y}{\partial \theta} \end{vmatrix} = \begin{vmatrix} \cos\theta & -r\sin 0 \\ \sin\theta & r\cos 0 \end{vmatrix} = r$$

Eq. (B.105) yields

$$f_{R\Theta}(r, \theta) = r f_{XY}(r\cos\theta, r\sin\theta) \tag{B.108}$$

B.5 Statistical Averages

A. Expectation:

The *expectation* (or *mean*) of a r.v. X, denoted by $E(X)$ or μ_X, is defined by

$$\mu_X = E(x) = \begin{cases} \displaystyle\sum_i x_i p_X(x_i) & X\text{: discrete} \\[4mm] \displaystyle\int_{-\infty}^{\infty} x f_X(x)\, dx & X\text{: continuous} \end{cases} \tag{B.109}$$

The expectation of $Y = g(X)$ is given by

$$E(Y) = E[g(X)] = \begin{cases} \displaystyle\sum_i g(x_i) p_X(x_i) & \text{(discrete case)} \\[4mm] \displaystyle\int_{-\infty}^{\infty} g(x) f_X(x)\, dx & \text{(continuous case)} \end{cases} \tag{B.110}$$

The expectation of $Z = g(X, Y)$ is given by

$$E(Z) = E[g(X, Y)] = \begin{cases} \sum_i \sum_j g(x_i, y_j) p_{XY}(x_i, y_j) & \text{(discrete case)} \\ \int_{-\infty}^{\infty} \int_{-\infty}^{\infty} g(x, y) f_{XY}(x, y) \, dx \, dy & \text{(continuous case)} \end{cases} \tag{B.111}$$

Note that the *expectation operation* is linear, that is,

$$E[X + Y] = E[X] + E[Y] \tag{B.112}$$

$$E[cX] = cE[X] \tag{B.113}$$

where c is a constant.

EXAMPLE B.11   If X and Y are independent, then show that

$$E[XY] = E[X]E[Y] \tag{B.114}$$

and
$$E[g_1(X)g_2(X)] = E[g_1(X)]E[g_2(Y)] \tag{B.115}$$

If X and Y are independent, then by Eqs. (B.79) and (B.111) we have

$$E[XY] = \int_{-\infty}^{\infty} \int_{-\infty}^{\infty} xy f_X(x) f_Y(y) \, dx \, dy$$

$$= \int_{-\infty}^{\infty} x f_X(x) \, dx \int_{-\infty}^{\infty} y f_Y(y) \, dy = E[X] E[Y]$$

Similarly,

$$E[g_1(X)g_2(Y)] = \int_{-\infty}^{\infty} \int_{-\infty}^{\infty} g_1(x) g_2(y) f_X(x) f_Y(y) \, dx \, dy$$

$$= \int_{-\infty}^{\infty} g_1(x) f_X(x) \, dx \int_{-\infty}^{\infty} g_2(y) f_Y(y) \, dy = E[g_1(X)E[g_2(Y)]$$

B. Moment:

The *n*th *moment* of a r.v. X is defined by

$$E(X^n) = \begin{cases} \sum_i x_i^n p_X(x_i) & X: \text{discrete} \\ \int_{-\infty}^{\infty} x^n f_X(x) \, dx & X: \text{continuous} \end{cases} \tag{B.116}$$

C. Variance:

The *variance* of a r.v. X, denoted by σ_X^2 or $\text{Var}(X)$, is defined by

$$\text{Var}(X) = \sigma_X^2 = E[(X - \mu_X)^2] \tag{B.117}$$

Thus,

$$\sigma_X^2 = \begin{cases} \sum_i (x_i - \mu_X)^2 p_X(x_i) & X: \text{discrete} \\ \int_{-\infty}^{\infty} (x - \mu_X)^2 f_X(x) \, dx & X: \text{continuous} \end{cases} \tag{B.118}$$

The positive square root of the variance, or σ_X, is called the *standard deviation* of X. The variance or standard variation is a measure of the "spread" of the values of X from its mean μ_X. By using Eqs. (B.112) and (B.113), the expression in Eq. (B.117) can be simplified to

$$\sigma_X^2 = E[X^2] - \mu_X^2 = E[X^2] - (E[X])^2 \tag{B.119}$$

Mean and variance of various random variables are tabulated in Table B-1.

TABLE B-1

RANDOM VARIABLE X	MEAN μ_X	VARIANCE σ_X^2
Bernoulli (p)	p	p(1-p)
Binomial (n, p)	np	np(1 − p)
Poisson (λ)	λ	λ
Uniform (a, b)	$\dfrac{a+b}{2}$	$\dfrac{(b-a)^2}{12}$
Exponential (λ)	$\dfrac{1}{\lambda}$	$\dfrac{1}{\lambda^2}$
Gaussion (normal)	μ	σ^2

D. Covariance and Correlation Coefficient:

The (k, n)th moment of a two-dimensional r.v. (X, Y) is defined by

$$m_{kn} = E(X^k Y^n) = \begin{cases} \displaystyle\sum_{y_j}\sum_{x_i} x_i^k y_j^n p_{XY}(x_i, y_j) & X:\ \text{discrete} \\[2mm] \displaystyle\int_{-\infty}^{\infty}\int_{-\infty}^{\infty} x^k y^n f_{XY}(x, y)\,dx\,dy & X:\ \text{continuous} \end{cases} \tag{B.120}$$

The $(1, 1)$th joint moment of (X, Y),

$$m_{11} = E(XY) \tag{B.121}$$

is called the *correlation* of X and Y. If $E(XY) = 0$, then we say that X and Y are *orthogonal*. The *covariance* of X and Y, denoted by $\mathrm{Cov}(X, Y)$ or σ_{XY}, is defined by

$$\mathrm{Cov}(X, Y) = \sigma_{XY} = E[(X - \mu_X)(Y - \mu_Y)] \tag{B.122}$$

Expanding Eq. (B.122), we obtain

$$\mathrm{Cov}(X, Y) = E(XY) - E(X)E(Y) \tag{B.123}$$

If $\mathrm{Cov}(X, Y) = 0$, then we say that X and Y are *uncorrelated*. From Eq. (B.123) we see that X and Y are uncorrelated if

$$E(XY) = E(X)E(Y) \tag{B.124}$$

Note that if X and Y are independent, then it can be shown that they are uncorrelated. However, the converse is not true in general; that is, the fact that X and Y are uncorrelated does not, in general, imply that they are independent. The *correlation coefficient*, denoted by $\rho(X, Y)$ or ρ_{XY}, is defined by

$$\rho(X, Y) = \rho_{XY} = \frac{\sigma_{XY}}{\sigma_X \sigma_Y} \tag{B.125}$$

It can be shown that (Example B.15)

$$|\rho_{XY}| \le 1 \qquad \text{or} \qquad -1 \le \rho_{XY} \le 1 \tag{B.126}$$

E.   Some Inequalities:

1.   Markov Inequality:
If $f_X(x) = 0$ for $x < 0$, then for any $\alpha > 0$,

$$P(X \ge \alpha) \le \frac{\mu_X}{\alpha} \tag{B.127}$$

2.   Chebyshev Inequality:
For any $\epsilon > 0$, then

$$P(|X - \mu_X| \ge \epsilon) \le \frac{\sigma_X^2}{\epsilon^2} \tag{B.128}$$

where $\mu_X = E[X]$ and σ_X^2 is the variance of X. This is known as the *Chebyshev inequality*.

3.   Cauchy-Schwarz Inequality:
Let X and Y be real random variables with finite second moments. Then

$$(E[XY])^2 \le E[X^2]\, E[Y^2] \tag{B.129}$$

This is known as the *Cauchy-Schwarz inequality*.

EXAMPLE B.12   Verify Markov inequality, Eq. (B.127).

From Eq. (B.54)

$$P(X \ge \alpha) = \int_\alpha^\infty f_X(x)\, dx$$

Since $f_X(x) = 0$ for $x < 0$,

$$\mu_X = E[X] = \int_0^\infty x f_X(x)\, dx \ge \int_\alpha^\infty x f_X(x)\, dx \ge \alpha \int_\alpha^\infty f_X(x)\, dx$$

Hence, $$\int_\alpha^\infty f_X(x)\, dx = P(X \ge \alpha) \le \frac{\mu_X}{\alpha}$$

EXAMPLE B.13   Verify Chebyshev inequality, Eq. (B.128).

From Eq. (B.54)

$$P(|X - \mu_X| \ge \epsilon) = \int_{-\infty}^{\mu_X - \epsilon} f_X(x)\, dx + \int_{\mu_X + \epsilon}^\infty f_X(x)\, dx = \int_{|x - \mu_X| \ge \epsilon} f_X(x)\, dx$$

By Eq. (B.118)

$$\sigma_X^2 = \int_{-\infty}^\infty (x - \mu_X)^2 f_X(x)\, dx \ge \int_{|x - \mu_X| \ge \epsilon} (x - \mu_X)^2 f_X(x)\, dx \ge \epsilon^2 \int_{|x - \mu_X| \ge \epsilon} f_X(x)\, dx$$

Hence,

$$\int_{|x - \mu_X| \ge \epsilon} f_X(x)\, dx \ge \frac{\sigma_X^2}{\epsilon^2}$$

or

$$P(|X - \mu_X| \geq \epsilon) \leq \frac{\sigma_X^2}{\epsilon^2}$$

EXAMPLE B.14 Verify Cauchy-Schwarz inequality Eq. (B.129).

Because the mean-square value of a random variable can never be negative,

$$E[(X - \alpha Y)^2] \geq 0$$

for any value of α. Expanding this, we obtain

$$E[X^2] - 2\alpha E[XY] + \alpha^2 E[Y^2] \geq 0$$

Choose a value of α for which the left-hand side of this inequality is minimum

$$\alpha = \frac{E[XY]}{E[Y^2]}$$

which results in the inequality

$$E[X^2] - \frac{(E[XY])^2}{E[Y^2]} \geq 0$$

or

$$(E[XY])^2 \leq E[X^2] \, E[Y^2]$$

EXAMPLE B.15 Verify Eq. (B.126).

From the Cauchy-Schwarz inequality Eq. (B.129) we have

$$\{E[(X - \mu_X)(Y - \mu_Y)]\}^2 \leq E[(X - \mu_X)^2] E[(Y - \mu_Y)^2]$$

or

$$\sigma_{XY}^2 \leq \sigma_X^2 \sigma_Y^2$$

Then

$$\rho_{XY}^2 = \frac{\sigma_{XY}^2}{\sigma_X^2 \sigma_Y^2} \leq 1$$

from which it follows that

$$|\rho_{XY}| \leq 1$$

APPENDIX C

Properties of Linear Time-Invariant Systems and Various Transforms

C.1 Continuous-Time LTI Systems

Unit impulse response: $h(t)$

Convolution: $y(t) = x(t) * h(t) = \int_{-\infty}^{\infty} x(\tau)h(t - \tau)\,d\tau$

Causality: $h(t) = 0, t < 0$

Stability: $\int_{-\infty}^{\infty} |h(t)|\, dt < \infty$

C.2 The Laplace Transform

The Bilateral (or Two-Sided) Laplace Transform:

Definition:

$$x(t) \xleftrightarrow{\mathscr{L}} X(s)$$

$$X(s) = \int_{-\infty}^{\infty} x(t)e^{-st}\, dt$$

$$x(t) = \frac{1}{2\pi j}\int_{c-j\infty}^{c+j\infty} X(s)e^{st}\, ds$$

Properties of the Bilateral Laplace Transform:

Linearity: $a_1 x_1(t) + a_2 x_2(t) \leftrightarrow a_1 X_1(s) + a_2 X_2(s), R' \supset R_1 \cap R_2$

Time shifting: $x(t - t_0) \leftrightarrow e^{-st_0} X(s), R' = R$

Shifting in s: $e^{s_0 t} x(t) \leftrightarrow X(s - s_0), R' = R + \text{Re}(s_0)$

Time scaling: $x(at) \leftrightarrow \dfrac{1}{|a|} X(s), R' = aR$

Time reversal: $x(-t) \leftrightarrow X(-s), R' = -R$

Differentiation in t: $\dfrac{dx(t)}{dt} \leftrightarrow sX(s), R' \supset R$

Differentiation in s: $-tx(t) \leftrightarrow \dfrac{dX(s)}{ds}, R' = R$

Integration: $\displaystyle\int_{-\infty}^{t} x(\tau)\,d\tau \leftrightarrow \dfrac{1}{s}X(s), R' \supset R \cap \{\operatorname{Re}(s) > 0\}$

Convolution: $x_1(t) * x_2(t) \leftrightarrow X_1(s)X_2(s), R' \supset R_1 \cap R_2$

Some Laplace Transforms Pairs:

$\delta(t) \leftrightarrow 1, \text{all } s$

$u(t) \leftrightarrow \dfrac{1}{s}, \operatorname{Re}(s) > 0$

$-u(-t) \leftrightarrow \dfrac{1}{s}, \operatorname{Re}(s) < 0$

$tu(t) \leftrightarrow \dfrac{1}{s^2}, \operatorname{Re}(s) > 0$

$t^k u(t) \leftrightarrow \dfrac{k!}{s^{k+1}}, \operatorname{Re}(s) > 0$

$e^{-at}u(t) \leftrightarrow \dfrac{1}{s+a}, \operatorname{Re}(s) > -\operatorname{Re}(a)$

$-e^{-at}u(-t) \leftrightarrow \dfrac{1}{s+a}, \operatorname{Re}(s) < -\operatorname{Re}(a)$

$te^{-at}u(t) \leftrightarrow \dfrac{1}{(s+a)^2}, \operatorname{Re}(s) > -\operatorname{Re}(a)$

$-te^{-at}u(-t) \leftrightarrow \dfrac{1}{(s+a)^2}, \operatorname{Re}(s) < -\operatorname{Re}(a)$

$\cos \omega_0 t u(t) \leftrightarrow \dfrac{s}{s^2 + \omega_0^2}, \operatorname{Re}(s) > 0$

$\sin \omega_0 t u(t) \leftrightarrow \dfrac{\omega_0}{s^2 + \omega_0^2}, \operatorname{Re}(s) > 0$

$e^{-at}\cos \omega_0 t u(t) \leftrightarrow \dfrac{s+a}{(s+a)^2 + \omega_0^2}, \operatorname{Re}(s) > -\operatorname{Re}(a)$

$e^{-at}\sin \omega_0 t u(t) \leftrightarrow \dfrac{\omega_0}{(s+a)^2 + \omega_0^2}, \operatorname{Re}(s) > -\operatorname{Re}(a)$

The Unilateral (or One-Sided) Laplace Transform:

Definition:

$$x(t) \xleftarrow{\;\mathscr{L}_I\;} X_I(s)$$

$$X_I(s) = \int_{0^-}^{\infty} x(t)e^{-st}\,dt \qquad 0^- = \lim_{\varepsilon \to \infty}(0 - \varepsilon)$$

Some Special Properties:

Differentiation in the Time Domain:

$$\frac{dx(t)}{dt} \leftrightarrow sX_I(s) - x(0^-)$$

$$\frac{d^2 x(t)}{dt^2} \leftrightarrow s^2 X_I(s) - sx(0^-) - x'(0^-)$$

$$\frac{d^n x(t)}{dt^n} \leftrightarrow s^n X_I(s) - s^{n-1}x(0^-) - s^{n-2}x'(0^-) - \cdots - x^{(n-1)}(0^-)$$

Integration in the Time Domain:

$$\int_{0^-}^{t} x(\tau)\, d\tau \leftrightarrow \frac{1}{s} X_I(s)$$

$$\int_{-\infty}^{t} x(\tau)\, d\tau \leftrightarrow \frac{1}{s} X_I(s) + \frac{1}{s}\int_{-\infty}^{0^-} x(\tau)\, d\tau$$

Initial value theorem : $x(0^+) = \lim_{s \to \infty} sX_I(s)$

Final value theorem : $\lim_{t \to \infty} x(t) = \lim_{s \to 0} sX_I(s)$

C.3　The Fourier Transform

Definition:

$$x(t) \xleftrightarrow{\ \mathscr{F}\ } X(\omega)$$

$$X(\omega) = \int_{-\infty}^{\infty} x(t)e^{-j\omega t}\, dt$$

$$x(t) = \frac{1}{2\pi}\int_{-\infty}^{\infty} X(\omega)e^{j\omega t}\, d\omega$$

Properties of the Fourier Transform:

Linearity: $a_1 x_1(t) + a_2 x_2(t) \leftrightarrow a_1 X_1(\omega) + a_2 X_2(\omega)$

Time shifting: $x(t - t_0) \leftrightarrow e^{-j\omega t_0} X(\omega)$

Frequency shifting: $e^{j\omega_0 t} x(t) \leftrightarrow X(\omega - \omega_0)$

Time scaling: $x(at) \leftrightarrow \dfrac{1}{|a|} X\left(\dfrac{\omega}{a}\right)$

Time reversal: $x(-t) \leftrightarrow X(-\omega)$

Duality: $X(t) \leftrightarrow 2\pi x(-\omega)$

Time differentiation: $\dfrac{dx(t)}{dt} \leftrightarrow j\omega X(\omega)$

Frequency differentiation: $(-jt)x(t) \leftrightarrow \dfrac{dX(\omega)}{d\omega}$

Integraton: $\displaystyle\int_{-\infty}^{t} x(\tau)\, d\tau \leftrightarrow \pi X(0)\, \delta(\omega) + \dfrac{1}{j\omega} X(\omega)$

Convolution: $x_1(t) * x_2(t) \leftrightarrow X_1(\omega) X_2(\omega)$

Multiplication: $x_1(t)x_2(t) \leftrightarrow \dfrac{1}{2\pi} X_1(\omega) * X_2(\omega)$

Real signal: $x(\mathrm{t}) = x_e(t) + x_o(t) \leftrightarrow X(\omega) = A(\omega) + jB(\omega)$

$$X(-\omega) = X^*(\omega)$$

Even component: $x_e(t) \leftrightarrow \mathrm{Re}\{X(\omega)\} = A(\omega)$

Odd component: $x_o(t) \leftrightarrow j\,\mathrm{Im}\{X(\omega)\} = jB(\omega)$

Parseval's Relations:

$$\int_{-\infty}^{\infty} x_1(\lambda) X_2(\lambda)\, d\lambda = \int_{-\infty}^{\infty} X_1(\lambda) x_2(\lambda)\, d\lambda$$

$$\int_{-\infty}^{\infty} x_1(t) x_2(t)\, dt = \frac{1}{2\pi}\int_{-\infty}^{\infty} X_1(\omega) X_2(-\omega)\, d\omega$$

$$\int_{-\infty}^{\infty} |x(t)|^2\, dt = \frac{1}{2\pi}\int_{-\infty}^{\infty} |X(\omega)|^2\, d\omega$$

Common Fourier Transforms Pairs:

$\delta(t) \leftrightarrow 1$

$\delta(t - t_0) \leftrightarrow e^{-j\omega t_0}$

$1 \leftrightarrow 2\pi\delta(\omega)$

$e^{j\omega_0 t} \leftrightarrow 2\pi\delta(\omega - \omega_0)$

$\cos \omega_0 t \leftrightarrow \pi[\delta(\omega - \omega_0) + \delta(\omega + \omega_0)]$

$\sin \omega_0 t \leftrightarrow j\pi[\delta(\omega - \omega_0) - \delta(\omega + \omega_0)]$

$u(t) \leftrightarrow \pi\delta(\omega) + \dfrac{1}{j\omega}$

$u(-t) \leftrightarrow \pi\delta(\omega) - \dfrac{1}{j\omega}$

$e^{-at}u(t) \leftrightarrow \dfrac{1}{j\omega + a}, a > 0$

$te^{-at}u(t) \leftrightarrow \dfrac{1}{(j\omega + a)^2}, \ a > 0$

$e^{-a|t|} \leftrightarrow \dfrac{2a}{a^2 + \omega^2}, \ a > 0$

$\dfrac{1}{a^2 + t^2} \leftrightarrow e^{-a|\omega|}$

$e^{-at^2} \leftrightarrow \sqrt{\dfrac{\pi}{a}}\, e^{-\omega^2/4a}, \ a > 0$

$p_a(t) = \begin{cases} 1 & |t| < a \\ 0 & |t| > a \end{cases} \leftrightarrow 2a\dfrac{\sin \omega a}{\omega a}$

$\dfrac{\sin at}{\pi t} \leftrightarrow p_a(\omega) = \begin{cases} 1 & |\omega| < a \\ 0 & |\omega| > a \end{cases}$

$\operatorname{sgn} t \leftrightarrow \dfrac{2}{j\omega}$

$\displaystyle\sum_{k=-\infty}^{\infty} \delta(t - kT) \leftrightarrow \omega_0 \sum_{k=-\infty}^{\infty} \delta(\omega - k\omega_0), \ \omega_0 = \dfrac{2\pi}{T}$

C.4 Discrete-Time LTI Systems

Unit sample response: $h[n]$

Convolution: $y[n] = x[n] * h[n] = \displaystyle\sum_{k=-\infty}^{\infty} x[k]h[n - k]$

Causality: $h[n] = 0, \ n < 0$

Stability: $\displaystyle\sum_{n=-\infty}^{\infty} |h[n]| < \infty$

C.5   The *z*-Transform

The Bilateral (or Two-Sided) *z*-Transform:

Definition:

$$x[n] \xleftrightarrow{\ \ 3\ \ } X(z)$$

$$X(z) = \sum_{n=-\infty}^{\infty} x[n] z^{-n}$$

$$x[n] = \frac{1}{2\pi j} \oint_c X(z) z^{n-1} \, dz$$

Properties of the z-Transform:

Lineartity: $a_1 x_1[n] + a_2 x_2[n] \leftrightarrow a_1 X_1(z) + a_2 X_2(z), \ R' \supset R_1 \cap R_2$

Time shifting: $x[n - n_0] \leftrightarrow z^{-n_0} X(z), \ R' \supset R_1 \cap \{ 0 < |z| < \infty \}$

Multiplication by z_0^n: $z_0^n x[n] \leftrightarrow X\left(\dfrac{z}{z_0}\right), \ R' = |z_0| R$

Multiplication by $e^{j\Omega_0 N}$: $e^{j\Omega_0 n} X[n] \leftrightarrow X(e^{-j\Omega_0} z), \ R' = R$

Time reversal: $x[-n] \leftrightarrow X\left(\dfrac{1}{z}\right), \ R' = \dfrac{1}{R}$

Multiplication by n: $nx[n] \leftrightarrow -z \dfrac{dX(z)}{dz}, \ R' = R$

Accumulation: $\displaystyle\sum_{k=-\infty}^{n} x[n] \leftrightarrow \dfrac{1}{1 - z^{-1}} X(z), \ R' \supset R \cap \{ |z| > 1 \}$

Convolution: $x_1[n] * x_2[n] \leftrightarrow X_1(z) X_2(z), \ R' \supset R_1 \cap R_2$

Some Common z-Transforms Pairs:

$\delta[n] \leftrightarrow 1, \ \text{all } z$

$u[n] \leftrightarrow \dfrac{1}{1 - z^{-1}} = \dfrac{z}{z - 1}, |z| > 1$

$-u[-n-1] \leftrightarrow \dfrac{1}{1 - z^{-1}} = \dfrac{z}{z-1}, \ |z| < 1$

$\delta[n - m] \leftrightarrow z^{-m}, \ \text{all } z \text{ except } 0 \text{ if } m > 0, \text{ or } \infty \text{ if } m < 0$

$a^n u[n] \leftrightarrow \dfrac{1}{1 - az^{-1}} = \dfrac{z}{z - a}, \ |z| > |a|$

$-a^n u[-n-1] \leftrightarrow \dfrac{1}{1 - az^{-1}} = \dfrac{z}{z - a}, \ |z| < |a|$

$na^n u[n] \leftrightarrow \dfrac{az^{-1}}{(1 - az^{-1})^2} = \dfrac{az}{(z - a)^2}, \ |z| > |a|$

$-na^n u[-n-1] \leftrightarrow \dfrac{az^{-1}}{(1 - az^{-1})^2} = \dfrac{az}{(z - a)^2}, \ |z| < |a|$

$(n+1)a^n u[n] \leftrightarrow \dfrac{1}{(1 - az^{-1})^2} = \left[\dfrac{z}{z - a}\right]^2, \ |z| > |a|$

$(\cos \Omega_0 n)u[n] \leftrightarrow \dfrac{z^2 - (\cos \Omega_0) z}{z^2 - (2\cos \Omega_0) z + 1}, \ |z| > 1$

$$(\sin \Omega_0 n)u[n] \leftrightarrow \frac{(\sin \Omega_0)\,z}{z^2 - (2\cos \Omega_0)\,z + 1}, \quad |z| > 1$$

$$(r^n \cos \Omega_0 n)u[n] \leftrightarrow \frac{z^2 - (r\cos \Omega_0)\,z}{z^2 - (2r\cos \Omega_0)\,z + r^2}, \quad |z| > r$$

$$(r^n \sin \Omega_0 n)u[n] \leftrightarrow \frac{(r\sin \Omega_0)\,z}{z^2 - (2r\cos \Omega_0)\,z + r^2}, \quad |z| > r$$

$$\begin{cases} a^n & 0 \le n \le N-1 \\ 0 & \text{otherwise} \end{cases} \leftrightarrow \frac{1 - a^N z^{-N}}{1 - az^{-1}}, \quad |z| > 0$$

The Unilateral (or One-Sided) z-Transform:

$$x[n] \xleftrightarrow{\ \Im_I\ } X_I(z)$$

$$X_I(z) = \sum_{n=0}^{\infty} x[n]\,z^{-n}$$

Some Special Properties:

Time-Shifting Property:

$$x[n-m] \leftrightarrow z^{-m}X_I(z) + z^{-m+1}x[-1] + z^{-m+2}x[-2] + \cdots + x[-m]$$

$$x[n+m] \leftrightarrow z^m X_I(z) - z^m x[0] - z^{m-1}x[1] - \cdots - zx[m-1]$$

Initial value theorem: $x[0] = \lim_{z \to \infty} X(z)$

Final value theorem: $\lim_{N \to \infty} x[N] = \lim_{z \to 1}(1 - z^{-1})X(z)$

C.6 The Discrete-Time Fourier Transform

Definition:

$$x[n] \xleftrightarrow{\ \mathscr{F}\ } X(\Omega)$$

$$X(\Omega) = \sum_{n=-\infty}^{\infty} x[n]e^{-j\Omega n}$$

$$x[n] = \frac{1}{2\pi}\int_{2\pi} X(\Omega)e^{j\Omega n}\,d\Omega$$

Properties of the Discrete-Time Fourier Transform:

Periodicity: $x[n] \leftrightarrow X(\Omega) = X(\Omega + 2\pi)$

Linearity: $a_1 x_1[n] + a_2 x_2[n] \leftrightarrow a_1 X_1(\Omega) + a_2 X_2(\Omega)$

Time shifting: $x[n - n_0] \leftrightarrow e^{-j\Omega n_0} X(\Omega)$

Frequency shifting: $e^{j\Omega_0 n}x[n] \leftrightarrow X(\Omega - \Omega_0)$

Conjugation: $x^*[n] \leftrightarrow X^*(-\Omega)$

Time reversal: $x[-n] \leftrightarrow X(-\Omega)$

Time scaling: $x_{(m)}[n] = \begin{cases} x[n/m] & \text{if } n = km \\ 0 & \text{if } n \ne km \end{cases} \leftrightarrow X(m\Omega)$

Frequency differentiation: $nx[n] \leftrightarrow j\dfrac{dX(\Omega)}{d(\Omega)}$

First difference: $x[n] - x[n-1] \leftrightarrow (1 - e^{-j\Omega})X(\Omega)$

Accumulation: $\displaystyle\sum_{k=-\infty}^{n} x[k] \leftrightarrow \pi X(0)\,\delta(\Omega) + \dfrac{1}{1 - e^{-j\Omega}}X(\Omega)$

Convolution: $x_1[n] * x_2[n] \leftrightarrow X_1(\Omega)\,X_2(\Omega)$

Multiplication: $x_1[n]x_2[n] \leftrightarrow \dfrac{1}{2\pi}X_1(\Omega) \otimes X_2(\Omega)$

Real sequence: $x[n] = x_e[n] + x_o[n] \leftrightarrow X(\Omega) = A(\Omega) + jB(\Omega)$

$$X(-\Omega) = X^*(\Omega)$$

Even component: $x_e[n] \leftrightarrow \mathrm{Re}\{x(\Omega)\} = A(\Omega)$

Odd component: $x_o[n] \leftrightarrow j\,\mathrm{Im}\{X(\Omega)\} = jB(\Omega)$

Parseval's Relations:

$$\sum_{n=-\infty}^{\infty} x_1[n]x_2[n] = \frac{1}{2\pi}\int_{2\pi} X_1(\Omega)X_2(-\Omega)\,d\Omega$$

$$\sum_{n=-\infty}^{\infty} |x[n]|^2 = \frac{1}{2\pi}\int_{2\pi} |X(\Omega)|^2\,d\Omega$$

Some Common Fourier Transform Pairs:

$\delta[n] \leftrightarrow 1$

$\delta[n - n_0] \leftrightarrow e^{-j\Omega n_0}$

$x[n] = 1 \leftrightarrow 2\pi\,\delta(\Omega),\ |\Omega| \le \pi$

$e^{-j\Omega_0 n} \leftrightarrow 2\pi\,\delta(\Omega - \Omega_0),\ |\Omega|,|\Omega_0| \le \pi$

$\cos\Omega_0 n \leftrightarrow \pi[\delta(\Omega - \Omega_0) + \delta(\Omega + \Omega_0)],\ |\Omega|,|\Omega_0| \le \pi$

$\sin\Omega_0 n \leftrightarrow -j\pi[\delta(\Omega - \Omega_0) - \delta(\Omega + \Omega_0)],\ |\Omega|,|\Omega_0| \le \pi$

$u[n] \leftrightarrow \pi\,\delta(\Omega) + \dfrac{1}{1 - e^{-j\Omega}},\ |\Omega_0| \le \pi$

$-u[-n-1] \leftrightarrow -\pi\,\delta(\Omega) + \dfrac{1}{1 - e^{-j\Omega}},\ |\Omega| \le \pi$

$a^n u[n] \leftrightarrow \dfrac{1}{1 - ae^{-j\Omega}},\ |a| < 1$

$-a^n u[-n-1] \leftrightarrow \dfrac{1}{1 - ae^{-j\Omega}},\ |a| > 1$

$(n+1)a^n u[n] \leftrightarrow \dfrac{1}{(1 - ae^{-j\Omega})^2},\ |a| < 1$

$a^{|n|} \leftrightarrow \dfrac{1 - a^2}{1 - 2a\cos\Omega + a^2},\ |a| < 1$

$x[n] = \begin{cases} 1 & |n| \le N_1 \\ 0 & |n| > N_1 \end{cases} \leftrightarrow \dfrac{\sin\!\left[\Omega\!\left(N_1 + \dfrac{1}{2}\right)\right]}{\sin(\Omega/2)}$

$\dfrac{\sin W_n}{\pi n}\,(0 < W < \pi) \leftrightarrow X(\Omega) = \begin{cases} 1 & 0 \le |\Omega| \le W \\ 0 & W < |\Omega| \le \pi \end{cases}$

$\displaystyle\sum_{k=-\infty}^{\infty} \delta[n - kN_0] \leftrightarrow \Omega_0 \sum_{k=-\infty}^{\infty} \delta(\Omega - k\Omega_0),\ \Omega_0 = \dfrac{2\pi}{N_0}$

C.7 Discrete Fourier Transform

Definition:

$$x[n] = 0 \qquad \text{outside the range } 0 \le n \le N - 1$$

$$x[n] \xleftrightarrow{\text{DFT}} X[k]$$

$$X[k] = \sum_{n=0}^{N-1} x[n] W_N^{kn} \qquad k = 0, 1, \ldots, N-1 \qquad W_N = e^{-j(2\pi/N)}$$

$$x[n] = \frac{1}{N} \sum_{n=0}^{N-1} X[k] W_N^{-kn} \qquad n = 0, 1, \ldots, N-1$$

Properties of the DFT:

Linearity: $a_1 x_1[n] + a_2 x_2[n] \leftrightarrow a_1 X_1[k] + a_2 X_2[k]$

Time shifting: $x[n - n_0]_{\text{mod } N} \leftrightarrow W_N^{kn_0} X[k]$

Frequency shifting: $W_N^{-kn_0} x[n] \leftrightarrow X[k - k_0]_{\text{mod } N}$

Conjugation: $x^*[n] \leftrightarrow X^*[-k]_{\text{mod } N}$

Time reversal: $x[-n]_{\text{mod } N} \leftrightarrow X[-k]_{\text{mod } N}$

Duality: $X[n] \leftrightarrow Nx[-k]_{\text{mod } N}$

Circular convolution: $x_1[n] \otimes x_2[n] \leftrightarrow X_1[k] X_2[k]$

Multiplication: $x_1[n] x_2[n] \leftrightarrow \frac{1}{N} X_1[k] \otimes X_2[k]$

Real sequence: $x[n] = x_e[n] + x_o[n] \leftrightarrow X[k] = A[k] + jB[k]$

$$X[-k]_{\text{mod } N} = X^*[k]$$

Even component: $x_e[n] \leftrightarrow \text{Re}\{X[k]\} = A[k]$

Odd component: $x_o[n] \leftrightarrow j \, \text{Im}\{X[k]\} = jB[k]$

Parseval's Relation:

$$\sum_{n=0}^{N-1} |x[n]|^2 = \frac{1}{N} \sum_{n=0}^{N-1} |X[k]|^2$$

Note

$$x_1[n] \otimes x_2[n] = \sum_{i=0}^{N-1} x_1[i] x_2[n - i]_{\text{mod } N}$$

C.8 Fourier Series

$$x(t + T_0) = x(t)$$

Complex Exponential Fourier Series:

$$x(t) = \sum_{k=-\infty}^{\infty} c_k e^{jk\omega_0 t} \qquad \omega_0 = \frac{2\pi}{T_0}$$

$$c_k = \frac{1}{T_0} \int_{-T_0/2}^{T_0/2} x(t) e^{-jk\omega_0 t} \, dt$$

Trigonometric Fourier Series:

$$x(t) = \frac{a_0}{2} + \sum_{k=1}^{\infty} (a_k \cos k\omega_0 t + b_k \sin k\omega_0 t)$$

$$a_k = \frac{2}{T_0} \int_{-T_0/2}^{T_0/2} x(t) \cos k\omega_0 t \, dt$$

$$b_k = \frac{2}{T_0} \int_{-T_0/2}^{T_0/2} x(t) \sin k\omega_0 t \, dt$$

Harmonic Form Fourier Series:

$$x(t) = C_0 + \sum_{k=1}^{\infty} C_k \cos(k\omega_0 t - \theta_k) \qquad \omega_0 = \frac{2\pi}{T_0}$$

Relations among Various Fourier Coefficients:

$$\frac{a_0}{2} = c_0 \qquad a_k = c_k + c_{-k} \qquad b_k = j(c_k - c_{-k})$$

$$c_k = \frac{1}{2}(a_k - jb_k) \qquad c_{-k} = \frac{1}{2}(a_k + jb_k)$$

$$C_0 = \frac{a_0}{2} \qquad C_k = \sqrt{a_k^2 + b_k^2} \qquad \theta_k = \tan^{-1} \frac{b_k}{a_k}$$

Parseval's Theorem for Fourier Series:

$$\frac{1}{T_0} \int_{T_0} |x(t)|^2 \, dt = \sum_{k=-\infty}^{\infty} |c_k|^2$$

C.9   Discrete Fourier Series

$$x[n + N_0] = x[n]$$

$$x[n] = \sum_{k=0}^{N_0-1} c_k e^{jk\Omega_0 n} \qquad \Omega_0 = \frac{2\pi}{N_0}$$

$$c_k = \frac{1}{N_0} \sum_{n=0}^{N_0-1} x[n] e^{-jk\Omega_0 n}$$

Parseval's Theorem for Discrete Fourier Series:

$$\frac{1}{N_0} \sum_{n=0}^{N_0-1} |x[n]|^2 = \sum_{k=0}^{N_0-1} |c_k|^2$$

APPENDIX D

Review of Complex Numbers

D.1 Representation of Complex Numbers

The complex number z can be expressed in several ways.
Cartesian or rectangular form:

$$z = a + jb \tag{D.1}$$

where $j = \sqrt{-1}$ and a and b are real numbers referred to the *real part* and the *imaginary part* of z. a and b are often expressed as

$$a = \mathrm{Re}\{z\} \qquad b = \mathrm{Im}\{z\} \tag{D.2}$$

where "Re" denotes the "real part of" and "Im" denotes the "imaginary part of."
Polar form:

$$z = re^{j\theta} \tag{D.3}$$

where $r > 0$ is the *magnitude* of z and θ is the *angle* or *phase* of z. These quantities are often written as

$$r = |z| \qquad \theta = \angle z \tag{D.4}$$

Fig. D-1 is the graphical representation of z. Using Euler's formula,

$$e^{j\theta} = \cos\theta + j\sin\theta \tag{D.5}$$

or from Fig. D-1 the relationships between the Cartesian and polar representations of z are

$$a = r\cos\theta \qquad b = r\sin\theta \tag{D.6a}$$

$$r = \sqrt{a^2 + b^2} \qquad \theta = \tan^{-1}\frac{b}{a} \tag{D.6b}$$

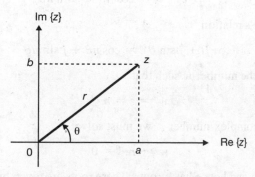

Fig. D-1

D.2   Addition, Multiplication, and Division

If $z_1 = a_1 + jb_1$ and $z_2 = a_2 + jb_2$, then

$$z_1 + z_2 = (a_1 + a_2) + j(b_1 + b_2) \tag{D.7}$$

$$z_1 z_2 = (a_1 a_2 - b_1 b_2) + j(a_1 b_2 + b_1 a_2) \tag{D.8}$$

$$\frac{z_1}{z_2} = \frac{a_1 + jb_1}{a_2 + jb_2} = \frac{(a_1 + jb_1)(a_2 - jb_2)}{(a_2 + jb_2)(a_2 - jb_2)}$$

$$= \frac{(a_1 a_2 + b_1 b_2) + j(-a_1 b_2 + b_1 a_2)}{a_2^2 + b_2^2} \tag{D.9}$$

If $z_1 = r_1 e^{j\theta_1}$ and $z_2 = r_2 e^{j\theta_2}$, then

$$z_1 z_2 = (r_1 r_2) e^{j(\theta_1 + \theta_2)} \tag{D.10}$$

$$\frac{z_1}{z_2} = \left(\frac{r_1}{r_2}\right) e^{j(\theta_1 - \theta_2)} \tag{D.11}$$

D.3   The Complex Conjugate

The *complex conjugate* of z is denoted by z^* and is given by

$$z^* = a - jb = re^{-j\theta} \tag{D.12}$$

Useful relationships:

1. $zz^* = r^2$
2. $\dfrac{z}{z^*} = e^{j2\theta}$
3. $z + z^* = 2\,\mathrm{Re}\{z\}$
4. $z - z^* = j2\,\mathrm{Im}\{z\}$
5. $(z_1 + z_2)^* = z_1^* + z_2^*$
6. $(z_1 z_2)^* = z_1^* z_2^*$
7. $\left(\dfrac{z_1}{z_2}\right)^* = \dfrac{z_1^*}{z_2^*}$

D.4   Powers and Roots of Complex Numbers

The nth power of the complex number $z = re^{j\theta}$ is

$$z^n = r^n e^{jn\theta} = r^n(\cos n\theta + j \sin n\theta) \tag{D.13}$$

from which we have De Moivre's relation

$$(\cos \theta + j \sin \theta)^n = \cos n\theta + j \sin n\theta \tag{D.14}$$

The nth root of a complex z is the number w such that

$$w^n = z = re^{j\theta} \tag{D.15}$$

Thus, to find the nth root of a complex number z, we must solve

$$w^n - re^{j\theta} = 0 \tag{D.16}$$

which is an equation of degree n and hence has n roots. These roots are given by

$$w_k = r^{1/n} e^{j[\theta + 2(k-1)\pi]/n} \qquad k = 1, 2, \ldots, n \tag{D.17}$$

Useful Mathematical Formulas

E.1 Summation Formulas

$$\sum_{n=0}^{N-1} \alpha^n = \begin{cases} \dfrac{1-\alpha^N}{1-\alpha} & \alpha \neq 1 \\ N & \alpha = 1 \end{cases}$$

$$\sum_{n=0}^{\infty} \alpha^n = \frac{1}{1-\alpha} \qquad |\alpha| < 1$$

$$\sum_{n=k}^{\infty} \alpha^n = \frac{\alpha^k}{1-\alpha} \qquad |\alpha| < 1$$

$$\sum_{n=0}^{\infty} n\alpha^n = \frac{\alpha}{(1-\alpha)^2} \qquad |\alpha| < 1$$

$$\sum_{n=0}^{\infty} n^2\alpha^n = \frac{\alpha^2 + \alpha}{(1-\alpha)^3} \qquad |\alpha| < 1$$

E.2 Euler's Formulas

$$e^{\pm j\theta} = \cos\theta \pm j\sin\theta$$

$$\cos\theta = \frac{1}{2}(e^{j\theta} + e^{-j\theta})$$

$$\sin\theta = \frac{1}{2j}(e^{j\theta} - e^{-j\theta})$$

E.3 Trigonometric Identities

$$\sin^2\theta + \cos^2\theta = 1$$

$$\sin^2\theta = \frac{1}{2}(1 - \cos 2\theta)$$

$$\cos^2\theta = \frac{1}{2}(1 + \cos 2\theta)$$

$$\sin 2\theta = 2\sin\theta\cos\theta$$

$$\cos 2\theta = \cos^2\theta - \sin^2\theta = 2\cos^2\theta - 1 = 1 - 2\sin^2\theta$$

$$\sin(a \pm \beta) = \sin\alpha\cos\beta \pm \cos\alpha\sin\beta$$

$$\cos(\alpha \pm \beta) = \cos\alpha\cos\beta \mp \sin\alpha\sin\beta$$

$$\sin\alpha\sin\beta = \frac{1}{2}\left[\cos(\alpha - \beta) - \cos(\alpha + \beta)\right]$$

$$\cos\alpha\cos\beta = \frac{1}{2}\left[\cos(\alpha - \beta) + \cos(\alpha + \beta)\right]$$

$$\sin\alpha\cos\beta = \frac{1}{2}\left[\sin(\alpha - \beta) + \sin(\alpha + \beta)\right]$$

$$\sin\alpha + \sin\beta = 2\sin\frac{\alpha + \beta}{2}\cos\frac{\alpha - \beta}{2}$$

$$\cos\alpha + \cos\beta = 2\cos\frac{\alpha + \beta}{2}\cos\frac{\alpha - \beta}{2}$$

$$a\cos\alpha + b\sin\alpha = \sqrt{a^2 + b^2}\,\cos\left(\alpha - \tan^{-1}\frac{b}{a}\right)$$

E.4 Power Series Expansions

$$e^{\alpha} = \sum_{k=0}^{\infty}\frac{\alpha^k}{k!} = 1 + \alpha + \frac{1}{2!}\alpha^2 + \frac{1}{3!}\alpha^3 + \cdots$$

$$(1 + \alpha)^n = 1 + n\alpha + \frac{n(n-1)}{2!}\alpha^2 + \cdots + \binom{n}{k}\alpha^k + \cdots + \alpha^n$$

$$\ln(1 + \alpha) = \alpha - \frac{1}{2}\alpha^2 + \frac{1}{3}\alpha^3 - \cdots + \frac{(-1)^{k+1}}{k}\alpha^k + \cdots \qquad |\alpha| < 1$$

E.5 Exponential and Logarithmic Functions

$$e^{\alpha}e^{\beta} = e^{\alpha + \beta}$$

$$\frac{e^{\alpha}}{e^{\beta}} = e^{\alpha - \beta}$$

$$\ln(\alpha\beta) = \ln\alpha + \ln\beta$$

$$\ln\frac{\alpha}{\beta} = \ln\alpha - \ln\beta$$

$$\ln\alpha^{\beta} = \beta\ln\alpha$$

$$\log_b N = \log_a N \log_b a = \frac{\log_a N}{\log_a b}$$

E.6 Some Definite Integrals

$$\int_0^{\infty} x^n e^{-ax}\,dx = \frac{n!}{a^{n+1}} \quad a > 0$$

$$\int_0^{\infty} e^{-ax^2}\,dx = \frac{1}{2}\sqrt{\frac{\pi}{a}} \quad a > 0$$

$$\int_0^{\infty} xe^{-ax^2}\,dx = \frac{1}{2a} \quad a > 0$$

Schaum's Signals and Systems Videos

1. Problem 1.46 – Signals and Systems – express a signal in terms of unit step functions
2. Problem 1.56 – Signals and Systems – determine if a signal is linear, time-invariant and/or causal
3. Problem 1.61 – Signals and Systems – determine if a system is invertible
4. Problem 2.46 – Linear Time-Invariant Systems – find the convolution of a pair of signals
5. Problem 2.58 – Linear Time-Invariant Systems – find the different equation for a 2nd order circuit
6. Problem 2.64 – Linear Time-Invariant Systems – find the output of a discrete-time system
7. Problem 3.43 – Laplace Transform and Continuous-Time LTI Systems – find the Laplace transform of a signal
8. Problem 3.49 – Laplace Transform and Continuous-Time LTI Systems – find the inverse Laplace transform of a signal
9. Problem 3.55 – Laplace Transform and Continuous-Time LTI Systems – find the transfer function of a system described by a block diagram
10. Problem 4.48 – The z-Transform and Discrete-Time LTI Systems – find the z-transform of a discrete-time system
11. Problems 4.53–4.54 – The z-Transform and Discrete-Time LTI Systems – find the inverse z-Transform of a discrete-time signal
12. Problem 4.56 – The z-Transform and Discrete-Time LTI Systems – use z-Transforms to find the transfer function and difference equation for a system
13. Problem 5.61 – Fourier Analysis of Continuous-Time Signals and Systems – find the trigonometric and complex exponential Fourier series of a continuous-time signal
14. Problems 5.67–5.69 – Fourier Analysis of Continuous-Time Signals and Systems – find the Fourier transform of a continuous-time signal
15. Problem 5.75 – Fourier Analysis of Continuous-Time Signals and Systems – find the frequency response and type of a filter
16. Problem 6.62 – Fourier Analysis of Discrete-Time Signals and Systems – find the discrete Fourier series for a periodic sequence
17. Problem 6.71 – Fourier Analysis of Discrete-Time Signals and Systems – find the frequency and impulse response of a causal discrete-time LTI system
18. Problem 7.65 – State Space Analysis – find the state space representation of a system
19. Problem 7.68 – State Space Analysis – find the state space representation of a system and determine whether it is asymptotically and/or BIBO stable
20. Problem 7.73 – State Space Analysis – use the state space method to solve a linear differential equation

Schaum's Signals and Systems
MATLAB Videos

1. Problems 1.1-1.2 – plot continuous-time and discrete-time signal transformations
2. Problem 1.5 – plot even and odd components of continuous-time and discrete-time signals
3. Problem 1.16 – determine if given are periodic and the fundamental period
4. Problem 1.20 – determine whether a signal is an energy or power signal
5. Problem 2.5 – compute y(t) by using convolution for a continuous-time LTI system
6. Problem 2.24 – find the impulse response and step response of a given system
7. Problem 2.30 – compute y[n] by using convolution for a discrete-time LTI system
8. Problem 3.43 – find the Laplace transform of given signals
9. Problem 3.49 – find the inverse Laplace transform of given signals
10. Problem 3.53 – find the output y(t) of a given C-T LTI system given h(t) and the input x(t)
11. Problem 4.41 – find the z-Transform of a given signal
12. Problem 4.53 – find the inverse z-Transform of a given signal
13. Problem 5.63 – find the Fourier series representation of a given signal
14. Problem 5.69 – find the inverse Fourier transform of a given signal
15. Problem 5.71 – find the Fourier transform of a given signal
16. Problem 6.62 – find the discrete Fourier series of a given signal
17. Problem 6.65 – find the Fourier transform of a given sequence
18. Problem 6.67 – find the inverse Fourier transform of a discrete-time system
19. Problem 7.62 – find the system function and whether a discrete-time LTI system is controllable of observable given a state space representation
20. Problem 7.74 – find the state space representation of a system and whether it is controllable or observable given a state space representation

MATLAB Prints for Online Videos

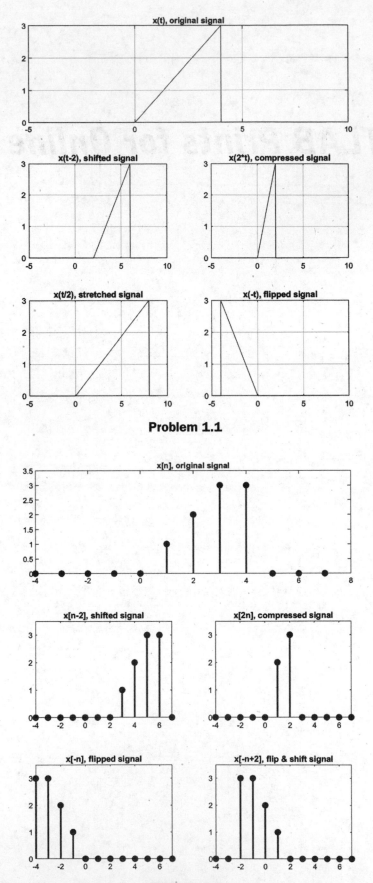

Problem 1.1

Problem 1.2

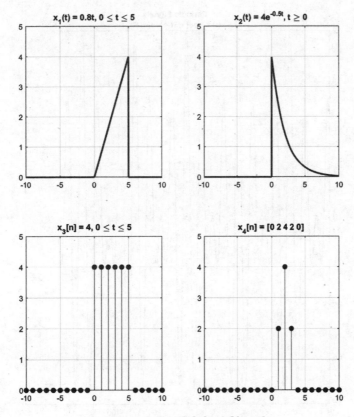

Problem 1.5a Original Signals

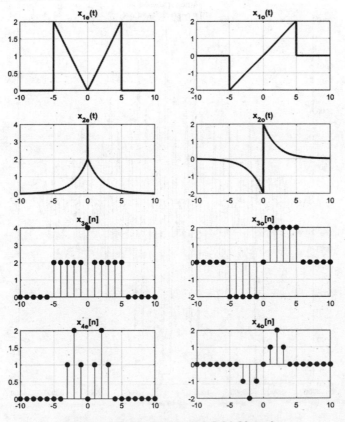

Problem 1.5b Even and Odd Signals

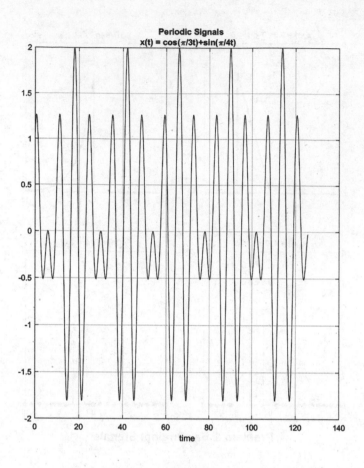

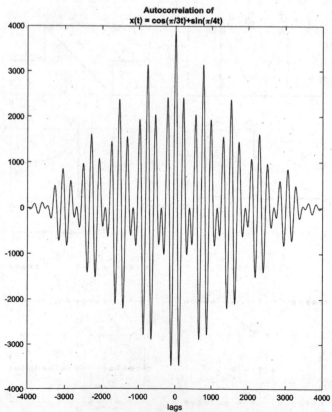

Problem 1.16c

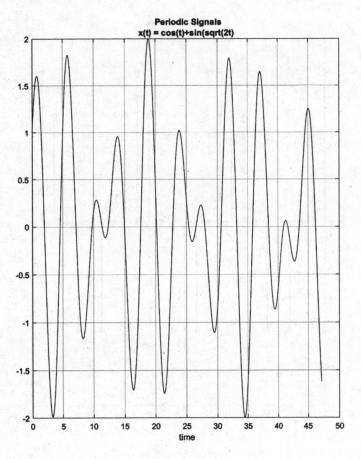

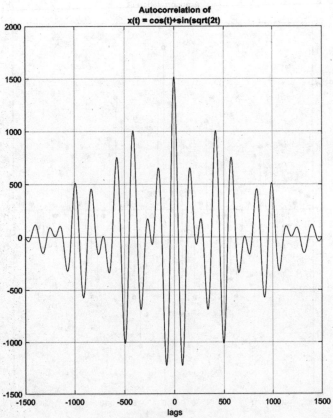

Problem 1.16d

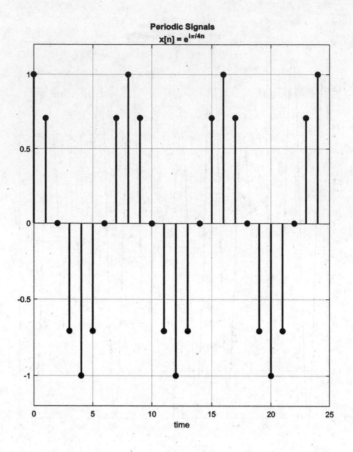

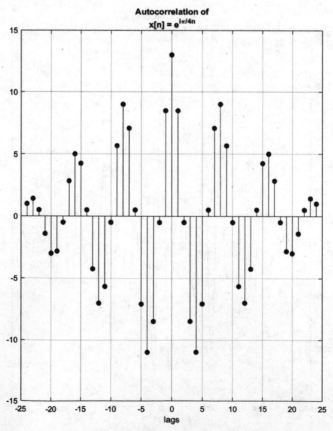

Problem 1.16g

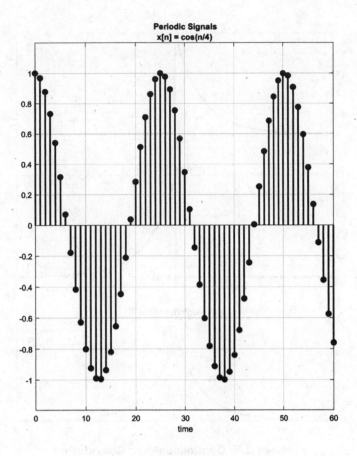

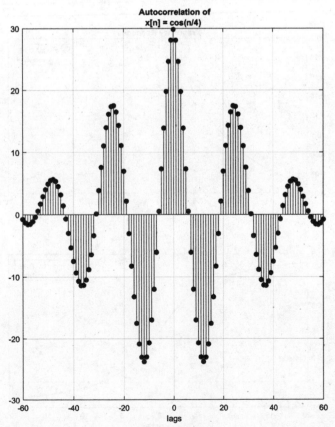

Problem 1.16h

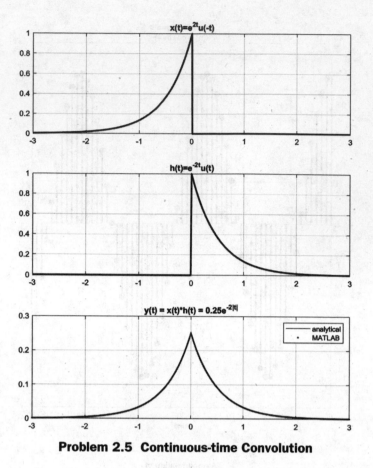

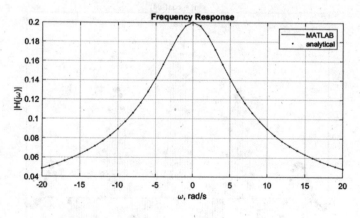

Problem 2.5 Continuous-time Convolution

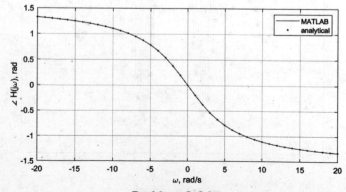

Problem 2.24a

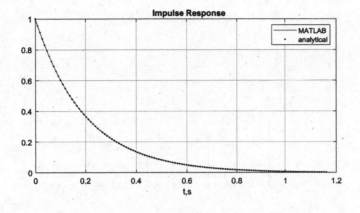

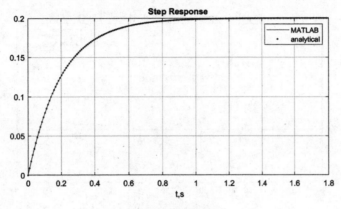

Problem 2.24b

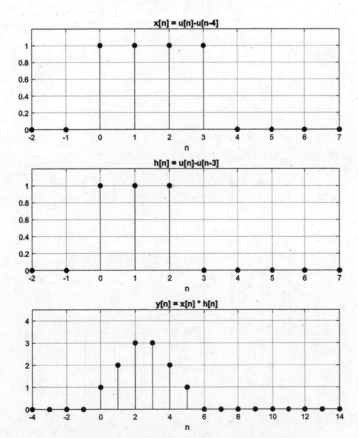

Problem 2.30 Discrete-time Convolution

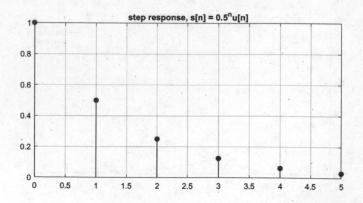

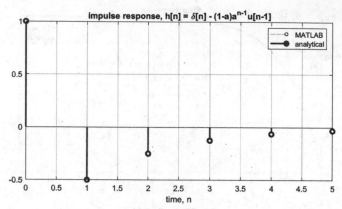

Problem 2.32

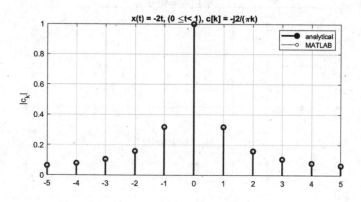

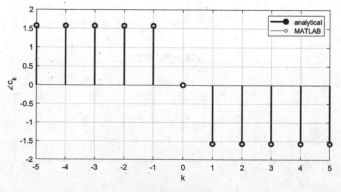

Problem 5.10a

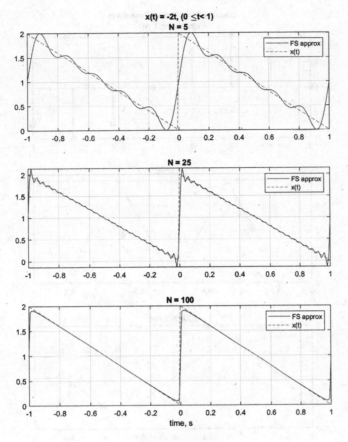

Problem 5.10b

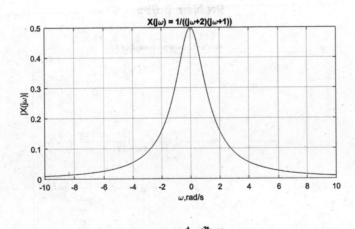

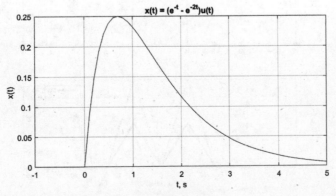

Problem 5.69a

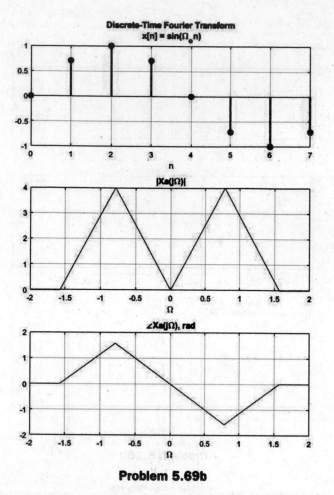

Problem 5.69b

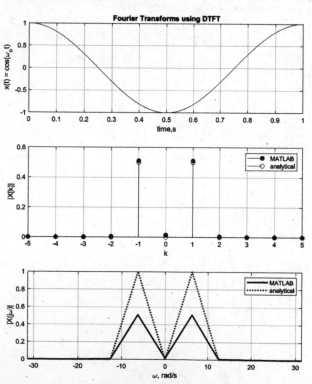

Problem 5.71

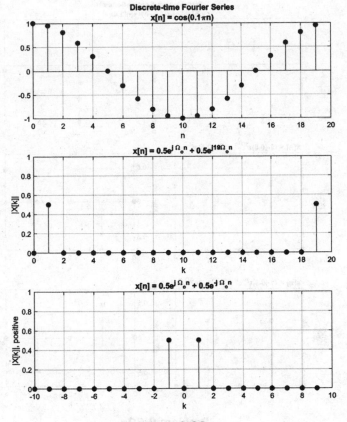

Problem 6.62a

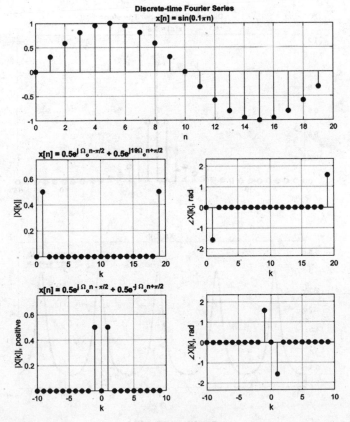

Problem 6.62b

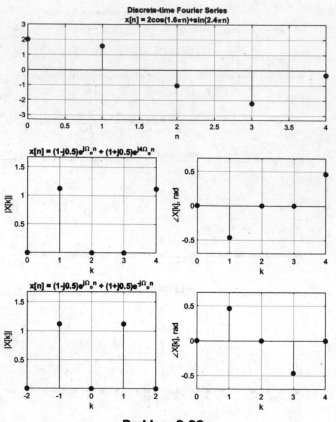

Problem 6.62c

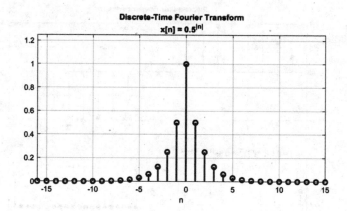

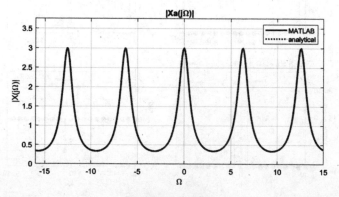

Problem 6.65a

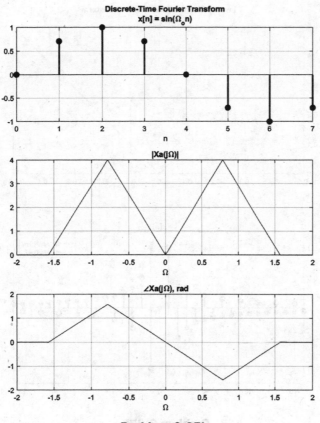

Problem 6.65b

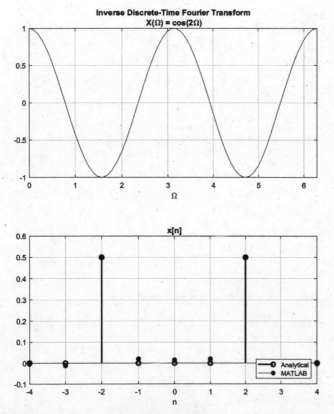

Problem 6.67a

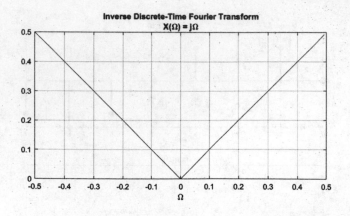

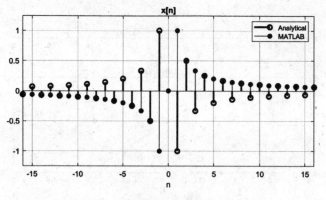

Problem 6.67b

INDEX